The Science of Construction Materials

Per Freiesleben Hansen
edited by Ole Mejlhede Jensen

The Science of Construction Materials

Springer

Prof. Ole Mejlhede Jensen
Technical University of Denmark
Dept. Civil Engineering
2800 Kgs. Lyngby
Brovej, Bldg. 118
Denmark
omj@byg.dtu.dk

This book is scientifically sponsored by the Educational Activities Committee, EAC, of the International Union of Laboratories and Experts in Construction Materials, Systems and Structures, RILEM.

ISBN: 978-3-642-42547-9 ISBN: 978-3-540-70898-8 (eBook)
DOI 10.1007/978-3-540-70898-8
Springer Heidelberg Dordrecht London New York

Translation from the Danish edition: Materialefysik for Bygningsingeniører - Beregningsgrundlag by Per Freiesleben Hansen, © Statens Byggeforskningsinstitut, Hørsholm, Denmark 1995. All rights reserved.

Photographs: The photos in this book come from different sources and are reproduced by permission. For a list of the origin of these illustrations see Appendix E.

Cover illustration: Paul Stutzman, Nat. Inst. of Sci. and Techn., USA; for details please refer to Appendix E.

Cover design: eStudio Calamar S.L, Figueres/Berlin

Printed on acid-free paper

Springer is part of Springer Science+Business Media (www.springer.com)

Contents

2. Thermodynamic concepts

3. First law

4. Second law

5. Calculations of equilibrium

6. Electrochemistry

APPENDIX A

Mathematical appendix

APPENDIX B

Tables

APPENDIX C

Solutions to check-up questions and exercises

APPENDIX D

Subject index

APPENDIX E

Acknowledgements for illustrations

List of symbols

The list contains symbols and indices used throughout the book. Note: in some cases the *same* symbol has been used for a **mole-specific** and a **mass-specific quantity**, for example for specific heat capacity (c). In such cases, the value concerned will be specified by the unit. Also, thermodynamic state functions such as U, H, G and S, which may occur as *both* **extensive** and **intensive** quantities, are described by the same symbol. Again, the applied unit shall be noted. These notation rules have been chosen to limit the use of indices and thus further the legibility. In the list, non-dimensional and abstract quantities are indicated by (–).

A	area of cross-section	(m^2)
A	nucleon number in atom nucleus (mass number)	(–)
A	the Helmholtz free energy	(J); (J/mol)
A_r	relative atomic mass m/m_u	(–)
a	activity of substance component	(–)
a	the van der Waals constant (correction of pressure)	($m^6 Pa/mol^2$)
a	stoichiometric coefficient in reaction equation	(–)
(aq)	ions or gases in aqueous solution	(–)
b	the van der Waals constant (correction of volume)	(m^3/mol)
b	stoichiometric coefficient in reaction equation	(–)
C	heat capacity of system	(J/K)
c	mole-specific heat capacity	(J/mol K)
c	mass-specific heat capacity	(J/kg K)
c_i	molar concentration of component (i)	(mol/m^3); (mol/ℓ)
c_p	heat capacity at constant pressure (isobaric)	(J/mol K)
c_V	heat capacity at constant volume (isochoric)	(J/mol K)
$c^{\ominus}$	standard concentration 1 mol/ℓ	(mol/ℓ)
c	stoichiometric coefficient in reaction equation	(–)
d	stoichiometric coefficient in reaction equation	(–)
$det()$	determinant of ()	(–)
$dim()$	physical dimension of ()	(–)
$\mathbf{E}$	electric field strength	(V/m)
E	modulus of elasticity	(MPa)
e	electric elementary charge $\sim 1.602 \cdot 10^{-19}$ C	(C)
eV	electron volt, energy unit $\sim 1.602 \cdot 10^{-19}$ J	(eV)
e^-	symbol for electron in reaction equation	(–)
$\mathbf{F}$	force (vector quantity)	(N)
F_x	force, component in x direction	(N)
$\mathcal{F}$	the Faraday constant ~ 96500 C/mol	(C/mol)
G	the Gibbs free energy in thermodynamic system	(J); (J/mol)
$G_T^{\ominus}$	standard free energy at temperature T	(J/mol)
$\Delta_r G$	free reaction energy	(J); (J/mol)
G	electric conductance	Ω^{-1}
(g)	gaseous state ("gas")	(–)
H	enthalpy content in thermodynamic system	(J); (J/mol)
$H_T^{\ominus}$	standard enthalpy at temperature T	(J/mol)
$\Delta_r H$	reaction enthalpy	(J); (J/mol)
I	electric current	(A); (C/s)
K_a	thermodynamic equilibrium constant	(–)
k	Boltzmann constant $R/\mathcal{N} \sim 1.38 \cdot 10^{-23}$ J/K	(J/K)

L	linear dimension	(m)
ℓ	litre	(ℓ)
(ℓ)	liquid state ("liquid")	(–)
ln()	natural logarithm of ()	(–)
$\log_{10}()$	base 10 logarithm of ()	(–)
M	molar mass	(kg/mol); (g/mol)
M_r	relative molecular mass m/m_u	(–)
m	mass	(kg)
m_u	atomic mass constant $\sim 1.6606 \cdot 10^{-27}$ kg	(kg)
$\widetilde{m}_i$	molality of component (i) in mixture	(mol/kg)
N	neutron number in atom nucleus	(–)
$\mathcal{N}$	the Avogadro constant $\sim 6.022 \cdot 10^{23}$ mol^{-1}	(mol^{-1})
n	amount of substance	(mol)
n	degree of polymerization	(–)
p	pressure	(Pa)
p_c	critical pressure for real gas	(Pa)
p_i	partial pressure of (i) in gas mixture	(Pa)
p_s	partial pressure of saturated water vapour	(Pa)
p_{tot}	total pressure in mixture of gases	(Pa)
$p^{\ominus}$	standard pressure 101325 Pa	(Pa)
pH	pH value: pH $= -\log_{10}(a(\mathrm{H}^+))$	(–)
Q	heat added to a system from its surroundings	(J); (J/mol)
Q	electric charge	(C); (A · s)
R	gas constant ~ 8.314 J/mol K	(J/mol K)
R	electric resistance	(Ω)
RH	relative humidity p/p_s	(%); (–)
r_0	equilibrium distance, Lennard-Jones potential	(m)
S	entropy content in system	(J/K); (J/mol K)
$S_T^{\ominus}$	standard entropy at temperature T	(J/mol K)
$\Delta_r S$	reaction entropy during process	(J/K); (J/mol K)
(s)	solid state ("solid")	(–)
(SHE)	standard hydrogen electrode	(–)
T	thermodynamic temperature $(273.15 + \theta)$	(K)
T_c	critical temperature for real gas	(K)
U	internal energy in a thermodynamic system	(J); (J/mol)
ΔU	increase in internal energy during process	(J); (J/mol)
u	absolute humidity	(g/m^3)
V	volume	(m^3); (ℓ)
V_{mix}	volume of a mixture or solution	(m^3); (ℓ)
V_a	electric potential at the point (a)	(V)
ΔV	increase in volume	(m^3)
ΔV	electric potential difference	(V)
$V^{\ominus}$	electrochemical standard potential against (SHE)	(V)
v_c	critical volume for real gas	(m^3/mol)
W	work done on a system by its surroundings	(J), (J/mol)
w_i	mass fraction of component (i) in a mixture	(–)
x_i	mole fraction of component (i) in a mixture	(–)
$x^{\ominus}$	mole fraction of pure solvent; $x^{\ominus} = 1$	(–)
Z	proton number in atom nucleus (atomic number)	(–)
z	transfer number for electrons	(–)

α	coefficient of volume expansion	(K^{-1})
β	coefficient of linear expansion	(K^{-1})
γ	heat capacity ratio c_p/c_V	(–)
γ	activity coefficient	(–)
Δ	increase in parameter value	(–)
ε	strain $\Delta L/L_0$	(–)
ε_u	ultimate strain for a material	(–)
ε	residual term, regression analysis	(–)
η	thermal efficiency, cyclic process	(–)
$\Phi(r)$	Lennard-Jones potential	(J); (J/mol)
Φ_0	Lennard-Jones constant	(J); (J/mol)
κ	compressibility	(Pa^{-1})
λ	thermal conductivity	(kJ/m h K); (W/m K)
λ	free mean path	(m)
μ	prefix micro-, corresponding to 10^{-6}	(–)
Π	non-dimensional parameter group	(–)
ϱ	density of a substance	(kg/m^3)
ϱ_i	mass concentration of component (i) in a mixture	(kg/m^3)
ρ	electric resistivity	(Ω m)
σ	electric conductivity	($\Omega^{-1}m^{-1}$)
σ	surface tension	(N/m)
θ	temperature, degrees Celsius	(°C)
Ω	number of micro states in a system	(–)

Preface

Preface to the English edition

This book on the *Science of Construction Materials* is the English version of the Danish textbook *Materialefysik for Bygningsingeniører* (Statens Byggeforskningsinstitut - Danish Building Research Institute, Hørsholm, 1995). The translation has been done by a team consisting of Professor Ole Mejlhede Jensen, Technical University of Denmark and Kirsten Aakjær, BA, Aalborg University. Comments on the English version were kindly supplied by Professor Sidney Diamond, Purdue University. Help with reproduction of the line figures was given by Trine Bay, MSc student, and help with the final text editing was given by Sara Laustsen, PhD student. Transfer of copyright was kindly granted by the family of Per Freiesleben Hansen. Financial support for the translation was donated by three Danish foundations: COWIfonden, Knud Højgaards fond, and Larsen & Nielsen fonden.

A number of modifications were made during the preparation of the English version of the textbook. Some of these have been necessitated by the translation, such as references to sources in Danish being changed to international references. Other corrections were made concerning a few apparent errors in the text. All line figures have been redrawn, and many photos have been substituted. However, in general, the English version is close to the Danish text. In addition to the textbook, a separate book of exercises exists. The exercises differ in extent and complexity and are organized to further the understanding and use of the subjects in the textbook. Based on this teaching concept, the subject of basic construction materials has been taught successfully in Denmark through 20 years.

It is more than two decades ago that Per Freiesleben Hansen initiated the writing of this textbook. This initiative was necessitated by the development within construction materials research from 1960 to 1985. In this period the research moved form being a simple study of the properties of different materials towards being a theoretically based specialist discipline. In the decade since the publication of the Danish textbook, the development has gone much further. In particular, complex computer simulation has become increasingly used as an alternative to measurement and direct calculation has been replaced by complex computer modelling. In this situation it is very important that the computer programmers build their algorithms on sound materials science. And likewise, it is very important that the user of the programs has an understanding of fundamental physics and chemistry that allows a critical interpretation of the output from the black box of complexity; the sharp knife of materials science that the textbook provides is more needed than ever.

It is our hope that this English version will contribute to enhancing the development of the science of construction materials internationally.

Lyngby, June 2009

O. Mejlhede Jensen

Per Freiesleben Hansen (1936-2002)

Per Freiesleben Hansen at his desk, October 2001. The Danish textbook was written during a 10-year period. It was his pedagogical intention that "the students should not be given a lunch packet, but be taught how to make one themselves - and then they should be given a sharp knife."

Extract from the preface to the Danish edition

In its classical form, the science of construction materials is a descriptive, empirical discipline related to certain types of materials, e.g. the study of the properties of wood, steel, concrete and plastics. This traditional division of the science into the separate studies of the different material types is appropriate as long as the purpose is to collect, disseminate and use knowledge about the simple physical and chemical properties of specific materials. Through generations, research, teaching and engineering practice have all functioned within this framework without problems.

However, during the last few decades the nature of the research being pursued within construction materials has changed. Increasingly, construction materials research is carried out by specialists within theoretical disciplines such as physics, chemistry and physical chemistry. The methods used for investigations have become more sophisticated and it is often necessary to "interpret" the results before they become meaningful for the practical civil engineer.

This development has given rise to an unfortunate gap between research and engineering practice in the field of construction materials. Fundamental research has gradually been fragmented into a number of narrow specialist disciplines, making it difficult for the researchers to communicate fundamental new knowledge in a form that can be utilized by the building materials engineer. This schism becomes more and more noticeable as research has resulted in development of a number of new building materials.

Within civil engineering education, an attempt to surmount these problems was the introduction of the specialist discipline "*Materials Science*", which aims at describing the macroscopic properties of materials based on their atomic and molecular structures. An early exponent of this specialist discipline was *van Vlack*. His "Elements of Materials Science" was introduced into the civil engineering education at the Technical University of Denmark in the late 1960'es.

Materials Science according to these principles has been shown to be readily accepted within the fields of electrical and mechanical engineering. In these fields well-defined materials are used, and well-defined requirements are established for their mechanical or electrical properties. However, within the field of construction materials, *Materials Science* has not been accepted to the same degree. Only in a few cases has this discipline lead to knowledge transfer of permanent significance to the practical civil engineer. The main reasons for this are the following:

- *Materials Science* is a theoretical science of materials which is based on the scientific disciplines of the researcher, and therefore reflects the reality on the building site only to a limited extent.
- The materials problems faced by civil engineers are as much related to a complex interaction of materials, structure and environments as they are to the properties of the pure materials themselves.

As used in the title of this book, the concept of *Materials Physics* [Editor: direct translation of the original Danish title] can be defined as having the following characteristics:

- *Materials Physics* is a theoretical science of materials formulated with respect to the special concerns of the *civil engineer.*

At a theoretical level, materials science is interdisciplinary. Also, Materials Physics includes elements of other sciences, e.g. chemical thermodynamics, mathematical analysis and numerical methods. An essential task in Materials Physics is to combine parts of these specialist fields into a practical set of calculation tools for building materials engineers.

The borderland of materials science where technical development and educational work meet is fascinating – and difficult – at the same time. Therefore, I express my great appreciation to many colleagues for their specialist inspiration, ideas and sound advice given to me during this work.

Aalborg, August 1994

P. Freiesleben Hansen

Introduction

The form of the textbook aims to make it fit for self-study. This has been achieved by the following additions to the theoretical subject matter.

- At the end of each theoretical section a number of *check-up questions* and *exercises* are given in order to make it possible for the reader to test his *understanding* of the text.

- The book contains a large number of thoroughly-prepared *examples*, showing the application of *formulas* and *calculation expressions*, and demonstrating at the same time the application of the theoretical text to specific *engineering* work.

- The book also contains a *mathematical appendix* with descriptions of the mathematical-physical methods that are often used to solve practical design problems and an *appendix of tables* that covers the application of the theoretical text to normal problems in materials science.

- The book is provided with an elaborate subject index making it possible to track *theory*, *examples*, *tables* and *exercises* separately.

This publication is both a *casebook and workbook*; however, it is the author's hope that it will also be used as a reference book by practising civil engineers, and thus serve as a source of inspiration during the work with technical tasks in materials science at different levels.

Contents of the book

Before working with this book, readers may find it useful to get an overview of the subjects dealt with in the different chapters. Therefore, readers should first read through the following overview and browse through the chapters concerned.

Chapter 1. Substance systems reviews a number of elementary, but fundamental *definitions* and *concepts* that are used to describe the *composition* and *properties* of substances. A large number of the addressed subjects are contained in the curriculum of mathematical-physical university entrance exams.

Chapter 2. Thermodynamic concepts contains an overview of the most essential *concepts* and *definitions* included in the thermodynamic *description of substances*. In the explanation of thermodynamics a number of precise terms concerning systems, state functions and process types are employed. The meaning of these concepts and their application to practical substance systems are explained.

Chapter 3. First law introduces the simple, but perfectly general principle of *the conservation of energy* as expressed through the first law of thermodynamics. The development of calculation rules for this principle is explained and the practical application of these rules is demonstrated to specific technical problems in material technology.

Chapter 4. Second law deals with the concept *entropy*, which is the basis for solving *equilibrium problems* in physical chemistry. Through the entropy concept of the second law, *equilibrium conditions* for substance systems is developed and the application of these conditions is illustrated by solution of practical problems within material science.

Chapter 5. Calculations of equilibrium deals with the concepts *free energy* and *thermodynamic equilibrium constants*, two quantities that form the basis of a number of essential *technical calculation methods* for *equilibrium systems*. The practical application of these quantities is demonstrated by a number of typical problems within materials technology.

Chapter 6. Electrochemistry explains the most essential *definitions* and *concepts* within the field that can be called *equilibrium electrochemistry*. The aim is, among others, to present the design basis of the subsequent treatment of *the science of corrosion*.

Appendix A. Mathematical appendix contains a survey of selected mathematical-physical *methods* that are often used for solution of practical problems. These include *numerical processing* and *uncertainty calculation*, *dimensional analysis*, *linear regression*, iterative *root finding*, etc.

Appendix B. Tables contains a selection of *tabular data* covering the most frequent types of problems within the science of construction materials: *physical constants*, overview of the *properties of elements*, including *table of the elements*, *thermochemical data* of inorganic and organic substances and ions, and a *table of the vapour pressure* of water.

Appendix C. Solutions to check-up questions and exercises contains a systematic overview of *solutions* to all *check-up questions* and *exercises* in the book.

Appendix D. Subject index refers to *theoretical sections*, *examples*, *problems*, *figures* and *tables* in the book.

Structure of the book

The layout of the book aims at making the study an *active* process. From experience, parts of the explained text may appear to be somewhat *abstract* – perhaps even *recondite* – *until* successfully applied to one's own specific problems and practical examples. In its layout and treatment of subjects the book reflects this philosophy of learning.

For a start, try to go through for example chapter 1 and familiarize yourself with the "signals" given in the text and with the organization of the subjects.

■ *Calculation expressions* are directly followed by an example showing how to apply the expression by solving a practical problem. These examples are preceded by a *black square*.

□ *Theoretical sections* are followed by a number of brief *check-up questions*, to check whether the text has been *understood* correctly. These control questions are preceded by a *white square* that can be checked off when the question has been answered. Solutions to control questions are given in *Appendix C*.

Definitions are framed in the text by a *bold* line. Such *bold-line* frame indicates the *definitions* and *concepts* that form the theoretical foundation of the text.

> Essential **calculation expressions** and **formulas** are framed by a *thin* line. Such thin-line frames indicate the expressions that form the practical calculation tools.

The individual chapters are concluded by a *list of key ideas* giving an overview of *definitions*, *concepts* and *calculation expressions* introduced within them.

■ After each chapter, a number of *examples* are given, typically six major examples with thoroughly prepared and discussed *calculation examples*. These examples are preceded by *black squares*. The examples have been carefully chosen so that they illustrate the application of the theory to practical engineering problems.

□ Minor *exercises* – typically 20 – are given at the end of each chapter to facilitate the *learning* of the theoretical subjects. The exercises are preceded by *white squares*; the exercises also show how to apply the explained theory by solving practical problems within materials science. Solutions to exercises are given in *Appendix C*.

A chemical laboratory from the 19th century; above and on top of the tiled tabletop, examples of the former, often homemade equipment for chemical tests can be seen.

CHAPTER 1

Systems of matter

The notion of a substance composed of small, indivisible particles, i.e. **atoms**, can be traced back to antiquity. In the fifth century BC, Greek philosophers such as **Leukippos** and **Democritus** were spokesmen for this point of view.

Through the Middle Ages up to the 17th and 18th centuries, scientists worked with an atomic notion of the composition of substances without, however, reaching a coherent physical and chemical description thereof. Thus, **Robert Boyle** (1626-91) formulated a concept of "elements" at a time where chemistry was not yet mature for this concept.

Today's knowledge of the composition of substances is highly dependent on the work of **Lavoisier** (1743-94). Lavoisier's merit was in particular that he was the first to base his work on careful measurements and weighings.

In 1808 the Englishman **John Dalton** (1766-1844) showed that elements combined with other elements in certain weight proportions. Through this - perhaps the most important observation in chemistry - the basis for modern, quantitative materials chemistry was created. **Atomic weights** could be ascribed to each of the elements, and it was possible to divide them into groups according to their chemical properties. These groups of elements could be arranged in the coherent, **periodic table of the elements**, that we know today.

This chapter gives an overview of a number of fundamental definitions and concepts used for describing the composition and properties of substances.

John Dalton (1766-1844)

English physicist and chemist, professor of mathematics and natural sciences at the College of Manchester.

Contents

1.1 Atoms

All solid matters, liquids, and gases are composed of atoms. At a crude level of understanding, the individual atoms are each composed of a number of elementary particles. Each of these elementary particles is in turn composed of a variety of sub-nuclear particles, but for purposes of ordinary materials science they may be treated as individual entities. Three of these are particularly significant to the science of the composition and properties of substances, viz. the **proton**, the **neutron** and the **electron**. Together protons and neutrons are denoted **nucleons**. These "elementary" particles can be characterized by their rest mass **m** and by their electric charge **e**.

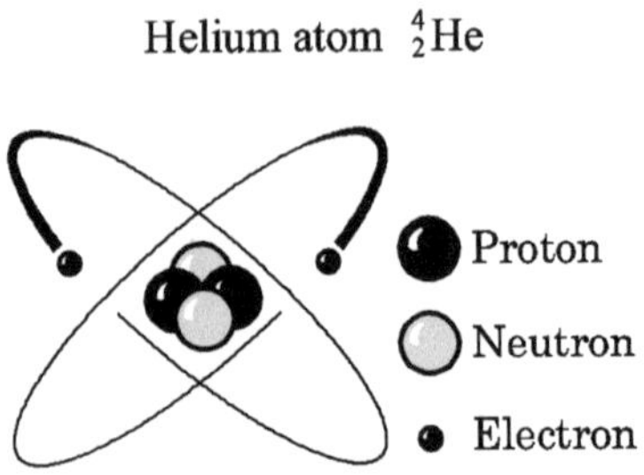

Figure 1.1. Schematic picture of the structure of the helium atom He. The atom nucleus contains two positively charged protons and two uncharged neutrons. The positive charge of the nucleus is neutralized by two negatively charged electrons so that the whole of the atom is electrically neutral.

Elementary particles	Rest mass (kg)	Charge (C)
Proton	$1.673 \cdot 10^{-27}$	$+1.602 \cdot 10^{-19}$
Neutron	$1.675 \cdot 10^{-27}$	0
Electron	$0.911 \cdot 10^{-30}$	$-1.602 \cdot 10^{-19}$

Table 1.1. Overview of the mass and electric charge of the elementary particles; the electric unit for charge is Coulomb: $C = A \cdot s$

Structure of the atom

The individual atom is formed by a positively charged nucleus composed of protons and neutrons surrounded by a negatively charged electron cloud. An electrically neutral atom contains equal numbers of electrons and protons. If an electrically neutral atom absorbs or emits one or more electrons, a **negative ion** or **positive ion**, respectively, is created.

The composition of an atom nucleus is determined by the number of protons and neutrons contained by the nucleus. To specify the composition of the atom nucleus the following quantities and symbols are used.

Quantity	Symbol	Definition
Proton number (atomic number)	Z	no. of protons in nucleus
Neutron number	N	no. of neutrons in nucleus
Nucleon number (mass number)	A	no. of nucleons in nucleus

Table 1.2. Normally used notation for the particles of atomic nuclei.

Elements

Atoms containing the same number of protons have the same **atomic number**. A substance in which all atoms have the same atomic number Z, is denoted an **element**. Atoms containing the same number of nucleons in the nucleus have the same **mass number** A. The following relation between the quantities proton number Z, neutron number N and nucleon number A applies:

$$\text{Nucleon number } A = \text{ Proton number } Z + \text{ Neutron number } N \quad (1.1)$$

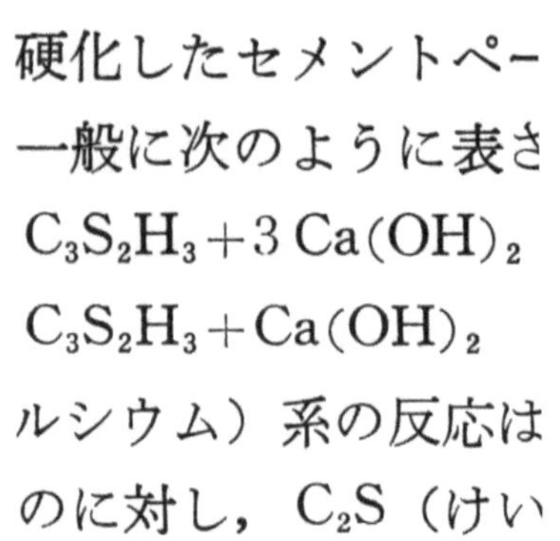
硬化したセメントペー
一般に次のように表さ
$C_3S_2H_3+3\,Ca(OH)_2$
$C_3S_2H_3+Ca(OH)_2$
ルシウム）系の反応は
のに対し，C_2S（けい

Figure 1.2. The names of the elements and thus the notation of the chemical equation is international. The figure shows a section of a page in a Japanese textbook on winter concreting.

In an **element** all atom nuclei contain the same number of protons. Therefore, all of the atoms in an element are able to form identical electron configurations and combine into identical chemical bonds – i.e. they are said to be **chemically identical**.

Most of the element names are international. As a symbol of the individual element the first or the first two letters of the international name of the element are used. For example, the symbol **Fe** is used for iron, which is derived from its Latin name *Ferrum*.

For each of the elements, mass number and atomic number are given based on the following convention: The mass number A is given as a left-hand superscript, and the atomic number Z is given as a left-hand subscript. Thus, carbon C with mass number $A = 12$ and atomic number $Z = 6$ is designated as: ${}^{12}_{6}\mathrm{C}$.

Isotopes

Atoms which are chemically identical and thus belong to the same element may contain different numbers of neutrons in the nucleus, and thus have different mass number. Chemically identical atoms with different mass number are said to be **isotopes** of the same element. For example, natural carbon is a mixture of three isotopes with the mass numbers 12, 13 and 14. According to the above rule these isotopes are denoted: ${}^{12}_{6}\mathrm{C}$, ${}^{13}_{6}\mathrm{C}$ and ${}^{14}_{6}\mathrm{C}$.

■ State: a) mass number A, b) proton number Z, c) neutron number N, d) number of electrons n_e and e) total charge of the nucleus Q for the potassium ion ${}^{39}_{19}\mathrm{K}^{+}$!

Solution. *a)* $A = \mathbf{39}$ *b)* $Z = \mathbf{19}$ *c)* $N = A - Z = \mathbf{20}$ *d)* 1 electron is released, i.e. $n_e = Z - 1 = \mathbf{18}$ e) $Q = Z \cdot e = \mathbf{19 \cdot e}$.

□ 1. Write a chemical symbol with mass number A and atomic number Z for the three isotopes of hydrogen: protium, deuterium and tritium!

□ 2. State the number of nucleons A, neutrons N and protons Z in the following three isotopes of chlorine, iron and lithium: ${}^{37}_{17}\mathrm{Cl}$, ${}^{56}_{26}\mathrm{Fe}$ og ${}^{7}_{3}\mathrm{Li}$!

□ 3. How many electrons pass through a cross-section of a metal conductor per hour at a current of 2 A?

□ 4. Write a chemical symbol stating mass number A and atomic number Z for: a) molybdenum with nucleon number 98, b) calcium with neutron number 20!

□ 5. Atom A1 has mass number 17 and contains 8 protons; atom A2 has nucleon number 17, and the nucleus contains 8 neutrons; Atom A3 has neutron number 10 and mass number 18. Which of these three atoms are chemically identical?

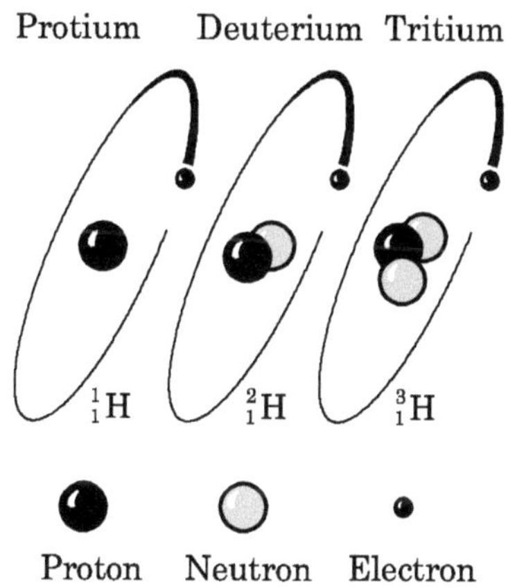

Figure 1.3. Hydrogen forms three isotopes: protium, deuterium and tritium with one, two and three nucleons, respectively, in the nucleus. Natural hydrogen contains more than 99.9 % protium.

1.2 Relative atomic mass

The order of magnitude of the mass of a single atom is 10^{-27} kg; therefore, it is not practical to use absolute mass units in calculations. Instead, according to international conventions, the **relative atomic mass** A_r, defined in the following way is normally used:

Relative atomic mass

$$A_r \stackrel{\text{def}}{=} \frac{m}{m_u} = \frac{\text{mass of atom}}{\text{atomic mass constant}} \tag{1.2}$$

The atomic mass constant $m_u = 1.6606 \cdot 10^{-27}$ kg is defined as $\frac{1}{12}$ of the rest mass of the carbon-12 atom ${}^{12}_{6}\mathrm{C}$.

Most natural elements are mixtures of isotopes with different mass numbers A. Generally, there is a constant proportion between the *contents* of the different isotopes. The relative atomic mass for elements is therefore conventionally given as a **weighted mean value**, where the mass contribution of the individual isotopes corresponds to their relative proportions in the substance in natural occurrences. Natural carbon, for example, has a relative atomic mass A_r which is slightly larger than 12, namely 12.011, because it contains small quantities of carbon-13 and carbon-14 in addition to the carbon-12 isotope.

The relative atomic mass A_r is often generally designated as the **atomic weight**. When this less tedious notation is used, it shall be borne in mind that it is a relative, i.e. non-dimensional, indication of weight.

Figure 1.4. Copper Cu is extracted from the frequently occurring mineral chalcopyrite $CuFeS_2$.

■ Natural copper Cu contains two isotopes; of a given number of atoms, 69.2% is made up of $^{63}_{29}Cu$ with relative atomic mass $A_r = 62.9296$ and 30.8% is made up of $^{65}_{29}Cu$ with $A_r = 64.9278$. Determine A_r for natural Cu!

Solution. $A_r = 0.692 \cdot 62.9296 + 0.308 \cdot 64.9278 = \mathbf{63.545}$

□ **1.** Calculate A_r of natural silver Ag, when 51.82% of the atoms are $^{107}_{47}Ag$ with A_r 106.905, and 48.18% of the atoms are $^{109}_{47}Ag$ with A_r 108.905!

□ **2.** Natural aluminium with atomic number 13 contains atoms with relative atomic mass 26.9815 only; state the mass of an Al atom given in kg!

□ **3.** An atom of the isotope $^{64}_{30}Zn$ weighs $1.0616{\cdot}10^{-25}$kg; what is the relative atomic mass A_r of this isotope?

□ **4.** State the relative atomic mass A_r and the absolute mass m (in kg) of a carbon-12 atom!

□ **5.** Natural hydrogen contains protium ^1_1H with relative atomic mass 1.007825 and deuterium ^2_1H with relative atomic mass 2.0140; A_r of natural hydrogen is 1.00797. Calculate how many % of the atoms are made up of protium or deuterium, respectively!

1.3 Relative molecular mass

Elements consist of chemically identical atoms, i.e. atoms with the same atomic number Z. **Compounds** consist of combinations of chemically different element atoms.

Compounds may exist as **molecules**, which are well-defined, discrete combinations of atoms. For example, water H_2O is formed by two H atoms and an O atom.

In other cases compounds are widespread three-dimensional, periodic arrangements of atoms; in these cases it does not make any sense to talk about molecular units. Common salt, i.e. sodium chloride is, for example, precipitated in a three-dimensional crystal lattice formed by positive Na^+ ions and negative Cl^- ions. Here the rather general concept **formula unit** is used to specify the stoichiometric composition of a given compound.

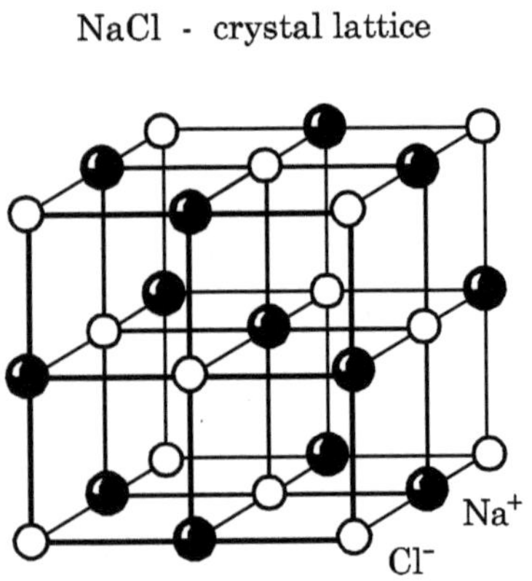

Figure 1.5. Sodium chloride forms a regular, cubic crystal lattice of positive ions Na^+ and negative ions Cl^-. The formula unit for this crystal lattice is NaCl; in this case, the concept "molecule" has no sense.

According to international rules the mass of molecules and formula units is given as **relative molecular mass** M_r, defined as follows:

Relative molecular mass

$$M_r \stackrel{\text{def}}{=} \frac{m}{m_u} = \frac{\text{mass of formula unit}}{\text{atomic mass constant}} \tag{1.3}$$

The atomic mass constant $m_u = 1.6606 \cdot 10^{-27}$ kg is *defined* as $\frac{1}{12}$ of the rest mass of the carbon-12 atom $^{12}_{6}C$.

Based on the definition of the relative molecular mass M_r, it is seen that it corresponds to the sum of the masses of the constituent atoms A_r. The relative molecular mass M_r, often denoted **molecular mass**, is a non-dimensional quantity.

■ When Portland cement hardens a hydration product called **ettringite** is formed, among others. The formula unit for ettringite is: $3CaO \cdot Al_2O_3 \cdot 3CaSO_4 \cdot 31H_2O$. The constituent elements of ettringite have the following relative atomic masses A_r; H: 1.00797, O: 15.9994, Al: 26.9815, S: 32.064 and Ca: 40.08. Calculate the relative molecular mass of ettringite!

Solution. A formula unit of ettringite is seen to contain 62 H atoms, 49 O atoms, 3 S atoms, 2 Al atoms and 6 Ca atoms. Thus, the relative molecular mass M_r is

$$M_r = 62 \cdot 1.00797 + 49 \cdot 15.9994 + 3 \cdot 32.064 + 2 \cdot 26.9815 + 6 \cdot 40.08 \quad = \mathbf{1237.1}$$

☐ 1. State the following substances for which no isolated molecular units exist: barium chloride $BaCl_2$, isoprene $C_5H_8(\ell)$, quartz SiO_2 and ice H_2O?

☐ 2. What is the relative molecular mass M_r of the carbon-12 isotope?

☐ 3. Calculate the relative molecular mass M_r of calcium sulphate dihydrate: $CaSO_4 \cdot 2H_2O$ and of tricalcium aluminate: $3CaO \cdot Al_2O_3$!

☐ 4. The relative molecular mass of water H_2O is 18.02; based on this, calculate the number of water molecules in 1 g of water!

☐ 5. The formula unit NaCl has a relative molecular mass M_r of 58.45; based on this, calculate the number of Na^+ ions in a sodium chloride crystal of 0.5 g!

Figure 1.6. Rods of ettringite, surrounding calcium-silicate-hydration products. These components are formed during hardening of Portland cement.

1.4 Amount of substance - the mole

A description of the composition and properties of substances will often refer to the **elementary parts** of which a substance is composed. Here, elementary parts mean well-defined units such as atoms, molecules, formula units or ions. In physical chemistry a special unit is applied for specifying the **amount of substance**, namely the unit **mole**. The formal, somewhat lengthy definition of the SI unit mole is:

> **Amount of substance - the mole** (1.4)
> The unit **mole** is *defined* as the **amount of a substance** that contains as many **elementary units** as the number of atoms in 0.012 kilogram of carbon 12; the symbol **n** is used to denote the amount of substance.

The Avogadro constant

This definition defines the number of elementary units $\mathcal{N}$ contained in a mole; the number $\mathcal{N}$ is denoted the **Avogadro constant**. The value of $\mathcal{N}$ can be derived as follows:

- the mass of 1 mole of carbon 12 is defined as : 0.012 kg
- the mass of 1 carbon 12 atom is defined as: $12 \cdot m_u$

Therefore, the number of elementary units, i.e. carbon atoms, in 1 mole of carbon 12 is:

$$\mathcal{N} = \frac{0.012 \text{ kg/mol}}{12 \cdot 1.6606 \cdot 10^{-27} \text{ kg}} = 6.022 \cdot 10^{23} \text{ mol}^{-1}$$

The Avogadro constant

$\mathcal{N} = 6.022 \cdot 10^{23} \text{ mol}^{-1}$

> **The Avogadro constant** (1.5)
> The Avogadro constant $\mathcal{N} = 6.022 \cdot 10^{23} \text{ mol}^{-1}$ defines the number of elementary units in an amount of substance of 1 mole.

The Avogadro constant $\mathcal{N}$ is a fundamental computational number within the field of materials chemistry.

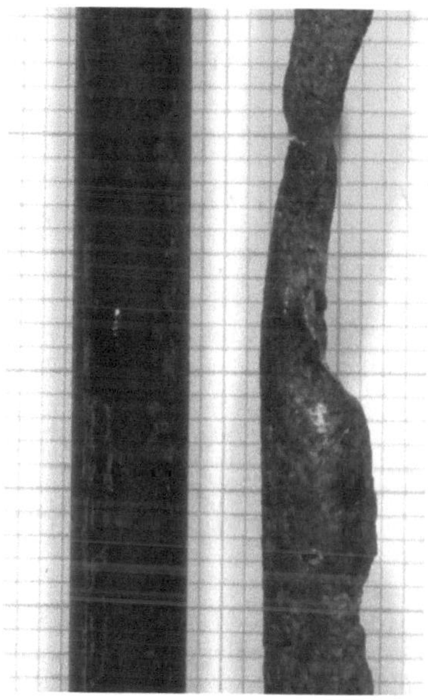

Figure 1.7. During electrolytic corrosion of iron Fe, the iron is dissolved by the anodic process:

$$Fe \rightarrow Fe^{++} + 2e^-$$

Electrolytic corrosion may result in extensive damage to steel structures. The figure shows how Ø10 reinforcing steel (left) can be corroded by this process (right).

■ The Faraday constant $\mathcal{F}$ with the unit C/mol, is defined as the electric charge per mole of elementary charges e^+, where e^+ indicates the electric charge of the electron proton $1.602 \cdot 10^{-19}$ C. Determine the numerical value of the Faraday constant $\mathcal{F}$!

Solution. $\mathcal{F} = \mathcal{N} \cdot e^+ = 6.022 \cdot 10^{23} \text{ mol}^{-1} \cdot 1.602 \cdot 10^{-19} \text{ C} \quad = \mathbf{96.5 \cdot 10^3 \text{ C/mol}}$

☐ 1. How many mol of: a) atoms, b) protons, c) neutrons, d) nucleons and e) electrons are found in 0.012 kg carbon 12?

☐ 2. How many mol of Fe atoms and S atoms are found in: a) 0.500 mol of pyrite FeS_2 and b) 0.500 kg of pyrite FeS_2?; the relative molecular mass of pyrite is 119.98

☐ **3**. How many mol electrons are liberated by electrolytic oxidation of 1 kg of iron Fe during the reaction: $Fe \rightarrow Fe^{++} + 2e^-$?

☐ **4**. The mean diameter of a grain of cement is approx. 10 μm, and the mass of the grain is approx. $1.6 \cdot 10^{-9}$ g; state the approximate mass of 1 mole of cement grains!

☐ **5**. How many moles of electrons are required to reduce 1.000 kg of aluminium by the reaction: $Al^{3+} + 3e^- \rightarrow Al$, when A_r for Al is 26.98?

1.5 Molar mass

The mass of a mole of elementary units, e.g. molecules or formula units, is named **molar mass** M. There is a simple relation between the molar mass M and the relative molecular mass M_r; if the mass m indicates the mass of a molecule or of a formula unit, the molar mass M is determined by:

$$M = \mathcal{N} \cdot m = \mathcal{N} \cdot m_u \cdot \frac{m}{m_u} = (\mathcal{N} \cdot m_u) \cdot M_r \tag{1.6}$$

Since the factor $(\mathcal{N} \cdot m_u)$ is a constant with the following numerical value:

$$\mathcal{N} \cdot m_u = \frac{0.012\,\text{kg/mol}}{12 \cdot m_u} \cdot m_u = 10^{-3}\ \text{kg/mol} = 1\ \text{g/mol}$$

then there is a simple relation between the molar mass M and the relative molecular mass M_r:

Molar mass (1.7)

Molar mass $M = M_r \cdot 10^{-3}$ kg/mol $= M_r$ g/mol

The molar mass M, with the unit g/mol, is equal to the numerical value of the relative molecular mass M_r.

When quantities are given in moles, the employed formula unit shall always be given as well. For example, it is unclear what is meant by an amount of substance of "1 mole of nitrogen" if it is not defined whether the formula unit is N or N_2.

Glucose unit $C_6H_{10}O_5$

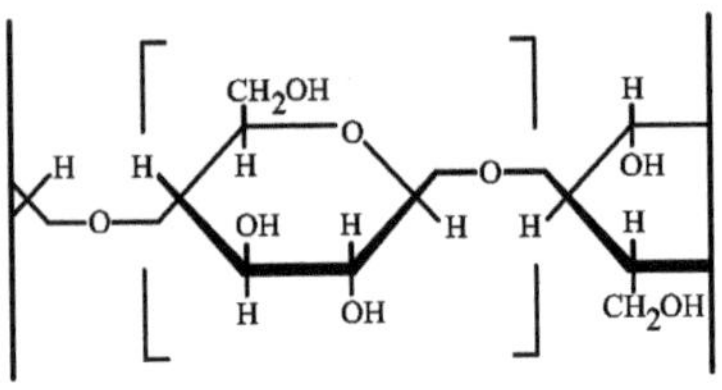

Figure 1.8. Wood cellulose is composed of molecular compounds. The compounds are formed by the glucose unit $-\lfloor C_6H_{10}O_5 \rfloor n-$. The chain molecules contain on average 10 000 units of this type.

■ The main constituent of xylem of wood is cellulose existing in the form of molecular compounds formed by glucose units composed as follows: $-\lfloor C_6H_{10}O_5 \rfloor n-$. The degree of polymerization n, i.e. the average number of glucose units in a cellulose molecule is approx. 10000. The relative atomic masses of carbon C, hydrogen H, and oxygen O, are respectively 12.01, 1.008 and 16.00. Calculate the average molar mass of wood cellulose!

Solution. $M(\text{glucose unit}) = 6 \cdot 12.01 + 10 \cdot 1.008 + 5 \cdot 16.00 = 162.14$ g/mol

$M(\text{cellulose}) \simeq n \cdot M(\text{glucose unit}) = 10000 \cdot 162.14$ g/mol $\simeq$ **1621 kg/mol**

☐ **1**. Explain why the following statement is inaccurate: "1 mol of oxygen"; "1 mol of ferric oxide" and "1 mol of stannic chloride"!

☐ **2**. A_r for hydrogen H is 1.008 and A_r for oxygen O is 16.00; determine the molar mass M of water H_2O based on the units kg/mol and g/mol!

☐ **3**. How many moles of O atoms are contained in 1 kg CO_2, when A_r(C) is 12.01 and A_r(O) is 16.00?

☐ **4**. Which of the following systems contain a) the highest number of molecules b) the highest number of O atoms: 1.0 g oxygen O_2 1.0 g ozone O_3?

☐ **5**. At a given oxidation process 100.0 g of iron Fe react with 42.97 g of oxygen O; find out whether hematite Fe_2O_3 or magnetite Fe_3O_4 is formed by the reaction!

1.6 Mixture of substances

A homogeneous substance may either be a **pure substance**, i.e. an element or a compound, or a **mixture** of different substances. Pure substances and mixtures may be **solid**, **liquid** or **gaseous**.

A homogeneous mixture is also called a **solution**. Often the content of one component in a solution is totally dominating; the dominating component of the solution is called the **solvent**. In chemistry the concept of solution includes solid, liquid and gaseous mixtures of substances. An example of a solid solution is steel made from technically pure iron Fe with a content of dissolved atomic carbon; at 721 °C ferrite Fe may contain approx. 0.025 wt-% of dissolved atomic carbon. In this case the solvent is solid iron Fe.

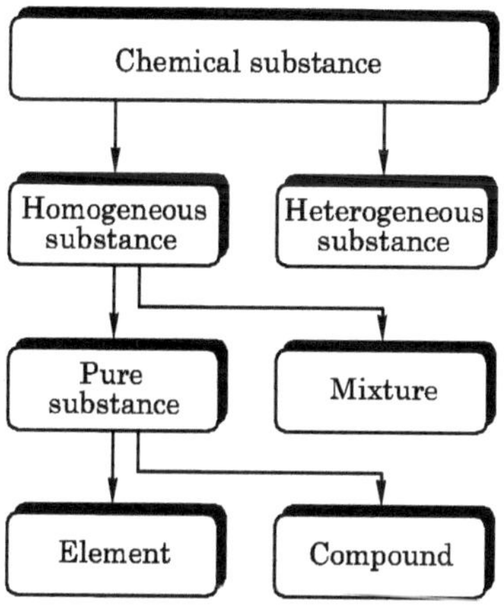

Figure 1.9. A homogeneous substance may be a pure substance or a mixture of several pure substances. A pure substance is an element or a compound.

Concentration

The composition of mixtures and solvents is described by certain, standardized concentration measures. The meaning of these measures is illustrated by a specific example where two substances, A and B, are mixed:

$$\left.\begin{array}{l}\textbf{SubstanceA}(V_A, n_A, m_A)\\ \textbf{SubstanceB}(V_B, n_B, m_B)\end{array}\right\} \rightarrow \textbf{MixtureAB}(V_{mix}, (n_A + n_B), (m_A + m_B))$$

where V denotes volumes, n denotes the amount of substances, and m denotes masses of the individual components.

The following table contains an overview of the most frequently used concentration measures of this binary system. The definitions may be directly extended to apply to multi-component systems.

Name	Symbol	Definition	Unit
Mass concentration	ϱ_B	$\frac{m_B}{V_{mix}}$	kg/m^3
Molar concentration	c_B; $[B]$	$\frac{n_B}{V_{mix}}$	mol/m^3; mol/ℓ
Molality	$\widetilde{m}_B$	$\frac{n_B}{m_A}$	mol/kg
Mole fraction	x_B	$\frac{n_B}{n_A + n_B}$	non-dimensional
Mass fraction	w_B	$\frac{m_B}{m_A + m_B}$	non-dimensional

Table 1.3. Normally used names, symbols, definitions and units of concentration measures to describe solution of substance B into substance A.

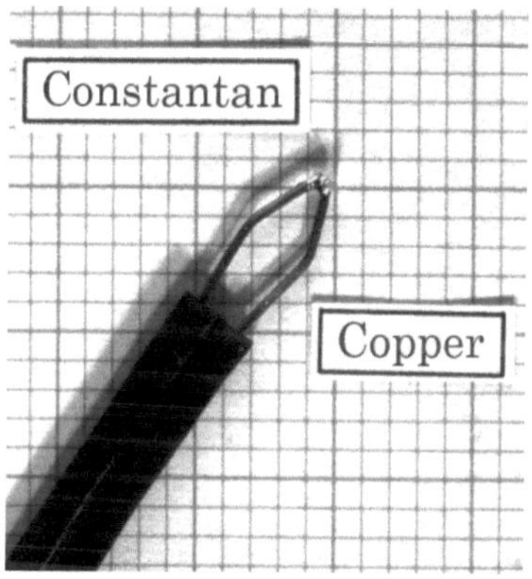

Figure 1.10. When alloying, a number of metals can form solid solutions in the form of homogeneous mixed crystals. An example is the Cu-Ni alloy constantan that, among others, is used to make thermocouples for measuring temperature.

In the definition of mass concentration ϱ_B, and of the molar concentration c_B the volume V_{mix} and *not* the sum of the component volumes $(V_A + V_B)$ is included. In mixtures, secondary chemical bonds between atoms or molecules will often produce the result that the volume of the mixture is less than the sum of the component volumes.

■ An ethanol/water mixture containing 44.00 wt-% of ethanol C_2H_5OH, has the density 0.93069 g/cm^3 at 15 °C. At the same temperature the density of undiluted ethanol and water is 0.79365 g/cm^3 and 0.99910 g/cm^3, respectively. Estimate the reduction in volume when ethanol and water are mixed in this proportion !

Solution. Evidently the mass of 1000.0 mℓ mixture is 930.69 g, of which 44.00 wt-%, i.e. 409.50 g, is ethanol and 56.00 wt-%, i.e. 521.19 g, is water. Therefore, the volume before mixing V_0 is:

$$V_0 = 409.50/0.79365 + 521.19/0.99910\,\text{cm}^3 = \mathbf{1037.6\,cm^3}$$

Intermolecular forces have caused a contraction of $\mathbf{37.6\,cm^3}$!

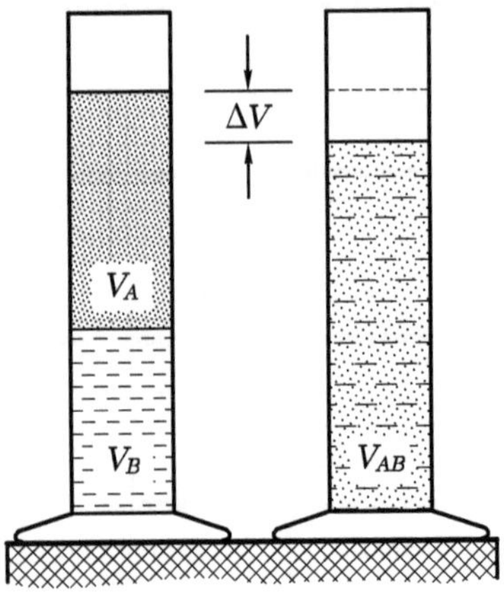

Figure 1.11. Normally, the volume of a mixture of two pure substances is less than the sum of the component volumes. Intermolecular forces may cause a significant reduction of volume when the substances are mixed.

☐ 1. A metal alloy contains 60 wt-% of lead Pb and 40 wt-% of tin Sn; determine the mole fractions x of Pb and Sn in the alloy!

☐ 2. Calculate the mass fraction w_{Fe} of iron Fe in iron ore hematite Fe_2O_3 and magnetite Fe_3O_4!

☐ 3. A solution contains 10.00 g of dissociated $CaCl_2$ per 500.0 mℓ; calculate the molar concentrations c of $CaCl_2$ and $[Cl^-]$ of chloride ions in the solution!

☐ 4. At 25 °C moisture saturated atmospheric air contains 23.1 g water vapour per m^3; calculate the molar concentration c and the mass concentration ϱ of water vapour H_2O!

☐ 5. An aqueous solution of sulphuric acid H_2SO_4 with density 1380 kg/m^3 contains 48.0 wt-% of H_2SO_4. Calculate the molar concentration and the molality of H_2SO_4 in the solution!

1.7 The ideal gas law

Dependent on pressure and temperature, most substances may exist in either a **solid**, **liquid**, or **gaseous** state. In the gaseous state the intermolecular forces are so weak that the molecules exist as freely movable particles. Therefore, gases will expand to fill up an arbitrarily large volume. Solid or liquid substances, on the contrary, have a well-defined volume because the space between the molecules are determined by intermolecular bonds.

In physical chemistry "real gases" are distinguished from "ideal gases". In a **real gas** the active intermolecular forces influence the gas properties; in an **ideal gas** the molecules behave as freely movable particles that interact through random, mutual, elastic impacts only.

It is general for gases that decreasing pressure and increasing temperature reduce the influence of intermolecular forces. At room temperature and atmospheric pressure many gases will obey the **ideal gas law** within a reasonable degree of approximation; the approximation gets better the higher the temperature and the lower the pressure.

States of matter

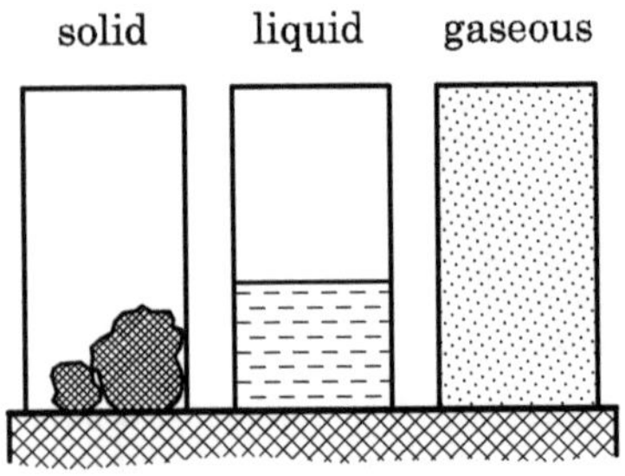

Figure 1.12. States: Solid substances have a certain volume and a certain shape. Liquids have a certain volume, but may assume any shape. Gases spontaneously fill up any volume.

The ideal gas law

$$p \cdot V = n \cdot R \cdot T = \frac{m}{M} \cdot R \cdot T \qquad (1.8)$$

p:	pressure	Pa	T:	temperature	K
V:	volume	m^3	m:	mass	kg
n:	amt. of substance	mol	M:	molar mass	kg/mol

The **universal gas constant** $R = 8.314$ J/mol · K in SI units. The **thermodynamic temperature** $T = \theta + 273.15$ K, where θ denotes the temperature measured in °C.

It is noteworthy that the composition and mass of the gas particles (i.e. atoms or molecules) is not included in the gas law; the relation between the gas pressure, volume and temperature is *solely* determined by the number of gas particles.

■ The **reduced volume** of a mass of air is defined as its volume in its standard state: 1 atm = 101325 Pa and 0 °C; determine the reduced volume of 1 mole of ideal gas!

Solution. $V = \dfrac{nRT}{p} = \dfrac{1\text{mol} \cdot 8.314\text{J/molK} \cdot 273.15\text{K}}{101325\text{Pa}} = \mathbf{22.4 \cdot 10^{-3}\ m^3}$

☐ 1. Calculate the volume V of a) 1.000 kg hydrogen $H_2(g)$ and b) 1.000 kg carbon dioxide $CO_2(g)$ at 20 °C, 1 atm, when the gases are assumed to be ideal!

☐ 2. The average molar mass of atmospheric air is 28.95 g/mol; calculate the density of dry air at 1 atm, 20 °C, when the air is assumed to obey the ideal gas law!

☐ **3.** At 1 atm, 20 °C, 1 m^3 nitrogen contains 1164.9 g $N_2(g)$; calculate the relative atomic mass A_r of the element N!

☐ **4.** Calculate the number of oxygen molecules in 1ℓ $O_2(g)$ at 100 °C, when the pressure p is 10.00 Pa!

☐ **5.** The coefficient of volume expansion α_v for solid substances, liquids and gases is defined by: $\alpha_v = V^{-1} \cdot (\partial V/\partial T)_p$; calculate α_v for an ideal gas at 0 °C!

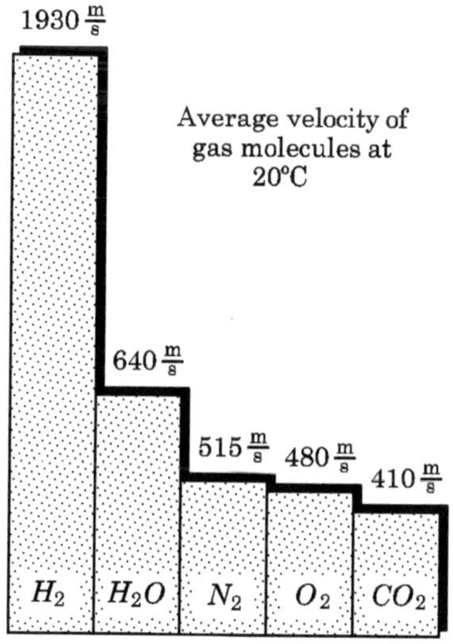

Figure 1.13. Molecules in a gas are freely movable particles that interact through random, mutual impacts. the figure shows the average velocity of the molecules of different gases at 20 °C.

1.8 Ideal gas mixture

In a mixture of ideal gases the individual gas component i exerts a **partial pressure** p_i, which is proportional to the **mole fraction** of the component in the mixture. Therefore, the total pressure in an ideal gas mixture is equal to the sum of the partial pressure of the components. This principle – **Dalton's law** – was formulated by the English physicist Dalton in 1801:

Dalton's law (1.9)

In an **ideal gas mixture** with the pressure p, the individual gas component i exerts a **partial pressure** p_i determined by:

$$p_i = x_i \cdot p \qquad \text{(a)}$$

where x_i denotes the mole fraction of the component in the mixture. The total pressure p in an ideal gas mixture is equal to the sum of the partial pressure of the components:

$$p = p_1 + p_2 + \cdots + p_n \qquad \text{(b)}$$

Dalton's law means that the ideal gas law can be applied to the individual gas component in an ideal gas mixture.

Atmospheric air

Atmospheric air is an example of a gas mixture that is often included in calculations in building technology. The natural content of **oxygen** and **water vapour** in the air, for example, is an important factor for corrosion processes, and the **carbon dioxide** in the air is an active component in the hardening of lime mortar. The composition of **dry** atmospheric air is given in the following table.

Component	Chem. symbol	M g/mol	vol-%	wt-%
Nitrogen	N_2	28.0134	78.09	75.51
Oxygen	O_2	31.9988	20.95	23.16
Carbon dioxide	CO_2	44.0100	0.03	0.05
Noble gases	—		0.93	1.28

Table 1.4. Composition of dry atmospheric air; the specified content of noble gases includes: Ne, He, Kr and Xe. Average molar mass for dry atmospheric air is 28.95 g/mol.

At normal pressures, atmospheric air obeys, as a good approximation, Dalton's law for ideal gas mixtures.

Figure 1.14. Atmospheric air has a natural content of carbon dioxide CO_2 of approx. 0.03 vol-%. This modest content of CO_2 is significant to the hardening of lime mortar due to the reaction:

$Ca(OH)_2 + CO_2 \rightarrow CaCO_3 + H_2O$

Humidity

Atmospheric air normally contains humidity in the form of water vapour. It is characteristic of the water vapour contained that the partial pressure p at any temperature has a maximum value – **the saturated vapour pressure** p_s – which corresponds to equilibrium between free water and the content of water vapour in air. The partial pressure p_s of saturated water vapour as a function of the temperature θ is given in the vapour pressure table in the appendix, page B.26.

At normal pressure and temperature conditions water vapour can, with a good approximation, be described by the ideal gas law, and humid air can in technical calculations be described as an ideal gas mixture.

Figure 1.15. Atmospheric air contains humidity in the form of invisible water vapour; when the air is cooled down this water vapour can condensate into microscopic drops, which can for example be seen when clouds are formed.

The humidity of the air is normally given in one of the following ways: as the actual **partial pressure** p of the water vapour, or as the **absolute humidity** u expressing the content of water vapour in g/m^3 air. For example, the partial pressure p_s of saturated water vapour at 20 °C is 2338.8 Pa, and the absolute humidity of the air u in this condition is 17.31 g/m^3.

The degree of moisture saturation of the air is described by specifying the **relative humidity** RH, which is defined as follows:

Relative humidity

$$RH \overset{\text{def}}{=} \frac{p}{p_s} = \frac{\text{actual partial pressure}}{\text{saturated vapour pressure}} \tag{1.10}$$

The relative humidity RH specifies the degree of moisture saturation of the air.

From the definition of relative humidity it is seen that the partial pressure of the water vapour p in fact assumes the value p_s when $RH = 100$ %. To calculate the absolute humidity of a mass of air it is necessary to know *both* the relative humidity RH *and* the temperature θ.

■ A mass of air of temperature 0 °C and relative humidity 90% is heated at constant pressure to 30 °C; calculate the relative humidity RH of the mass of air after heating!

Solution. The partial pressure of the water vapour does not change by heating. According to the vapour pressure table, p_s is 611.3 Pa at 0 °C and 4245.5 Pa at 30 °C. Thus, after heating:

$$RH \overset{\text{def}}{=} \frac{p}{p_s} = \frac{0.90 \cdot 611.3\text{Pa}}{4245.5\text{Pa}} \cdot 100\ \% = \mathbf{13.0}\ \%$$

Figure 1.16. Moisture saturated air contains approximately 9 g of water vapour per m^3 at 10 °C (left); at 50 °C the content is approximately 83 g of water vapour per m^3 (right).

☐ **1.** Dry atmospheric air has an average molar mass M of 28.95 g/mol; calculate how many g of oxygen O_2(g) are contained in 1 m^3 air at atmospheric pressure and 20 °C!

☐ **2.** A container with the volume 100 ℓ contains at 25 °C an ideal gas mixture of 90.41 g N_2(g) and 0.807 mol CO_2(g); what is the pressure p in the container?

☐ **3.** Determine from the vapour pressure table the partial pressure p of water vapour in atmospheric air at 30 °C, 80 % RH!

☐ **4.** The dimensions of a tunnel plant for hardening of concrete roofing tiles are: 30m ×8m ×5m; how many kg of water vapour are contained in the chamber air at 40 °C, 95 % RH?

☐ **5.** The definition of the compressibility κ of solid substances, liquids and gases is: $\kappa = -V^{-1} \cdot (\partial V/\partial p)_T$ (Pa^{-1}); calculate κ for air at atmospheric pressure!

1.9 Real gases

The ideal gas law describes, to a good approximation, the behaviour of real gases at room temperature and atmospheric pressure. At high pressures or low temperatures, however, it may be necessary to use an extended gas law to get a satisfactory description of the behaviour of gases.

When real gases deviate from ideal gases this is in particular due to two conditions:

- attractive intermolecular forces act between the gas molecules; this attraction entails a **reduction** of the pressure exerted by the molecules on the surroundings.
- the gas molecules themselves take up part of the available volume; this reduction of the available volume entails an **increase** of the pressure exerted by the molecules on the surroundings.

The van der Waals equation

Adjustment for these two opposite effects has been made in the important **van der Waals equation** (1873) for real gases:

The van der Waals equation

$$(p + a \cdot (\frac{n}{V})^2) \cdot (V - n \cdot b) = n \cdot R \cdot T \qquad (1.11)$$

p:	pressure	Pa	T:	temperature	K
V:	volume	m^3	n	amt. of substance	mol
a:	constant	m^6Pa/mol^2	b:	constant	m^3/mol

The **universal gas constant** $R = 8.314$ J/mol · K

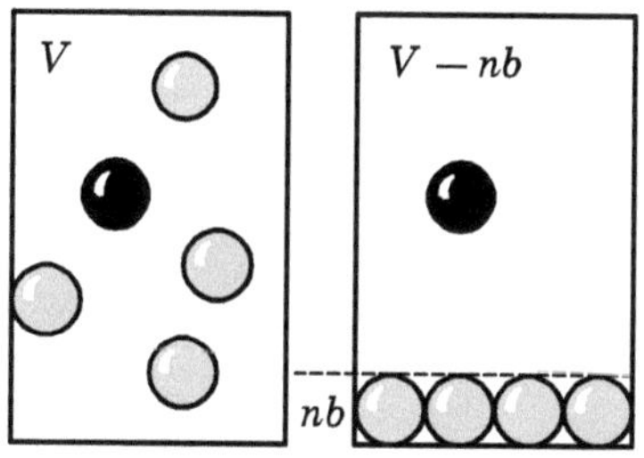

Figure 1.17. In the van der Waals equation the term $(n \cdot b)$ corrects for the reduction of available volume occurring when the concentration of gas molecules is increased in a system; the constant b is a measure of the molar "volume" of the actual molecules in the gas.

In the van der Waals equation the term $a \cdot (n/V)^2$ corrects for the reduction of pressure due to attractive forces between the gas molecules. The quantity $n \cdot b$ corrects for the reduction of the available volume due to the gas molecules. The physical meaning of these correction terms can be seen by comparison with the ideal gas law (1.8):

$$p_{ideal} = p_{real} + a \cdot (\frac{n}{V})^2 = p_{real} + \Delta p$$

$$V_{ideal} = V_{real} - n \cdot b \qquad = V_{real} - \Delta V$$

The pressure according to the ideal gas law p_{ideal} is **larger**, than the actual pressure p_{real} exerted by the system when attractive forces act between the molecules. The actual volume V_{ideal} in which the gas molecules can move is **less** than the apparent volume of the system V_{real}.

The van der Waals constants

At one point the ideal gas law deviates fundamentally from the van der Waals equation. The ideal gas law is of general validity; the real gas law is dependent on substance since the magnitude of the **van der Waals constants** a and b depends on the molecular properties of the gas considered. The constants have been experimentally determined for a large number of gases and are available in the literature. The following table contains the van der Waals constants for some chosen gases; in Appendix B, page B.5 more comprehensive numerical information is given.

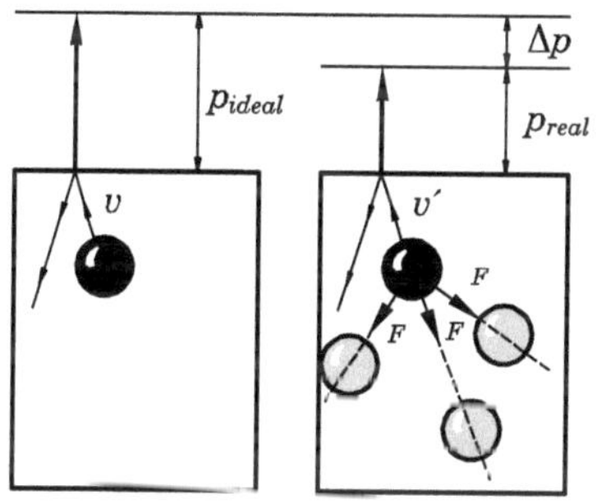

Figure 1.18. In the van der Waals equation the term $a \cdot (n/V)^2$ denotes the reduction in pressure which is due to attractive forces between the gas molecules in a system. With increasing concentration the velocity of the molecules, and thus the impulse on the wall, is reduced.

Substance	gas	**a** m^6Pa/mol^2	**b** m^3/mol
Helium	He(g)	0.0035	$23.70 \cdot 10^{-6}$
Hydrogen	H_2(g)	0.0248	$26.61 \cdot 10^{-6}$
Nitrogen	N_2(g)	0.1408	$39.13 \cdot 10^{-6}$
Oxygen	O_2(g)	0.1378	$31.83 \cdot 10^{-6}$
Carbon dioxide	CO_2(g)	0.3640	$42.67 \cdot 10^{-6}$
Water vapour	H_2O(g)	0.5536	$30.49 \cdot 10^{-6}$

Table 1.5. Overview of the van der Waals constants for some chosen gases; Appendix B contains a more comprehensive table of these constants.

The significance of the van der Waals equation is based on the fact that apart from expressing the properties of real gases over a large pressure and temperature range, it also expresses the **liquefaction** and transformation to **liquid state** of real gases as can be seen from the following section.

Figure 1.19. Industrial gas such as nitrogen N_2 and oxygen O_2 that cannot be brought to liquefy at room temperature are sold in steel bottles at a pressure of approx. 150 atm; the density of N_2 at this pressure is approx. 175 kg/m^3.

■ At 25 °C a 40.0ℓ pressure bottle contains 16.0 mol of oxygen $O_2(g)$; calculate the relation p_r/p_p between the gas pressure p_r based on the van der Waals equation and the gas pressure p_p based on the ideal gas law!

Solution. The ideal gas law (1.8) and the van der Waals equation (1.11) give:

$$\frac{p_r}{p_p} = (\frac{nRT}{V-nb} - a \cdot (\frac{n}{V})^2)/(\frac{nRT}{V}) = \frac{V}{V-nb} - \frac{na}{RTV}$$

$$\frac{p_r}{p_p} = \frac{0.040}{0.040 - 16.0 \cdot 31.83 \cdot 10^{-6}} - \frac{0.1378 \cdot 16.0}{8.314 \cdot 298.15 \cdot 0.040} = \mathbf{0.991}$$

Thus the difference in calculated pressure is approx. 1 %.

□ **1.** The van der Waals constant a is 0.5536 m^{6}Pa/mol^2 for water vapour $H_2O(g)$ and 0.0035 m^{6}Pa/mol^2 for helium He(g); explain why $a(\text{He}) \ll a(H_2O)$!

□ **2.** Calculate the average space in Å between the gas molecules in an ideal gas of argon Ar(g) at 25 °C for the following pressures: a) 1 atm b) 10 atm and c) 30 atm!

□ **3.** Calculate the reduction in pressure Δp, caused by the intermolecular forces in saturated water vapour at 25 °C?

□ **4.** Calculate the pressure of $CO_2(g)$ at 25 °C, when the molar gas concentration c is 50 mol/m^3!

□ **5.** At 25 °C, a volume of 1 m^3 contains 40.00 mol $N_2(g)$; determine the deviation in % between the gas pressure p_{ideal}, calculated according to the ideal gas law, and the gas pressure p_{real}, calculated according to the van der Waals equation!

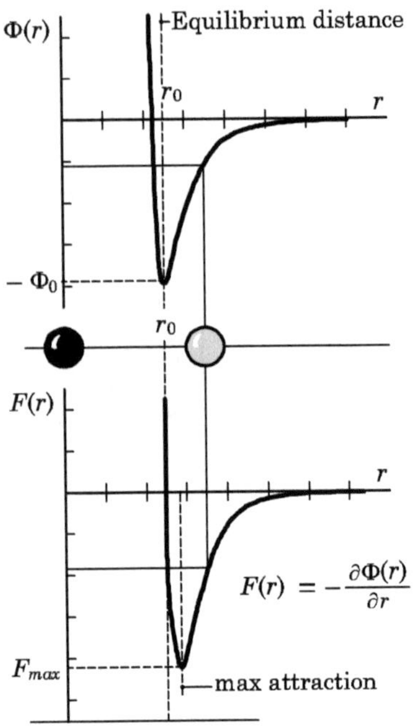

Figure 1.20. The Lennard-Jones potential $\Phi(r)$ and acting force $F(r)$ between two gas molecules. Note that the force $F(r)$ is zero and that the potential energy Φ is at a minimum in the equilibrium distance $r = r_0$.

1.10 Intermolecular forces

Any atom is formed by a **positively charged** nucleus surrounded by a **negatively charged** electron cloud. This structure entails that all atoms and molecules in a system of matter are provided with a surrounding, electrostatic field. By interaction between fields of this kind, **attractive** or **repulsive**, intermolecular forces arise.

Attractive forces arise when atoms and molecules by mutual action form electric **dipoles**; the attractive dipole forces are **weak**, and the distances over which they act correspond to a few times the atom diameter. On the other hand, when two atoms are forced to a distance which is less than the characteristic atom diameter, very **strong** repulsive **nucleus forces** arise.

Dipole forces of the kind mentioned here entail that gases at low temperatures may liquefy because the molecules are mutually bound into a state with low potential energy. It must be emphasized, however, that the bonds between adjacent atoms or molecules in a liquid are so-called **secondary bonds** which are much weaker than the **primary bonds** that hold the atoms together in molecules.

The Lennard-Jones potential

The **potential energy** $\Phi(r)$ of two mutually interacting molecules is a function of the distance r between the molecules. Mathematically the function $\Phi(r)$ can be approximately described by the so-called **Lennard-Jones** potential:

The Lennard-Jones potential (1.12)

$$\Phi(r) = \Phi_0 \cdot \left[(\tfrac{r_0}{r})^{12} - 2 \cdot (\tfrac{r_0}{r})^6 \right] \qquad \text{(J/mol)}$$

The function $\Phi(r)$ denotes the **potential energy** between two molecules as a function of the **distance** r between the centres of the molecules; Φ_0 and r_0 are substance-dependent constants.

The acting **force** $\mathbf{F}(r)$ between two molecules is determined by: $\mathbf{F}(r) = -\partial\Phi(r)/\partial r$; therefore, the intermolecular forces $\mathbf{F}(r)$ corresponding to the Lennard-Jones potential (1.12) are:

$$\mathbf{F}(r) = -\frac{\partial \Phi(r)}{\partial r} = \Phi_0 \cdot \frac{12}{r_0} \cdot \left[(\tfrac{r_0}{r})^{13} - (\tfrac{r_0}{r})^7 \right] \qquad (1.13)$$

The **equilibrium distance** between two molecules is achieved when the intermolecular force $\mathbf{F}(r)$ is 0; from (1.13) it is seen that the constant r_0 denotes the equilibrium distance between two molecules. Further it is seen that the Lennard-Jones potential (1.12) has **minimum** $\Phi(r) = -\Phi_0$ for $r = r_0$, i.e. when the molecules are at their mutual equilibrium distance. For $r \to \infty$, $\Phi(r)$ and $F(r) \to 0$.

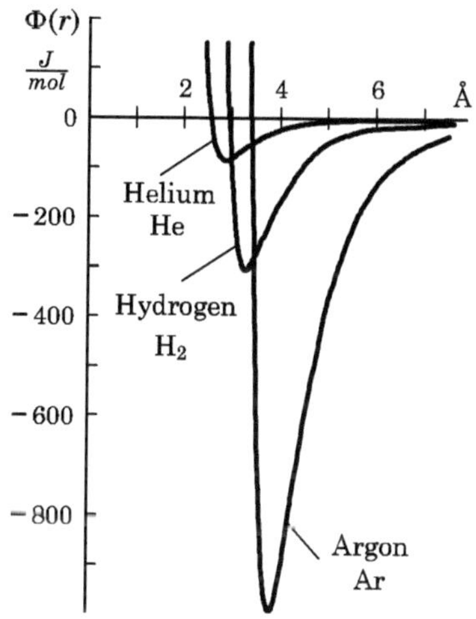

Figure 1.21. The Lennard-Jones potential for helium He, hydrogen H_2 and argon Ar, calculated from the parameters in table 1.6. An increasing bond energy in the series: He < H_2 < Ar is also reflected in the increasing values of the boiling points of the substances, which are: 4 K, 20 K and 87 K, respectively.

The Lennard-Jones parameters

The following table shows numerical values of the Lennard-Jones parameters Φ_0 and r_0 for some simple gases.

Substance	Gas	Φ_0 (J/mol)	r_0 (Å)	Boiling point (K)
Helium	He(g)	85	2.87	4
Hydrogen	H_2(g)	308	3.29	20
Argon	Ar(g)	996	3.82	87
Nitrogen	N_2(g)	791	4.15	77
Oxygen	O_2(g)	984	3.88	90

Table 1.6. Overview of the Lennard-Jones parameters for some chosen gases (Hill,T.L.: "Lectures on Matter & Equilibrium").

It should be noted that dipole forces introduce bond energies in the range of 0.1-1.0 kJ/mol; for comparison the primary chemical bonds in molecules and crystals typically have a bond energy of 100-1000 kJ/mol.

Hydrogen bond

Molecules in which the hydrogen ion H^+ is bonded to strongly electronegative ions such as fluorine F^-, oxygen O^{--}, or nitrogen N^{---} normally have an asymmetric charge distribution; thus, this kind of molecules are **polar**, and exist as permanent dipoles. Therefore, between these molecules a particularly strong **dipole bond**, the so called **hydrogen bond**, can be formed.

A hydrogen bond is formed between a bound hydrogen atom, with a positive charge surplus, and a nearby F, O or N atom with local net negative charge adjacent to the bound hydrogen atom. Examples of hydrogen bonds ($\cdots$) are.

$$\mathrm{F-H\cdots F} \qquad \mathrm{O-H\cdots O} \qquad \mathrm{N-H\cdots N} \qquad \mathrm{F-H\cdots O}$$

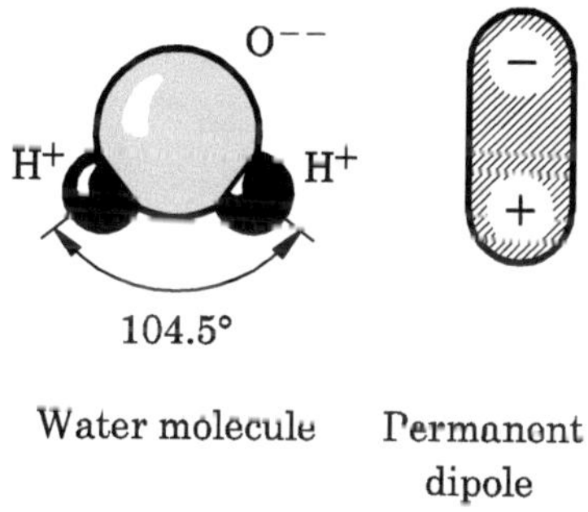

Figure 1.22. The molecule H_2O has an asymmetric charge distribution causing the water molecule to form a permanent electrical dipole. Due to this, water gets special properties as a solvent for salts and the molecules can be adsorbed by hydrogen bonding on the surface of solid substances.

The hydrogen bond is a significantly stronger bond than the induced dipole bonds occurring in non-polar substances such as Ar and O_2. Typical bond energies for the hydrogen bond are 20-30 kJ/mol.

The water molecule H_2O belongs to the group of substances that readily form hydrogen bonds. This property of the H_2O molecule entails that water has a number of special physical and chemical properties in relation to closely related chemical compounds.

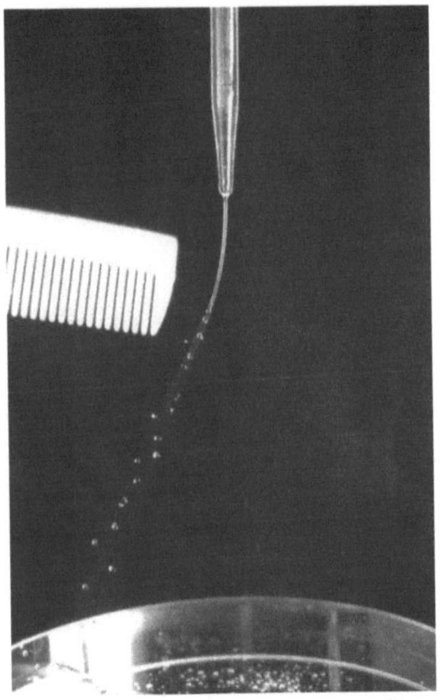

Figure 1.23. The polar structure of the H_2O molecule entails that a water jet is influenced by an electrical field. The photo shows this effect caused by the electrical field around a comb which has been charged by rubbing it against a cloth.

■ The potential energy between two argon atoms is assumed to be described by the Lennard-Jones potential (1.12); according to table 1.6 the equilibrium distance r_0 is then 3.82 Å . Determine the distance r_m between the atoms, where the attractive force $\mathbf{F}(r)$ assumes its maximum value $\mathbf{F}_m$!

Solution. The acting force $\mathbf{F}(r)$ is determined by (1.13); maximum of the function is determined by the condition:

$$\frac{\partial \mathbf{F}}{\partial r} = -\Phi_0 \cdot \frac{12}{r_0^2} \cdot \left[13 \cdot (\frac{r_0}{r_m})^{14} - 7 \cdot (\frac{r_0}{r_m})^8\right] = 0$$

By reduction: $r_m = (13/7)^{1/6} \cdot r_0 = 1.11 \cdot 3.82$ Å= **4.24 Å**

□ **1**. State the approximate value of: a) dipole bond, non-polar molecules, b) hydrogen bond and c) primary chemical bonds; unit kJ/mol.

□ **2**. In the water molecule H_2O the valence angle between the two O—H bonds is 104.5°; show how water molecules can form mutual hydrogen bonds!

□ **3**. The potential energy $\Phi(r)$ between the molecules in an argon gas is assumed to comply with the Lennard-Jones potential; find the distance $r > r_0$ where $|\Phi(r)/\Phi_0| \approx 1$ %!

□ **4**. Based on the ideal gas law, find the mean distance $\bar{r}$ between the molecules in N_2(g) at 25 °C for $p = 1$, 10 and 50 atm and calculate $\Phi(\bar{r})$ between two molecules at these $\bar{r}$ values!

□ **5**. Molecules complying with the Lennard-Jones potential have a potential energy $\Phi(r^*) = 0$ at the distance $r^* < r_0$; determine r^* for He(g) and N_2(g) with the unit Å!

1.11 Critical temperature

The states of matter – solid, liquid, or gaseous – is determined by a balance between two opposite tendencies. The molecules of any system of matter are affected by:

- **Intermolecular forces** furthering formation of a dense, organized structure with a low potential energy, i.e. a **solid** or **liquid** state.
- **Thermal molecular movements** furthering the formation of a dispersed state with freely movable molecules, that is, a **gaseous state**.

The ideal gas law (1.8) describes the macroscopic properties of a system of gas particles that *solely* interact through random, elastic impacts; in the ideal gas the thermal molecular movements dominate, and the gas properties are *independent* of intermolecular forces.

The van der Waals equation (1.11) describes the macroscopic properties of a real gas where the intermolecular forces and thermal molecular movements are in mutual balance. From a molecular point of view the real gas is a **transition state** between the ideal gas and the condensed, liquid and solid states.

The van der Waals equation illustrates in a remarkably simple way the physical phenomena that determine changes of state of matter.

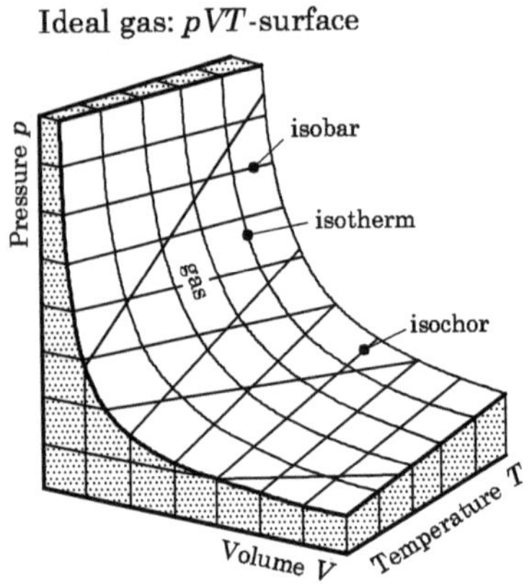

Figure 1.24. pVT surface for an ideal gas. In this picture the isotherms form a cluster of regular hyperbolas: $p = k \cdot V^{-1}$. Isobars and isochors, respectively, run as straight lines on the pVT-surface.

The van der Waals isotherms

Intermolecular forces entail that a real gas will exert a smaller pressure than is predicted by the ideal gas law. At low temperatures and high pressures, these forces can cause gases to liquefy – change state – because the molecules are mutually bonded in a condensed liquid phase with low potential energy.

The graphical depiction of the pressure p as a function of volume V at constant temperature T is called an **isotherm**. For an ideal gas the isotherms form a cluster of **hyperbolas**, corresponding to the equation:

$$p = \frac{nRT}{V} = \text{constant} \cdot (\tfrac{1}{V}) \qquad (T \text{ constant, ideal gas}) \qquad (1.14)$$

For a real gas the isotherms run in a more complicated way. According to the van der Waals equation (1.11), an isotherm is described by:

$$p = \frac{nRT}{V - nb} - a \cdot (\tfrac{n}{V})^2 \qquad (T \text{ constant, real gas}) \qquad (1.15)$$

At sufficiently high temperatures, isotherms for a real gas are hyperbolic in accordance with (1.14). At lower temperatures, deviations occur due to intermolecular forces between the gas molecules, and below a certain temperature, a horizontal discontinuity of the curve (fig. 1.25) is seen.

The discontinuity where the pressure p is independent of V, shows that the gas is **liquefied**; in this field there is **equilibrium** between two phases: gas$\rightleftharpoons$liquid. The pressure p corresponds to the **saturated vapour pressure** p_s of the liquid phase. By compression in this state, gas will liquefy so that the pressure in the system is kept constant. After liquefaction of the gas, the isotherm increases steeply since a liquefied phase is only to a small degree influenced by the pressure.

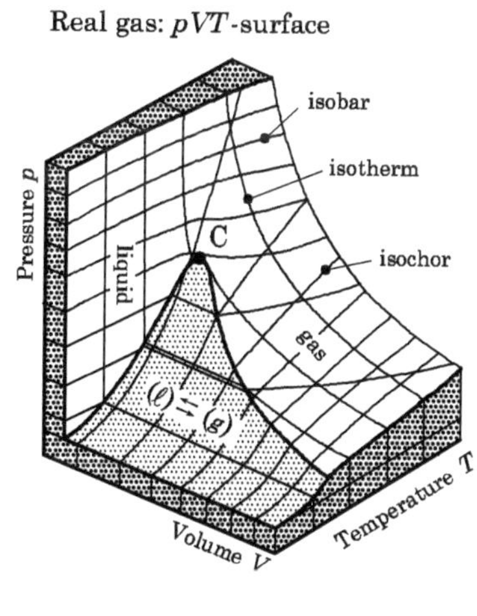

Figure 1.25. The pVT surface for a real gas with the critical point: (p_c, V_c, T_c). Below the critical temperature T_c the gas can be brought to liquefy. Above the critical temperature T_c the substance can only exist in the gaseous phase irrespective of the magnitude of the pressure.

The critical point

With rising temperature the discontinuous area of the isotherm, where there is equilibrium between two phases, will narrow; at a certain temperature T_c the area will decrease to a point, the so-called **critical point**. Above this temperature T_c the substance is in a gaseous state at all pressures and densities. Therefore, the condition for liquefaction of a gas is that the gas temperature is below T_c.

On the pVT-surface the critical point for a gas is related to the **critical temperature** T_c, the **critical pressure** p_c and the **critical volume** V_c. At the critical point, the isotherm T_c will fulfil the following two conditions: a) it has horizontal tangent at the point: $\partial p/\partial V = 0$ and b) it has inflection at the point: $\partial^2 p/\partial V^2 = 0$. For a van der Waals gas the condition for horizontal tangent at the critical point is

$$\left(\frac{\partial p}{\partial V}\right)_T = -\frac{nRT}{(V - nb)^2} + \frac{2an^2}{V^3} = 0 \qquad \text{(horizontal tangent)} \qquad (1.16)$$

Correspondingly, the condition for inflection at the critical point is

$$\left(\frac{\partial^2 p}{\partial V^2}\right)_T = +\frac{2nRT}{(V - nb)^3} - \frac{6an^2}{V^4} = 0 \qquad \text{(inflection)} \qquad (1.17)$$

If these equations and the equation for the van der Waals gas (1.11) are solved, the critical temperature T_c, the critical pressure p_c and the molar, critical volume $v_c = V_c/n$ can be expressed by the van der Waals constants a and b:

The critical point (1.18)

For a **real gas** complying with the van der Waals equation the **critical temperature** T_c, the **critical pressure** p_c and the molar **critical volume** v_c are :

$$T_c = \frac{8a}{27bR} \text{ (K)}; \qquad p_c = \frac{a}{27b^2} \text{ (Pa)}; \qquad v_c = 3b \text{ (m}^3\text{/mol)}$$

where a and b are substance dependent van der Waals constants.

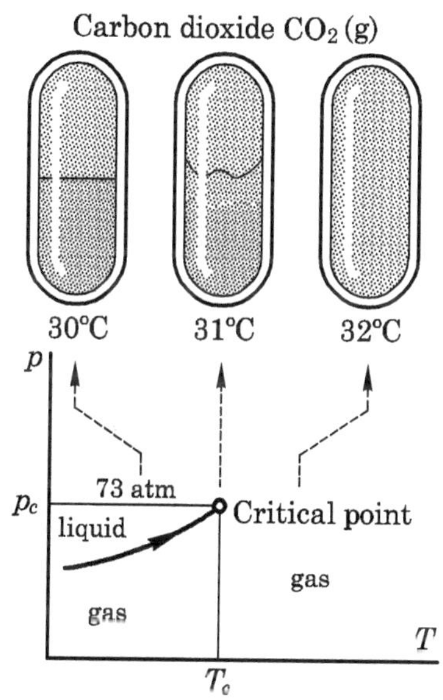

Figure 1.26. At 30°C a glass ampoule (left) contains carbon dioxide CO_2 at the critical volume V_c; the interface between the liquid and the gaseous state is clearly seen. If the ampoule is heated slowly, the interface vanishes when the critical temperature, which is approx. 31°C (right), is exceeded.

Critical constants

The following table shows experimentally determined critical constants for some chosen gases; in Appendix B a more comprehensive numerical material is shown.

Substance	Gas	T_c K	p_c MPa	v_c cm^3/mol
Helium	He(g)	5.2	0.23	57.8
Hydrogen	$H_2(g)$	33.2	1.30	65.0
Argon	Ar(g)	150.7	4.86	75.3
Nitrogen	$N_2(g)$	126.3	3.40	90.1
Oxygen	$O_2(g)$	154.8	5.08	78.0
Carbon dioxide	$CO_2(g)$	304.2	7.38	94.0

Table 1.7. Experimentally determined values of critical temperature T_c, critical pressure p_c, and critical molar volume v_c for some chosen gases; Data converted to SI units (Based on Atkins: "Physical Chemistry").

Figure 1.27. Technical carbon dioxide CO_2 is sold in pressure bottles. At room temperature CO_2 exists in a liquid and a gaseous state, depending on pressure. Above the critical temperature of approx. 31 °C, CO_2 is transformed into a homogeneous gaseous phase.

By experience, critical temperature T_c and critical pressure p_c can be predicted with a good accuracy from the van der Waals equation via (1.18); the calculated value of critical volume V_c according to (1.18) may, however, deviate somewhat from test data.

For a number of common gases the critical temperature T_c is above room temperatures; examples include carbon dioxide $CO_2(g)$, sulphur dioxide $SO_2(g)$ and ammonia $NH_3(g)$. These gases can be liquefied by compression alone and can therefore be stored in pressure bottles in their **liquid state.** For other gases such as oxygen $O_2(g)$, nitrogen $N_2(g)$, and hydrogen $H_2(g)$, the critical temperature T_c is considerably lower; these gases can only be liquefied after **cooling.**

■ Calculate the **critical temperature** T_c (°C) and the **critical pressure** p_c (atm) for water vapour, and comment on the result!

Solution. The targeted critical constants can be determined from (1.18). The van der Waals constants for water vapour $H_2O(g)$ are cf. table 1.5: $a = 0.5536$ m^6Pa/mol^2 and $b = 30.49 \cdot 10^{-6}$ m^3/mol

$$T_c = \frac{8a}{27bR} = \frac{8 \cdot 0.5536}{27 \cdot 30.49 \cdot 10^{-6} \cdot 8.314} = 647 \text{ K} = \mathbf{374\ °C}$$

$$p_c = \frac{0.5536}{27 \cdot (30.49 \cdot 10^{-6})^2} = 2.21 \cdot 10^7 \text{ Pa} = \mathbf{218\ atm}$$

The critical temperature T_c for water vapour is considerably higher than room temperature; accordingly, water exists at room temperature as a **two-phase system** with **equilibrium** between liquid phase (water) and gaseous phase (water vapour). The vapour pressure curve for water shows how this equilibrium depends on the temperature.

Figure 1.28. Steam turbines in modern power plants are operated by water vapour at a pressure of 110-115 bar and a temperature of approx. 540 °C. This is above the critical temperature of water 374 °C; therefore, H_2O can only exist in a gaseous form under these operational conditions.

☐ **1.** Specify which of the gases: nitrogen $N_2(g)$, oxygen $O_2(g)$, and carbon dioxide $CO_2(g)$ can be liquefied at room temperature!

☐ **2.** Rewrite the equations of conditions (1.16) and (1.17), so that the molar volume v m^3/mol is used instead of the absolute volume $V m^3$!

☐ **3.** Mercury Hg(g) has: $a = 0.820$ m^6Pa/mol^2 and $b = 16.96 \cdot 10^{-6}$ m^3/mol; calculate the critical temperature T_c (°C) and the critical pressure p_c (atm) for Hg(g)!

☐ **4.** For monosilane SiH_4 the critical temperature is −3.46 °C and the critical pressure is 47.8 atm; based on this, determine the van der Waals constants a and b for SiH_4!

☐ **5.** Butane C_4H_{10} that is sold in pressure bottles for technical heating purposes has the van der Waals constants: $a = 1.466$ m^6Pa/mol^2 and $b = 122.6 \cdot 10^{-6}$ m^3/mol; can butane be delivered in the liquid state at room temperature?

1.12 SI units

The International Standards Association has in the series ISO 31 published a number of **standard sheets** specifying physical quantities, measurement units and symbols used within the essential areas of science and technology. The ISO 31 series describes the international system of units **SI**, which is short for *"Systéme international d'Unité"*. The following overview comprises a range of SI units, symbols and terms that are frequently used to describe systems of substance.

A comprehensive overview of standards for SI units, physical quantities, measurement units and symbols can be found in *Tables of physical and chemical constants* by G.W.C. Kaye and T.H. Laby.

Figure 1.29. Laboratory scales in a glass box from around 1850. Weighing has played an essential role for the development of the chemistry of substances; accurate weighing could be carried out by such balances since relatively early times.

Base units

The international SI unit system is based on seven **base quantities**: *length, mass, time, electric current, temperature, amount of substance* and *luminous intensity*. For each base quantity a practical unit of measurement – the so-called **base unit** is defined. The SI system determines the following seven base quantities and base units:

SI base units (1.19)

The SI unit system os based on the following **seven** base units:

Base quantity	Base unit	Symbol	Dimension
length	metre	m	L
mass	kilogram	kg	M
time	second	s	T
electric current	ampere	A	I
thermodynamic temperature	kelvin	K	Θ
amount of substance	mole	mol	N
luminous intensity	candela	cd	J

As an example, the base unit of **measurement** for mass is defined as the mass of the **international kilogram prototype**; kg is used as symbol for the base unit kilogram. As symbol for the base quantity mass, the dimension M is used.

Figure 1.30. The Danish prototype for mass is a 1 kilogram weight made from platinum alloyed with 10 % iridium. The weight is kept at Danish National Metrology Institute at the Technical University of Denmark.

Derived units

Based on the seven base units of the SI system, **derived units** for other physical quantities are formed. Often a special name and symbol are used for the derived units; however, these can *always* be expressed by the base units of the SI system. Thus, the derived pressure unit pascal Pa can be expressed by the base units of the SI system as: $kg \cdot m^{-1} \cdot s^{-2}$. Derived units that are often used within physical chemistry are:

Quantity	Unit	Symbol	Definition	SI base unit
Force	newton	N	$1\,N = 1\,kg \cdot m \cdot s^{-2}$	$kg \cdot m \cdot s^{-2}$
Energy	joule	J	$1\,J = 1\,N \cdot m$	$kg \cdot m^2 \cdot s^{-2}$
Power	watt	W	$1\,W = 1\,J \cdot s^{-1}$	$kg \cdot m^2 \cdot s^{-3}$
Pressure	pascal	Pa	$1\,Pa = 1\,N \cdot m^{-2}$	$kg \cdot m^{-1} \cdot s^{-2}$
El. charge	coulomb	C	$1\,C = 1\,A \cdot s$	$A \cdot s$
Potential diff.	volt	V	$1\,V = 1\,W \cdot A^{-1}$	$kg \cdot m^2 \cdot s^{-3} \cdot A^{-1}$
El. resistance	ohm	Ω	$1\,\Omega = 1\,V \cdot A^{-1}$	$kg \cdot m^2 \cdot s^{-3} \cdot A^{-2}$

Table 1.8. Derived SI units often used within physical chemistry.

Prefixes

So-called **prefixes** can be used in front of SI units; they can be used to form appropriate numerical values, names and symbols for the decimal multiples of SI units. The following prefixes are given in the SI system:

Factor	Prefix	Symbol	Factor	Prefix	Symbol
10^{18}	exa	E	10^{-1}	**deci**	d
10^{15}	peta	P	10^{-2}	**centi**	c
10^{12}	tera	T	10^{-3}	**milli**	m
10^{9}	**giga**	G	10^{-6}	**micro**	μ
10^{6}	**mega**	M	10^{-9}	**nano**	n
10^{3}	**kilo**	k	10^{-12}	**pico**	p
10^{2}	hecto	h	10^{-15}	femto	f
10	deca	da	10^{-18}	atto	a

Table 1.9. Overview of the prefixes of the SI system; the most frequently used prefixes are emphasized in the table.

Note: The base unit for mass – kilogram – contains the SI prefix *"kilo"*, so in this special case multiples of the SI unit are formed by adding the prefix to the unit **gram**. For example, notation is **milligram** (mg), and *not* microkilogram (μkg).

Special units

Some previously used units may be used together with the SI units; therefore, it is recommended that the time units **minute** (min), **hour** (h) and **day** (d) as well as volume **litre** (ℓ) are used where practical.

Further, because of their use within special areas the following special units can be used together with the SI units and multiples of thereof:

Prefix	**Symbol**	**Factor**
giga	G	10^{9}
mega	M	10^{6}
kilo	k	10^{3}
...	...	...
milli	m	10^{-3}
micro	μ	10^{-6}
nano	n	10^{-9}

Quantity	Name	Symbol	Definition
length	ångström	Å	$1 = 10^{-10}\,\text{m} = 0.1\,\text{nm}$
mass	atomic mass constant	m_u	$1\,m_u = 1.6606 \cdot 10^{-27}\,\text{kg}$
energy	electron volt	eV	$1\,\text{eV} = 1.602 \cdot 10^{-19}\,\text{J}$
pressure	bar	bar	$1\,\text{bar} = 10^{5}\,\text{Pa}$

Table 1.10. Overview of units applied to special areas and conventionally used with SI units.

■ In tables, thermal conductivity λ is frequently based on the unit: kJ/m · h °C. Determine the corresponding SI base unit!

Solution. According to table 1.8, kJ corresponds to the base unit: kg · m^2 · s^{-2}; The time unit h corresponds to the base unit s. Therefore, thermal conductivity λ has the unit:

$$\text{kJ/m} \cdot \text{h} \cdot {}^\circ\text{C} \;\rightarrow\; (\text{kg} \cdot \text{m}^2 \cdot \text{s}^{-2}) \cdot (\text{m} \cdot \text{s} \cdot \text{K})^{-1} \;\rightarrow\; \mathbf{kg \cdot m \cdot s^{-3} \cdot K^{-1}}$$

☐ **1.** Rewrite the numerical value $5.4 \cdot 10^{-10}$ s to an SI unit in prefix form and rewrite the numerical value 0.023 GJ to a decimal multiple of the unit J!

☐ **2.** The permeability coefficient K for porous materials is often denoted by the unit: kg/Pa · m · s; rewrite this unit to SI base units!

☐ **3.** An aqueous solution contains 0.050 g NaCl per litre. Calculate the molar concentration of NaCl in μmol/ℓ; M(NaCl) is 58.5 g/mol!

☐ **4.** The Lennard-Jones parameter Φ_0 for oxygen is J/mol (table 1.6); calculate Φ_0 for the O_2 molecule based on the unit electron volt eV!

☐ **5.** The typical mean grain diameter d for Portland cement is 10 μm; State d based on the units: Å, nm, mm, and m!

List of key ideas

The following overview lists the key definitions, concepts and terms introduced in Chapter 1.

Nucleon number A .... denotes the total number of **nuclear particles**, that is, protons and neutrons contained in an atom nucleus; the nucleon number A is also called **mass number**.

Proton number Z .... denotes the number of **protons** in an atom nucleus; the proton number Z is also called **atomic number**.

Neutron number N .... denotes the number of **neutrons** in an atom nucleus.

Element .... is a substance where all atoms have the **same atomic number**, i.e. contain the same number of protons in the nucleus.

Isotope .... chemically identical atoms with different **mass number** A is said to be isotopes of the same **element**.

Atomic mass constant ..... $m_u = 1.6606 \cdot 10^{-27}$ kg is defined as 1/12 of the rest mass of the carbon 12 atom.

Relative atomic mass ...... $A_r = m/m_u$ is defined as the relation between the mass of an atom and the **atomic mass constant** m_u.

Relative molecular mass ... $M_r = m/m_u$ is defined as the relation between the mass of a molecule or a formula unit and the **atomic mass constant** m_u.

Amt. of substance, mole ... the unit **mole** denotes the **amount of substance** that contains the same number of elementary parts as the number of atoms in 0.012 kg of carbon 12; symbol n.

The Avogadro constant .... $\mathcal{N} = 6.022 \cdot 10^{23}\ \text{mol}^{-1}$ denotes the number of **elementary parts** in an amount of substance of 1 mole.

Molar mass .... M g/mol denotes the mass of a mole of elementary parts such as atoms, molecules, formula units or ions; $M = M_r \cdot 10^{-3}$ kg/mol $= M_r$ g/mol.

Mass concentration .... $\rho_B = m_B/V_{mix}$ kg/m³.

Molar concentration .... c_B; $[B] = n_B/V_{mix}$ mol/ℓ.

Molality .... $\tilde{m}_B = n_B/m_A$ mol/kg.

Mole fraction .... $x_B = n_B/(n_B + n_A)$ (non-dimensional)

Mass fraction .... $w_B = m_B/(m_A + m_B)$ (non-dimensional)

The ideal gas law .... $p \cdot V = n \cdot R \cdot T = (m/M) \cdot R \cdot T$ (ideal gas)

Dalton's law .... $p_i = x_i \cdot p_{total}$ (ideal gas mixture)

Relative humidity .... $RH = p/p_s$ %

Van der Waals equation ... $(p + a \cdot (\frac{n}{V})^2) \cdot (V - nb) = n \cdot R \cdot T$ (real gas)

Lennard-Jones potential ... $\Phi(r) = \Phi_0 \cdot [(\frac{r_0}{r})^{12} - 2 \cdot (\frac{r_0}{r})^6]$ denotes the **potential energy** between two molecules as a function of the distance r between the molecules.

Hydrogen bond .... molecules in which H is bonded to strongly electronegative atoms such as fluorine F, oxygen O or nitrogen N and thus forms permanent **dipoles**; this kind of molecules may form a particularly strong **dipole bond**, the so-called **"hydrogen bond"**.

Critical point .... the point on the pVT surface for a **real gas** determined from the critical temperature T_c, the critical pressure p_c and the critical volume V_c of the gas.

Critical temperature ...... $T_c = 8a/27bR$ (K); the critical temperature for a real gas complying with the van der Waals equation. At temperatures below T_c the gas can be **liquefied**; at higher temperatures it is not possible to distinguish between the liquid and gaseous states.

Critical pressure .... $p_c = a/27b^2$ (Pa); the critical pressure for a real gas complying with the van der Waals equation.

Critical volume .... the critical molar volume is $v_c = 3b$ (m^3/mol), for a real gas complying with the van der Waals equation.

Examples

The following examples illustrate how subject matter explained in Chapter 1 can be applied in practical calculations.

Example 1.1

■ Corrosion of iron – stray current

Figure 1.31. Oxygen take-up during electrolytic corrosion of iron Fe: A glass that contains steel wool is placed with the open end immersed in water (left); in the corrosion process according to reaction equation (a), O_2 is consumed in the confined quantity of air and the water table rises in the glass in a couple of hours (right).

When the insulation of electric underground cables is damaged, so-called "stray currents" may arise, i.e. electric currents that are conducted through nearby water or gas pipes. This may cause serious **electrolytic corrosion** of the installations concerned.

The corrosion-affected installations are decomposed by a so-called anodic process acting where the positive stray current leaves the metal. The overall corrosion process includes the following transformations:

$$\textbf{Metal} + \textbf{Oxygen} + \textbf{Water} \rightarrow \textbf{Corrosion product} \quad \text{(a)}$$

This process contains two sub-reactions: a metal-dissolving **anode process** and an oxygen-consuming **cathode process**. During the corrosion of iron, Fe, these processes can be described by the reactions below:

$$2Fe \rightarrow 2Fe^{++} + 4e^- \quad \text{(anode process)} \quad \text{(b)}$$

$$O_2 + 2H_2O + 4e^- \rightarrow 4OH^- \quad \text{(cathode process)} \quad \text{(c)}$$

The combination of these reactions leads to the corrosion process for iron Fe that corresponds to (a):

$$2Fe + O_2 + 2H_2O \rightarrow 2Fe(OH)_2 \quad \text{(gross reaction)} \quad \text{(d)}$$

For calculation of electrochemical corrosion processes, the **Faraday constant** $\mathcal{F}$ is an essential quantity. The Faraday constant states the electric charge transmission Q of one mole of elementary charges e, where e denotes the electric charge of the electron or proton in coulombs C.

The Faraday constant

$\mathcal{F}$ denotes the electric charge Q of 1 mole of elementary charges e.

Problem. Determine the value of the Faraday constant $\mathcal{F}$, and calculate how many kg of iron Fe can be corroded by the anode reaction (b) of a direct current of 1 ampere acting for 1 year !

Conditions. In calculations the following numerical constants are used:

Elementary charge $e = 1.602 \cdot 10^{-19}\,\mathrm{C}$; $\mathrm{M_{Fe}} = 55.847\,\mathrm{g/mol}$

Solution. From the definition of the Faraday constant $\mathcal{F}$ it follows that:

$$\mathcal{F} = \mathcal{N} \cdot e = \text{ the Avogadro constant} \cdot \text{elementary charge} \qquad (1)$$

$$\mathcal{F} = 6.022 \cdot 10^{23}\,\mathrm{mol^{-1}} \cdot 1.602 \cdot 10^{-19}\,\mathrm{C} = \mathbf{96472\,C/mol}$$

During one year, a direct current of 1 A will entail a charge transmission Q determined by:

$$Q = \text{ amperage} \cdot \text{time} = 1\mathrm{A} \cdot 365 \cdot 24 \cdot 60 \cdot 60\,\mathrm{s} = 31.5 \cdot 10^6\,\mathrm{C}$$

The quantity of iron Fe that is corroded in the process can now be determined. It is noted that 1 mole of elementary charges corresponds to $\frac{1}{2}$ mol Fe^{++} ions in the anode process (b), so that:

$$\mathrm{m_{Fe}} = \frac{Q}{2 \cdot \mathcal{F}} \cdot \mathrm{M_{Fe}} = \frac{31.5 \cdot 10^6\,\mathrm{C}}{2 \cdot 96472\,\mathrm{C/mol}} \cdot 55.847\,\mathrm{g/mol} = 9118\,\mathrm{g}$$

Thus, the total quantity of iron Fe that is decayed per year is: **9.12 kg**

Discussion. The internationally determined value of the Faraday constant $\mathcal{F}$ is 96484.56 ± 0.27 C/mol; therefore, the applied tabular values for the elementary charge e and the Avogadro constant $\mathcal{N}$ determine the internationally accepted value of $\mathcal{F}$ with a deviation of approx. 0.01 %.

Example 1.2

■ Air hardening of lime mortar

Historically, a mortar mainly consisting of **calcium hydroxide** $Ca(OH)_2$, water and sand has been used for brick-laying. This lime mortar hardens when the carbon dioxide CO_2 of the air reacts with the $Ca(OH)_2$ and forms sparingly soluble **calcium carbonate** $CaCO_3$. The total binder reaction can be written as:

$$Ca(OH)_2 + CO_2 \rightarrow CaCO_3 + H_2O \qquad (a)$$

Thus, hardening of lime mortar requires access to carbon dioxide from the air. During hardening and strength development, water – 1 mol of water per mol of $Ca(OH)_2$ – is formed as a by-product and must be removed by drying.

It should be understood that actually the reaction of $Ca(OH)_2$ with CO_2 can take place at a reasonable speed only in a **humid environment**. Waterless calcium hydroxide can be exposed to carbon dioxide without forming $CaCO_3$, even though the equilibrium for (a) is strongly shifted towards carbonate formation. In order to react the CO_2 molecules must be dissolved in water.

For brick-laying with lime mortar in closed structures the requirement for:

- access to carbon dioxide CO_2 from the air
- removal of chemically liberated water by drying

Figure 1.32. Lime mortar for brick-laying consists of calcium hydroxide $Ca(OH)_2$, sand and water. The mortar hardens by uptake of the carbon dioxide CO_2 of the air, since sparingly soluble calcium carbonate $CaCO_3$ is formed after the binder reaction (a). Normally, lime mortar is strengthened by addition of small quantities of Portland cement.

will require a number of special measures. In practice it is a question of exchange of considerable quantities of substance in the gaseous state.

Figure 1.33. When lime mortar is hardening, the reaction strongly tends towards formation of $CaCO_3$. An airtight metal can (left) contains CO_2 gas at 1 atm and approx. 100 g lime mortar. Within a few hours the confined CO_2 is consumed at the reaction (a) and the pressure in the surrounding atmosphere crushes the can (right).

Problem. Calculate how many m^3 of atmospheric air are theoretically required to provide the quantity of carbon dioxide CO_2 necessary to transform 1 kg $Ca(OH)_2$ in the binder reaction (a), and determine how many kg of water are liberated at the reaction!

Conditions. The calculation shall be based on the following conditions regarding substances and states of substance:

Substance	$Ca(OH)_2$	CO_2	$CaCO_3$	H_2O	Unit
M	74.10	44.01	100.09	18.02	g/mol

Atmospheric air: ideal gas mixture, 20 °C, 1 atm (101325 Pa), with a natural CO_2 content of 0.033 vol-%.

Solution. According to **Dalton's law** (1.9) the molecules in an ideal gas mixture are equivalent; the CO_2 content of 1 m^3 atmospheric air, therefore, corresponds to the content $0.033 \cdot 10^{-2}\,m^3$ of pure CO_2 gas at 20 °C, 1 atm. From **the ideal gas law** (1.8), the CO_2 amount of substance n is now determined:

$$n(\mathrm{CO_2}) = \frac{101325\mathrm{Pa} \cdot 0.033 \cdot 10^{-2}\mathrm{m}^3}{8.314\mathrm{J/molK} \cdot 293.15\mathrm{K} \cdot 1\mathrm{m}^3} = 1.37 \cdot 10^{-2}\,\mathrm{mol}$$

The amount of substance in 1 kg $Ca(OH)_2$ is:

$$n(\mathrm{Ca(OH)_2}) = \frac{m}{M} = \frac{1000\mathrm{g}}{74.10\mathrm{g/mol}} = 13.5\,\mathrm{mol}$$

Therefore, the necessary air volume for hardening of 1 kg $Ca(OH)_2$ is:

$$V = n(\mathrm{Ca(OH)_2})/n(\mathrm{CO_2}) = 13.5\,\mathrm{mol}/1.37 \cdot 10^{-2}\mathrm{mol/m}^3 \simeq \mathbf{985\,m^3}$$

During hardening of 1 mol $Ca(OH)_2$, 1 mol H_2O is liberated; the quantity of water m formed can be determined by:

$$m = n(\mathrm{Ca(OH)_2}) \cdot M(\mathrm{H_2O}) = 13.5\mathrm{mol} \cdot 18.02 \cdot 10^{-3}\mathrm{kg/mol} = \mathbf{0.24\,kg\,H_2O}$$

Discussion. The mortar consumption for a one-brick wall is approx. 80 ℓ/m^2 wall surface. When lime mortar is used, this corresponds to a lime content of approx. 15 kg $Ca(OH)_2$ per m^2 wall surface. A partial carbonation and hardening of the lime mortar in a brick structure, therefore, requires a considerable air change to ensure access to CO_2 and removal of the chemically liberated water.

Figure 1.34. Aerated concrete is a foamed concrete with large content of air. It is among others available as building blocks and precast elements. The large content of air voids gives the concrete particularly good heat-insulating properties. The density of aerated concrete is typically 500-800 kg/m^3.

Example 1.3

■ Foaming of aerated concrete

Aerated concrete can be made by chemical foaming of a mixture of **Portland cement**, **lime** and ground **coarse aggregates.** Foaming of the fresh mixture is induced by adding a gas-forming agent. Because of its large content of **air voids**, aerated concrete has particularly good **heat-insulating** properties.

A very frequently used foaming agent in the manufacture of aerated concrete is **aluminium** in the form of a fine powder. Aluminium reacts in a strongly basic environment by formation of **aluminate ions** AlO_3^{---}; by the reaction, free hydrogen $H_2(g)$ is developed:

$$2\mathrm{Al} + 6\mathrm{OH}^- \rightarrow 2\mathrm{AlO}_3^{---} + 3\mathrm{H_2(g)} \qquad \text{(a)}$$

A mix of Portland cement, lime and water is alkaline with **pH** values of the order of 13.5. When aluminium powder is added to this mix the reaction (a) is induced, which will result in heavy foaming within a few hours.

After foaming and pre-curing, the aerated concrete is dimensionally stable and stiff. In this state the foamed mass is cut into **blocks** or **elements**, after which the final curing is performed with steam in **autoclave** at approximately 180 °C. At this temperature the added lime CaO reacts with quartz SiO_2 to form **calcium silicate hydrates**, which increase the strength significantly.

Typically, the density of the the finished aerated concrete is 500-800 kg/m^3 and the air void content is approximately 70-80 vol-%.

Problem. Calculate how many g of aluminium powder are consumed for development of 1.00 m^3 H_2(g) at 20 °C, 1 atm!

Conditions. In the calculations it is assumed that H_2(g) is an ideal gas at 20 °C and 1 atm; molar mass for Al: 26.98 g/mol.

Solution. According to the **ideal gas law** (1.8) at the pressure 1 atm = 101325 Pa and the temperature 20 °C, a volume of 1.00 m^3 H_2(g) contains the following amount of substance:

$$n(\mathrm{H_2}) = \frac{p \cdot V}{R \cdot T} = \frac{101325\mathrm{Pa} \cdot 1.00\mathrm{m}^3}{8.314\mathrm{J/molK} \cdot 293.15\mathrm{K}} = 41.6\,\mathrm{mol}$$

In accordance with the reaction equation (a), 1 mol H_2 is liberated by $\frac{2}{3}$ mol Al; development of 1.00 m^3 H_2(g), therefore, requires transformation of:

$$m = \tfrac{2}{3} \cdot n(\mathrm{H_2}) \cdot M(\mathrm{Al}) = \tfrac{2}{3} \cdot 41.6\mathrm{mol} \cdot 26.98\mathrm{g/mol} = \mathbf{748\,g\,Al}$$

Discussion. In practice, foaming of aerated concrete is a difficult and critical process. The setting of the concrete must be exactly adjusted to the progress of the hydration so that the concrete stiffens at the time when the hydrogen development ceases; otherwise the foamed mass will collapse again or its air void content will be too low.

Figure 1.35. The final strength development in aerated concrete takes place during steam curing in approx. 50 m long pressure chambers – so-called autoclaves. The curing period is 10-20 hours at approx. 180 °C; the vapour pressure in the autoclave is approx. 10 atm.

Example 1.4

■ Accelerated testing of concrete – carbonation properties

During hardening of concrete, **calcium hydroxide** $Ca(OH)_2$ is precipitated and alkali hydroxides are formed within the mix water. The pore water in concrete is thus strongly **alkaline** with pH $\simeq$ 13.5. This high pH value of the pore water counteracts corrosion attack on embedded reinforcement bars.

Iron Fe is a **base** metal, which in a neutral, humid environment can easily corrode. In a strongly alkaline environment, however, iron forms a thin protective surface layer of iron hydroxide that inhibits further corrosion: the iron is **"passivated"**.

In a hardened, reinforced concrete the high pH value induced by dissolved alkali hydroxide is in a sense "backed up" by the presence of solid calcium hydroxide, which maintains a "floor pH value" of 12.6 as long as it is present. Thus, a pH state is normally maintained that passivates the iron and prevents harmful corrosion of the reinforcement.

The dissolved alkali hydroxides are subject to reaction, and further reaction can slowly transform the solid calcium hydroxide to calcium carbonate. The concrete is said to **"carbonate"**. The reaction equation for the transformation of the calcium hydroxide to calcium carbonate is:

$$\mathrm{Ca(OH)_2 + CO_2 \rightarrow CaCO_3 + H_2O} \qquad \text{(a)}$$

In the outer parts of the concrete where $Ca(OH)_2$ has been totally transformed into $CaCO_3$, the pH decreases to approximately 8. The passivation of the iron cannot be maintained and the reinforcement is exposed to destructive corrosion. The carbonation rate and thus the time for development of damage, if any, depends on the concrete quality; the denser the concrete, the more slowly the carbonation rate.

Problem. An accelerated test method for comparing the carbonation properties of concrete shall be developed according to the following principle: a geometrically well-defined concrete specimen is placed in a pressure tank of steel. Carbon dioxide $CO_2(g)$ is added under pressure at 20 °C. The free gas volume in the tank is 4.00 ℓ. The progress of the carbonation is recorded by measuring the drop of pressure in the tank.

At the time t_1 a CO_2 pressure of 28.5 atm is measured; at a later time t_2 the pressure has decreased to 21.8 atm. Calculate according to the van der Waals equation:

- how many grams $CO_2(g)$ are consumed during the measuring period,
- how many grams $Ca(OH)_2$ are transformed during the measuring period.

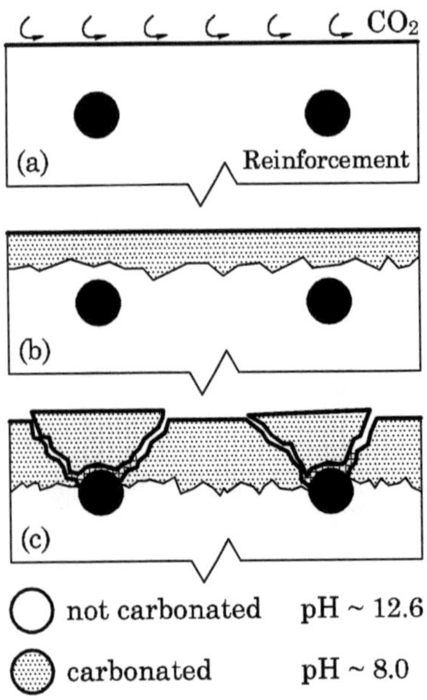

Figure 1.36. When a hardened concrete is influenced by the CO_2 in the air, the content of $Ca(OH)_2$ in the concrete will slowly be transformed into $CaCO_3$. Hereby the pH of the concrete decreases from approx. 12.6 to approx. 8; this neutralizes the protective passivation of the embedded reinforcement and can result in harmful corrosion damages.

Conditions. The following conditions apply to the calculations: the concrete carbonates according to the reaction (a); constant gas volume 4.00 ℓ at 20 °C during the reaction; molar mass CO_2 = 44.01 g/mol; the gas complies with the van der Waals equation at the change of state considered. The van der Waals constants are, cf. table 1.5:

$$CO_2(g): \quad a = 0.3640\,m^6Pa/mol^2; \qquad b = 42.67 \cdot 10^{-6}\,m^3/mol$$

Solution. At the time t_1 the CO_2 pressure is p = 28.5 atm = $2.89 \cdot 10^6$Pa. Insertion into the van der Waals equation (1.11) gives:

$$(2.89 \cdot 10^6 + 0.3640 \cdot (\frac{n}{4.00 \cdot 10^{-3}})^2) \cdot (4.00 \cdot 10^{-3} - n \cdot 42.67 \cdot 10^{-6}) = n \cdot R \cdot T$$

which by reduction gives the following third degree equation for the amount of substance n of $CO_2(g)$:

$$n^3 - 93.74 \cdot n^2 + 2.638 \cdot 10^3 \cdot n - 1.191 \cdot 10^4 = 0$$

By iterative solution the amount of substance CO_2 is found: $n_1 = \mathbf{5.54\,mol}$

At the time t_2, insertion of p = 21.8 atm = $2.21 \cdot 10^6$Pa gives a corresponding third degree equation for n:

$$n^3 - 93.74 \cdot n^2 + 2.608 \cdot 10^3 \cdot n - 0.9106 \cdot 10^4 = 0$$

By iterative solution the amount of substance CO_2 is found: $n_2 = \mathbf{4.06\,mol}$

Thus, during the considered measuring period the following amount is consumed:

$$m(CO_2) = (n_2 - n_1) \cdot M(CO_2) = (5.54 - 4.06)\text{mol} \cdot 44.01\,\text{g/mol} = \mathbf{65.1\,g\,CO_2}$$

Carbonation

100 g $Ca(OH)_2$ + 59 g CO_2
forms
135 g $CaCO_3$ + 24 g H_2O

By the reaction (a)

$$m(CaOH)_2) = \frac{M(Ca(OH)_2)}{M(CO_2)} \cdot m(CO_2) = \frac{74.10}{44.01} \cdot 65.1\,g = \mathbf{109.6\,g\,Ca(OH)_2}$$

has been transformed.

Discussion. If the reacted quantity of $CO_2(g)$ is calculated from the ideal gas law a consumption of approx. 49 g CO_2 is found; at the pressures and temperatures used in the test, $CO_2(g)$ deviates noticeably from an ideal gas.

Newton-Raphson iteration is useful for iterative solution of third degree equations frequently found when the van der Waals equation is invoked; the method is described in the Mathematical appendix, subchapter A3.

Example 1.5

■ Molar mass of atmospheric air

Atmospheric air mainly consists of **nitrogen** $N_2(g)$ and **oxygen** $O_2(g)$; in addition the air contains small quantities of gases such as **carbon dioxide** $CO_2(g)$, **argon** Ar(g) and **neon** Ne(g). In the *"CRC Handbook of Chemistry and Physics"* it is specified that the density of dry atmospheric air at $\theta = 24.00\,°C$ and $p = 730$ mm Hg is $\varrho = 1.142\,kg/m^3$. The pressure $p = 1.000\,atm = 101325\,Pa$ corresponds to 760 mm Hg.

Problem. Calculate from the above information the molar mass $\overline{M}$ for dry atmospheric air assuming that the air is an ideal gas mixture!

Conditions. In the calculations the **ideal gas law** (1.8) and **Dalton's law** (1.9) are assumed to apply to the considered mixture of gases.

Atmospheric air

Nitrogen N_2	*78.09 vol-%*
Oxygen O_2	*20.95 vol-%*
Noble gases	*0.93 vol-%*
Carb. dioxide CO_2	*0.03 vol-%*

Solution. The thermodynamic temperature of the gas $T = 273.15 + 24.00 = 297.15\,K$. The gas pressure is converted into the SI unit Pa as follows:

$$p = 730\,\text{mm Hg} = \frac{730}{760} \cdot 101325\,\text{Pa} = 97325\,\text{Pa}$$

Now, the following applies to the assumed ideal gas mixture:

$$p_{\text{tot}} = p_1 + \cdots + p_i = (n_1 + \cdots + n_i) \cdot \frac{RT}{V} = n \cdot \frac{RT}{V} \tag{1}$$

In (1) the total pressure p_{tot} has been determined as the sum of the partial pressure of the components in accordance with **Dalton's law**. The average molar mass $\overline{M}$ satisfies the condition:

$$m = n_1 \cdot M_1 + \cdots + n_i \cdot M_i = n \cdot \overline{M} \tag{2}$$

where m denotes the mass of the gas and n denotes the total amount of substance in the system. Comparing (1) and (2), the average molar mass $\overline{M}$ can be determined from:

$$\overline{M} = \frac{mRT}{pV} = \frac{1.142\text{kg} \cdot 8.314\text{J/mol K} \cdot 297.15\text{K}}{1\text{m}^3 \cdot 97325\text{Pa}} = 0.02899\,\text{kg/mol}$$

Average molar mass for atmospheric air: **28.99 g/mol**

Discussion. According to the *"CRC Handbook of Chemistry and Physics"*, dry atmospheric air of normal composition has an average molar mass $\overline{M}$ of 28.95 g/mol. This value deviates by approximately 0.1% from the estimated value of 28.99 g/mol calculated above. In this case the condition of ideal gas mixture is a fair approximation in connection with practical calculations.

Atmospheric air

Average molar mass of dry atmospheric air is 28.95 g/mol

Example 1.6

■ Chemical shrinkage by hardening of Portland cement

Portland cement is manufactured by burning until partial melting of an intimate mixture of substances that mainly contain **calcium** Ca, **silicon** Si, **aluminium** Al and **iron** Fe. During burning, so-called **clinkers** are formed. When ground, these clinkers are able to react with water - to hydrate - and to deposit hydrates with binder properties. A typical clinker mineral composition of Portland cement is:

Name	Chemical composition	Cement chemical name	Typical content
Tricalcium silicate	$3CaO \cdot SiO_2$	C_3S	55 %
Dicalcium silicate	$2CaO \cdot SiO_2$	C_2S	20 %
Tricalcium silicate	$3CaO \cdot Al_2O_3$	C_3A	7 %
Tetracalcium aluminoferrite	$4CaO \cdot Al_2O_3 \cdot Fe_2O_3$	C_4AF	9 %

During grinding of these clinkers 3-5 % **gypsum** $CaSO_4 \cdot 2H_2O$ is added to regulate the setting of the C_3A mineral. For concise cement chemical descriptions the following "cement chemist's shorthand" names or abbreviations are used:

"C"$= CaO$; "S"$= SiO_2$; "A"$= Al_2O_3$; "F"$= Fe_2O_3$; "$\bar{S}$"$= SO_3$; "H"$= H_2O$

The hydration of Portland cement results in an overall **reduction of volume**: that is, the hydration products formed have less volume than the volume of solid substances

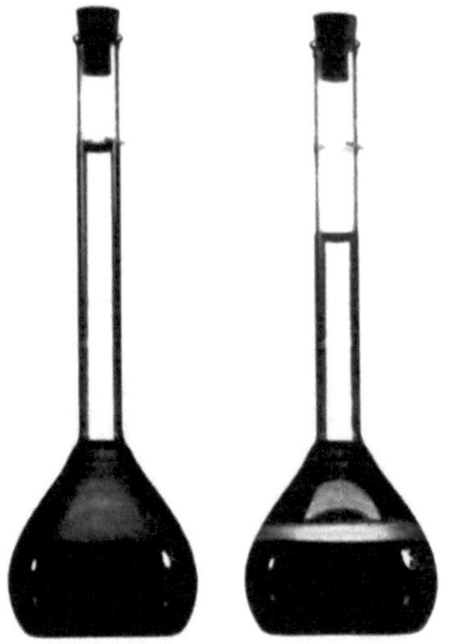

Figure 1.37. The chemical shrinkage by hydration of cement paste can be measured by the simple setup shown: the water level in the narrow neck of the flask over a specimen of cement paste directly indicates the shrinkage of volume. The specimen is photographed 0 hours and 48 hours, respectively, after casting.

Figure 1.38. A Danish equipment for measuring chemical shrinkage in hardening cement paste. With this equipment, which is thermostatically controlled, simultaneous registration of chemical shrinkage in up to 6 specimens of cement paste is possible (Conometer of the make Lolk).

and water from which they are formed. This reduction of volume is called the chemical shrinkage, and is mainly assumed to be caused by changes of the state of the water.

- Chemical bonding of water in hydrates results in a reduction of the molar volume of the water.
- Physical adsorption of water on the surface of the reaction products results in a reduction of the molar volume of the water.

The chemical shrinkage has the consequences that a mixture of Portland cement and water – cement paste – has a natural tendency to imbibe water from the environments during hydration. If this water imbibition is prevented, capillary tensile stresses will arise in the pore water.

In practice this "chemical shrinkage" may cause problems; at low w/c ratios where the quantity of water is small compared to the quantity of cement the chemical shrinkage may result in cracking during hardening. This phenomenon is for example encountered in connection with casting of fibre-reinforced cement-bound thin plates with a low w/c ratio.

Problem. A Portland cement contains 65 wt-% C_3S, 28 wt-% C_2S and 7 wt-% C_3A; during hydration of the cement the following chemical reactions are assumed to take place (ignoring the effects of the sulphate present):

$$2C_3S + 6H \rightarrow C_3S_2H_3 + 3CH \qquad (C_3S\text{-reaction}) \qquad (a)$$
$$2C_2S + 4H \rightarrow C_3S_2H_3 + CH \qquad (C_2S\text{-reaction}) \qquad (b)$$
$$2C_3A + 27H \rightarrow C_2AH_8 + C_4AH_{19} \qquad (C_3A\text{-reaction}) \qquad (c)$$

In these idealized hydration reactions **C-S-H gel** $C_3S_2H_3$, **calcium hydroxide** CH and the two **aluminate hydrates**: C_2AH_8 and C_4AH_{19} are formed. For these assumed hydration reactions, determine:

- The chemical shrinkage due to hydration by the reactions (a), (b) and (c).
- The chemical shrinkage due to hydration by full hydration of the actual Portland cement; unit (mℓ/100 g cement).

Conditions. In the hydration reactions (a), (b) and (c), full hydration of the individual clinkers is assumed and the reactions are assumed to be mutually independent in the composite system. Only changes of volume caused by chemical reactions in the system are considered.

Densities ϱ (g/cm^3) for reactants and products from: *Lea,F.M.: "The Chemistry of Cement and Concrete"* and calculated molar masses M (g/mol) given in the following table shall be used for calculations:

Substance	C_3S	C_2S	C_3A	$C_3S_2H_3$	C_2AH_8	C_4AH_{19}	CH	H	Unit
ϱ	3.13	3.28	3.00	2.63	1.95	1.80	2.23	1.00	g/cm^3
M	228.3	172.2	270.2	342.4	358.1	668.3	74.1	18.0	g/mol

Solution. For the individual reactants and products the molar volume v can be determined by: $v = M/\varrho$ (cm^3/mol). For the C_3S reaction, the following changes of volume due to the chemical binding of water in hydrates are obtained per mole of reaction equation (a):

$$V_{prod} = 1 \cdot (342.4/2.63) + 3 \cdot (74.1/2.23) = 230.0 \text{ cm}^3$$
$$V_{react} = 2 \cdot (228.3/3.13) + 6 \cdot (18.0/1.00) = 253.9 \text{ cm}^3$$
$$\Delta V = V_{prod} - V_{react} = 230.0 \text{ cm}^3 - 253.9 \text{ cm}^3 = -23.9 \text{ cm}^3$$

Change of volume in % for C_3S: $\varepsilon = (\Delta V/V_{react}) \cdot 100\% = \mathbf{-9.4}$ **%**

Change of volume per 100 g C_3S: $\Delta V = -23.9 \cdot (100/(2 \cdot 228.3)) = \mathbf{-5.2\ cm^3}$

Similarly, a calculation of the chemical shrinkage for the C_2S reaction and for the C_3A reaction give the following results per mole of reaction equation:

Quantity	C_3S	C_2S	C_3A	Unit
V_{prod}	230.0	163.4	554.9	cm^3
V_{react}	253.9	177.0	666.1	cm^3
ΔV	-23.9	-13.5	-111.2	cm^3
Change of volume ε	**-9.4**	**-7.6**	**-16.7**	**%**
Change of volume per 100 g	**-5.2**	**-3.9**	**-20.6**	**mℓ/100 g**

For 100 g cement with the given mix proportions: 65 wt-% C_3S, 28 wt-% C_2S and 7 wt-% C_3A, the calculated chemical shrinkage due to full hydration becomes

$$\varepsilon_{tot} = -(0.65 \cdot 5.2 + 0.28 \cdot 3.9 + 0.07 \cdot 20.6)\ \mathrm{m}\ell/100\,\mathrm{g} = \mathbf{5.9\,m\ell/100\,g}$$

Discussion. Experimental investigations show that the chemical shrinkage of Portland cement is typically of the order of magnitude 6 mℓ/100 g cement. The chemical shrinkage due to hydration calculated above is close to this empirical value. Therefore, it can be assumed that the chemical shrinkage is mainly caused by reduction of volume by chemical binding of water in the hydrates formed.

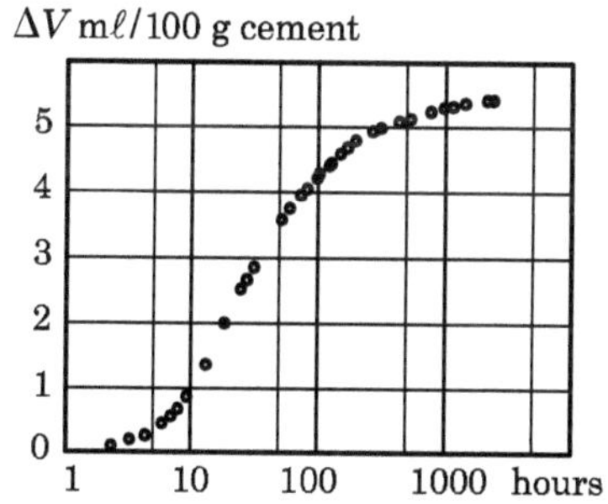

Figure 1.39. Chemical shrinkage measured on hydrating cement paste. Cement type: PC(R/HS/EA/W), 20°C, w/c ratio 0.50 (Geiker M: Studies of Portland Cement Hydration).

Exercises

The following exercises can be used when learning the subjects addressed in Chapter 1. The times given are suggested guidelines for use in assessing the effectiveness of the teaching.

Exercise 1.1

☐ Molar mass of isoprene *(2 min.)*

Isoprene is a so-called **poly-functional monomer**; the molecule contains two unsaturated double bonds between carbon atoms that are able to form cross-linking to other molecules by a polymerization process. Natural rubber, for example, is a **polyisoprene** compound formed by polymerization of isoprene molecules of the chemical formula: C_5H_8. Calculate the molar mass M of isoprene!

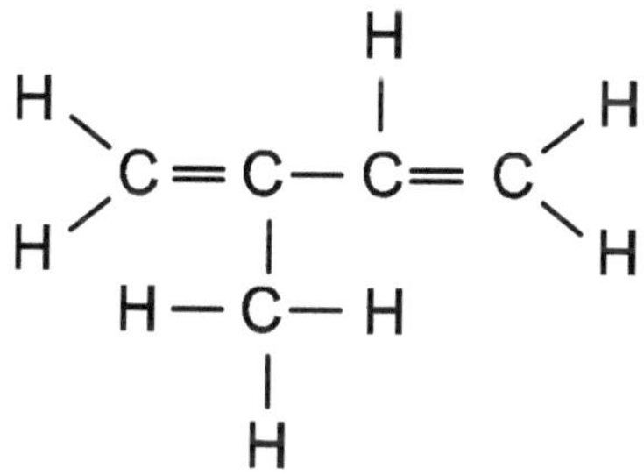

Figure 1.40. Configuration of an isoprene molecule C_5H_8. Natural rubber is poly-isoprene; furthermore, isoprene is an essential ingredient of many types of artificial rubber.

Exercise 1.2

☐ Calculation of amount of substance, tricalcium silicate *(3 min.)*

Tricalcium silicate is an essential ingredient of **Portland cement**. Tricalcium silicate has the formula unit: $3CaO \cdot SiO_2$. The molar masses of the components are: M(Ca): 40.08 g/mol; M(O): 16.00 g/mol and M(Si): 28.09 g/mol. Calculate the amount of substance n in 1.00 kg tricalcium silicate!

Exercise 1.3

☐ Relative atomic mass, natural iron Fe *(3 min.)*

Natural iron Fe is composed of four **isotopes**: $^{54}_{26}Fe$ with $A_r = 53.9396$ constitutes 5.8% of the atoms; $^{56}_{26}Fe$ with $A_r = 55.9349$ constitutes 91.8 % of the atoms; $^{57}_{26}Fe$ with $A_r = 56.9354$ constitutes 2.1 % of the atoms and $^{58}_{26}Fe$ with $A_r = 57.9333$ constitutes 0.3 % of the atoms. Calculate the relative atomic mass A_r of natural iron Fe!

Exercise 1.4

☐ Purity of iron ore *(6 min.)*

Crude iron is made from natural ferric oxides (ores) which contain compounds such as **hematite** Fe_2O_3 and **magnetite** Fe_3O_4. Calculate the mass fraction w and the mole fraction x of Fe in these two ore compounds!

Exercise 1.5

☐ Absolute humidity in air *(4 min.)*

At 0°C the partial pressure of saturated water vapour is 611.3 Pa. The molar mass of water M is 18.02 g/mol. Calculate, based on the **ideal gas law** how many g of water vapour m at this temperature are contained in a room with the dimensions: 6.0 m ×2.5 m ×10.0 m, when the **relative humidity** RH is 80 %!

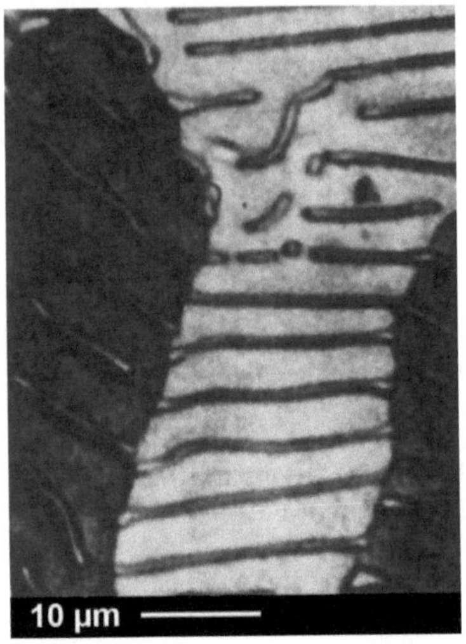

Figure 1.41. Typical micro structure of a carbon steel. Deposits of dark laminae of cementite Fe_3C can be seen in a light groundmass of ferrite, which is low-carbon Fe; the shown structure is called perlite.

Exercise 1.6

☐ Mass fraction and mole fraction of carbon in cementite *(3 min.)*

At room temperature carbon steel contains a compound of iron Fe and carbon C, which in metallurgy is denoted "cementite". Cementite is an **iron carbide** with formula unit Fe_3C. Calculate the mass fraction $w(C)$ and the molar fraction $x(C)$ of carbon in cementite! The molar masses of the components are: $M(Fe)$ 55.847 g/mol and $M(C)$ 12.011 g/mol.

Exercise 1.7

☐ Molar concentration of chloride ions *(4 min.)*

Under certain circumstances, chloride exposure of concrete may result in serious corrosion of the embedded steel reinforcement. In a paper on laboratory tests describing penetration of chloride ions into concrete, reference is made to an aqueous 1.00 molal NaCl solution. The density of the solution ϱ is 1.036 g/cm^3. Calculate the molar concentration $[Cl^-]$ of chloride ions in the solution used! The molar masses of the components: $M(NaCl)$ 58.44 g/mol; $M(H_2O)$ 18.02 g/mol.

Exercise 1.8

☐ Conversion of technical units *(5 min.)*

In *CRC Handbook of Chemistry and Physics* the following data for the wood fibre product "*Insulite*" is given: density $\varrho = 16.2$ pounds/cubic feet and thermal conductivity $\lambda = 0.34\,\mathrm{BTU \cdot inch/h \cdot ft^2 \cdot {}^\circ F}$. The following conversion values are given: 1 BTU ("British Thermal Unit") = 1055.9 J; 1 pound = 0.4536 kg; 1 inch = 0.0254 m; 1 ft = 0.3048 m and $1\,^\circ\mathrm{F} = \frac{5}{9}\,^\circ\mathrm{C}$. Calculate the density ϱ and the thermal conductivity λ for this wood fibre product in SI units kg/m^3 and W/mK!

Exercise 1.9

☐ The gas constant with unit of pressure torr *(6 min.)*

In the literature the **ideal gas law** is sometimes given with other units of pressure and volume than those standardized in the SI system. Determine the numerical value for the gas constant R with the unit $\ell \cdot \mathrm{torr/mol \cdot K}$, and calculate with this value of R the pressure p (torr) in 30 g of an ideal gas given the following conditions: $V = 15.0\,\ell$; $\theta = 22.0\,^\circ\mathrm{C}$; $M = 42.00\,\mathrm{g/mol}$. The unit of pressure 1 torr = 1 mm Hg = 1/760 atm.

Exercise 1.10

☐ Relative and absolute humidity *(8 min.)*

At 25 °C the partial pressure of saturated water vapour is $p_s = 3169.1$ Pa. Calculate a) the relative humidity RH in air with an absolute humidity of 12.0 g/m^3 at 25 °C, b) the partial pressure p of water vapour in a 25 °C hot air with a relative humidity RH of 10 %, and c) the absolute humidity in air with a relative humidity RH of 30 % at 25 °C! In the calculations water vapour in air shall be assumed to comply with the ideal gas law.

Relative humidity

$$RH = \frac{\textit{actual partial pressure}}{\textit{saturated vapour pressure}}$$

Exercise 1.11

☐ Calculation of gas quantity by permeability tests *(12 min.)*

In a laboratory test to measure the permeability of a concrete, pressurized **nitrogen** N_2 is applied from a 40.0 ℓ pressure bottle. The gas pressure in the bottle is 124 atm at 20 °C. Calculate a) based on the **ideal gas law** and b) based on the **van der Waals equation**, how many kg $N_2(g)$ can be expected to be in the pressure bottle! $M(N_2) = 28.02\,\mathrm{g/mol}$.

Exercise 1.12

☐ Electrolytic galvanizing *(8 min.)*

Steel specimens can be effectively protected against corrosion by surface coating with metallic zinc Zn. The protective effect is based on the fact that Zn is a less noble metal than iron Fe; in a corrosive environment Zn will form the **anode** and the subjacent Fe will act as **cathode**. By galvanic corrosion the anode metal – here Zn – corrodes, while the cathode metal remains intact. The zinc coating thus functions as a so-called "sacrificial anode".

Galvanizing of small steel specimens can be made by **electrolysis** in a solution of zinc salts. During the electrolysis $Zn^{++}(aq)$ is **reduced** and is deposited as metallic Zn on the cathode.

$$Zn^{++}(aq) + 2e^- \quad \rightarrow \quad Zn(s) \qquad (a)$$

In a process it has been measured that 1500 g Zn per hour is deposited at constant current in a galvanic bath. Calculate the current i (A) in the bath! $M(Zn) = 65.39$ g/mol.

Figure 1.42. Examples of steel specimens that are corrosion protected by electrolytic galvanizing. The electrolytic method is particularly used for galvanizing of small mass-produced articles.

Exercise 1.13

☐ Critical temperature and pressure for industrial gas *(15 min.)*

For industrial heating purposes combustible gases such as **propane** C_3H_8, **butane** C_4H_{10} or a mixture of these gases are often used. In the *CRC Handbook of Chemistry and Physics* the following van der Waals constants are given for propane and butane:

Gas	Chem. name	**a** $\ell^2 \cdot atm/mol^2$	**b** ℓ/mol
Propane	C_3H_8	8.664	0.08445
Butane	C_4H_{10}	14.47	0.1226

a) convert these van der Waals constants into SI units, and b) calculate critical temperature (°C) and c) critical pressure p (atm) for propane and butane. d) Indicate whether propane and butane gases can be kept in their liquid form in pressure bottles at room temperature!

Exercise 1.14

☐ Cracking by heat curing of concrete *(15 min.)*

When concrete is heated quickly, e.g. in connection with heat curing, an **overpressure** is developed in the air voids of the concrete. In freshly cast concrete, this overpressure may cause harmful **cracking** if the temperature rise is not adjusted to the strength development of the concrete.

Estimate changes of pressure in the following ideal model of an air void in concrete: A rigid, enclosed volume in a tank V contains saturated atmospheric air at 20 °C; the total pressure in the tank is $p_1 = 1$ atm $= 101325$ Pa. At the bottom of the tank there is enough of water to keep the air in the tank saturated during heating. The temperature in the tank is increased to 50 °C. Calculate the equilibrium pressure p_2 (atm) and the increase of pressure $\Delta p = p_2 - p_1$ (atm) in this state. The mixture of atmospheric air and water vapour is assumed to be an ideal gas.

Figure 1.43. Example of cracking in freshly cast, hardening concrete caused by overpressure in air voids during heating.

Exercise 1.15

☐ Volume contraction by dissolution of NaOH in water *(8 min.)*

In solutions **intermolecular** forces between the dissolved substance and the solvent may affect the volume of the solution. This molecular effect can e.g. be seen when concentrated saline solutions are made: the volume of the mix will normally be considerably less than the sum of the component volumes.

The *CRC Handbook of Chemistry and Physics* specifies that the density of a 30 wt-% solution of sodium hydroxide NaOH in water is $\varrho = 1.3277$ g/cm³ at 20 °C. At this temperature the density ϱ of H_2O and NaOH are 0.9982 g/cm³ and 2.130 g/cm³, respectively. The molar mass of NaOH is 40.01 g/mol. Assume a 1000 mℓ 30 wt-% NaOH solution. Calculate the sum of $V(H_2O)$ and $V(NaOH)$ mℓ prior to mixing and calculate the contraction of volume in (mℓ) and (%) caused by intermolecular forces in the solution!

Exercise 1.16

☐ Spalling of concrete cover due to corrosion of reinforcement *(8 min.)*

In hardened concrete an excess of calcium hydroxide $Ca(OH)_2$ together with alkali hydroxide normally ensures that the pore water is strongly **alkaline** with pH ≃13.5. At this high pH value, corrosion of the embedded reinforcement is prevented because a protective layer of **ferric oxides** is deposited on the surface of the reinforcement – the iron is "passivated". Under the influence of carbon dioxide CO_2 in the air, $Ca(OH)_2$ can gradually be transformed into calcium carbonate $CaCO_3$ – particularly in porous concretes. In such a **carbonated** concrete pH has been reduced to approx. 8; at this pH value the protective passivation layer on the reinforcement can decay. When exposed to

Figure 1.44. If reinforcing bars are exposed to moisture and oxygen in a pH neutral environment, they will quickly corrode. Because the corrosion products formed have a larger volume than the transformed iron, the concrete over the reinforcement spalls.

oxygen and moisture, reinforcing bars in a carbonated concrete can decay by electrolytic corrosion in accordance with the following reaction

$$2Fe + O_2 + 2H_2O \quad \rightarrow \quad 2Fe(OH)_2 \qquad \text{(a)}$$

The **ferrous hydroxide** $Fe(OH)_2$ thus formed is a solid substance with larger volume than the iron, Fe, from which it has been transformed; therefore, due to the corrosion process (a) the concrete cover over the attacked reinforcement spalls.

In the *CRC Handbook of Chemistry and Physics* the density ϱ of iron Fe and ferrous hydroxide $Fe(OH)_2$ is specified as 7.86 g/cm^3 and 3.4 g/cm^3, respectively. Calculate the increase in volume of solid substance in % due to the transformation (a)!

Exercise 1.17

☐ Manufacture of NaCl solution for test purposes *(10 min.)*

In a laboratory test of the **alkali silica reactivity** an aqueous solution of NaCl with a molar concentration of $c = 4.00$ mol/ℓ is required. The solution is made in the following way: An **amount of substance** of 4.00 mol NaCl is weighed out and placed in a 1000 mℓ measuring cylinder. Water is added and the solution is adjusted to 1000 mℓ at 20 °C. The density of the finished solution is $\varrho = 1.150$ g/cm^3. Determine a) how many g NaCl shall be weighed out for the solution, b) the NaCl content of the solution in wt-% and c) the molality $\widetilde{m}$ of NaCl in the solution!

Exercise 1.18

☐ Electric charge quantity in a solution of ions *(5 min.)*

1.00 ℓ water containing 100 g dissolved and fully dissociated $CaCl_2$ is assumed. Calculate the magnitude of the **positive** electric charge Q^+ (C) constituted by Ca^{++} ions in the solution and the **negative** electric charge Q^- (C) constituted by Cl^- ions in the solution. Calculate the period of time for which a current of 1 amp must run through a conductor to transmit an electric charge corresponding to Q^- (C)! $M(CaCl_2) = 111.0$ g/mol.

Exercise 1.19

☐ Typical molecular distances in humid atmospheric air *(10 min.)*

Assume 1 m^3 humid atmospheric air at 20 °C and a relative humidity of $RH = 80\%$. The atmospheric air and the water vapour are assumed to be an ideal gas mixture. At 20 °C the partial pressure of saturated water vapour is 2338.4 Pa. Calculate a) the average distance in (Å) between the gas molecules in the air mixture, b) the average distance in (Å) between the water molecules in the air and c) the percentage of H_2O molecules in the gas!

Literature

The following literature and supplementary readings are recommended for further reference on the subjects covered in Chapter 1.

References

In chapter 1 a number of references to the literature have been made; the following list contains a complete listing of the references used.

- Hill, T.L.: *Lectures on Matter & Equilibrium*, W.A.Benjamin, Inc., New York 1966.
- Atkins, P.W.: *Physical Chemistry*, Oxford University Press 1986.
- Kaye, G.W.C. and Laby, T.H.: *Tables of Physical and Chemical Constants*, 5th ed., Longman, Harlow, 1992.
- Weast, R.C. (ed.): *CRC Handbook of Chemistry and Physics*, CRC Press, Inc., Florida 1983.
- Lea, F.M.: *The Chemistry of Cement and Concrete*, Edward Arnold (Publishers) Ltd., London 1956.

- Geiker, M.: *Studies of Portland Cement Hydration*, Institute of Mineral Industry, Technical University of Denmark, Lyngby 1983.
- Alexanderson, J.: *Strength Losses in Heat Cured Concrete*, CBI Handlingar Nr. 43, Stockholm 1972.

Supplementary literature

A fundamental collection of tables with very comprehensive and updated data for the chemical and physical properties of substances is:

- Weast, R.C. (ed.): *CRC Handbook of Chemistry and Physics*, CRC Press, Inc., Florida.

These tables are continuously updated and published; compared to its volume and contents, it is a very inexpensive work of reference.

A recommended chemical work of reference, with an extensive explanation of compounds and their chemical and physical properties, can be found in the classical work:

- Holleman-Wiberg: *Textbook of Anorganic Chemistry*. In German: *Lehrbuch der Anorganischen Chemie*, Watter de Gruyter, Berlin.

A unique key to substance data for cement-chemical compounds and to the work with cement-chemical calculations is available in the work:

- Babushkin, Matveyev & Mchedlov-Petrossyan: *Thermodynamics of Silicates*, Springer-Verlag, Berlin.

which is a heavily revised and extended editon of the well-known work: *Thermodynamik der Silikate* by Mcedlov-Petrossyan.

A steam-powered Newcomen machine from around 1770; the cylinder is to be assumed filled with water vapour – by injection of cold water into the cylinder the water condenses whereby the piston is driven down into the cylinder due to the outer air pressure.

CHAPTER 2

Thermodynamic concepts

The term "thermodynamics" has been translated as "The science of the moving force of heat"; the name identifies the classical origin of thermodynamics, namely the study of the functions of heat engines from the beginning of the 19th century.

The classical foundations of thermodynamics were formulated in a number of papers written by the American physicist and chemist **J.W. Gibbs** during the years 1875 – 1878. The foundations of Gibbs' thermodynamics included earlier works by among others **Sadi Carnot** (France), **J.P. Joule** (England), **R. Clausius** (Germany) and **W. Thomson**, later Lord Kelvin (England).

The classical thermodynamics – energetics – can be briefly described as the science of the energy content of macroscopic systems, and the transformation of the different forms of energy in macroscopic systems. In this form thermodynamics is **axiomatic**. The basis is formed by four central postulates – the so-called **laws of thermodynamics**. From these laws mathematical derivation of a number of general rules for the physical and chemical properties of macroscopic systems is possible.

In the description of thermodynamics a number of exact terms are used. This chapter contains an overview of the main concepts and definitions within thermodynamics.

J.W. Gibbs (1839-1903)

American physicist and chemist, professor of theoretical physics at Yale University.

Contents

2.1 Thermodynamic systems

All thermodynamic considerations refer to a *predefined* **system**. A thermodynamic system can be a quantity of substance, a specimen, or a working machine which in a well-defined way is set apart from its **surroundings**. The boundary between the system and its surroundings can be real, as for example a container wall or the surface of a specimen, or it can be an imaginary mathematical envelope that separates the system from its environments. Collectively a system and its surroundings are denoted as the thermodynamic **universe**.

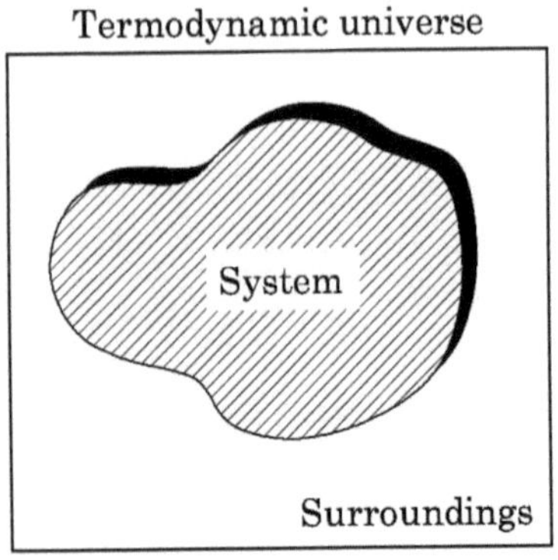

Figure 2.1. Thermodynamic considerations refer to a system which is separated from its surroundings by an imaginary or real boundary. System and surroundings constitute the thermodynamic universe.

> **Thermodynamic system** 2.1
>
> Any thermodynamic consideration refers to a well-defined **system** that is separated from its **surroundings** by a real or imaginary boundary: collectively system and surroundings constitute the thermodynamic **universe**.

A thermodynamic system is not necessarily bound to a predefined geometry. A system can consist, for example, of a given amount of gas which is enclosed by a cylinder and a movable piston. When the piston is moved the geometry of the system is changed: the original quantity of gas – within the changed volume – still constitutes the system.

The chemical composition as well as the physical state of a thermodynamic system can change. If, for example, a system consists of one kg of water $H_2O(\ell)$, the *same* system can for example be changed by a process to one kg of water vapour $H_2O(g)$, or to one kg of ice $H_2O(s)$.

System types

Thermodynamic systems can be divided into three types: **open** systems, **closed** systems and **isolated** systems. An open system can exchange **matter** and **energy** with its surroundings. A closed system can exchange **energy**, but *not* matter with its surroundings. An isolated system can exchange *neither* energy *nor* matter with its surroundings.

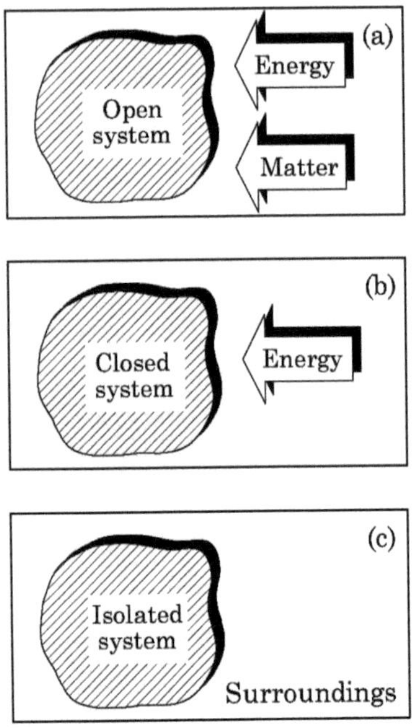

Figure 2.2. An open thermodynamic system (a) can exchange both matter and energy. A closed system (b) can only exchange energy, and an isolated system (c) can exchange neither energy nor matter with its surroundings.

> **Types of thermodynamic systems** (2.2)
>
> - An **open** system can exchange **matter** and **energy** with its surroundings.
> - A **closed** system can only exchange **energy** with its surroundings.
> - An **isolated** system can exchange *neither* energy *nor* matter with its surroundings.

Open thermodynamic systems are, for example, used to describe the energy conditions in combustion engines. Here the exchange of matter is represented by consumption of fuel and the exhaust of combustion products. In the following description of the physical and chemical substances, however, we shall confine ourselves to considering **closed** and **isolated** systems.

■ A tension rod is loaded to failure in a tensile test machine. During testing the stress-strain curve of the steel is registered by measuring paired values of load and elongation of the rod. After the measurement we want to analyse the fracture mechanical properties of the steel rod. For this purpose, define a thermodynamic system and its surroundings and specify the type of system concerned!

Solution. The thermodynamic system consists of the **steel rod**; the boundary between the system and its surroundings is the surface of the rod. Everything beyond the surface of the rod is considered to be surroundings. This is a **closed** system, because the system is able to exchange **energy**, but *not* matter with its surroundings.

☐ **1.** A system consists of 1 mole of gas, confined in a cylinder with a movable, closely fitting piston – specify the type of system!

☐ **2.** The temperature rise in a hardening concrete specimen insulated from heat and moisture is measured in an adiabatic calorimeter – which type of system is the concrete specimen?

☐ **3.** A system consisting of 1 kg of liquid ethanol C_2H_5OH is evaporated into the atmosphere – specify which type of system the specified amount of ethanol is?

☐ **4.** A system is defined as a unit length of a feed duct for a ventilation plant – define the type of system when the plant is operating!

☐ **5.** In its initial state, a system consists of 2 moles of hydrogen gas H_2; by combustion the system absorbs 1 mole of O_2 from the air, and is transformed into a system consisting of 2 moles of water H_2O – specify the type of system concerned!

Figure 2.3. A thermodynamic system can be a concrete cylinder tested in compression; in this case, the surroundings consist of a testing machine exchanging energy with the system.

2.2 Description of state

We have now given the principles explaining how to separate and define a thermodynamic system; the next task is to determine how to describe the state of the system in an unambiguous way.

Consider a simple system consisting of one, point **particle** with a known mass exposed to the gravitational field of the earth. At a given time this system is fully described by specifying the location of the particle in relation to the ground (x, y, z) and by specifying the particle velocity (v_x, v_y, v_z). When these six coordinates are known, the **state** of the system is described.

If we consider a more interesting – and more complex – system, for example 1 mole of an **ideal gas**, this form of description of state becomes unmanageable. A detailed description of the state of all gas particles would in this case require specification of $6 \cdot 6.022 \cdot 10^{23}$ quantities.

We have now reached the first, important limitation of the description of systems in classical thermodynamics. Classical thermodynamics can *only* describe the **macroscopic state** of a system. For a complete description of the macroscopic state of 1 mole of ideal gas in a stationary, closed container, only **two** of three variables that describe an ideal gas (T, p, V) need be stated.

> **Description of state** (2.3)
> Classical thermodynamics can *only* describe the **macroscopic state** of systems and **changes** in them.

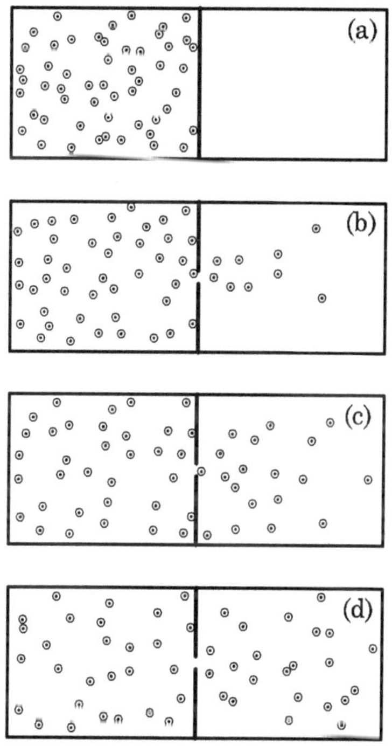

Figure 2.4. Expansion of a gas into a vacuum. In the initial state (a) all particles are confined in the chamber volume (left). The partition is opened and the gas particles expand freely into a vacuum (b). During the expansion (b) and (c) the system is in a non-equilibrium state where the gas cannot be associated with a particular pressure p or temperature T. In the final state (d) the system has regained equilibrium. Classical thermodynamics only deals with equilibrium states like (a) and (d).

The quantities used for describing the macroscopic state of systems are called the **state functions** or **state variables**. The name indicates that these quantities only depend on the state of the system, and are *independent* of the way in which the actual state has been reached.

A simple example of the concept of state variables is encountered in the gas law: $pV = nRT$; this law describes the state of an ideal gas by specifying three state variables: pressure p, volume V and absolute temperature T. If two of these state variables are known, the system state has been unambiguously determined. Later in this section we shall learn that a number of important energy quantities can be defined in such a way that they become state quantities.

The concepts **"state variables"** and **"state functions"** express the following essential property. If a system is changed from state 1 into state 2, the *change* of a given state variable or state function only depends on the **initial state** and the **final state** of the system; in other words, the change is *independent* of the path linking these states.

> **State parameters** (2.4)
>
> A **state variable** or **state function** only depends on the state of the system and is *independent* of the way in which the state has been achieved. The change of state variables and state functions is *solely* determined by the **initial state** and **final state** of the system.

Figure 2.5. The gas flame represents a state of non-equilibrium. In spite of this, the attainable temperature in the flame can be calculated by the special state parameters of thermodynamics.

We have now reached the second important limitation of the description of systems in classical thermodynamics. The thermodynamic description of the macroscopic state of systems can only be applied to systems in **thermodynamic equilibrium**. If, for example, we consider a gas expanding into a vacuum, no definite pressure p or temperature T can be ascribed to the gas during the expansion. During the process the gas is in a state of **non-equilibrium** and therefore, the state cannot be described by simple, macroscopic variables as p and T.

> **Equilibrium condition** (2.5)
>
> Classical thermodynamics can *only* describe systems in **thermodynamic equilibrium** and processes connecting **states of equilibrium.**

Offhand, the conditions (2.3) and (2.5) appear to be serious limitations on the practical application of thermodynamics. The reason that classical thermodynamics – in spite of this – is characterized by an extraordinary generality and practical applicability, is that the analysis is made with **state variables** and **state functions** such as item (2.4). This means that changes of state of non-equilibrium processes can be described by a series of imaginary equilibrium processes connecting the same starting and final conditions and thus bypasses the limitations mentioned above.

■ A system consists of 1 kg of pure copper Cu. State at least 10 different macroscopic state variables that can be applied to this system!

Solution. Mass m, amount of substance n, volume V, pressure p, temperature T, density ϱ, compressibility κ, coefficient of thermal expansion β, heat capacity c, specific electric resistance ρ.

□ **1.** A system consists of 1 mole of ideal gas in equilibrium. How many macroscopic state variables must be given to determine the gas density ϱ unambiguously?

□ **2.** Assume an aqueous solution of NaCl in equilibrium; state at least 8 macroscopic state variables for this system!

□ **3.** A bowl of water is placed in a closed chamber; the temperature is 20 °C everywhere. The partial pressure of the water vapour is 1705 Pa. Is this a system in equilibrium?

□ **4.** A given amount of calcium hydroxide $Ca(OH)_2$ is added to 1 ℓ of water. Describe the state when the system water + salt has reached a state of equilibrium!

□ **5.** A beaker with hot water (20 °C) is placed in a freezer at −18 °C; describe states of equilibrium and non-equilibrium during the subsequent process!

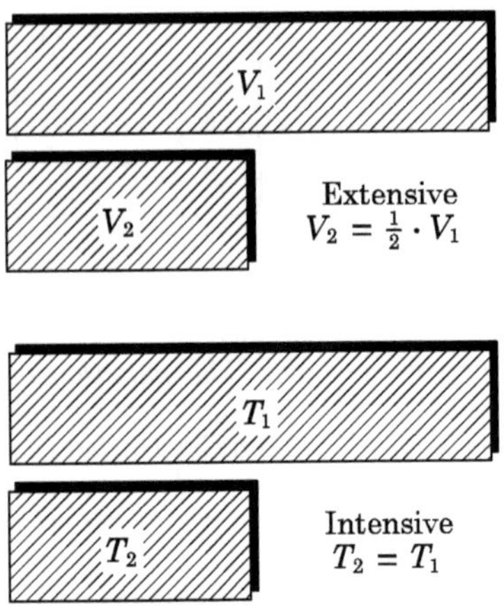

Figure 2.6. If a system of substance is assumed to be reduced by one-half, the value of an extensive state variable will also be reduced by one-half, while the value of an intensive state variable will remain the same.

2.3 Thermodynamic variables

The **state** of a thermodynamic system can be described by stating a number of measurable, physical quantities of the system – so-called state variables. Examples of such state variables are the pressure p, temperature T, density ϱ and volume V of the system. Normally the state of the system will be fully described by a limited number of variables.

Thermodynamic state variables can be **extensive**, such as the system volume V, or they can be **intensive**, such as the pressure p and temperature T of the system. The importance of this distinction is as follows. Extensive variables are proportional to the size of a chosen system; intensive variables are independent of the size of the chosen system. Intensive variables are sometimes described as **point variables**, because this type of variable – for example, the temperature T – can be applied to any "point" in the system.

Normally the relation between two extensive variables will form an intensive variable. This special property is often utilized in the description of systems of substance; here intensive variables are formed by division by the **amount of substance** n of the system, by division by the system **volume** V or by division by the system **mass** m. Hereby **mole**-specific, **volume**-specific and **mass**-specific quantities are obtained. A well-known example of a volume-specific state variable is the density ϱ, which is the system mass m divided by the system volume V.

Types of thermodynamic variables (2.6)

- **Intensive variable**: a state variable whose value is **independent** of the size of the chosen system *(example: temperature T, pressure p)*.
- **Extensive variable**: a state variable whose value is **proportional** to the size of the chosen system *(example: amount of substance n, volume V)*.

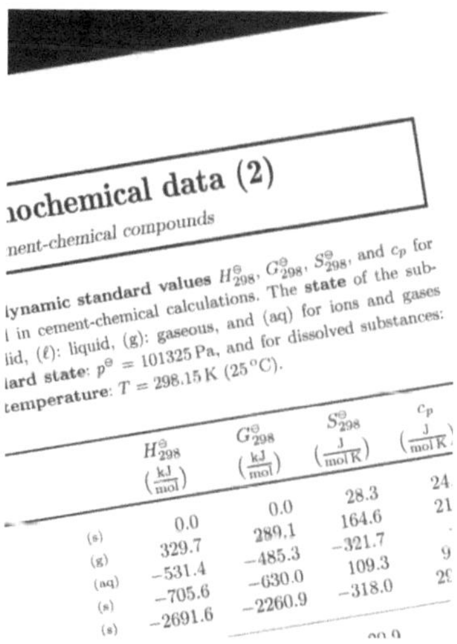

Figure 2.7. Tables of the physical and chemical properties of substances list intensive quantities of the substances that are independent of the system.

Use of specific, intensive state variables can in many cases simplify calculations and analyses. Usually, tables of the physical properties of substances list specific substance values, i.e. intensive values that are independent of the size of the chosen system.

For state variables that may occur as intensive as well as extensive state variables, one should specify the nature of the variable to avoid confusion. If for example a specific heat capacity c is given it should always be made clear from units, indices or notation whether a **mole-specific** or a **mass-specific** heat capacity is considered.

■ The ideal gas law in the form $pV = nRT$ contains two **intensive** state variables: the pressure p and the temperature T as well as two **extensive** state variables: amount of substance n and volume V. Convert the ideal gas law into a form that contains intensive variables only!

Solution. Conversion can either be made by division by the system volume V or by division by the amount of substance n of the system. The first case gives

$$p = c_m RT \qquad c_m = n/V \text{ denotes the molar concentration of the gas}$$

Conversion by division by the amount of substance n gives the following form of the ideal gas law:

$$pv_m = RT \qquad v_m = V/n \text{ denotes the molar volume of the gas}$$

Both of the above formulations of the ideal gas law are often used in the literature.

☐ **1.** Divide the quantities: density ϱ, viscosity η, amount of substance n, mass m, compressibility κ, volume V and mole fraction x into extensive and intensive variables, respectively!

☐ **2.** A physical quantity k is given based on the following unit: (kJ/mh°C); specify whether it is an intensive or extensive quantity!

☐ **3.** Convert the van der Waals equation (1.11) into a form where the equation only contains intensive state variables!

☐ **4.** Specify what physical quantity should be multiplied by the mass-specific heat capacity to convert it into a mole-specific heat capacity?

☐ **5.** State at least 5 intensive and 5 extensive state variables for an ideal gas mixture consisting of nitrogen N_2(g) and oxygen O_2(g)!

2.4 Temperature

In the description of substance the **thermodynamic temperature** T is included as a fundamental quantity. In 1854 the English physicist *William Thomson*, later *Lord Kelvin*, proved that based on the 2nd law of thermodynamics it was possible to define an absolute temperature quantity that was independent of any thermometric substance or measuring method. The unit for this thermodynamic temperature is named after its inventor the **kelvin** (K).

Figure 2.8. The temperature in a system of substance is a measure of the average kinetic energy of the system's molecules.

Thermodynamic temperature scale

The temperature of a system of matter, for example a gas, is a measure of the average **kinetic energy** of its molecules. The higher the temperature in a system of matter, the higher the average kinetic energy of its molecules. This simple physical interpretation of the concept of temperature illustrates at the same time the existence of a lowest temperature, **absolute zero**, where the molecules are in a state of rest with a minimum of kinetic energy.

The thermodynamic temperature T can be experimentally determined with a **gas thermometer**. This thermometer utilizes that the pressure p in an **ideal gas** at constant volume V is proportional to the thermodynamic temperature T

$$p = \frac{n \cdot R}{V} \cdot T = \text{constant} \cdot T \qquad (\textit{ideal gas})$$

When hydrogen $H_2(g)$ or helium He(g) is used at low pressure, the ideal gas state can be approximated with close accuracy. For a number of years the gas thermometer was the basic instrument for fundamental measurements in the thermodynamic temperature scale.

For practical applications the thermodynamic temperature scale was standardized by international agreement in 1990 ("ITS-90"); this agreement forms the basis of the practical temperature definition of the SI unit system.

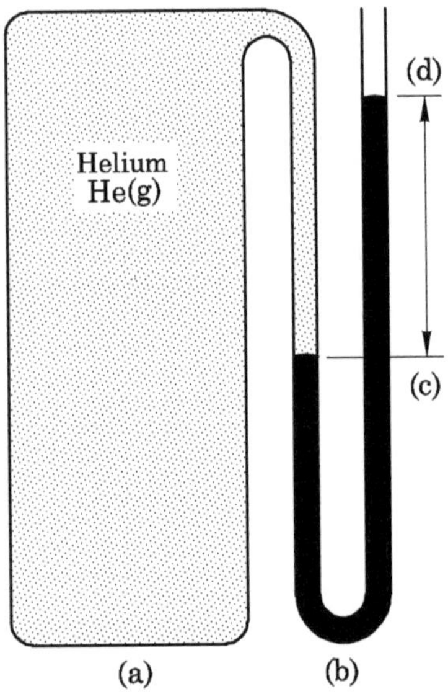

Figure 2.9. The structure of a gas thermometer shown schematically. A container (a) with helium He(g) is connected to a manometer (b). At (c) the liquid column is kept at a constant level. When in equilibrium the gas pressure is determined by the head of pressure (c)-(d).

Thermodynamic temperature T (2.7)

The unit **kelvin** is *defined* as the fraction 1/273.16 of the **triple point** of the thermodynamic temperature of water.

The calibration point of the thermodynamic temperature unit, kelvin, is the **triple point** of water, i.e. the temperature at which the pure phases **ice** $H_2O(s)$, **water** $H_2O(\ell)$ and **water vapour** $H_2O(g)$ are in mutual equilibrium. The temperature of the triple point is +0.01 °C.

This definition ensures that temperature intervals in the kelvin scale are *identical* with the temperature intervals in the Celsius scale. Thus, conversion from thermodynamic temperature T to Celsius temperature θ is:

$$\theta = T - T_0 = T - 273.15\ (^\circ\text{C}); \qquad T = \theta + T_0 = \theta + 273.15\ (\text{K}) \qquad (2.8)$$

By this definition **absolute zero** in the thermodynamic temperature scale is determined as −273.15 °C. Note that $T_0 = 273.15\,\text{K}$ is defined as an **exact** quantity.

Other temperature units

In practical calculations it is sometimes necessary to refer to English works of reference, where the temperature unit **Fahrenheit** °F is still in use. By conversion of temperature units from °F into °C and vice versa the following relations apply

Conversion of temperature values (2.9)

$$\text{celsius value} = \tfrac{5}{9} \cdot (\text{fahrenheit value} - 32)$$

$$\text{fahrenheit value} = \tfrac{9}{5} \cdot \text{celsius value} + 32$$

Conversion of temperature intervals is carried out by the relationships $\Delta C = \frac{5}{9} \cdot \Delta F$ and $\Delta F = \frac{9}{5} \cdot \Delta C$. These factors are used, for example, for conversion of physical units containing temperature quantities.

In American literature the temperature unit **Rankine** (°R) can often be encountered. The Rankine scale is an absolute thermodynamic temperature scale with the *same* unit as the Fahrenheit scale, i.e. the *unit* (°R) is *identical* with the unit (°F). These two units are mutually coordinated in the same way as the **kelvin** and **Celsius** units. In particular, the absolute thermodynamic zero 0 °R corresponds to −459.67 °F.

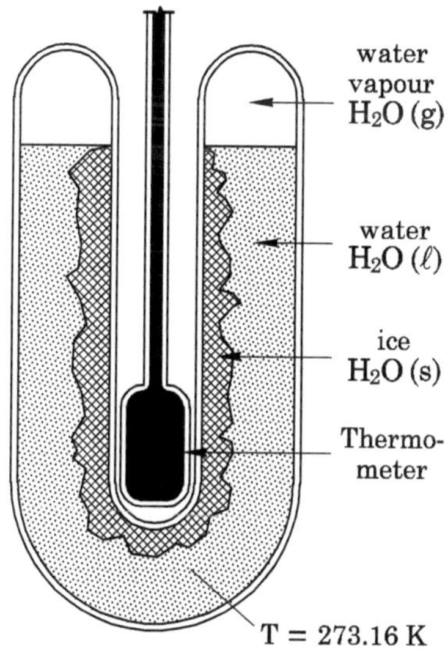

Figure 2.10. The calibration point in the thermodynamic temperature scale is the triple point of water where phase equilibrium exists between ice, water and water vapour in a closed system; these three phases can only coexist in equilibrium at one particular temperature: 273.16 K.

■ A system consists of a closed helium-filled container with constant volume V. In state 1 the pressure is $p_1 = 10205$ Pa and the temperature is 62.5 °F. After a change of temperature a pressure of $p_2 = 10304$ Pa is measured in state 2. Helium He(g) is assumed to comply with the ideal gas law. Calculate the thermodynamic temperature T_2 (K) in state 2!

Solution. First we calculate the thermodynamic temperature T_1 (K) using the relations (2.8) and (2.9)

$$T_1 = \left(\tfrac{5}{9} \cdot (62.5 - 32)\right) + 273.15 = \mathbf{290.1\,K}$$

The ideal gas law (1.8) shows that the thermodynamic temperature T is proportional to the pressure p for a change of state at a constant volume; then we can determine the requested temperature T_2 by the following calculation

$$T_2 = T_1 \cdot \frac{p_2}{p_1} = 290.1\,\mathrm{K} \cdot \frac{10304\,\mathrm{Pa}}{10205\,\mathrm{Pa}} = \mathbf{292.9}\,\mathrm{K}$$

☐ **1.** Convert the temperature values: 70.5 °F, -63.5 °F and 896 °F into the units (°C) and (K) using the relations (2.8) and (2.9)!

☐ **2.** Convert the temperature values: 0.00 °C, 25.0 °C and 100.0 °C into the units (°F) and (K) using the relations (2.8) and (2.9)!

☐ **3.** Specify at which temperature the temperature value given in (°F) and the temperature value given in (°C) are identical?

☐ **4.** Given the tabular value: $\alpha = 24 \cdot 10^{-6}\,{}^{\circ}\mathrm{C}^{-1}$ for the coefficient of thermal expansion of aluminium, calculate the value of the coefficient based on the units $^{\circ}\mathrm{F}^{-1}$ and K^{-1}!

☐ **5.** Analogous to (2.8), pose an expression for conversion of the thermodynamic temperature unit (°R) into (°F) and (°F) into (°R)!

2.5 Work

Energy in the form of either **work** or of **heat** can be transferred to a thermodynamic system. As shown in the following, work and heat represent two fundamentally different forms of energy exchange.

Figure 2.11. A thermodynamic system may consist of a given amount of substance in the form of water H_2O. This amount of water still constitutes the system after the water evaporates and is transformed into water vapour in the atmosphere.

Work and heat

By a **work process** energy can be transferred to a system through an **organized** action such as moving a piston. In contrast, **heat transfer** is a **non-organized** exchange of kinetic energy between atoms at the boundary between system and its surroundings. Energy-rich atoms can stimulate less energy-rich atoms by random impacts. If, for example, the surroundings have a higher temperature than the system, these random impacts will result in a net transfer of molecular kinetic energy to the system.

> **Work and heat** (2.10)
>
> **Work** and **heat** are quantities that represent the exchange of **energy** between a system and its surroundings; work and heat are assumed to be **positive** when energy is **transferred to** the system.

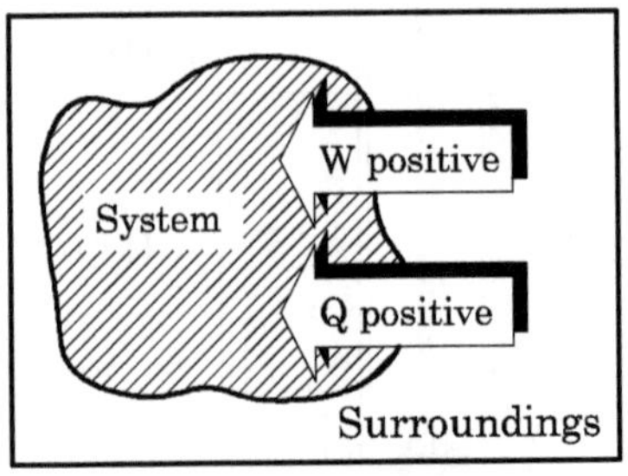

Figure 2.12. Work W and heat Q are each an exchange of energy between a thermodynamic system and its surroundings; in all calculations it is assumed that W and Q are positive when energy is transferred to the system.

Work and heat describe different forms of energy exchange between a system and its surroundings – in a thermodynamic context, therefore, it makes no sense to talk about the "content" of heat or work of a system. The thermodynamic concept of **heat** is an accurately-defined quantity that does not always correspond to the use of the word "heat" in everyday language. Correspondingly, the thermodynamic concept of **work** has a broader meaning than the usual mechanical work concept in physics.

The form that the energy assumes in the system is not necessarily determined by the way of transfer. A temperature rise in a system – i.e. an increase of the average molecular kinetic energy – can be caused by energy transfer in the form of heat as well as in the form of work.

The concept of work

Work is connected with a **process** that causes **energy exchange** between a system and its surroundings; the work is considered **positive** when the energy is **transferred to** the system from its surroundings.

In thermodynamics there are different forms of work such as **mechanical work**, **volume work**, **electrical work** and **surface work**. The total work δW done by a system can formally be expressed as

$$\delta W = \sum Y_i \, dX_i \tag{2.11}$$

where Y_i represents a generalized **force** and dX_i represents a generalized **displacement**. By summation, contributions from all active work processes are added. As a rule, the generalized forces Y_i are **intensive** variables ("potentials") and the corresponding displacements dX_i are **extensive** variables ("quantities").

The most important forms of work included in the description of systems of matter will be described in the following.

Mechanical work

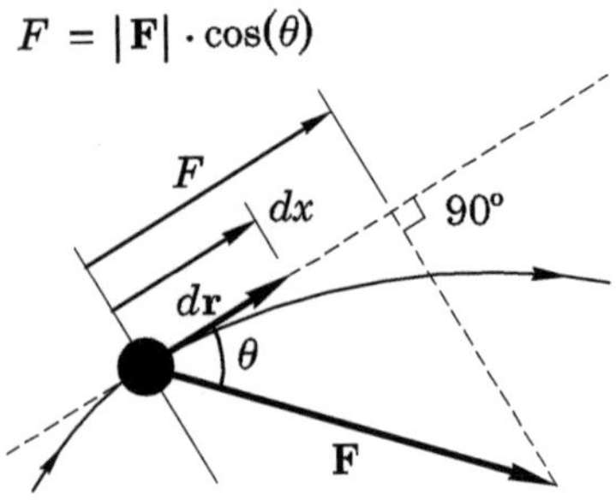

*Figure 2.13. Force **F** and displacement **dr** are vector quantities. The work δW is the scalar product $\mathbf{F} \cdot \boldsymbol{dr}$. When the magnitude of the displacement is denoted dx and the force component in the direction of the displacement is denoted F, then $\delta W = F\,dx$.*

Mechanical work δW is defined by a **force** F multiplied by an infinitesimal **displacement** dx in the force direction; the unit for work is joule, defined as $(\mathrm{J}) = (\mathrm{N} \cdot \mathrm{m})$.

As physical quantities, force **F** and displacement **dr** are **vectors**. Formally, therefore, work δW is produced as the **scalar product** of an acting force **F** and a displacement **dr**; if the angle between the vector **F** and the vector **dr** is denoted θ, then the scalar product expresses that

$$\delta W = \mathbf{F} \cdot \boldsymbol{dr} = |\mathbf{F}| \cos(\theta) \cdot |\boldsymbol{dr}| = F \cdot dx \tag{2.12}$$

where $F = |\mathbf{F}| \cos(\theta)$ denotes the force component in the direction of the displacement dx.

In cases where the vector notation is not essential to the description, we shall denote mechanical work as the product of a force F and a displacement dx in the force direction: $\delta W = F\,dx$.

The expression (2.12) gives the work contribution for an *infinitesimal* displacement dx in the force direction. For a *finite* work process where the force $F = F(x)$ is a function of the displacement x from x_1 to x_2, the total work $W_{1.2}$ is produced by integration of these infinitesimal contributions, i.e.

$$W_{1.2} = \int_{x_1}^{x_2} F(x)\, dx \tag{2.13}$$

If, for example, we consider the elongation of a steel rod in a tensile test machine where $F(x)$ denotes the force curve of the rod, then the force F is a function of the elongation x of the rod. The work $W_{1.2}$ done on the rod during its elongation from x_1 to x_2 is determined by the integral (2.13). The magnitude of the work $W_{1.2}$ is numerically equal to the area below the covered part of the force curve $F(x)$ from x_1 to x_2.

Mechanical work (2.14)

The work done by an external **concentrated force** F acting on a system is

$$\delta W = F dx; \qquad W_{1.2} = \int_{x_1}^{x_2} F(x)\, dx \qquad \text{(J)}$$

where δW denotes the work produced by an **infinitesimal** displacement dx in the force direction and $W_{1.2}$ denotes the work by a **finite** work process where the force $F = F(x)$ is a function of the displacement from x_1 to x_2. The work done **by** surroundings **on** a system is considered **positive**.

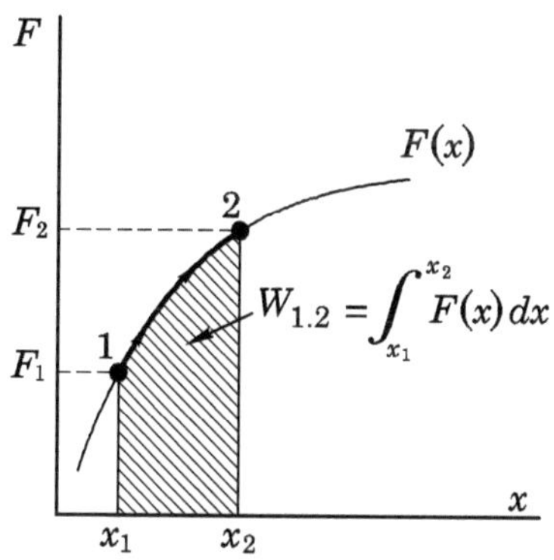

Figure 2.14. In a work process, if the force $F(x)$ is a function of the displacement x, the work $W_{1.2}$ is numerically equal to the area below the part of the force curve $F(x)$ between x_1 and x_2.

Volume work

In thermodynamics, work contributions are often seen in connection with compression or expansion of systems of matter – the so-called **volume work**. Assume a system consisting of n moles of an ideal gas confined in a cylinder under a frictionless piston. The gas pressure is denoted p and the piston area is denoted A. To maintain equilibrium in this system the piston is subjected to an external force $F = p \cdot A$.

Now the piston is pushed slowly a differential distance dx into the cylinder; the force F is assumed to be constant during the movement. By this displacement the surroundings do *positive* work δW on the system as the displacement occurs in the force direction. The work δW done on the system is

$$\delta W = F \cdot dx = p \cdot A \cdot dx = -p \cdot dV \tag{2.15}$$

using that $dV = -A \cdot dx$.

The expression (2.15) is universal for **hydrostatic** systems, irrespective of its change of form during the process; a hydrostatic system is a system of gas or liquid where the pressure p is the same in all directions.

The work W done by a *finite* process is calculated by integration of (2.15); if the equilibrium pressure $p = p(V)$ as a function of the system volume V is known, the work $W_{1.2}$ due to a change of the system volume from V_1 to V_2 is determined by

$$W_{1.2} = -\int_{V_1}^{V_2} p(V)\, dV \tag{2.16}$$

If this work process is illustrated in a pV coordinate system, the work $W_{1.2}$ is numerically equal to the area below the covered part of the curve $p(V)$ from V_1 to V_2.

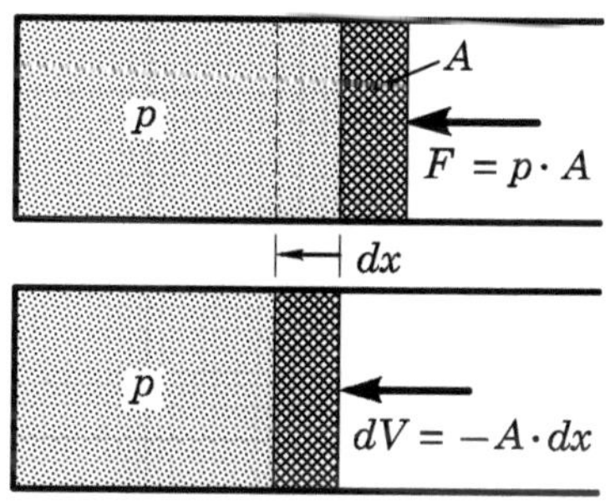

Figure 2.15. Volume work may occur by compression or expansion of a gas confined in a cylinder below a frictionless piston. The gas was in equilibrium at the pressure p when an external force $F = A \cdot p$ was applied to the piston. When the piston was displaced differentially dx, the work $\delta W = -p \cdot dV$ was done on the gas.

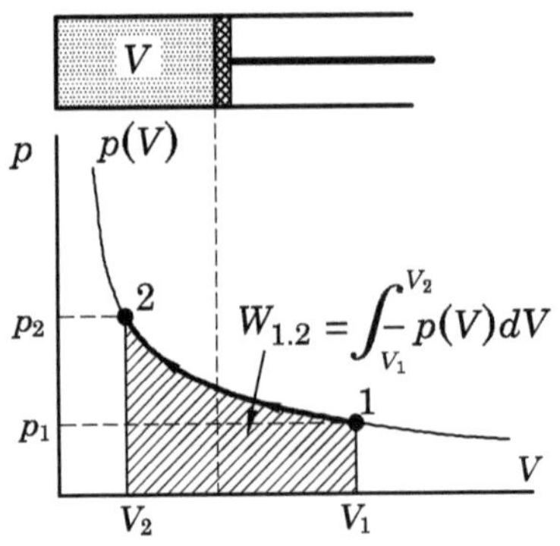

Figure 2.16. Where the equilibrium pressure p is a function of the volume $p(V)$, the numerical value of the volume work $W_{1.2}$ done corresponds to the area under the $p(V)$ curve between V_1 and V_2.

> **Volume work** (2.17)
>
> The work done on a **hydrostatic** system with internal equilibrium pressure p is
>
> $$\delta W = -pdV; \qquad W_{1.2} = -\int_{V_1}^{V_2} p(V)\, dV \qquad \text{(J)}$$
>
> where δW denotes the work due to an **infinitesimal** volume change dV and $W_{1.2}$ denotes the work due to a **finite** work process, where the equilibrium pressure $p = p(V)$ is a function of the volume V from V_1 to V_2. The work done **by** surroundings **on** a system is considered **positive**.

Volume work of the kind mentioned here is generally seen in connection with the change of state in gas systems.

Surface work

In liquids and solid substances the energy level of the molecules is higher at the surface of the substance than further inside the substance. Therefore, **positive** work needs to be done on a system to increase its surface area. This surface effect is seen both in liquids and in solid substances. In liquid systems the effect is visible in the form of a **surface tension** σ; in solid substances the effect is not directly visible. This work contribution is an essential quantity in the thermodynamic description of systems of matter with a large specific surface.

Assume a system where a film of soapy water is stretched across a U-shaped frame and a movable wire of the length L. At the surface of the liquid film a **surface tension** σ tends to contract the film. To balance the surface tension at the surfaces of the liquid film, the surroundings act on the movable wire with a force $F = 2\sigma\ L$. Now, assume that the movable wire is displaced differentially dx in the force direction; this results in a **positive** work δW on the system

$$\delta W = F \cdot dx = 2 \cdot \sigma \cdot L \cdot dx = \sigma \cdot dA \tag{2.18}$$

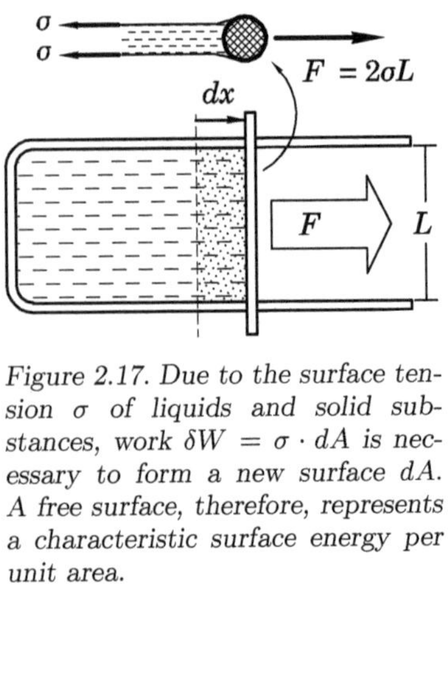

Figure 2.17. Due to the surface tension σ of liquids and solid substances, work $\delta W = \sigma \cdot dA$ is necessary to form a new surface dA. A free surface, therefore, represents a characteristic surface energy per unit area.

It is seen above that the total surface A of the system has been increased by $dA = 2 \cdot L \cdot dx$ due to the displacement of the wire; the factor 2 results because the soapy film has two boundary surfaces to the surrounding air.

The expression (2.18) can be shown to be universal for the work in the formation of a new surface in systems of matter; thus, in conclusion

> **Surface work** (2.19)
>
> The work done on a system with constant **surface tension** σ is
>
> $$\delta W = \sigma \cdot dA; \qquad W_{1.2} = \int_{A_1}^{A_2} \sigma\, dA = \sigma \cdot (A_2 - A_1) \qquad \text{(J)}$$
>
> where δW denotes the work due to an **infinitesimal** increase of the surface of the system dA and $W_{1.2}$ denotes the work due to a **finite** process by which the surface area A of the system is increased from A_1 to A_2. Work done **by** surroundings **on** a system is considered **positive**.

Electrical work

If an **electric charge** dQ (C) is placed in an **electrostatic field**, the charge will be subjected to a **force** F whose magnitude is determined by the field strength (volts/metre) at the location. Therefore, a displacement dx of the charge in the field will be connected with a work process – **electrical work** $F\,dx$.

In the field of electrochemistry that deals with electrochemical corrosion among other matters, electrical work constitutes an essential computational quantity. The description of this thermodynamic quantity is formulated in chapter 6: *Electrochemistry.* In the following, only the final expression for electrical work is given.

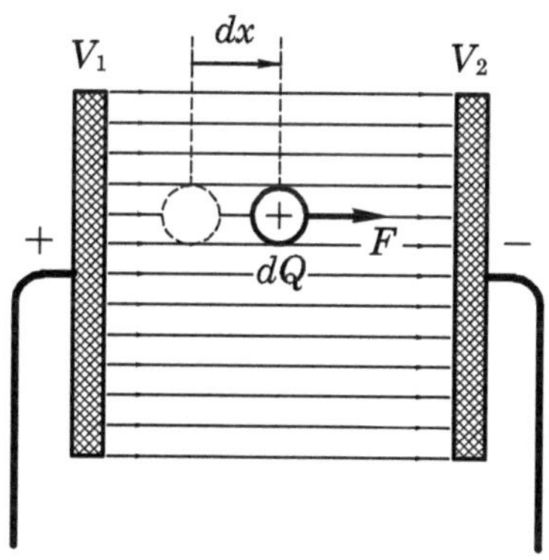

Figure 2.18. An electrostatic field acts on any electric charge dQ with a force F. A displacement dx of the charge in the field is therefore induced, with the work F dx being performed as the so-called electrical work.

Electrical work (2.20)

Electrical work is done on an **electric charge** dQ, which within an **electric field** is transferred from the potential V_a to the potential V_b. The work done is

$$\delta W_{ab} = (V_b - V_a)dQ; \qquad W_{ab} = (V_b - V_a)Q \qquad \text{(J)}$$

where dQ is the electric charge with a positive or negative *sign*. Work done **by** surroundings **on** a system is considered **positive**.

■ A system consists of 2 moles of an ideal gas. In the initial state the gas volume is $V_1 = 20.0\,\ell$. The gas is slowly compressed to the volume $V_2 = 10.0\,\ell$. During compression the gas is assumed to have an evenly distributed hydrostatic pressure $p(V)$ and a constant temperature $T = 293.15$ K. Calculate the magnitude of the volume work $W_{1.2}$ done on the system!

Solution. Using the ideal gas law (1.8), $p(V)$ can be determined during the process

$$p(V) = \frac{nRT}{V}$$

The work done $W_{1.2}$ can be calculated from (2.17)

$$W_{1.2} = -\int_{V_1}^{V_2} \frac{nRT}{V}\,dV = -\int_{V_1}^{V_2} nRT\,d\ln(V) = -nRT \cdot \ln(\frac{V_2}{V_1})$$

The volume work is obtained by inserting numerical values

$$\mathrm{W}_{1.2} = -2\,\mathrm{mol} \cdot 8.314\,\mathrm{J/molK} \cdot 293.15\,\mathrm{K} \cdot \ln(\frac{10.0}{20.0}) = \mathbf{3377\,J}$$

☐ **1.** A system with the mass $m = 100.0$ kg is lifted 10.0 m in the gravitational field; calculate the work $W_{1.2}$ done on the system! ($g = 9.81\,\mathrm{m/s^2}$)

☐ **2.** At constant temperature 20.0 °C, 3 moles of ideal gas are compacted from $V_1 = 70.0\,\ell$ to $V_2 = 50.0\,\ell$; calculate the volume work $W_{1.2}$ done on the system!

☐ **3.** A system of 1.00 kg of ideal He(g) at $\theta = 0.0\,°\mathrm{C}$ and $p_1 = 101325$ Pa is compacted at constant temperature to $p_2 = 250000$ Pa. Calculate the volume work $W_{1.2}$!

☐ **4.** A negative electric charge $Q = -100$ C is moved from the potential $V_a = -10.0$ V to $V_b = 15.0$ V; calculate the electrical work W done on the charge!

☐ **5.** A linear-elastic steel rod of the length 1.00 m and diameter $d = 10.0$ mm is subjected to compression F which is slowly increased from $F_1 = 0$ N to $F_2 = 8000$ N. The E modulus is $2.1 \cdot 10^5$ MPa. Calculate the work $W_{1.2}$ during the process!

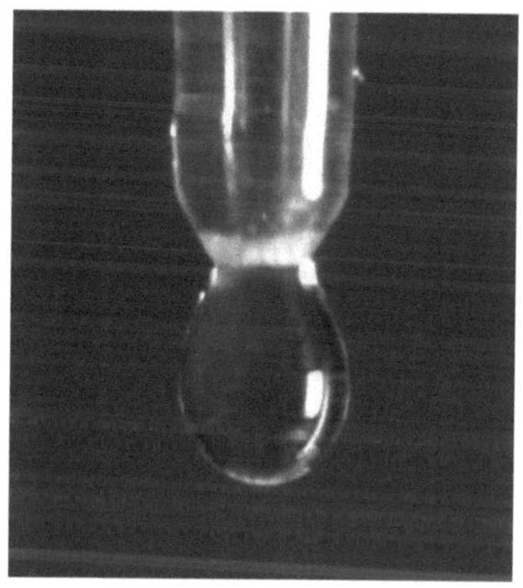

Figure 2.19. The molecular energy level is higher at the surface of a liquid than within the liquid; this surface effect is made manifest by the existence of a surface tension σ, which, among other things, produces the characteristic equilibrium form of drops.

2.6 Heat

Heat is a form of energy transfer caused by the difference in temperature between a system and its surroundings. Within thermodynamics, **heat** and **work** are two different forms of **energy exchange** between a system and its surroundings.

The **work process** transfers energy to or from a system by an **organized** action. Volume work, for example, is connected with a concurrent and organized displacement of a boundary between a system and its surroundings corresponding to the movement of a piston in a cylinder.

Heat transfer is a random, **non-organized** exchange of kinetic energy between individual atoms in the boundary between a system and its surroundings. Energy-rich atoms may, through random impacts, stimulate less energy-rich atoms. If, for example, the surroundings have a higher temperature than the system, such random impacts will result in net transfer of molecular kinetic energy to the system.

It is important to bear in mind that heat and work describe two different forms of energy exchange between a system and its surroundings – *not* two different forms of energy. In a thermodynamic context, therefore, it does not make sense to talk about the "content" of heat or the "content" of work of a system.

Heat exchange between a system and its surroundings is described by the following sign convention: heat **added** to a system is considered **positive.**

Figure 2.20. The typical mean dimension of Portland cement particles is 10 μm, which is achieved by grinding cement clinkers in tube mills as shown. During the grinding process, part of the mechanical energy added to the clinker charge is consumed by the work connected with the increase of the surface area of the solid clinker minerals.

Heat capacity of a system

If a heat quantity δQ is transferred to a system of substances from its surroundings the temperature of the system will rise and the temperature change dT will depend on the **substance properties** of the system as well as the actual **process conditions.** This connection is expressed by the **heat capacity** C of the system, which for a differential process is defined by

Heat capacity of a system (2.21)

The **heat capacity of a system** C is *defined* as the ratio between added heat δQ and the temperature increment dT of the system

$$C \stackrel{\text{def}}{=} \frac{\delta Q}{dT} = \frac{\text{added heat}}{\text{temperature increment}} \qquad \text{(J/K)}$$

The heat capacity of a system at **constant volume** is denoted C_v and the heat capacity at **constant pressure** is denoted C_p. The heat capacity C is an *extensive* quantity, C being dependent on the size of the system. The unit for the heat capacity of a system C is (J/K).

The definition of the heat capacity of a system C distinguishes between heat capacity at constant volume C_v and heat capacity at constant pressure C_p. For **solid substances** and **liquids** C_v and C_p are almost identical; for **gases** C_p is significantly greater than C_v. The difference in C_p and C_v for gases is due to the following:

If a gas is heated at constant volume V, the added heat δQ is consumed *only* to heat the gas. If, on the other hand, a gas is heated at constant pressure p, part of the added heat δQ will be consumed to heat the gas and the other part will be transformed to **volume work** during the thermal expansion of the gas.

Normally, in practical calculations C_v and C_p are only distinguished by heating or cooling of gases.

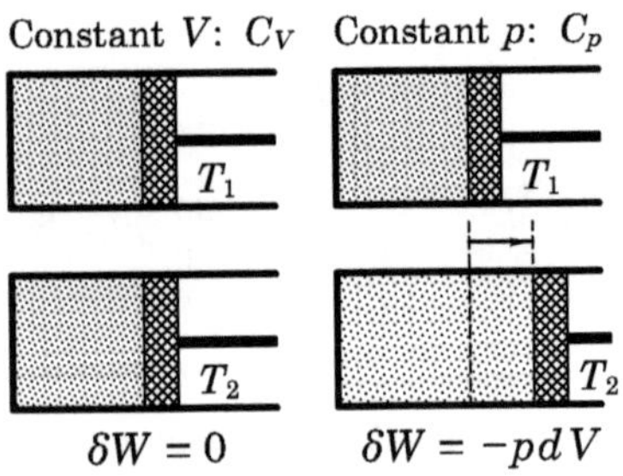

Figure 2.21. If a gas is heated at constant volume V, the added heat δQ is consumed only to heat the gas. If a gas is heated at constant pressure p, part of the added heat δQ will be transformed into volume work δW during the thermal expansion of the gas. Therefore, the same temperature rise requires a larger quantity of heat at constant p than at constant V, i.e. $C_p > C_v$.

Specific heat capacity

Tabulating substance values it is appropriate to use the heat capacity per mol of substance or per kg of substance – the so-called **specific heat capacity** c as (J/mol · K) or (J/kg · K). The specific heat capacity c is an **intensive** variable that is independent of the size of the system.

Generally, in thermodynamic calculations the **mole-specific** heat capacity is used, defined by

> **Mole-specific heat capacity** (2.22)
>
> The **mole-specific heat capacity** c of a substance is *defined* as the ratio between added heat δQ per mole of substance and the temperature increment dT
>
> $$c \stackrel{\text{def}}{=} \frac{1}{n} \cdot \frac{\delta Q}{dT} = \frac{C}{n} = \frac{\text{added heat per mole of substance}}{\text{temperature increment}} \qquad (\text{J/mol K})$$

Mole-specific heat capacity at **constant volume** is denoted c_v and mole-specific heat capacity at **constant pressure** is denoted c_p. Often, in technical calculations the so-called **mass-specific** heat capacity is used, defined by

> **Mass-specific heat capacity** (2.23)
>
> The **mass-specific heat capacity** c of a substance is *defined* as the ratio between added heat δQ per kg of substance and the temperature increment dT
>
> $$c \stackrel{\text{def}}{=} \frac{1}{m} \cdot \frac{\delta Q}{dT} = \frac{C}{m} = \frac{\text{added heat per kg of substance}}{\text{temperature increment}} \qquad (\text{J/kg K})$$

Mass-specific heat capacity at **constant volume** is denoted c_v and mass-specific heat capacity at **constant pressure** is denoted c_p.

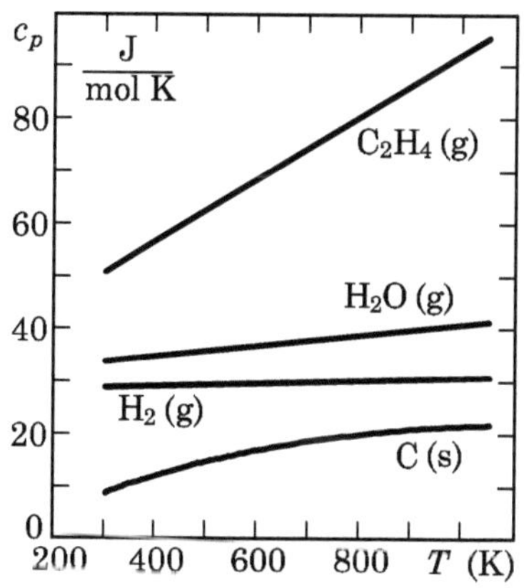

Figure 2.22. The diagram shows examples of temperature dependence of the specific heat capacity c of various substances. (Barin & Knacke: Thermochemical properties of inorganic substances).

Heating of a system of substances

When heating a system of substances over a finite temperature interval from T_1 to T_2, the necessary heat $Q_{1.2}$ is calculated by integration. If the heat capacity c is assumed to be **constant** in the temperature interval considered, the following is obtained using (2.22) for 1 mole of substance

$$Q_{1.2} = c \int_{T_1}^{T_2} dT = c \cdot (T_2 - T_1) \qquad (\text{J/mol}) \qquad \text{constant } c \qquad (2.24)$$

The specific heat capacity c of substances is slightly **temperature dependent** so that $c = c(T)$. If the temperature is changed over a large interval it may be necessary to take this dependence into consideration. The necessary heat $Q_{1.2}$ for heating 1 mole of substance is in this case determined by

$$Q_{1.2} = \int_{T_1}^{T_2} c(T)\, dT \qquad (\text{J/mol}) \qquad \text{temperature dependent } c \qquad (2.25)$$

The temperature dependence $c(T)$ of the specific heat capacity is determined experimentally for a large number of substances and can be found in reference tables.

For temperature changes of approximately 100 °C or below, c may, to a good approximation, be assumed constant in technical calculations

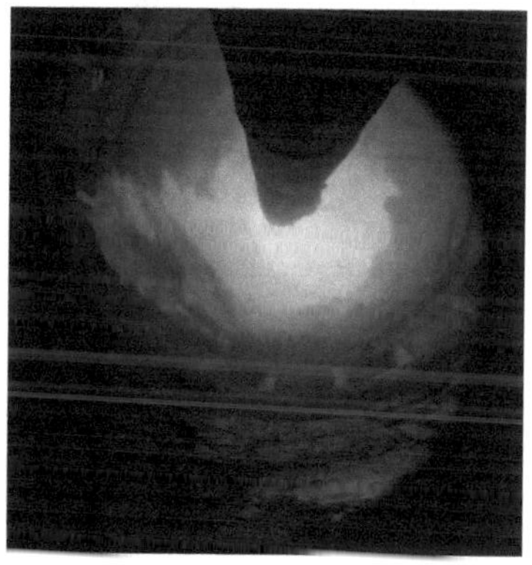

Figure 2.23. The fire zone in a rotary kiln for cement production has a temperature of approximately 1500 °C. In thermodynamic calculations for high-temperature processes of this kind it is necessary to take into account the fact that the heat capacity is dependent on temperature.

Symbols and units

No standardized or incorporated list of symbols distinguishes between **mole-specific** and **mass-specific** heat capacity. In calculations, therefore, one should always specify by **units** or **notation** which kind of heat capacity is applied.

In the present book the heat capacity of systems is always denoted C with the unit (J/K); mole-specific and mass-specific heat capacity are both denoted c with unit (J/mol K) and (J/kg K), respectively.

■ A system consists of $m = 20.0$ kg of water in temperature equilibrium at 18.00 °C. From the surroundings the system is supplied with a heat quantity $\delta Q = 41800$ J at constant pressure, whereby the water temperature rises to 18.50 °C. Calculate the heat capacity C_p (J/K) of the system, the mole-specific heat capacity c_p (J/mol K) and the mass-specific heat capacity c_p (J/kg K) of water. Molar mass for water is 18.02 g/mol.

Solution. The heat capacity C_p is found from the expression (2.21); in the calculation the heat capacity is assumed to be constant

$$C_p = \frac{\delta Q}{dT} = \frac{41800\,\mathrm{J}}{0.50\,\mathrm{K}} = \mathbf{83600\,J/K}$$

The mole-specific and mass-specific heat capacity is determined from (2.22) and (2.23), respectively

$$c_p = \frac{C_p}{n} = \frac{C_p}{m} \cdot M = \frac{83600\,\mathrm{J/K}}{20.0\,\mathrm{kg}} \cdot 18.02 \cdot 10^{-3}\,\mathrm{kg/mol} = \mathbf{75.3\,J/mol\,K}$$

$$c_p = \frac{C_p}{m} = \frac{83600\,\mathrm{J/K}}{20.0\,\mathrm{kg}} = \mathbf{4180\,J/kg\,K}$$

☐ **1.** Explain why c_p is not distinguished from c_v when the specific heat capacity of metals is listed in reference tables!

☐ **2.** Given the mole-specific heat capacity c of a substance (J/mol K); how is this value converted into a mass-specific heat capacity c (J/kg K)?

☐ **3.** A system of substances with 500.0 g iron *Fe* has a heat capacity of $C = 224.7$ J/K; calculate mole-specific and mass-specific heat capacity for *Fe*!

☐ **4.** A concrete has the heat capacity $c = 1.10$ kJ/kgK and density $\varrho = 2350\,\mathrm{kg/m^3}$; how many kJ must be added to 1 m^3 of concrete to increase the temperature by 50.0 °C?

☐ **5.** Given that for nitrogen N_2(g): $c_v(T) = 0.7137 + 0.000067 \cdot T$ (J/gK). Calculate $Q_{1.2}$ for reversible, isochoric heating of 1.00 kg N_2(g) from 20 °C to 500 °C!

2.7 Thermodynamic process

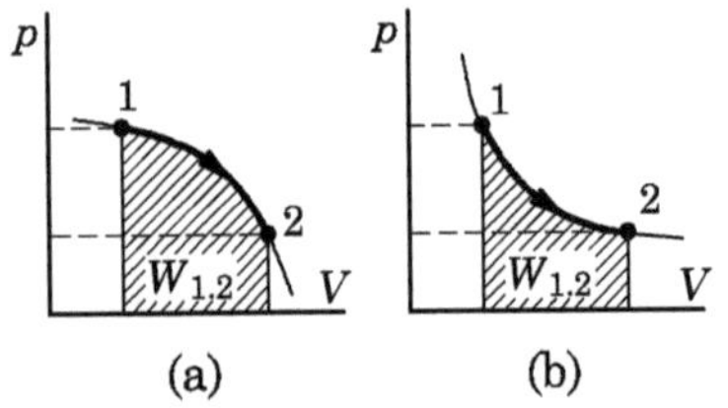

Figure 2.24. Work W is not a state function. For the same change of state, W will depend on the chosen process path. Numerically the volume work $W_{1.2}$ for this process (a) – the covered area below the process curve – is clearly larger than that for process (b), although the change of state is the same.

In a thermodynamic process bringing a system from state 1 to state 2 *both* the **work** done *and* the added **heat** will depend on the chosen process path. This important experience shows that work W and heat Q are *not* thermodynamic state variables, cf. (2.4).

The **volume** V of a system is a **state variable**. By a change of state a system volume can change from V_1 to V_2; for this process it makes physical sense to refer to the change of volume $\Delta V = V_2 - V_1$. Correspondingly, an infinitesimal change of volume can be denoted dV.

Heat Q and **work** W are *not* state variables. For a change of state one *cannot* say that a system's "content" of heat Q or work W is changed. Therefore, in thermodynamic calculations it makes no sense to use designations as ΔQ and ΔW for a process.

Reminding that heat and work are *not* state variables, we shall denote infinitesimal heat and work quantities δQ and δW, respectively, and *not* dQ or dW. Correspondingly, the notations $Q_{1.2}$ and $W_{1.2}$, respectively, are used for finite heat and work quantities and *not* ΔQ or ΔW.

Work and heat (2.26)

Work and heat are *not* thermodynamic **state variables**; for a **change of state** in a system it applies that

- The **work** done $W_{1.2}$ depends on the chosen process path.
- The added **heat** $Q_{1.2}$ depends on the chosen process path.

Process conditions

Work and heat are *not* state variables. Therefore, the work done $W_{1.2}$ and the added heat $Q_{1.2}$ can only be determined when the process path is known. A number of special designations apply to the most frequently used process conditions in thermodynamics.

Process terminology (2.27)

Isothermal process	$\Delta T = 0\,;$	process at **constant temperature** T.
Isobaric process	$\Delta p = 0\,;$	process at **constant pressure** p.
Isochoric process	$\Delta V = 0\,;$	process at **constant volume** V.
Adiabatic process	$Q = 0\,;$	process **without heat exchange**.

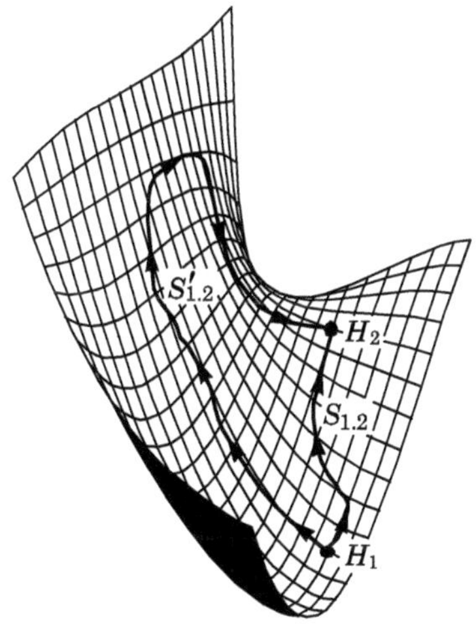

Figure 2.25. The concept of state variable can be illustrated by an example. Let the states (1) and (2) denote two points in a terrain. When we move from (1) to (2) the change of elevation is independent of the route: $\Delta H = H_2 - H_1$. The elevation H is a state variable. The distance $S_{1.2}$ is clearly dependent on the chosen route, i.e. S is not a state variable (see Mathematical appendix, subchapter A6).

Comparing this notation with the definition of heat capacity (2.22) and (2.23) it is seen that

$c_v \Rightarrow$ specific heat capacity at an **isochoric** process.

$c_p \Rightarrow$ specific heat capacity at an **isobaric** process.

Process types

It is a fundamental principle of classical thermodynamics that it *only* addresses systems in thermodynamic balance and processes *connecting* thermodynamic states of equilibrium. This principle entails that all processes connecting thermodynamic states of equilibrium can be divided into two types: **reversible** processes and **irreversible** processes. Therefore, the meaning is as follows:

A **reversible** process is an *idealized* process in which the system undergoes states with only infinitesimal deviation from **equilibrium**. During reversible heating the heat flow is assumed to be caused by an infinitesimal temperature difference dT between the system and its surroundings; during the process the system is assumed to be in **quasi-stationary** equilibrium. Similarly, for a reversible compression or expansion of gas, there are only infinitesimal differences in pressure dp between the system and its surroundings.

During the reversible process the system undergoes a sequence of states all of which are infinitely near a true, thermodynamic equilibrium. Therefore, the reversible process can be unambiguously illustrated in a diagram, for example as a curve depicting a reversible isothermal change of state for a gas in a pV diagram.

All non-reversible processes are called **irreversible**. An example of an irreversible process is expansion of a gas into a vacuum; during the expansion process the system is in a state of **non-equilibrium** and cannot be described by the usual macroscopic state variables such as temperature T and pressure p. The irreversible expansion of a gas into a vacuum can therefore *not* be shown as a process curve in a pV diagram.

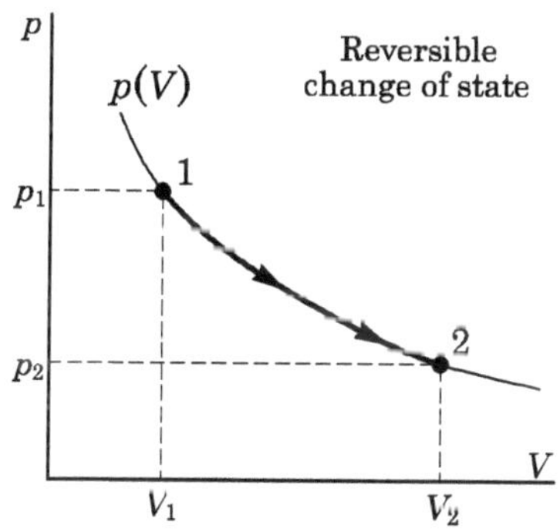

Figure 2.26. The development of a reversible change of state in a gas can be uniquely illustrated in a pV diagram. At each stage of the process, the gas is in equilibrium and the state can be fully described by the general macroscopic state variables.

Process types (2.28)

A process connecting thermodynamic **states of equilibrium**, can be

- a **reversible** process developing through successive states of equilibrium
- an **irreversible** process developing through states of non-equilibrium

In practical systems irreversible processes will generally appear as **spontaneously** developing processes that change a system towards a state of equilibrium.

■ A system consists of 2 moles of ideal gas. In state (1) the gas volume is $V_1 = 20.0\,\ell$ and $T_1 = 293.15\,\mathrm{K}$. In a **reversible, isobaric** compression the gas volume is changed to $V_2 = 10.0\,\ell$; this state is denoted (x). Finally, the gas temperature is raised again to $T = 293.15\,\mathrm{K}$ through a **reversible, isochoric** heating. Calculate the work $W_{1.2}$ for the whole process: $(1) \rightarrow (x) \rightarrow (2)$!

Solution. First we calculate the state variables of the gas (p, V, T) in the states (1), (x) and (2); using the ideal gas law (1.8) we get

State variable:	p (Pa)	V (m^3)	T (K)
state (1)	243725	**0.0200**	**293.15**
state (x)	243725	**0.0100**	146.58
state (2)	487450	0.0100	**293.15**

In the table the supplied data are written in bold digits; in the calculation $p_x = p_1$ is applied, cf. the given process conditions.

By **isobaric** compression $(1) \rightarrow (x)$, $p = 243725\,\mathrm{Pa}$ is constant. Thus, from (2.17),

$$W_{1.2} = -p\int_{V_1}^{V_2} dV = -p\cdot(V_2 - V_1) = -243725\cdot(0.0100 - 0.0200) = 2437\,\mathrm{J}$$

By **isochoric** heating $(x) \rightarrow (2)$, dV, and thus the volume work $W_{x.2}$, is zero. In this case, the total work in the process is

$$W_{1.2} = W_{1.x} + W_{x.2} = 2437\,\mathrm{J} + 0\,\mathrm{J} = \mathbf{2437\,J}$$

Note: In section **2.5 Work**, the work $W_{1.2}$ for the *same* change of state was determined as $W_{1.2} = 3377\,\mathrm{J}$ for an **isothermal** process. From this it is seen that the work done depends on the process path, i.e. W is *not* a state variable!

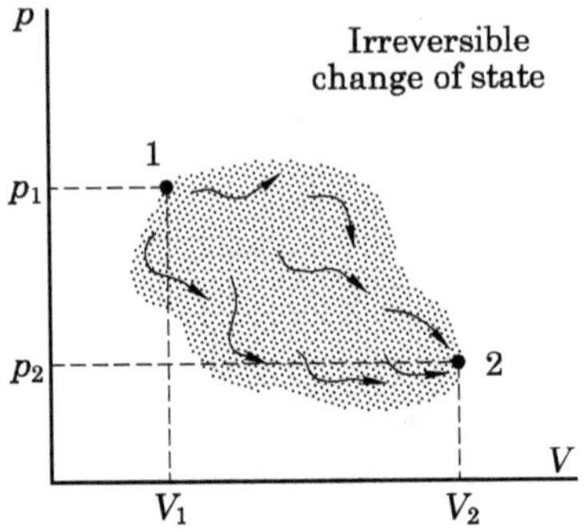

Figure 2.27. An irreversible change of state in a gas cannot be depicted in a pV diagram. During the process the gas is in a state of non-equilibrium: the gas is located, so to speak, at several points in the pV diagram at the same time. Therefore, the process cannot be described by the usual macroscopic state variables (see also fig. 2.4).

☐ **1.** (a) is an isolated, thermodynamic system; is it also necessarily an adiabatic system? (b) is an adiabatic, thermodynamic system; is it also necessarily an isolated system?

☐ **2.** Given a lump of ice in a cup of water. Explain the process conditions for: (a) a reversible melting of the ice! – (b) an irreversible melting of the ice!

☐ **3.** 200 grams of NaCl is put into a $1\,\ell$ beaker of water; on standing the salt slowly dissolves in the water. Is this a reversible or an irreversible process?

☐ **4.** Outline the development of an *isobaric* process for an ideal gas in: (a) a VT diagram, (b) a pV diagram and, (c) a pT diagram!

☐ **5.** Assume 5 moles of ideal gas at 20.0 °C; the gas volume is $V = 120.0\,\ell$. Calculate the work $W_{1.2}$ during a reversible, isobaric heating of the gas to 100.0 °C!

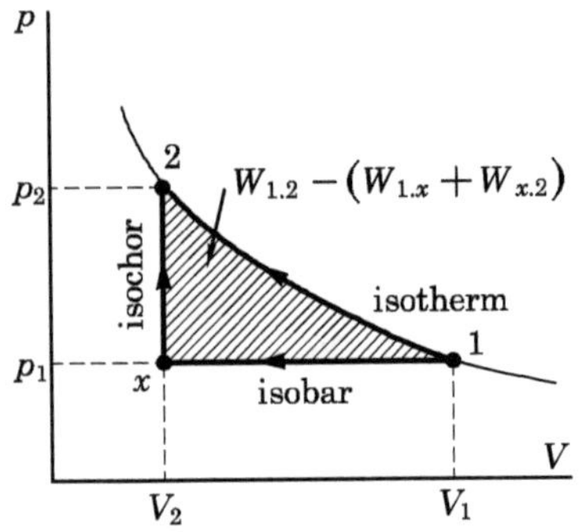

Figure 2.28. The examples calculated in the sections 2.5 Work and 2.7 Thermodynamic process describe the same change of state by two different processes. The difference of the volume work $W_{1.2}$ thus found corresponds to the area shown in the figure.

List of key ideas

The following overview lists the key definitions, concepts and terms introduced in Chapter 2.

Open system .... A thermodynamic system that can exchange **substance** as well as **energy** with its surroundings.

Closed system .... A thermodynamic system that can exchange **energy**, but *not* substance with its surroundings.

Isolated system .... A thermodynamic system that can exchange *neither* substance *nor* energy with its surroundings.

State variable .... A physical quantity *solely* dependent on the state of a system; in a **change of state** the change of state variable is *independent* of the chosen process path (for example temperature T, volume V).

State function .... A physical quantity that can be expressed as a **function** of one or more **state variables**.

Intensive variable	A system variable whose value is *independent* of the size of the chosen system (for example temperature T).
Extensive variable	A system variable whose value is *proportional* to the size of the chosen system (for example volume V).
Temperature	The temperature of a system of matter is a measure of the average **kinetic energy** of the molecules.
Kelvin	The thermodynamic temperature scale is given in the unit **kelvin** (K), which is defined as the fraction 1/273.16 of the **triple point** of the thermodynamic temperature of water.
Triple point	The **triple point** of water indicates the state in which **ice** $H_2O(s)$, **water** $H_2O(\ell)$ and **water vapour** $H_2O(g)$ are in thermodynamic equilibrium; these three phases can only coexist at one particular temperature: 273.16 K.
Fahrenheit	Celsius $= \frac{5}{9} \cdot (\text{Fahrenheit} - 32)$ Fahrenheit $= \frac{9}{5} \cdot \text{Celsius} + 32$
Work	In a thermodynamic process **work** W indicates an *organized* exchange of **energy** between a system and its surroundings; work done **on** a system is considered **positive**.
Heat	In a thermodynamic process **heat** Q indicates a *non-organized* exchange of **molecular kinetic energy** between a system and its surroundings; heat **added to** a system is considered **positive**.
Mechanical work	$\delta W = F \cdot dx; \quad W_{1.2} = \int_{x_1}^{x_2} F(x)\, dx$ (J)
Volume work	$\delta W = -p \cdot dV; \quad W_{1.2} = -\int_{V_1}^{V_2} p(V)\, dV$ (J)
Surface work	$\delta W = \sigma \cdot dA; \quad W_{1.2} = \sigma \cdot (A_2 - A_1)$ (J)
Electrical work	$\delta W = (V_b - V_a)dQ; \quad W = (V_b - V_a)Q$ (J)
Heat capacity of a system	$C = \frac{\delta Q}{dT} = \frac{\text{added heat}}{\text{temperature increment}}$ (J/K)
Specific heat capacity	$c = \frac{1}{n} \cdot \frac{\delta Q}{dT} = \frac{C}{n}$ (mole-specific) (J/mol K)
Specific heat capacity	$c = \frac{1}{m} \cdot \frac{\delta Q}{dT} = \frac{C}{m}$ (mass-specific) (J/kg K)
Isothermal process	Thermodynamic process developing at **constant temperature** T, i.e. $\Delta T = 0$ during process.
Isobaric process	Thermodynamic process developing at **constant pressure** p, i.e. $\Delta p = 0$ during process.
Isochoric process	Thermodynamic process developing at **constant volume** V, i.e. $\Delta V = 0$ during process.

Adiabatic process .... Thermodynamic process *without* **heat exchange** between a system and its surroundings.

Reversible process .... An idealized, thermodynamic process developing through **states of equilibrium**.

Irreversible process .... A spontaneously developing thermodynamic process during which the system is in states of **non-equilibrium**.

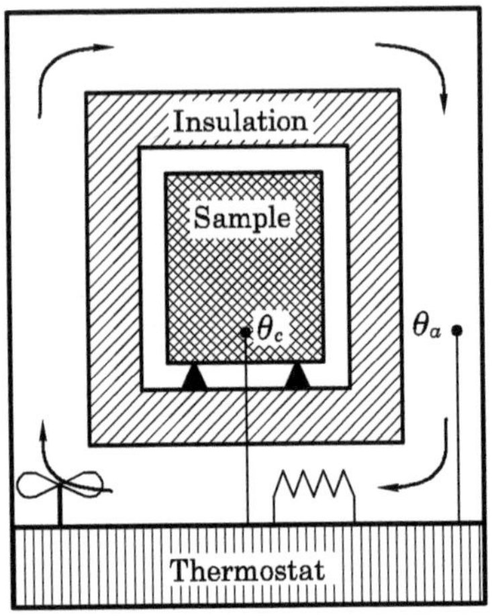

Figure 2.29. Schematic illustration of the composition of an adiabatic calorimeter. A sample of concrete is placed in a thermally insulated container encased in a thermostatically controlled chamber. During the measurement, a thermostat ensures that the air temperature θ_a in the outer chamber is equal to the concrete temperature θ_c. The hydration process will thus develop adiabatically.

Examples

The following examples illustrate how subject matter explained in Chapter 2 can be applied in practical calculations.

Example 2.1

■ Measuring the adiabatic heat development of concrete

Concrete consists of a mixture of **cement**, **water** and coarse and fine **aggregates**. During the reaction between the cement and the added water, the concrete hardens and gains strength. The chemical reactions between cement and water – the so-called hydration – are marked by considerable heat development. If concrete hardens under **adiabatic** conditions, i.e. without exchanging heat with its surroundings, the concrete temperature will typically increase by 50-70 °C due to the heat developed.

When casting solid concrete structures, the developed heat may cause considerable **temperature differences** within the hardening concrete; since the coefficient of thermal expansion α is approximately $1 \cdot 10^{-5}\,\mathrm{K}^{-1}$, this may give rise to **stresses** and **cracking** in the concrete.

In the work specifications for all major concrete work it is specified that temperature differences in the hardening concrete must be kept within certain limits. Therefore, it is in many cases necessary to make advance calculations of the temperature development in the hardening cross-sections under different conditions, and then, based on this, decide upon a method of execution that safeguards the concrete against crack damages during hardening. In such calculations the data of the **adiabatic** heat development of the applied concrete are used.

An **adiabatic calorimeter** is used for measuring heat development of concrete under adiabatic hardening conditions. Samples of fresh concrete are used for the measurement. Heat exchange with its surroundings is prevented during the hardening of the concrete, so that the heat developed is fully converted to a temperature rise in the test sample. The temperature rise is measured as a function of time. Finally, with knowledge of the heat capacity and cement content of the sample, the quantity of heat Q (in kJ per kg of cement) can be calculated.

Figure 2.30. An actual adiabatic calorimeter for measuring heat of hydration of concrete. For the measurement a sample quantity of 4-5 ℓ of concrete is used, and the measurement is typically carried out over 24 to 48 hours.

Problem. In an adiabatic calorimeter the temperature θ_c in a sample of hardening concrete is measured. The cement content C of the concrete is 350 kg/m^3. The concrete density ϱ is 2350 kg/m^3 and the mass-specific heat capacity c_p of the concrete is calculated as 1.10 kJ/kg K. The measurement shows the following relation between hardening time t and the adiabatic concrete temperature θ_c

t	(h)	0	4	7	11	14	16	18	26	80
θ_c	(°C)	22	28	38	51	57	61	63	69	72

Calculate and draw a graph showing the adiabatic heat development Q as a function of time t; the heat development Q should be given in kJ per kg of cement.

Conditions. In the calculations the mass-specific heat capacity of the concrete c_p is assumed to be constant at 1.10 kJ/kg K, independently of the degree of hardening. The measurement is assumed to be exactly adiabatic.

Solution. We consider a system of V m^3 concrete with density $\varrho = 2350\,\mathrm{kg/m^3}$. The heat developed ΔQ (kJ) at a given temperature ΔT (K) is according to (2.21) and (2.23) determined by

$$\Delta Q = C_p \cdot \Delta T = m \cdot c_p \cdot \Delta T = V \cdot \varrho \cdot c_p \cdot \Delta T \qquad \text{(kJ)} \qquad \text{(a)}$$

If the concrete contains C kg of cement per m^3, the adiabatic heat development Q in kJ per kg of cement can be expressed as a function of the temperature rise ΔT by

$$Q = \frac{\Delta Q}{C \cdot V} = \frac{V \cdot \varrho \cdot c_p \cdot \Delta T}{C \cdot V} = \frac{\varrho c_p}{C} \cdot \Delta T \qquad \text{(kJ/kg)} \qquad \text{(b)}$$

Inserting the known quantities, the following numerical relation is obtained

$$Q = \frac{2350\mathrm{kg/m^3} \cdot 1.10\mathrm{kJ/kgK}}{350\mathrm{kg/m^3}} \cdot \Delta T = 7.39 \cdot \Delta T \qquad \text{(kJ/kg)} \qquad \text{(c)}$$

With the expression (c) the measured temperature rise in the specimen can be converted to adiabatic heat development

t	(h)	0	4	7	11	14	16	18	26	80
θ_c	(°C)	22	28	38	51	57	61	63	69	72
ΔT	(K)	0	6	16	29	35	39	41	47	50
Q	(kJ/kg)	0	44	118	214	259	288	303	347	370

The target graph $Q(t)$ can then be drawn. By graphical representation of the property development of concrete it is normally appropriate to use a logarithmic time axis as shown in figure 2.31.

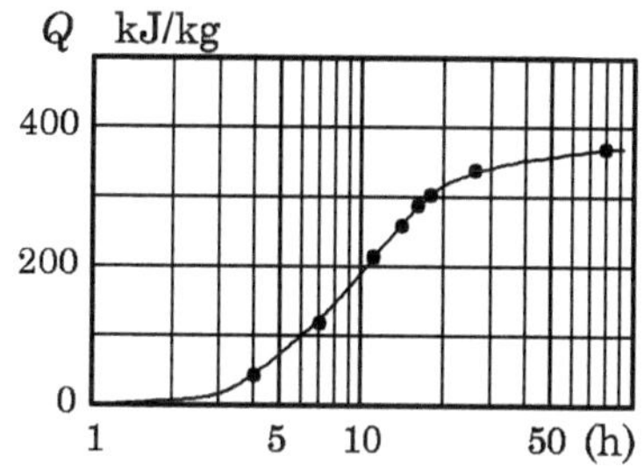

Figure 2.31. Graphical representation of adiabatic heat development Q as a function of hardening time t.

Discussion. The adiabatic heat development Q of 370 kJ/kg of cement found after 80 hours of adiabatic hardening is of normal magnitude for Portland cement. Depending on cement type and w/c ratio, Portland cements typically develop 350 – 450 kJ/kg under such conditions.

For practical mapping of heat development of concrete, Q is normally stated as a function of the concrete maturity M (h), which is the equivalent hardening time at 20 °C. The theory for calculating concrete maturity M is addressed within concrete technology.

Example 2.2

■ Mechanical work in tensile testing of a steel rod

For determining the needed dimensions of load-bearing structures it is necessary to know the strength and deformation properties of the materials used. A frequently used method to describe the deformation properties is the specification of the so-called **stress-strain relation**.

The stress-strain relation is a graphical or analytical representation of the relation between **stress** σ and **strain** ε when the material is loaded. The stress-strain relation is determined by measurement on well-defined specimens subjected to tension or compression in special test machines.

To ensure comparable and reproducible results, detailed **standard methods** have been worked out for determining the strength and deformation properties of different materials. These standards prescribe for example, the design of test specimens and the method of applying loads, and specify requirements for the accuracy and calibration of the test machines used. Examples of international standards for this purpose include *EN 12390* for testing hardened concrete, and *EN ISO 15630: Steel for reinforcement and prestressing of concrete*.

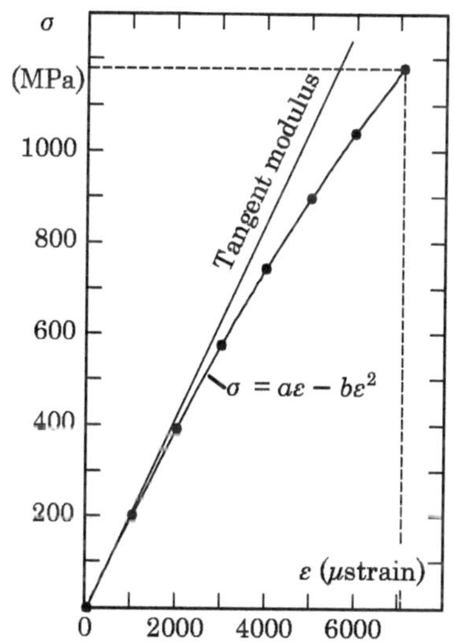

Figure 2.32. Graphical representation of the tensile stress-strain relation (a) for hardened steel. Note that the stress-strain relation $\sigma = \sigma(\varepsilon)$ contains intensive quantities only and therefore is independent of the size or shape of the test specimen.

Problem. A round rod of length 1.00 m and Ø 10.0 mm of hardened steel is subjected to tension in a test machine. During tensioning corresponding values of the applied force F and the elongation $\Delta\ell$ are registered. The force applied is increased until failure in the bar. In a subsequent analysis of the results measured, the following connection between **tensile stress** σ (MPa) and **strain** ε in the steel rod is found.

$$\sigma = a \cdot \varepsilon - b \cdot \varepsilon^2 \qquad \text{(MPa)} \qquad \text{(a)}$$

Figure 2.33. Normally, tensile failure in hardened steel is a sudden, brittle fracture without previous plastic deformation (left). Typically, tensile failure in mild steel is a ductile fracture with considerable elongation taking place prior to fracture. Around the place of fracture a severe necking and reduction of cross-section is seen (right). The materials samples shown are made of amorphous metal.

where : $a = 0.21 \cdot 10^6$ MPa; $b = 6.15 \cdot 10^6$ MPa.

The ultimate stress of the steel is $\sigma_u = 1180$ MPa at the strain $\varepsilon_u = 7.1 \cdot 10^{-3}$. Based on this information, calculate the magnitude of the **mechanical work** W (kJ) done on the steel rod at the instant where fracture occurs.

Conditions. The deformed length of the steel rod is assumed to be $\ell = 1.00$ m; the length ℓ does not include the part of the rod which is held by the jaws of the test machine.

In calculations the closed system constituted by the loaded part of the steel rod will be considered.

Solution. The work W done on the steel rod is calculated from the expression for mechanical work (2.14)

$$\delta W = F \cdot dx; \qquad W_{1.2} = \int_{x_1}^{x_2} F(x) dx \tag{b}$$

Denoting the cross-sectional area of the steel rod A and its length ℓ, we have the following relation between the stress σ and force F, and between strain ε and elongation in the force direction dx

$$F = \sigma \cdot A; \qquad dx = \ell \cdot d\varepsilon \tag{c}$$

Inserting these relations into the work expression we have the total work at failure W_u as

$$W_u = \int_0^{\varepsilon_u} A \cdot (a \cdot \varepsilon - b \cdot \varepsilon^2) \cdot \ell \, d\varepsilon \tag{d}$$

Integration of this expression gives

$$W_u = A\ell \cdot (\tfrac{a}{2} \cdot \varepsilon_u^2 - \tfrac{b}{3} \cdot \varepsilon_u^3) \tag{e}$$

Inserting the measured value of the deformation at rupture $\varepsilon_u = 7.1 \cdot 10^{-3}$ and the constants a and b, the targeted work at failure W_u can be determined

$$W_u = \tfrac{\pi}{4} \cdot (0.0100)^2 \cdot 1.00 \cdot \left(\frac{0.21 \cdot 10^6}{2} \cdot (7.1 \cdot 10^{-3})^2 - \frac{6.15 \cdot 10^6}{3} \cdot (7.1 \cdot 10^{-3})^3\right)$$

$$W_u = 3.6 \cdot 10^{-4} \quad (\mathrm{MPa} \cdot \mathrm{m}^3) \tag{f}$$

As the unit $(\mathrm{Pa} \cdot \mathrm{m}^3)$ corresponds to (J), the final result is:

Work W_u done in the failure condition: **= 360 J**

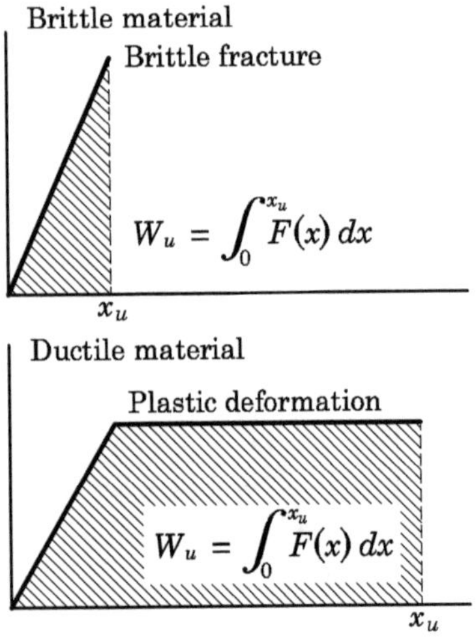

Figure 2.34. The work to failure in tension W in a brittle fracture (top) and in a ductile fracture (bottom). For comparable test specimens the work to failure in a ductile fracture is large compared to the work to failure in a brittle fracture.

Discussion. In load-bearing structures it is necessary to distinguish between **brittle** materials and **ductile** materials; the W_u is much larger for ductile materials.

A **ductile** material – such as construction steel *S235JR (EuroNorm 10025)* – shows large plastic deformation before the actual parting of a specimen takes place. A **brittle** material – such as glass – shows sudden failure without appreciable previous plastic deformation. Generally, the work at failure is large for a ductile material while it is modest for a brittle material.

For reasons of safety, ductile materials are preferred in a load-bearing structure to reduce the risk of sudden failure and collapse due to excessive loading of the structure.

The hardened steel described in the example has a high ultimate strength; at the same time because the steel is brittle, the work at failure W_u is modest.

Finally it is interesting to note the following: the total energy added to the steel rod up to the point of failure, approximately 360 J, corresponds to the energy that is released in the form of heat by combustion of about half a match!

Example 2.3

■ Volume work by evaporation of water

To transform water into water vapour at atmospheric pressure – for example by boiling – a considerable quantity of energy in the form of heat must be added. Evaporation of 1 kg of water at 100 °C consumes approx. 2300 kJ – the so-called **heat of evaporation**.

The heat energy added to the liquid water partly serves to increase the **potential energy** of the molecules by formation of gas phase, and partly serves to do **volume work** on the surrounding atmosphere. This latter contribution arises because the change of state:

$$H_2O(\ell) \quad \rightarrow \quad H_2O(g) \tag{a}$$

entails a considerable volume increase. During the boiling process a volume of the surrounding atmospheric air corresponding to the volume of water vapour generated is necessarily displaced: thus the system does volume work W on its surroundings.

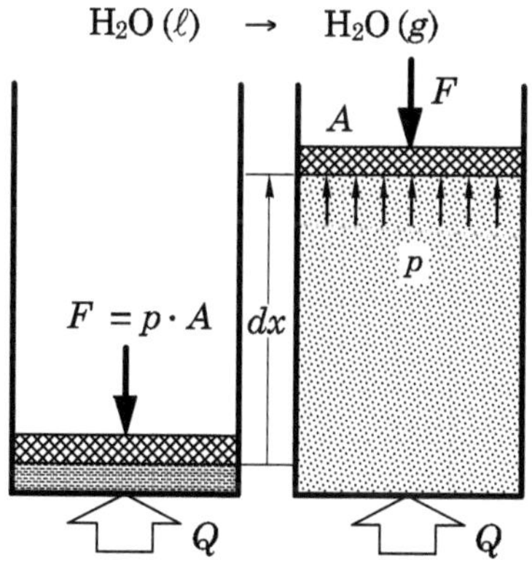

Figure 2.35. Evaporation of water results in a considerable increase of volume. Evaporation of water into a surrounding atmosphere requires that volume work W shall be done to displace some of the gas molecules previously present in the atmosphere around the liquid water.

Problem. Assume a system consisting of 1.000 kg of water $H_2O(\ell)$ at 100 °C. The water evaporates during boiling and is fully transformed into water vapour at 100 °C. During the transformation the pressure p is constant at 1 atm. Calculate the magnitude of the volume work W done by the surroundings on the system during the evaporation process.

Conditions. We consider a **reversible**, **isobaric** and **isothermal** change of state transforming the system from water into water vapour

$$\left\{\begin{matrix} 1.000\,\text{kg}\,H_2O(\ell) \\ \theta = 100\,^\circ\text{C} \\ p = 101325\,\text{Pa} \end{matrix}\right\} \rightarrow \left\{\begin{matrix} 1.000\,\text{kg}\,H_2O(g) \\ \theta = 100\,^\circ\text{C} \\ p = 101325\,\text{Pa} \end{matrix}\right\} \tag{b}$$

Formally this process can be performed by confining the actual quantity of water in a cylinder under a frictionless piston. By adding heat reversibly $H_2O(\ell)$ is transformed into $H_2O(g)$; under the concurrent isobaric expansion the system does volume work on its surroundings.

In calculations the water vapour will be assumed to be an ideal gas. The density ϱ of liquid water at 100 °C has been determined from reference tables as 958.4 kg/m³. The molar mass of water is M = 18.02 g/mol.

Solution. In its initial state the system has a volume V_1 determined by

$$V_1 = \frac{m}{\varrho} = \frac{1.000\,\text{kg}}{958.4\,\text{kg/m}^3} = 1.043 \cdot 10^{-3}\,\text{m}^3 \tag{c}$$

In its final state the water vapour formed has a volume V_2 which can be determined using the **ideal gas law** (1.8); insertion gives

$$V_2 = \frac{mRT}{Mp} = \frac{1.000 \cdot 8.314 \cdot 373.15}{18.02 \cdot 101325} = 1.699\,\text{m}^3 \tag{d}$$

The volume work $W_{1.2}$ done on the system during the process can now be calculated from (2.17)

$$W_{1.2} = -\int_{V_1}^{V_2} p(V)dV = -p_{atm} \cdot \int_{V_1}^{V_2} dV = -p_{atm} \cdot (V_2 - V_1) \tag{e}$$

$$W_{1.2} = -101325 \cdot (1.699 - 1.043 \cdot 10^{-3})\,\text{Pa} \cdot \text{m}^3 = -1.720 \cdot 10^5\,\text{J} \tag{f}$$

Thus the volume work $W_{1.2}$ done by evaporation of 1.000 kg of water is = $\mathbf{-1.720 \cdot 10^5}$ J

Discussion. The negative sign for W shows that the system has done positive work on its surroundings during the process. This means that part of the added heat Q was consumed in this work process. The heat of evaporation for 1 kg of water at 100 °C is approximately $23 \cdot 10^5$ J; approximately $1.7 \cdot 10^5$ J or about 7 % of this heat of evaporation was consumed for the volume work.

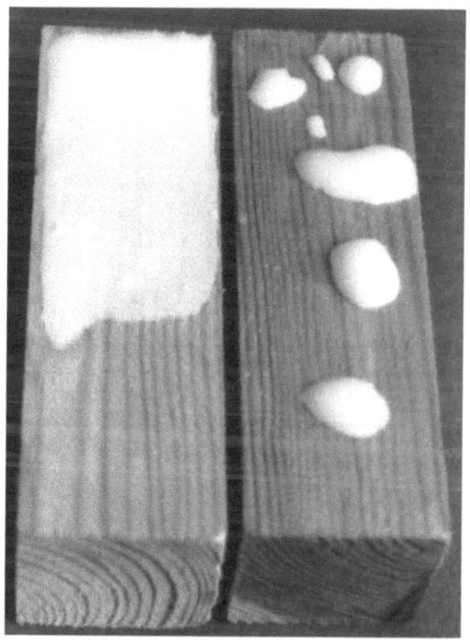

Figure 2.36. To achieve adherence and high strength in a glued joint, it is necessary that the glue completely wets the surface to be glued (left). If the surface tension σ of the glue exceeds the surface tension of the surface to be glued, wetting is not possible (right).

Example 2.4

■ Surface work by atomization of water

In liquids and solids the molecules at the surface of the substance have larger potential energy than those inside the substance. Therefore, *positive* work on a system must be done to increase its surface area. A free surface existing in a system thus involves the presence of an amount of **surface energy** in the system.

This surface effect is seen both in liquids and in solid substances. In liquids the effect becomes visible in the form of a **surface tension** σ, which, for example, produces the characteristic form of equilibrium for drops.

The magnitudes of the quantities of energy connected with surface work are normally rather modest. In spite of this, surface work is a decisive parameter in many connections. Within the field of **glue technology**, changes of energy connected with surface work are vital to the glue's ability to wet a specimen; without completely wetting the substance to be glued adherence and joint strength cannot be achieved. Surface work is also important for many fracture mechanical phenomena.

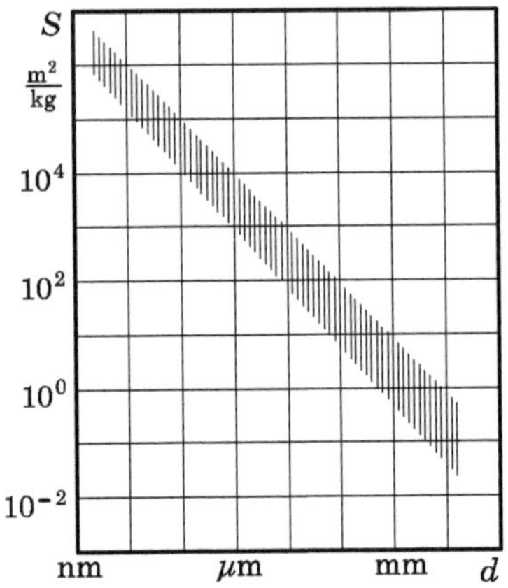

Figure 2.37. The specific surface S (m^2/kg) of a system of identical particles is inversely proportional to the particle dimension d. This relation implies that surface energy is an important parameter in the description of colloid systems such as gels, emulsions and aerosols (fogs) due to the large specific surface of these systems.

Problem. Assume a system consisting of 1.000 kg of water at 25 °C. The water is atomized into mist droplets which are assumed to be identical spherical droplets of radius $r = 0.005$ mm. Calculate the magnitude of the surface work W (J) connected with this atomization of the water!

Conditions. In this calculation, *only* the work required to increase the surface area of the water is to be determined; all mechanical work losses during the atomization are to be disregarded.

The *Handbook of Chemistry and Physics* lists the **density** ϱ and the **surface tension** σ of water at 25.0 °C as:

$$\varrho = 0.99705 \text{ g/cm}^3 \qquad \sigma = 71.97 \text{ dyn/cm} \tag{a}$$

The force unit (dyn) is equal to $1 \cdot 10^{-5}$ (N).

Solution. First we calculate the total volume V of the system

$$V = \frac{m}{\varrho} = \frac{1.000\,\text{kg}}{997.05\,\text{kg/m}^3} = 1.003 \cdot 10^{-3}\,\text{m}^3 \tag{b}$$

The volume V_d of a single drop is determined by

$$V_d = \tfrac{4}{3}\pi r^3 = \tfrac{4}{3}\pi(0.005 \cdot 10^{-3})^3\,\text{m}^3 = 5.24 \cdot 10^{-16}\,\text{m}^3 \tag{c}$$

Correspondingly, the surface area A_d of one single drop is determined by

$$A_d = 4\pi r^2 = 4\pi(0.005 \cdot 10^{-3})^2\,\text{m}^2 = 3.14 \cdot 10^{-10}\,\text{m}^2 \tag{d}$$

Thus, after atomizing the total surface area A of the system is

$$A = \frac{V}{V_d} \cdot A_d = \frac{1.003 \cdot 10^{-3}}{5.24 \cdot 10^{-16}} \cdot 3.14 \cdot 10^{-10}\,\text{m}^2 = 601\,\text{m}^2 \tag{e}$$

For calculation of the surface work $W_{1.2}$ during atomizing, the expression (2.19) is used; $\sigma = 71.97 \cdot 10^{-3}$ N/m is inserted as surface tension. In the calculation the surface area of the system in its initial state is disregarded, since $A_1 \ll A_2$

$$W_{1.2} = \sigma \cdot (A_2 - A_1) = 71.97 \cdot 10^{-3} \cdot (601 - 0) = 43.3 \text{ J} \tag{f}$$

Surface work $W_{1.2}$ during atomizing of 1.000 kg water = **43 J**

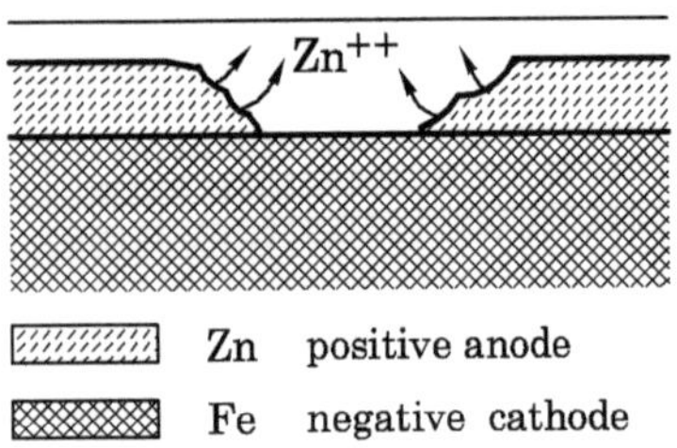

Figure 2.38. A zinc cover on steel counteracts electrolytic corrosion of the steel even if there are holes in the cover. Zn is a less noble metal than Fe; therefore, the zinc cover acts as sacrificial anode and corrodes, while the steel as cathode remains intact.

Discussion. Formation of new surface of a substance is a physical process – a **change of phase** – and can therefore, to a certain extent, be compared with processes such as **melting** or **evaporation** of substances. Therefore, it is illustrative to compare the relative scales of surface work with those of typical processes of energy changes due to melting or evaporation.

Melting of 1 kg of ice requires an energy supply of approximately 340 000 J. Evaporation of 1 kg of water requires an energy supply of approximately 2 400 000 J. In comparison, a very modest supply of energy – ca. 43 J – is required for atomizing 1 kg of water into fog droplets.

Example 2.5

■ Electrical work by galvanizing of steel

An effective corrosion protection for steel can be obtained by depositing a cover of metallic zinc Zn. The corrosion-preventing effect of the zinc layer is due to the following circumstances: Zn is a less noble metal than iron Fe; in electrolytic corrosion of a zinc-covered steel specimen, Zn will therefore become anodic with respect to the steel. In electrolytic corrosion, the **anode** oxidizes and loses metal; the **cathode**, steel in this case, remains intact. The zinc cover acts as a **sacrificial anode**; corrosion of the underlying steel can take place only when all of the anodic zinc has been oxidized.

In practice, galvanizing of structural steel is carried out in one of several ways:

- *Hot-galvanizing*: steel is immersed in a melt of liquid zinc; typical cover thicknesses attained in this process range from 30 to about 150 μm
- *Spray galvanizing*: atomized, molten zinc is sprayed onto the surface of the steel; typical cover thicknesses vary between 80 and 200 μm
- *Electro-galvanizing*: metallic zinc is deposited on the surface of the steel by electrolysis; typical cover thicknesses are only 10-20 μm

For small products in particular, such as fittings, bolts, nuts, etc., electro-galvanizing is widely used; using this method, the steel is not subjected to the intense heating which is necessary when using the other two methods.

Problem. A set of steel specimens is to be galvanized in a galvanic bath containing an aqueous solution of zinc sulphate $ZnSO_4$. The specimens are suspended at the negative pole and thus function as **cathodes**; the positive **anode** is formed by solid zinc plates immersed in the bath. During the process the anode metal is **oxidized**

$$\mathrm{Zn(s)} \quad \rightarrow \quad \mathrm{Zn^{++}(aq)} + 2e^- \tag{a}$$

and the zinc ions pass into solution. At the cathode the $Zn^{++}(aq)$ ions are **reduced** to metallic zinc and plated out on the surfaces of the specimens. The cathode process is:

$$\mathrm{Zn^{++}(aq)} + 2e^- \quad \rightarrow \quad \mathrm{Zn(s)} \tag{b}$$

By the process the anode is given an electric voltage $V_a = +4.5$ volt, whereas the cathode with the steel specimens has the potential $V_b = 0.0$ volt.

Calculate the magnitude of the electrical work W which needs to be done on the system in order to deposit 1000 g of metallic zinc on the steel specimens!

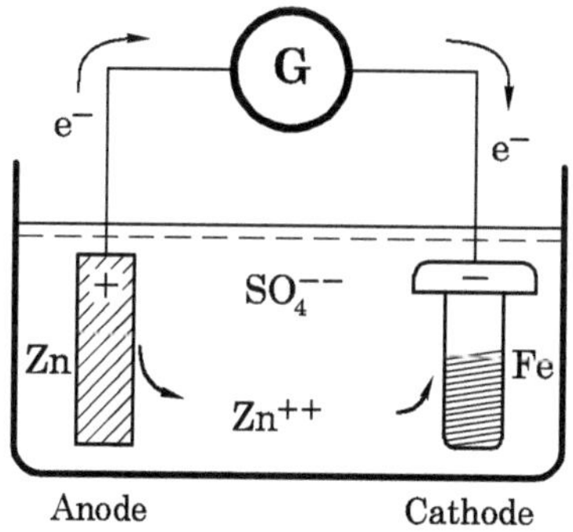

Figure 2.39. Electrolytic galvanizing of a steel specimen is shown schematically. At the positive zinc anode, Zn atoms are oxidized to Zn^{++} ions which are dissolved. At the negative cathode formed by the specimen, Zn^{++} ions are reduced to Zn atoms which are deposited on the surface of the steel. The process can be carried out in a solution of zinc sulphate $ZnSO_4$ and may be driven by an external generator at a voltage of 4-6 volts.

Conditions. In the calculations a system consisting of a galvanic bath and two electrodes is considered. Electrical work is done as follows. Using an external generator, electrons e^- are transferred from the anode with an electric potential of $V_a = 4.5$ volt to the cathode with a potential of $V_b = 0.0$ volt.

The following tabular data are used: molar mass of zinc Zn $M = 65.38$ g/mol; electric elementary charge $e^+ = 1.602 \cdot 10^{-19}$ C.

Solution. Depositing of 1000 g of metallic Zn corresponds to an amount of substance determined by

$$n(\mathrm{Zn}) = \frac{m}{M} = \frac{1000\,\mathrm{g}}{65.38\,\mathrm{g/mol}} = 15.3\,\mathrm{mol} \tag{c}$$

From equation (b) describing the reduction process it is seen that he number of electrons e^- that need to be transferred is twice the number of zinc ions needed to deposit the specified mass of zinc, i.e., $n(e^-) = 30.6$ mol. Thus, the total quantity of electric charge Q(C) transferred by the process is

$$Q = n(e^-) \cdot \mathcal{N} \cdot Q(e^-) \tag{d}$$

$$Q = 30.6\mathrm{mol} \cdot 6.022 \cdot 10^{23}\mathrm{mol}^{-1} \cdot (-1.602 \cdot 10^{-19}\mathrm{C}) = -2.95 \cdot 10^6\ \mathrm{C} \tag{e}$$

The electrical work W *done by* the surroundings *on* the system is determined by the expression (2.20) in integrated form

$$W = (V_b - V_a)Q = (0.0 - 4.5) \cdot (-2.95 \cdot 10^6)\ \mathrm{J} = 1.33 \cdot 10^7\ \mathrm{J} \tag{f}$$

Electrical work W done by depositing of 1000 g of Zn = $\mathbf{1.33 \cdot 10^4\ kJ}$

Figure 2.40. Electrolytic galvanizing of steel is often used for large numbers of small steel specimens; the photo shows galvanic baths for galvanizing of bolts and nuts.

Discussion. The positive sign of the work W shows that work has been done on the system. The value found corresponds to an energy consumption of approx. 3.7 kWh.

Example 2.6

■ Heating of mortar by employing hot mixing water

Figure 2.41. Winter concreting requires a number of special measures to protect the fresh and hardening concrete against freezing. Typical winter measures are the use of hot concrete and provision of heat insulation of concrete structures during hardening.

During winter it is often necessary to use hot concrete or mortar for concreting to ensure fast hardening and strength development; if not, the concrete may be destroyed by premature freezing.

Hot concrete or mortar can be produced by using **preheated** aggregates or by blowing **steam** into the mixer during the mixing process. In other cases the required concrete or mortar temperature can be achieved simply by using hot mixing water.

Problem. **Cement mortar** is required for casting joints between concrete units. The work is to be done during winter, and a mortar temperature of at least 20 °C is required in the fresh mix. Preheated water is to be used to meet the temperature requirement. The mortar is made according to the following mix design:

Component	Subscripts	m (kg)	θ (°C)	c (kJ/kg K)
Sand	(s)	140	−1	0.84
Cement	(c)	50	+5	0.71
Water	(w)	25	+50	4.19

In the table m, θ and c denote the mass, temperature and mass-specific heat capacity, respectively, of the constituent materials. With the given subscripts, the mass of sand is denoted m_s, temperature of cement is denoted θ_c, etc.

- Set up an analytic expression to determine the mixing temperature of the mortar as a function of the masses m (kg), temperatures θ (°C) and specific heat capacities c (kJ/kg K) of the constituent materials.
- Calculate the actual temperature θ of the mortar mix from the this expression.

Conditions. In the calculations any loss of heat during weighing and mixing of materials should be disregarded. Furthermore, the initial heat development normally occurring during mixing of the cement with water should also be disregarded.

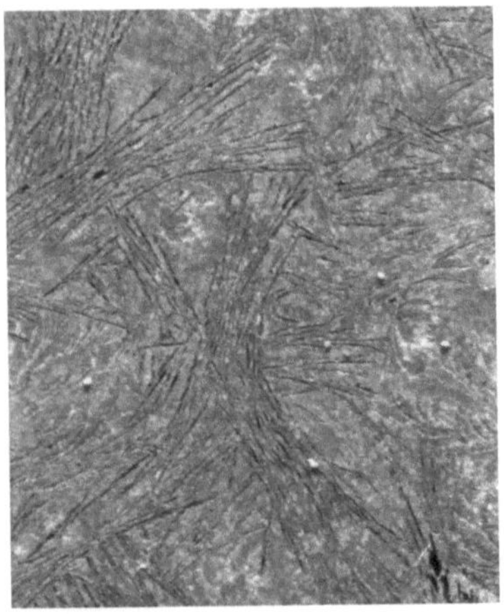

Figure 2.42. Freezing of fresh or modestly hardened concrete can produce ice formations that may cause permanent destruction of the concrete. The photo shows ice impressions developed after ice formation in a fresh concrete exposed to premature freezing.

Solution. Assume a system formed by **adiabatic** mixing of the three constituents. After the temperature equilibrium has occurred in the mix, the following will apply:

$$Q_{system} = Q_s + Q_c + Q_w = 0 \tag{a}$$

where Q denotes the heat added to the individual systems. Denoting the temperature of the mix θ, this relation between heat quantities can be expressed by:

$$m_s c_s \cdot (\theta - \theta_s) + m_c c_c \cdot (\theta - \theta_c) + m_w c_w \cdot (\theta - \theta_w) = 0 \tag{b}$$

Isolating the temperature of the mix θ, the targeted expression is obtained as follows:

$$\theta = \frac{m_s c_s \theta_s + m_c c_c \theta_c + m_w c_w \theta_w}{m_s c_s + m_c c_c + m_w c_w} = \frac{\Sigma(m \cdot c \cdot \theta)_i}{\Sigma(m \cdot c)_i} \qquad i = s, c, w \tag{c}$$

By introduction of the given material data into (c) the temperature of the actual mortar mix is determined by:

$$\theta = \frac{140 \cdot 0.84 \cdot (-1) + 50 \cdot 0.71 \cdot 5 + 25 \cdot 4.19 \cdot 50}{140 \cdot 0.84 + 50 \cdot 0.71 + 25 \cdot 4.19} = 21\,^\circ\mathrm{C} \tag{d}$$

The mixing temperature of the mortar is $\theta =$ **ca. 21 °C**

Discussion. The required mixing temperature of at least 20 °C can be obtained by adding 50 °C hot mixing water. It shall be emphasized, however, that in practice it is difficult to control mixing temperatures exactly; normally the temperature of the aggregates used is not very well defined, and in winter they may contain moisture in the form of ice.

Exercises

The following exercises can be used when learning the subjects addressed in Chapter 2. The times given are suggested guidelines for use in assessing the effectiveness of the teaching.

Exercise 2.1

☐ Extensive and intensive variables *(10 min.)*

Divide the following physical variables into **intensive** and **extensive** variables, respectively. Then show by examples how to convert the extensive variables into intensive variables. 1) The mass m (kg) of a system of substances. 2) The modulus of elasticity E (MPa) of a steel specimen. 3) The viscosity η (Pa s) of a given saline solution. 4) The electric charge Q (C) on a charged condenser plate. 5) The molar mass M (g/mol) of a chemical compound. 6) The cement content C (kg) in a given concrete specimen. 7) The elongation $\Delta\ell$ (m) of a loaded test steel rod. 8) The coefficient of thermal expansion α (K^{-1}) of pyrex glass.

Exercise 2.2

☐ Conversion of temperature units *(10 min.)*

Perform the following conversions of temperature units: 1) 57.5 °F into unit (°C); 2) 385.2 K into unit (°F); 3) 33.5 °R (*Rankine*) into unit (°F) and (°C); 4) 1052 °F into unit (K); 5) 852.4 K into unit (°R) and 6) 4.2 K into unit (°C).

Exercise 2.3

☐ Specific heat capacity of alloy stainless steel *(4 min.)*

Stainless steel is the designation for steel that has been made particularly corrosion proof by adding **chromium** Cr and **nickel** Ni. The table reference *Engineering Materials Handbook* specifies that **austenitic** stainless steel with 18 wt-% Cr and 8 wt-% Ni has a specific heat capacity c of 0.12 Btu/(°F · lb). The unit 1 Btu (*"British Thermal Unit"*) = 1054 J and the unit 1 lb (*"libra": pound*) = 0.454 kg. Calculate the mass-specific heat capacity c for the steel concerned using the SI unit kJ/kg K!

Figure 2.43. Stainless steel alloys are widely used in the manufacture of vessels, pipelines, facings and fittings for buildings. A frequently used stainless steel alloy is "stainless steel 18/8", which is a steel alloy with 18 wt-% Cr and 8 wt-% Ni.

Exercise 2.4

☐ Thermal properties of zinc Zn *(8 min.)*

In the reference book *Handbook of Chemistry and Physics* the following thermal data are given for the pure metal zinc Zn: **Melting point** $\theta_m = 785\,°F$; **coefficient of thermal expansion** $\alpha = 18\,\mu$inch/(inch ·° F); **heat conductivity** $\lambda = 114\,$Btu/(h · ft ·° F). The following conversion factors are given: 1 Btu = 1054 J and 1 ft = 0.3048 m. Based on this, determine θ_m, α and λ for Zn using the SI units (°C), (K^{-1}) and (W/m K), respectively!

Exercise 2.5

☐ Calculation of heat capacity and adiabatic heat development *(15 min.)*

The heat development of a concrete is investigated by measurement in an **adiabatic calorimeter**. The cement content of the tested concrete is $C = 310\,kg/m^3$ and the concrete density is $\varrho = 2320\,kg/m^3$. A concrete specimen with the mass $m = 12.5$ kg is used, and the specific heat capacity of the concrete is $c_p = 1.12$ kJ/kg K. Based on this, determine 1) the heat capacity C of the specimen in (kJ/K); 2) the heat (kJ) developed in the specimen at an adiabatic temperature rise of 56.5 °C and 3) the heat Q in kJ per kg of cement developed at this temperature rise.

Mechanical work

$$W_{1.2} = \int_{x_1}^{x_2} F(x)\,dx$$

Exercise 2.6

☐ Mechanical work in tensile testing of a steel bar *(6 min.)*

A steel rod is subjected to tension in a test machine. The load is slowly increased from $F_1 = 0$ N to $F_2 = 30000$ N; the elongation of the rod due to this load, $\Delta\ell$, is 1.4 mm. Linear-elastic deformation is assumed in the steel rod, i.e. $\Delta\ell = k \cdot F$. Calculate the amount of **mechanical work** $W_{1.2}$ done on the steel rod when the load is increased from F_1 to F_2!

Volume work

$$W_{1.2} = -\int_{V_1}^{V_2} p(V)\,dV$$

Exercise 2.7

☐ Volume work by compression of atmospheric air *(10 min.)*

Consider a system, which in its initial state consists of 1.00 m^3 dry atmospheric air at 20.0 °C and an initial pressure of $p_1 = 1.00$ atm. By a **reversible**, **isothermal** compression the air pressure is changed to $p_2 = 2.00$ atm. The atmospheric air is assumed to behave in accordance with ideal gas laws (1.8). Calculate the amount of volume work $W_{1.2}$ (J) done on the air by the compression described!

Exercise 2.8

☐ The van der Waals equation with intensive variables *(10 min.)*

The van der Waals equation for real gases (1.11) contains two **intensive** variables, (pressure p and temperature T), and two **extensive** variables, (volume V and amount of substance n). Rewrite the van der Waals equation to a form solely containing the following three intensive variables: pressure p, temperature T and the molar gas concentration c_m. From the state equation set up, calculate the pressure p (Pa) in water vapour with a molar concentration c_m of 33.2 mol/m^3 at $\theta = 100\,^\circ$C!

Exercise 2.9

☐ Water vapour in humid atmospheric air *(6 min.)*

The partial pressure of saturated water vapour in atmospheric air at 10.0 °C is $p =$ 1228.1 Pa; in this state the air contains 9.41 g water vapour per m^3. The water vapour is assumed to comply with the ideal gas law. Based on these data, calculate the **molar mass** M of H_2O (g/mol) and the **molar concentration** c_m of saturated water vapour in atmospheric air at 10.0 °C!

Electrical work

$$\delta W = (V_b - V_a)\,dQ$$

Exercise 2.10

☐ Electrical work by a storage battery *(4 min.)*

A system consists of a storage battery (or accumulator) with terminals marked a and b, respectively. When the accumulator is charged, an external generator transmits an electric charge $Q = +1.0$ C from terminal a to terminal b; the electric potentials are: $V_a = 0.0$ volt and $V_b = +12.0$ volt. Calculate the amount of the **electrical work** W (J) done on the system during this process!

Exercise 2.11

☐ Work to failure during tensile test of brittle materials *(12 min.)*

A 1.00 m long round rod of hardened steel with diameter $d = 15$ mm is loaded in tension until failure in a testing machine. The steel is assumed to show **linear-elastic** deformation until failure, i.e. that **Hooke's law** $\sigma = \mathrm{E} \cdot \varepsilon$ is assumed to be valid for $0 \leq \sigma \leq \sigma_u$. The ultimate stress of the steel $\sigma_u = 850$ MPa is determined by testing. The **modulus of elasticity** of the steel is $\mathrm{E} = 2.10 \cdot 10^5$ MPa. Calculate the amount of the **mechanical work** W (J) done on the test rod when failure occurs! In the calculation the deformed length ℓ is assumed to be 1.00 m.

Figure 2.44. When large, solid concrete members are cast, the concrete hardens under near-adiabatic conditions near the middle of the cross-section. The maximum temperature rise in this part of the concrete will correspond approximatively to that measured in an adiabatic calorimeter.

Exercise 2.12

☐ Measuring of the heat capacity of a metal alloy *(10 min.)*

The thermal properties of a metal alloy is to be determined in a laboratory. Measurements are made on a specimen of the metal with the mass $m = 1050.0$ g. The specimen is heated to the temperature $\theta = 45.5\,^\circ$C and then transferred to an **adiabatic calorimeter** containing 410 g water at a temperature of 18.2 °C. Any heat loss from the specimen during transfer to the calorimeter is disregarded. After temperature equilibrium has been reached in the calorimeter, a common temperature of 25.1 °C is measured for the water and the specimen. The heat capacity of the calorimeter and the water is $C = 1.72$ kJ/K. Calculate the heat capacity C (kJ/K) of the metal specimen and the mass-specific heat capacity c (kJ/kg K) of the metal!

Exercise 2.13

☐ Adiabatic temperature rise in hardening concrete *(8 min.)*

A structural concrete with density $\varrho = 2350$ kg/m^3 has a **cement content** of 410 kg/m^3. At full hydration the heat developed by the cement is $Q = 405$ kJ/kg. The specific heat capacity of the concrete is $c = 1.10$ kJ/kg K. Calculate the maximum **adiabatic** increase of temperature ΔT that can arise in the concrete concerned!

Exercise 2.14

☐ Surface work by mist-atomization of water *(10 min.)*

5.0 kg of water is mist-atomized at 50 °C; the process is **isothermal**. At 50 °C the density of water is $\varrho = 998$ kg/m^3 and the surface tension is $\sigma = 0.0679$ N/m. The mist droplets are assumed to have identical spherical shape with the radius $r = 0.01$ mm. Calculate the **surface work** W connected with the atomization process!

Exercise 2.15

☐ Heat capacity as an extensive and intensive quantity *(6 min.)*

A system consists of 5.00 kg of water in temperature equilibrium at 25.0 °C. A heat quantity of $Q = 14630$ J at constant pressure is added to the system from its surroundings so that the water temperature is increased to 25.7 °C. Calculate the **heat capacity of the system** C_p (J/K), the **mole-specific heat capacity** of the water c_p (J/mol K) and the **mass-specific heat capacity** c_p (J/kg K) of the water. The molar mass of H_2O is 18.02 g/mol.

Exercise 2.16

☐ Heating of metallic zinc Zn *(12 min.)*

Zinc Zn is a greyish white metal with the density 7100 kg/m^3. Zinc is won from ores such as zinc blende ZnS and calamine $ZnCO_3$. Barin, I. & Knache, O.: *Thermochemical Properties of Inorganic Substances* specify the following temperature-dependent heat capacity for the metal Zn

$$c(T) = 22.40 + 10.05 \cdot 10^{-3} \cdot T \qquad \text{(J/mol K)}$$

Calculate from the expression (2.25) the quantity of heat Q necessary to add to 1.000 kg of Zn in order to increase the metal temperature from 25 °C to the melting temperature 419 °C!

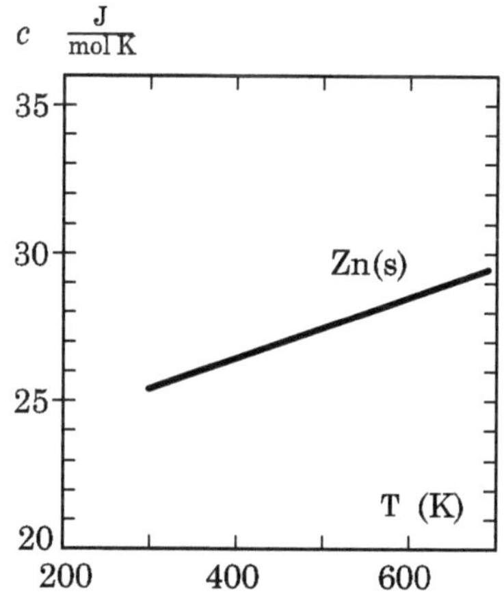

Figure 2.45. Normally, the molar heat capacity c of solid substances, liquids and gases all increase with temperature. The figure shows $c = c(T)$ for the metal zinc Zn in the temperature range from 20 °C to the melting temperature for Zn 419 °C (Barin & Knacke: Thermochemical properties of inorganic substances).

Exercise 2.17

☐ Change of state of atmospheric air *(15 min.)*

A cylinder, whose volume is delimited by a frictionless piston contains 5 g of dry atmospheric air. The initial state of the air is: temperature $\theta_1 = 20.0$ °C, pressure $p_1 = 101325$ Pa and volume $V_1 = 4.15\,\ell$. An external force F slowly moves the piston outwards until the gas volume V_2 is 5.00 ℓ; the process is carried out **isothermally** at 20.0 °C. During the process the confined quantity of air is assumed to be in equilibrium. The air is assumed to comply with the ideal gas law. Calculate a) the **average molar mass** M of atmospheric air, b) the **volume work** $W_{1.2}$ done on the system during the process and c) the **pressure** p (Pa) in the cylinder in the final state!

Exercise 2.18

☐ Electrical work in the production of aluminium *(10 min.)*

Industrial production of aluminium is done by **electrolysis** according to the so-called *Hall*-process. **Aluminium oxide** Al_2O_3 is used as the raw material in this process. The electrolysis is carried out in a melt of **cryolite** Na_3AlF_6 at approximately 1000 °C. At this temperature Al_2O_3 is partly soluble into Na_3AlF_6, and liquid, metallic Al can be deposited from the melt due to the following **cathode reaction**

$$Al^{+++} + 3e^- \quad \rightarrow \quad Al(\ell)$$

Given: the electric elementary charge $e = 1.602 \cdot 10^{-19}$ C; the relative atomic mass of aluminium Al is 27.0; 1 C = 1 As (ampere second). Calculate the **electrical work** W connected with manufacture of 1000 g Al(ℓ) when an anode voltage of +6.0 volt and a cathode voltage of 0 volt are used for the electrolysis; specify the work in the units (kJ) and (kWh)!

Figure 2.46. The milk-white mineral cryolite Na_3AlF_6 is used as a fluxing agent in the electrolytic production of aluminium. The cryolite itself is not involved in the reaction; the raw material is the Al-containing bauxite Al_2O_3. Cryolite is a rare mineral. A well-known occurrence is found at Ivigtut in Greenland.

Exercise 2.19

☐ Heating of homogeneous SiO_2 system of matter *(20 min.)*

For accurate energy calculations where the substance is heated or cooled over a wide temperature interval, it is necessary to take account of the temperature dependence of the heat capacity: $c = c(T)$. In *Thermodynamik der Silikate* by Mcodlov-Petrosjan the molar heat capacity $c(T)$ of **quartz** β-SiO_2 is given as

$$c(T) = A + B \cdot T + C \cdot T^{-2} \quad \text{cal/mol K} \qquad (298\,\text{K} \leq T \leq 848\,\text{K})$$

where A = 11.22 cal/mol K; B = $8.2 \cdot 10^{-3}$ cal/mol K^2 and C = $-2.7 \cdot 10^5$ cal K/mol. The molar mass for SiO_2 is 60.06 g/mol; 1 cal = 4.186 J. Based on this information, calculate a) the mole-specific heat capacity c of β-SiO_2 at 25 °C given in SI unit (J/mol K) and b) determine by integration, cf. (2.25), the heat quantity Q necessary to add to a specimen of 1000 g β-SiO_2 to rise the temperature from 25 °C to 575 °C and calculate c) the heat quantity Q for the same heating of the specimen if the heat capacity is assumed to be constant $c = c(298\,\text{K})$, cf. (2.24)!

Figure 2.47. Copper Cu is a good electric conductor and is therefore widely used for production of electric cables. For this use electrolytic copper that has been electrolytically refined to high purity is preferred.

Exercise 2.20

□ Electrolytic refining of copper Cu *(12 min.)*

Copper is produced to a purity of approx. 99.8 % by reduction of Cu in natural ores such as **chalcopyrite** $CuFeS_2$ and **chalcocite** Cu_2S. Particularly pure copper can be achieved by electrolytic refining of copper; this so-called **electrolytic copper**, which is made industrially to a purity of approx. 99.95 %, is universally used for production of electric conductors. The electrolysis is made in a bath into which copper sulphate $CuSO_4$ is dissolved. During the process crude copper is dissolved at the **positive anode** and deposited at the **negative cathode** as electrolytic copper of high purity. The cathode reaction leading to the deposition of the metallic Cu is

$$Cu^{++}(aq) + 2e^{-} \quad \rightarrow \quad Cu(s)$$

The process is driven by an external generator that transfers electrons from the anode to the cathode. Calculate the magnitude of the **electrical work** W that is necessary to produce 1000 g of electrolytic copper, when an anode voltage of $V_1 = 4.8$ volt and a cathode voltage of $V_2 = 0.0$ volt are used in the electrolysis; the work W shall be given in (kJ) and in (kWh)!

Literature

The following literature and supplementary readings are recommended for further reference on the subjects covered in Chapter 2.

References

In chapter 2 a number of references to the literature have been made; the following is a complete list of citations of the references used.

- Mantell, L.C. (ed.): *Engineering Materials Handbook*, McGraw-Hill Book Company, Inc., New York 1958.
- Weast, R.C. (ed.): *CRC Handbook of Chemistry and Physics*, CRC Press, Inc. Florida 1983.
- Barin, I. & Knacke O.: *Thermochemical properties of inorganic substances*, Springer-Verlag Berlin 1973.
- Mcedlov-Petrosjan: *Thermodynamik der Silikate*, VEB Verlag für Bauwesen, Berlin 1966.
- Danish Building Research Institute: *Winter Concreting*, SBI Instruction 125, 1982. In Danish: Statens Byggeforskningsinstitut: *Vinterstøbning af beton*, SBI-anvisning 125, Statens Byggeforskningsinstitut 1982.

Supplementary literature

Well-written and easily comprehensible explanations of the concepts and definitions of thermodynamics are provided in the following books:

- Whittaker, G., Mount, A. & Heal, M.: *Instant Notes in Physical Chemistry*, Taylor & Francis, 2000.
- Atkins, P.W.: *The Elements of Physical Chemistry*, W.H. Freeman, 1996.

An advanced explanation of thermodynamic concepts, definitions and system types is given in the following textbooks:

- Atkins, P.W. & de Paula, J.: *Physical Chemistry*, 7th ed., W.H. Freeman, 2001.

- Morse, P.M.: *Thermal Physics*, W.A. Benjamin, Inc., New York.

An overshot water wheel from the end of the 18th century. The hydraulic power machine utilizes the work produced by falling water. The potential energy of the water in the gravity field is transformed into mechanical work which can be secured from the axle of the wheel.

CHAPTER 3

First law

The first law of thermodynamics expresses the simple and universal principle of the **energy conservation**. The formulation of this principle took its final form in the middle of the 19th century; the **concept of heat** and the **concept of energy** are summarized in a very fruitful form using this principle.

The principle of energy conservation was a breakaway from the opinion of that time that heat was a weightless substance – a view which was then inhibitory of the development of a universal concept of energy. Works of **J.P. Joule** in the 1840's, in particular, gave conclusive clarification of the equivalence between work and heat which forms the basis of classical thermodynamics.

The first law of thermodynamics is a **postulate**, which is based on experience; the statement *cannot* be derived or proved. The first law forms the basis of a number of energy relations in thermodynamics. In this chapter, the most important relations are presented and the use of the calculation expressions is illustrated by problems within materials technology.

J.P. Joule (1818-1889)

English brewer, who gained major recognition as a physicist and a scientist.

Contents

3.1 Energy

Thermodynamic considerations refer to a predefined **system** separated from its **surroundings** by an imaginary or real boundary; as a whole, the system and its surroundings form the thermodynamic **universe** (see chapter 2). The following contains a brief explanation of the concept of energy and the energy forms of the thermodynamic system of matter.

Figure 3.1. Energy is a measure of the capacity to do work. An energy content is ascribed to a tight metal spring: the spring can do work on its surroundings during relaxation.

Energy and energy conservation

The notation "energy" is a measure of the capacity to do **work** or to release **heat**. An energy content is ascribed to a tight elastic spring: the spring is able to do work on its surroundings during relaxation. An energy content is ascribed to fuel: by combustion the fuel is able to emit heat to its surroundings or generate a motor that is able to do work.

In these examples, the presence of energy – the capacity to do work or to release heat – is evident. However, we *cannot* base the description of the composition of systems of matter on such more or less intuitive energy considerations. What is for example the energy content in a loaded tensile steel bar? – and what does this energy content mean in relation to the physical and chemical properties of the steel, e.g. the resistance to corrosion. Such questions are addressed in the thermodynamic description of substances.

In classical thermodynamics, general rules for the physical and chemical properties are derived from a few central **principles of conservation of energy**. These principles are empirical **postulates** that *cannot* be proved.

The basis of thermodynamic energy considerations is the following definition of energy and energy conservation

Energy and energy conservation (3.1)

- **Energy** is a measure of the capacity to do **work** or the capacity to release **heat**.
- **Energy** *cannot* be created or destroyed; the energy in an isolated thermodynamic system is **constant**.

Figure 3.2. Energy is a measure of the capacity to release heat. An energy content is ascribed to a combustible substance: by combustion it is able to release heat to its surroundings or to generate a motor that is able to do work.

Energy forms

An energy content can be ascribed to any thermodynamic system of matter – the **internal energy** U of the system – which is bound up with phenomena in the electronic, atomic and molecular structure of the system. For systems of matter at room temperature, changes of the internal energy ΔU will mainly be caused by activation of the following energy forms in the system:

- **Kinetic energy** related to different forms of *movement* of atoms or molecules in the system.
- **Potential energy** related to acting *Coulomb forces* between electrically charged particles in the system.
- **Chemical energy** related to the development of *chemical reactions* between atoms or molecules in the system.

Under more extreme temperature and pressure conditions, energy forms such as **radiation** and **nuclear energy** are considered to be essential quantities in the description of changes of the internal energy ΔU of the system. These energy forms, however, are beyond the scope of the present description.

Kinetic energy

The **internal energy** U in a system of matter can be **kinetic energy** E_k connected with movement of atoms or molecules. Kinetic energy exists in two qualitatively different forms

- **Molecular kinetic energy** connected with *disordered*, random thermal movements of atoms and molecules. The **temperature** T of the system of matter is therefore a measure of the average molecular kinetic energy in the system.
- **Macroscopic kinetic energy** created by an *ordered* simultaneous movement of a system of matter, for example in the form of a weight falling down in a gravitational field. The macroscopic kinetic energy of the system of matter is an often used quantity in calculations within mechanical physics.

The molecular kinetic energy E_k is an important quantity in the thermodynamic description of systems of matter; any temperature change therefore indicates a change in the average kinetic energy of the molecules.

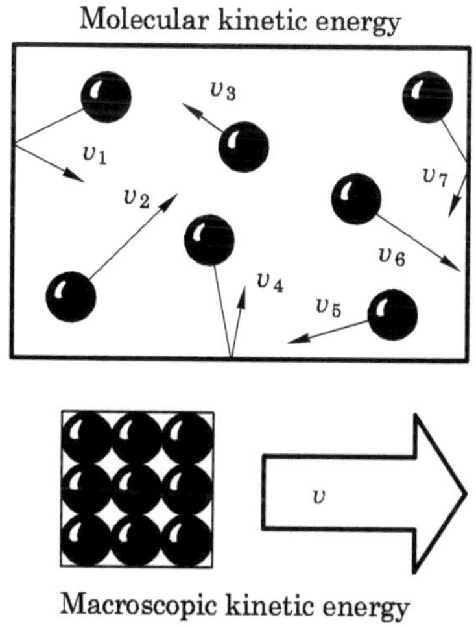

Figure 3.3. Molecular kinetic energy is disordered thermal movement of atoms and molecules in a system. Macroscopic kinetic energy is an ordered, simultaneous movement of a system of matter. The temperature of a system of matter is a measure of the average molecular kinetic energy of the atoms and molecules of the system.

Potential energy

All atoms and molecules in a system of matter are surrounded by an electrostatic field. By interaction between these fields, *attractive* or *repulsive* electrostatic forces – so-called **Coulomb forces** – between atoms and molecules are generated (see section 1.10). If two particles are displaced from their equilibrium state determined by their charge, they are given a mutual **potential energy**, i.e. a capacity to do work.

This molecular potential energy is visibly reflected in the macroscopic behaviour of the substances. Assume, for example, an elastic spring exerting force on its environments. The tight spring is a macroscopic representation of a potential energy. The spring force **F** is preserved because a large number of atoms in the substance have been displaced from their mutual equilibrium state during deformation of the spring.

By relaxation the spring has the capacity to do work $\delta W = F\,dx$ on its environments; this work is connected with a decrease of potential binding energy between the atoms of the substance when they regain their original state of equilibrium due to relaxation. Also, this interaction between macroscopic potential energy and molecular potential energy gives a fine illustration of the concept of energy as a *capacity* to do work.

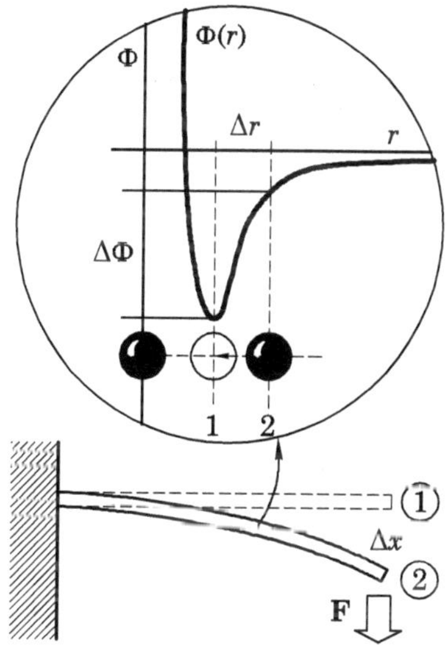

*Figure 3.4. (Bottom) A tight spring represents potential energy. The spring force **F** is preserved because a large number of atoms in the spring are displaced from their mutual equilibrium state and thus have obtained increased potential energy $\Delta\phi$.*

Chemical energy

A **chemical reaction** is defined as a transformation that causes changes in the molecular structure of the substances and thus formation of new chemical compounds. During a chemical reaction, strong **primary bonds** between atoms in the system are formed or destroyed. In contrast, a **phase transformation** will normally be connected with formation or destruction of far weaker **secondary bonds** between atoms or molecules. Examples of chemical reactions and phase transformations are

Chemical reaction:	$H_2(g) + \frac{1}{2}O_2(g)$	$\rightarrow$	$H_2O(\ell)$
Phase transformation	$H_2O(g)$	$\rightarrow$	$H_2O(\ell)$

A chemical reaction during which a substance is transformed is normally connected with a considerable change of the energy content of the substance: we say that the **chemical energy** of the substance is changed. A change in the chemical energy of a substance is registered, for example by the reaction heat released by the chemical transformation of the substance.

Summary

In **classical thermodynamics** we do *not* in principle distinguish between different forms of internal energy: States and changes of state are described by general, macroscopic variables such as **internal energy** U, **temperature** T and **pressure** p. In classical thermodynamics, a system is only a delimited macroscopic object *without* molecular structure (a "black box").

Thus, the qualitative overview of energy forms given above is beyond the scope of classical thermodynamics. The detailed description of energy states and forms in the molecular structure of the substances is covered by the discipline of **statistical thermodynamics**.

The author recommends that when reading the following chapters the student should always ask the question: *why?* – and thus seek a certain qualitative understanding of the physical phenomena affecting the molecular structure that determine the properties of substances.

Figure 3.5. Due to stress concentration at notches and other imperfections, large local tensile stresses can arise around cracks in tension. In these loaded areas, local stress corrosion can lead to development of failure in apparently intact specimens. The photo shows stress corrosion in a bellows of austenitic stainless steel.

■ A system (A) consists of an unloaded coil spring of steel immersed in hydrochloric acid. A system (B) consists of a similar, but compressed coil spring of steel in hydrochloric acid. After some time, both of the steel springs have been dissolved in the acid. What has happened to the potential energy E_p that was originally present in the spring (B)?

Solution. The potential energy in the spring (B) has been transformed into heat in the system, i.e. into molecular kinetic energy. In its loaded state, steel corrodes faster and thus releases a slightly larger amount of heat during the dissolution.

☐ **1.** When substances are divided into fine particles their energy content is increased due to the surface work done – in what form will the increased internal energy ΔU be present in the system?

☐ **2.** In an isolated system 1 mol of H_2(g) and $\frac{1}{2}$ mol of O_2(g) are combined by combustion to form 1 mol of H_2O(g) – which energy transformation will dominate during this process?

☐ **3.** Two moles of ideal gas with $c_v = 20.0$ J/mol K are subjected to isochoric heating from 0 °C to 100 °C; specify a) the increase of internal energy ΔU and b) which energy form is supplied to the system.

☐ **4.** Which energy form is dominating in ideal gas at room temperature? – and which energy forms are dominating in real gas at room temperature?

☐ **5.** An isolated system contains water vapour H_2O(g). Part of the water vapour condenses into water $H_2O(\ell)$; what happens with a) the total energy E_{total}; b) the potential energy E_p and c) the molecular kinetic energy E_k of the system?

3.2 First law

A thermodynamic system can be given energy in the form of **work** W done *on* the system or in the form of **heat** Q *supplied* to the system. Work and heat are *not* thermodynamic state variables (section 2.7); for a given change of state in a system

- the **work** W done depends on the chosen process path,
- the supplied **heat** Q depends on the chosen process path.

However, experience shows that *irrespective* of the chosen process path, the **sum** of the **work** W done on the system and the **heat** Q supplied to the system will remain constant for a given thermodynamic process 1 → 2. This experience is expressed in the **first law** of thermodynamics, which for a closed system reads

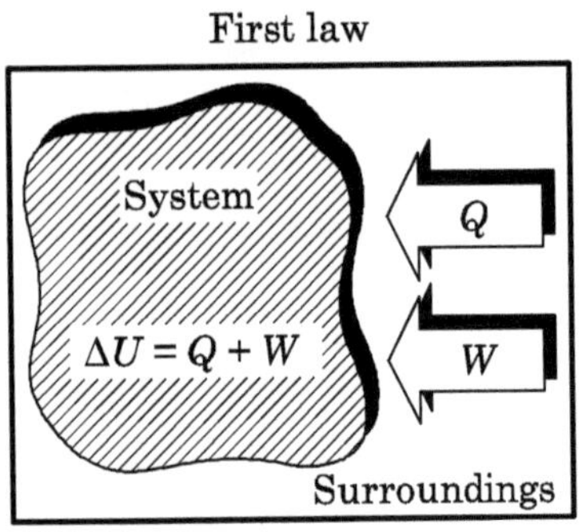

Figure 3.6. The first law postulates the existence of a state function: the internal energy U. The increase in the internal energy ΔU in a closed system is equal to the sum of the heat Q supplied to the system and the work W done on the system.

The first law of thermodynamics (3.2)

$$dU = \delta Q + \delta W; \qquad \Delta U = Q + W$$

– the increase in the **internal energy** dU in a closed system is the *sum* of the **heat** δQ supplied to the system and the **work** δW done on the system.

The first law expresses a **postulate** that cannot be proved; phenomena contradicting the law, however, have never been observed.

During a process connecting two states of equilibrium, the *change* of the internal energy ΔU of a system will *solely* depend on the **initial state** and the

final state of the system; the internal energy U is said to be a **state function** (section 2.2). In contrast, *neither* work W *nor* heat Q are state functions as both W and Q depend on the chosen process path. This essential property of the internal energy U will be explained in further detail in the following.

Internal energy as a state function

The differentials of Q and W in (3.2) are given as δQ and δW, respectively, while the differential of the internal energy U is given as dU. This special notation has been used to illustrate the fundamental difference between **state variables** such as U, and energy quantities such as Q and W.

In the state of equilibrium a certain internal energy U can be ascribed to any thermodynamic system; for a thermodynamic process connecting two states of equilibrium (1) and (2), therefore, it applies that

$$\Delta U = \int_{(1)}^{(2)} dU = U_2 - U_1 \tag{3.3}$$

Independently of the chosen process path, the increase (or decrease) in internal energy ΔU can *always* be expressed as the internal energy of the system in its **final state** U_2 minus the internal energy of the system in its **initial state** U_1. For a **cyclic process** where the final state and the initial state of the system are identical, the change in the internal energy ΔU of the system will be zero

$$\Delta U = \oint dU = 0 \qquad \text{(cyclic process)} \tag{3.4}$$

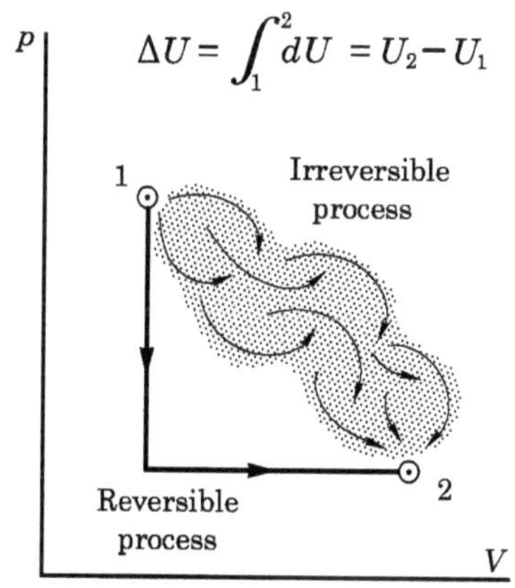

Figure 3.7. Independently of the chosen process path, the increase in the internal energy ΔU of a system is equal to the internal energy U_2 of the system in its final state minus the internal energy U_1 of the system in its initial state.

Relations such as (3.3) and (3.4) do *not* apply to the energy quantities Q and W. Work W and heat Q express transfer of energy from surroundings to system. In a thermodynamic context, therefore, it makes no sense to talk about the "content" of work or heat of a system – we can only talk about the content of **energy** of a system. For a final thermodynamic process connecting two states of equilibrium (1) and (2), we can therefore not talk about the increase in work "ΔW" or the increase in heat "ΔQ" of the system; but for an *actual* process we can specify the work $W_{1.2}$ done and the heat $Q_{1.2}$ supplied, i.e.

$$Q_{1.2} = \int_{(1)}^{(2)} \delta Q; \qquad W_{1.2} = \int_{(1)}^{(2)} \delta W \tag{3.5}$$

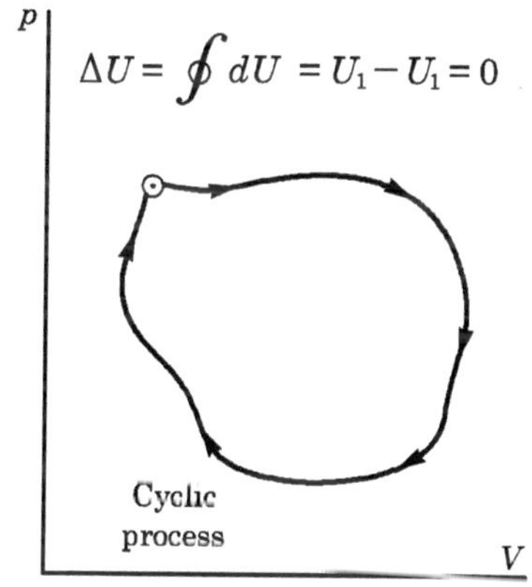

Figure 3.8. In any cyclic process during which a system is restored to its initial state, the increase in the internal energy of the system is $\Delta U = 0$.

The mathematical difference between dU and δQ, or between dU and δW is as follows: The differential of the internal energy dU is an **exact** (or **total**) differential; δQ and δW are *not* exact differentials (see *Mathematical appendix, subchapter A6*).

Use of the **first law** requires that the initial state (1) and the final state (2) are thermodynamic **equilibrium states**. The processes connecting the states (1) and (2) can be arbitrarily chosen. Thus, no assumption of equilibrium has been made for these processes; this is a central point in the thermodynamic description of substances, as will be explained later.

■ Assume a system containing 5.00 mol of ideal gas at the pressure 101325 Pa. In its initial state (1), the gas temperature is 100 °C. By an **isobaric** and **reversible** cooling, the gas temperature is reduced to 0 °C in the final state (2). The molar heat capacity of the gas $c_p = 20.8$ J/mol K is assumed to be constant. Calculate the change in the internal energy ΔU of the system during the process!

Solution. The heat $Q_{1.2}$ supplied during the isobaric process is calculated from the expression (2.24)

$$Q_{1.2} = n \cdot c_p \cdot (T_2 - T_1) = 5.00 \cdot 20.8 \cdot (273.15 - 373.15)\,\text{J} = -10.40\,\text{kJ}$$

The ideal gas law (1.8) determines the gas volumes V_1 and V_2 as

$$V_1 = \frac{nRT_1}{p} = \frac{5.00 \cdot 8.314 \cdot 373.15}{101325} \text{m}^3 = 0.1531\,\text{m}^3$$

$$V_2 = \frac{nRT_2}{p} = \frac{5.00 \cdot 8.314 \cdot 273.15}{101325} \text{m}^3 = 0.1121\,\text{m}^3$$

The volume work $W_{1.2}$ done on the system is determined from (2.17)

$$W_{1.2} = -\int_{(1)}^{(2)} p(V)dV = -p \cdot (V_2 - V_1) = -101325 \cdot (0.1121 - 0.1531) = 4.15\,\text{kJ}$$

The **first law** (3.2) determines the desired change of internal energy ΔU as

$$\Delta U = Q_{1.2} + W_{1.2} = -10.40 + 4.15\,\text{kJ} = \mathbf{-6.25\,kJ}$$

☐ **1.** By an explosion in an isolated thermodynamic system, 2 mol of $H_2(g)$ and 1 mol of $O_2(g)$ are transformed into 2 mol of $H_2O(g)$. Specify the increase in the internal energy ΔU of the system!

☐ **2.** An amount of heat $Q = 1025$ kJ is supplied to a system; during heating, the work done by the system on its surroundings is $W = 625$ kJ. Calculate ΔU (kJ)!

☐ **3.** Two mol of ideal gas with $c_p = 20.8$ J/molK are heated in an isobaric, reversible process from 20.0 to 30.0 °C at $p = 1\,\text{atm}$. Calculate ΔU (kJ) for the process!

☐ **4.** During an isochoric cooling of an ideal gas in a closed system, the system emits an amount of heat of 840 kJ; specify the increase in the internal energy ΔU (kJ) of the system!

☐ **5.** By a reversible adiabatic process a 5.00 m long bar of steel of circular cross-section with a diameter of 20.0 mm is subjected to tension of $F = 32000$ N. The E-modulus of the steel is $2.10 \cdot 10^5$ MPa. Calculate the increase in the internal energy ΔU (J) of the bar due to the applied tension!

Heat and work

Q and W depend on the process path; therefore, heat and work are not state variables.

3.3 Internal energy U

The **first law** of thermodynamics (3.2) introduces the quantity **internal energy** U as a measure of the energy content in a system. In the following, some important properties of this new quantity are summarized.

The internal energy U of a thermodynamic system is a **state function**. When a system is changed from the state of equilibrium (1) to the state of equilibrium (2), the increase of the internal energy of the system ΔU will *solely* depend on the terminal states (1) and (2); it will be independent of the process connecting these states. Therefore, in a **cyclic process** that leads the system back to its initial state, ΔU is equal to zero.

Figure 3.9. State functions and state variables are fundamental concepts in explanation of thermodynamics. The following analogy can help illustrate these concepts. If we move from one point (1) to another (2) in the terrain, the change of height level $\Delta H = H_2 - H_1$ is independent of the chosen route ("process path"): the height level is a state variable. The distance S from (1) to (2), on the other hand, is dependent on the chosen route in the terrain, and therefore, the distance S is not a state variable.

Internal energy U is a state function (3.6)

$$\Delta U = \int_{(1)}^{(2)} dU = U_2 - U_1 \qquad \text{(process } 1 \rightarrow 2\text{)}$$

$$\Delta U = \oint dU = 0 \qquad \text{(cyclic process)}$$

The increase of the **internal energy** U of a system is *independent* of the process path; in a **cyclic process**, $\Delta U = 0$.

It is important to note the following: In tests one can *only* determine the *changes* of the internal energy ΔU of a system, and therefore, it is *not* possible to determine absolute values of the internal energy U by measurement on a system.

Symbols and units

The first law, equation (3.2), introduces the internal energy U as an **extensive** variable with the unit (J) (see section 2.3). In the description of systems of pure substances, however, it is expedient to use **intensive**, rather than extensive variables. Therefore, the corresponding **mole-specific** quantity of the unit (J/mol) is often used.

Mole-specific internal energy (3.7)

$$U_m = \frac{U}{n}; \qquad \Delta U_m = \frac{\Delta U}{n} \qquad \text{(J/mol)}$$

The **mole-specific** internal energy U_m of a **pure substance** denotes the internal energy per mole of the substance.

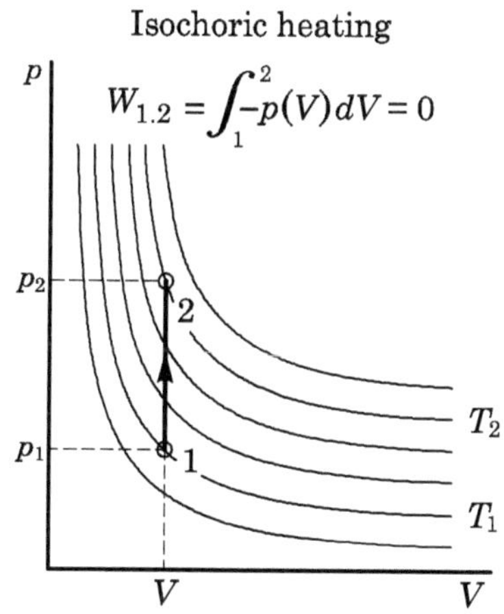

Figure 3.10. If an ideal gas is heated at constant volume (i.e., undergoes an isochoric transformation), the quantity $-pdV$, and thus the volume work W, is zero. Therefore, for isochoric heating, the increase in the internal energy ΔU is equal to the supplied heat Q.

There is no standardized or universally-accepted notation that distinguishes the internal energy (J) of a system from the mole-specific internal energy of a pure substance (J/mol). In the following sections, the common symbol U will be used for both of these quantities. Which quantity is of concern should be clear from the context, units and other designations in the text.

Internal energy and heat capacity

If an amount of heat δQ is supplied to a closed thermodynamic system at *constant volume* (isochoric process), the temperature increment dT is, according to (2.22), determined from

$$\delta Q = C_V \cdot dT \tag{3.8}$$

When dV, and thus the volume work $-p \cdot dV$, is zero during the process, the first law (3.2) indicates that

$$dU = \delta Q + \delta W = \delta Q + 0 = C_V \cdot dT \qquad (V, n \text{ constant}) \tag{3.9}$$

Thus, the previously defined quantity, the heat capacity C_V at constant volume, is intimately connected with the internal energy through the following relation

Heat capacity at constant volume (3.10)

$$\text{Heat capacity, system:} \qquad C_V = \left(\frac{\partial U}{\partial T}\right)_V \qquad \text{(J/K)}$$

$$\text{Mole-specific heat capacity:} \qquad c_V = \left(\frac{\partial U}{\partial T}\right)_V \qquad \text{(J/mol K)}$$

The heat capacity at **constant volume** denotes the increase in **internal energy** per kelvin (K).

When measuring on closed systems at **constant volume**, the internal energy U is a convenient parameter as the volume work $\delta W = -p \cdot dV$ on the system is zero. In this case, the increment in the internal energy dU is equal to the supplied heat δQ, if there are no contributions from other work processes.

Figure 3.11. A bomb calorimeter for measuring the heat release for processes occurring at constant volume, i.e. isochoric processes. The increase in internal energy ΔU is equal to the heat Q supplied to the system. Bomb calorimeters are e.g. used for measuring the calorific value of substances.

■ A closed system contains 5.00 mol of ideal gas. By isochoric heating, $Q = 1300$ J is supplied; the gas temperature then increases from 20.0 °C to 32.5 °C. Calculate a) the increase in internal energy ΔU of the system; b) the heat capacity C_V (J/K) of the system and c) the mole-specific heat capacity of the gas c_V (J/mol K)!

Solution. By integration of (3.9) for constant (V, n), the following expression for ΔU is obtained

$$\Delta U = \int_{(1)}^{(2)} dU = \int_{(1)}^{(2)} \delta Q = Q_{1.2} = \mathbf{1300\,J}$$

$$C_V = \frac{\Delta U}{\Delta T} = \frac{1300\,\mathrm{J}}{305.65\,\mathrm{K} - 293.15\,\mathrm{K}} = \mathbf{104\,J/K}$$

after which the mole-specific heat capacity c_V can be determined from (2.22)

$$c_V = \frac{1}{n} \cdot C_V = \frac{1}{5.00\,\mathrm{mol}} \cdot 104\,\mathrm{J/K} = \mathbf{20.8\,J/mol\,K}$$

☐ **1.** 100 g of water vapour is heated so that the temperature is increased by 15 °C at constant volume; $c_V = 25.3$ J/mol K for H_2O(g). Calculate the increase in mole-specific internal energy ΔU (J/mol)!

☐ **2.** Does it make sense to talk about mole-specific internal energy U (J/mol) for a) a saturated NaCl solution in water; b) crystalline NaCl(s); c) a melt of NaCl(ℓ)?

☐ **3.** A concrete hardens at constant volume in a moisture-proof adiabatic calorimeter; find the increase in the internal energy ΔU (J) of the system?

☐ **4.** $Q = 1500$ kJ is supplied to a closed system; during the process the internal energy is increased by $\Delta U = -300$ kJ. Calculate the work W done on the system!

☐ **5.** By isochoric heating of 18.5 mol of gas, $Q = 3510$ J is supplied; the gas temperature is increased by 7.5 °C. Calculate the quantities: ΔU (J), ΔU (J/mol), C_V (J/K), and c_V (J/mol K)!

3.4 Enthalpy H

The **internal energy** U is an appropriate state function in the description of processes developing at **constant volume** V, since contributions from volume work $\delta W = -pdV$ can be omitted. In everyday systems of matter, the majority of reactions, however, develop at a constant atmospheric pressure rather than at constant volume. Therefore, it is an advantage to introduce a modified state function, **enthalpy** H, which is adjusted to the description of processes developing at constant pressure, i.e. isobaric processes.

For a pV-system of the volume V and the pressure p, the **enthalpy** of the system is defined by the following equation

Enthalpy H (3.11)

$$H \stackrel{\mathrm{def}}{=} U + pV$$

The **enthalpy** H of the system is the sum of the **internal energy** U and the product pV of **pressure** p and **volume** V.

Figure 3.12. When exposed to the weather, copper will form the well-known green layer of verdigris, which is an alkaline copper carbonate. This process, in which Cu reacts with O_2, CO_2 and H_2O, is a typical example of an isobaric process.

The **enthalpy** H of a system is produced by supplying the additional energy contribution pV (J) to the **internal energy** U of the system; the quantity pV – the "**spatial energy**"of the system – is a measure of the work involved in displacing the system volume V against an "external pressure" $p_e = p$.

The enthalpy H as a state function

The enthalpy H is defined as a function of three state quantities: U, p, and V; the enthalpy H, therefore, is a **state function** itself. When a system is changed from the state of equilibrium (1) to the state of equilibrium (2) the increase of the enthalpy of the system ΔH *solely* depends on the initial and final states (1) and (2); ΔH is *independent* of the chosen process path. Therefore, the increase of the enthalpy ΔH of a system in a **cyclic process**, during which the system is brought back to its initial state, is always zero.

The enthalpy H is a state function (3.12)

$$\Delta H = \int_{(1)}^{(2)} dH = H_2 - H_1 \qquad \text{(process } 1 \to 2\text{)}$$

$$\Delta H = \oint dH = 0 \qquad \text{(cyclic process)}$$

The increase of the **enthalpy** H of a system is *independent* of the process path; in a **cyclic process**, $\Delta H = 0$.

It is important to note the following: In a test it is *only* possible to determine *changes* in the enthalpy ΔH of a system; it is *not* possible by measurement to determine absolute values of the enthalpy content H of a system.

Definition of enthalpy

Enthalpy: $H \overset{\text{def}}{=} U + pV$

Symbols and units

In practical calculations and measurements, the enthalpy H is found as an **extensive** system-dependent quantity of unit (J), and as an **intensive** mole-specific quantity of unit (J/mol); see section (2.3). Since there are no standardized symbols distinguishing these cases, the common symbol H is used universally; the meaning in each individual case will be seen from the units and other designations used in the context.

The **mole-specific** enthalpy H, specifying the enthalpy content per mole of a **pure** substance, is used in description of systems of matter.

Mole-specific enthalpy (3.13)

$$H_m = \frac{H}{n}; \qquad \Delta H_m = \frac{\Delta H}{n} \qquad \text{(J/mol)}$$

– the **mole-specific** enthalpy H_m of a **pure substance** specifies the enthalpy content per mole of the substance.

The enthalpy H is a state function with the same basic properties as the internal energy U. For a description of processes that develop at constant pressure p, it is, however, simpler to use the H function.

Isobaric heating

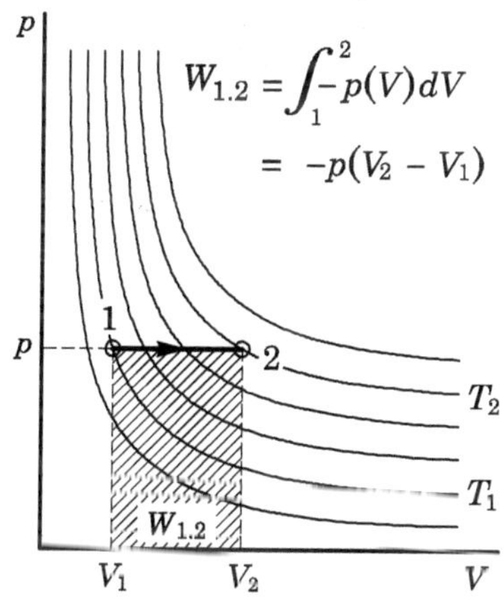

Figure 3.13. When an ideal gas is heated at constant pressure, i.e. isobarically, the volume work is $W = -p(V_2 - V_1)$. For isobaric heating, the increase of enthalpy ΔH, therefore, is equal to the supplied heat Q.

Enthalpy and heat capacity

Assume a **closed** system where the internal pressure p is equal to a constant external pressure p_e; this assumption is typically true for processes developing at the atmospheric pressure $p = 1$ atm. During a process in this system, the differential dH fulfils the condition

$$dH = dU + d(pV) = \delta Q + \delta W + pdV + Vdp \tag{3.14}$$

where the last rewriting uses the law (3.2). If *only* **volume work** $\delta W = -pdV$ is done on the system, the differential dH for the isobaric process is

$$dH = \delta Q - pdV + pdV + V \cdot 0 = \delta Q \qquad (p, n \text{ constant}) \tag{3.15}$$

The increment of the enthalpy dH of a system during an isobaric process is exactly equal to the supplied **heat** δQ, if *only* volume work is done on the system. However, according to (2.21) and (2.22), the supplied amount of heat δQ at constant pressure is $\delta Q = C_p \cdot dT$. Therefore, the heat capacities C_p and c_p are related to the H function in the following simple way

The heat capacity at constant pressure (3.16)

$$\text{Heat capacity, system}: \quad C_p = \left(\frac{\partial H}{\partial T}\right)_p \quad (\text{J/K})$$

$$\text{Mole-specific heat capacity}: \quad c_p = \left(\frac{\partial H}{\partial T}\right)_p \quad (\text{J/mol K})$$

– the heat capacity at **constant pressure** denotes the increase of **enthalpy** per kelvin (K).

Normally, as can be seen from the calculation examples, the difference between ΔH and ΔU is only important for **gas phase reactions**. For reactions in liquids and solid matters, $\Delta(pV)$ usually is a modest correction quantity.

Heat capacities

$$c_V = \left(\frac{\partial U}{\partial T}\right)_V \quad (\text{J/mol K})$$

$$c_p = \left(\frac{\partial H}{\partial T}\right)_p \quad (\text{J/mol K})$$

■ The mole-specific heat capacity c_V for an ideal gas, $c_V = 12.5$ J/molK, has been determined by measurement in a volume-constant calorimeter. Calculate the mole-specific heat capacity c_p (J/molK) of the gas!

Solution. The difference between c_V and c_p is based on the fact that gas expands when heated at constant pressure; hereby a volume work δW is done on the surroundings during heating (see figure 2.21). Therefore, in order to obtain the *same* temperature rise of the gas, a larger amount of heat than that at constant volume must be supplied, i.e. $c_p > c_V$. We aim at an analytical answer through the following considerations: according to (3.11), increment of enthalpy dH and increment of internal energy dU fulfil the condition

$$dH = dU + d(pV) \qquad \text{(a)}$$

According to the ideal gas law (1.8), the term for one mole of ideal gas is $d(pV) = d(RT)$. Next, using (3.10) and (3.15), we find for one mole of gas

$$c_p \cdot dT = c_V \cdot dT + R \cdot dT \quad \Rightarrow \quad c_p = c_V + R \qquad \text{(b)}$$

$$c_p = c_V + R = 12.5 + 8.314 \text{ J/molK} = \mathbf{20.8\,J/mol\,K}$$

☐ **1.** State the quantity $(H - U)$ in unit (J) at $p = 1$ atm, $\theta = 25\,°\text{C}$ for: a) 1 mol of ideal gas; b) 1 mol of water $H_2O(\ell)$, and c) 1 mol of iron Fe(s).

☐ **2.** By isobaric heating of 1.000 kg of water $H_2O(\ell)$, $Q = 1000$ J is supplied. Calculate the enthalpy increase for the system ΔH (J), and the increase of molar enthalpy ΔH (J/mol)!

☐ **3.** Explain, under what process conditions the increase of the enthalpy ΔH is equal to the increase of the internal energy ΔU of a system?

☐ **4.** A concrete hardens at constant pressure p in a moisture-proof adiabatic calorimeter. State the increase of the enthalpy of the system ΔH!

☐ **5.** By isobaric heating of 18.5 mol of ideal gas, $Q = 4660$ J is supplied, increasing the gas temperature by 7.5 °C. Calculate the quantities: ΔH (J), ΔH (J/mol), C_p (J/K), and c_p (J/molK)!

Figure 3.14. Light, carbonated surface zone and dark non-carbonated core in a cut concrete beam; the colour difference is caused by a pH-sensitive indicator phenolphthalein that has been applied to the cut surface. In the description of the penetration and reaction of gases into porous materials such as concrete, ideal gases are normally assumed.

3.5 Ideal gas

For a number of technical calculations, knowledge of the properties of gases is necessary. For example, this applies to treatment of subjects such as hygroscopic **moisture absorption**, **carbonation** of concrete, **gas permeability** of porous materials, and **moisture diffusion** in materials. For technical calculations, **ideal gases** are often assumed, which in most cases is an acceptable assumption. For these kinds of calculations, the concept of ideal gas is briefly explained in the following portions of this section.

Assume a **closed** system consisting of 1 mol of gas. The state of the gas can be described by two variables, e.g. the **temperature** T and the **volume** V, which determine the internal energy $U = U(T,V)$. In this case, a differential change in the internal energy of the system is uniquely determined by the change of temperature dT and the volume dV through the differential

$$dU = \left(\frac{\partial U}{\partial T}\right)_V \cdot dT + \left(\frac{\partial U}{\partial V}\right)_T \cdot dV \tag{3.17}$$

In this expression, two partial differential coefficients are included: $(\frac{\partial U}{\partial T})_V$ and $(\frac{\partial U}{\partial V})_T$ as coefficients of dT and dV, respectively. These coefficients contain information of the physical properties of the gas (see Mathematical appendix, subchapter A8 *Maxwell's relations*).

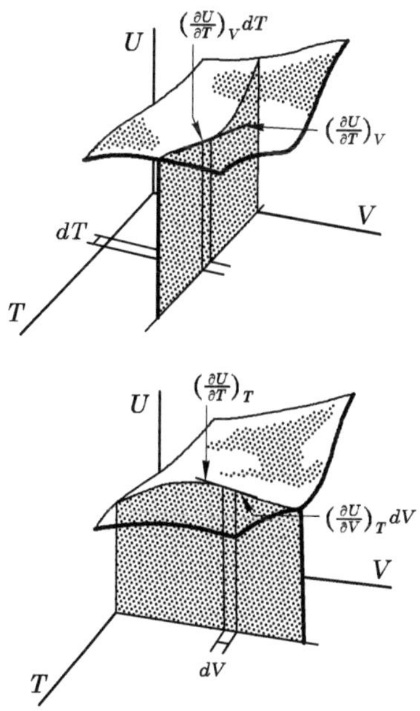

Figure 3.15. The graphical meaning of differential coefficients for the internal energy $U = U(T,V)$.

Joule's law

The quantity $(\frac{\partial U}{\partial T})_V$ denotes the molar **heat capacity** c_V of the gas at constant volume; this quantity has been introduced into (3.10). The expression (3.17) can thus be rewritten in the form

$$dU = c_V \cdot dT + \left(\frac{\partial U}{\partial V}\right)_T \cdot dV \tag{3.18}$$

The coefficient $(\frac{\partial U}{\partial V})_T$ expresses the increment of the internal energy of the gas at **isothermal** change of volume. For **ideal gases** it was found experimentally ("*Joule's experiment*" 1843) that $(\frac{\partial U}{\partial V})_T = 0$. Thus, the internal energy U of an ideal gas is *solely* a function of the gas temperature $U = U(T)$. This important property of ideal gases is expressed in **Joule's law** (1845).

Joule's law for ideal gases (3.19)

$$U = U(T); \qquad \left(\frac{\partial U}{\partial V}\right)_T = 0$$

The **internal energy** U of **ideal gas** is *solely* a function of the gas **temperature** T.

"Joule's experiment" 1843

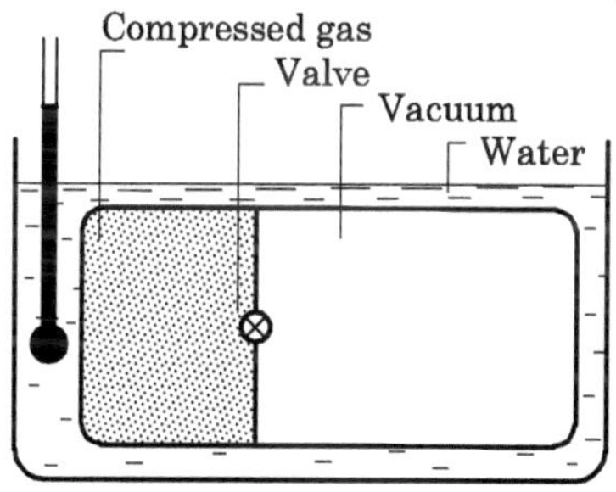

Figure 3.16. Schematic presentation of Joule's experiments. Joule's experiments verify that the temperature T of an ideal gas does not change when the gas expands into a vacuum; since Q and W are zero during the process, the experiment shows that ΔU, and thus $(\partial U/\partial V)_T$, are zero during the expansion of the gas.

According to Joule's law, the **enthalpy** H of an ideal gas is solely a function of the gas temperature; this is directly seen when applying the ideal gas law (1.8) to the definition equation (3.11). For 1 mol of gas we obtain

$$H = U(T) + pV = U(T) + RT = H(T) \tag{3.20}$$

Especially, the fact that U and H for ideal gas only depend on the gas temperature T implies that the difference between c_p and c_V is constant for an ideal gas:

$$c_p - c_V = \left(\frac{\partial H}{\partial T}\right)_p - \left(\frac{\partial U}{\partial T}\right)_V = \frac{dH}{dT} - \frac{dU}{dT} = \frac{d(pV)}{dT} = R \tag{3.21}$$

For rewriting the partial differential coefficients, it has been used that the assumption of constant p and V, respectively, can be omitted for ideal gas; therefore, we have

$$c_p - c_V = R = \text{constant} \tag{3.22}$$

Summing up, the following important properties of ideal gases are

Ideal gases (3.23)

$pV = nRT$	complies with the **ideal gas law**
$U = U(T)$	the **internal energy** U *solely* depends on the **temperature** T
$H = H(T)$	the **enthalpy** H *solely* depends on the **temperature** T
$c_p - c_V = R$	the difference $c_p - c_V$ is **constant**

Offhand, it may seem surprising that the internal energy U and the enthalpy H of ideal gas *only* depend on the gas **temperature** T, i.e. that irrespective of the gas **pressure** p and **volume** V, all ideal gases have the *same* U and H, respectively, at the same temperature.

The explanation is that the internal energy of ideal gas solely has the form of **molecular kinetic energy** E_k; the potential energy of the gas is zero because *no forces act between gas* particles of ideal gas. At the same time, this also makes it clear that fulfilment of such ideal state can only be *approximate* in real gases.

■ Calculate the amount of heat Q (J) that must be supplied to 100 g of water vapour $H_2O(g)$ to heat the vapour 2.0 °C at constant volume V. Given data for $H_2O(g)$: $c_p = 33.6$ J/molK; $M = 18.02$ g/mol; the water vapour is assumed to be an ideal gas.

Solution. In calculation of Q, the molar heat capacity c_V at constant volume is used; according to (3.23), c_V is determined from

$$c_V = c_p - R = 33.6 - 8.314 \text{ J/molK} = 25.3 \text{ J/mol}$$

The necessary supply of heat Q can now be determined from

$$Q = n \cdot c_V \cdot \Delta T = \frac{100\,\text{g}}{18.02\,\text{g/mol}} \cdot 25.3\,\text{J/molK} \cdot 2.0\,\text{K} = \mathbf{281\ J}$$

Ideal gas state

$pV = nRT$
$U = U(T)$
$H = H(T)$
$c_p - c_V = R$

☐ **1.** In an isothermal process with an ideal gas, the pressure p is reduced from 101325 Pa to 75994 Pa. State ΔU and ΔH for the process!

☐ **2.** 10 mol of ideal gas are supplied with $Q = 520$ J at constant pressure $p = 101325$ Pa; hereby the gas temperature is raised by 2.5 °C. Calculate ΔH (J/mol) and c_V (J/molK) for the gas!

☐ **3.** At constant V, 5 mol of ideal gas are supplied with $Q = 187.5$ J; thus, the increase of the gas temperature is $\Delta T = 2.0$ K. Determine c_p (J/molK), c_V (J/molK) and ΔH (J/mol) for the gas!

☐ **4.** In a closed container, 6 mol of ideal gas are heated 2.0 °C. Calculate the numerical value of the quantity $(\Delta H - \Delta U)$ of unit (J) for the system!

☐ **5.** In a reversible, isothermal process, 2 mol of ideal gas are compressed from $p =$ 1000 Pa to $p =$ 2000 Pa at 25 °C. Calculate the heat Q (J) supplied during the process!

3.6 Isothermal change of state

According to Joule's law (3.19), the internal energy U of **ideal gas** is *solely* a function of the gas temperature: $U = U(T)$. For an **isothermal** change of state with ideal gas, this means that dT, and thereby dU, is zero. From the first law (3.2) we then get

$$dU = \delta Q + \delta W = 0 \qquad \Rightarrow \qquad Q_{1.2} + W_{1.2} = 0 \tag{3.24}$$

For an **isothermal** change of state of ideal gas, the sum of the supplied heat $Q_{1.2}$ and the work done $W_{1.2}$ is zero.

For a **reversible** and **isothermal** change of state (1) → (2) with an ideal gas, the volume work done on the system can be determined by

Reversible isothermal process

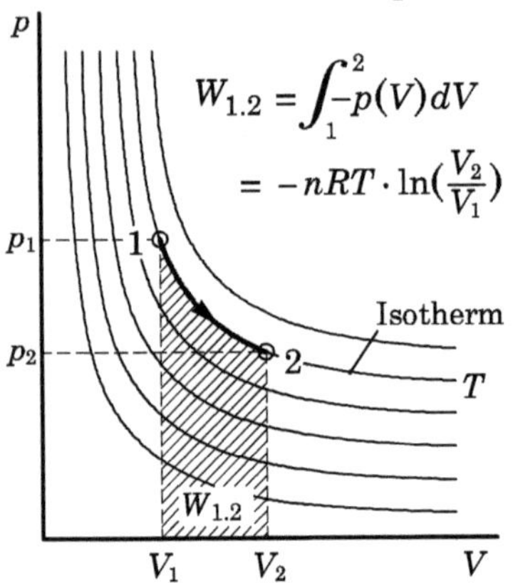

Figure 3.17. During a reversible, isothermal process with an ideal gas, the work W is proportional to the area under the shaded part of the isotherm $V_1 \to V_2$.

$$W_{1.2} = \int_{(1)}^{(2)} \delta W = -\int_{(1)}^{(2)} p(V)dV = -\int_{(1)}^{(2)} \frac{nRT}{V}dV = -nRT \cdot \ln\left(\frac{V_2}{V_1}\right) \tag{3.25}$$

The heat $Q_{1.2}$ supplied during the process is determined from (3.24)

$$Q_{1.2} = -W_{1.2} = +nRT \cdot \ln\left(\frac{V_2}{V_1}\right) \tag{3.26}$$

Summing up, we have

Isothermal change of state, ideal gas (3.27)

(a) *Generally*, for an **isothermal** change of state,

$$Q_{1.2} + W_{1.2} = 0$$

(b) *Specifically*, for a **reversible**, **isothermal** change of state,

$$W_{1.2} = -nRT \cdot \ln\left(\frac{V_2}{V_1}\right); \qquad Q_{1.2} = +nRT \cdot \ln\left(\frac{V_2}{V_1}\right);$$

Note: The assumption of a **reversible**, isothermal change of state is necessary for a unique calculation of the volume work $W_{1.2}$ done, since the work W is *not* a state function. However, as state functions, ΔU and ΔH, are *independent* of the process path.

■ Assume a system that in its initial state (1) contains $V_1 = 1.000\,\mathrm{m}^3$ of atmospheric air at 20.0 °C; the pressure of the gas is $p_1 = 1\,\mathrm{atm} = 101325\,\mathrm{Pa}$. The air is assumed to be an ideal gas with molar mass $M = 28.98$ g/mol. By a **reversible, isothermal** process the air is compressed to a volume of $V_2 = 0.500\,\mathrm{m}^3$. Calculate the amount of heat (J) that has to be conducted away from the air during the process and determine the pressure p_2 in the final state (2)!

Ideal gas state

The internal energy U is constant by an isothermal process.

Solution. The considered system contains an amount of substance n, determined from the gas law (1.8)

$$n = \frac{pV}{RT} = \frac{101325 \cdot 1.000}{8.314 \cdot 293.15}\,\mathrm{mol} = 41.57\,\mathrm{mol}$$

The amount of heat Q *supplied* during the process is determined from (3.27b)

$$Q_{1.2} = +nRT \cdot \ln\left(\frac{V_2}{V_1}\right) = 41.57 \cdot 8.314 \cdot 293.15 \cdot \ln\left(\frac{0.500}{1.000}\right)\,\mathrm{J} = -7.02 \cdot 10^4\,\mathrm{J}$$

The negative sign of Q means that $7.02 \cdot 10^4$ J is **conducted away** from the system during the process. The pressure p_2 after the reversible, isothermal process is calculated from the ideal gas law (1.8)

$$p_2 = \frac{nRT_2}{V_2} = \frac{41.57 \cdot 8.314 \cdot 293.15}{0.500}\,\mathrm{Pa} = \mathbf{2.03 \cdot 10^5\,Pa}$$

☐ **1.** In an isolated system, an ideal gas expands irreversibly into a vacuum (see figure 2.4). State and explain: Q, W, ΔT, ΔU and ΔH for the process!

☐ **2.** By a reversible, isothermal process with 5 moles of ideal gas, p is changed from 0.25 to 0.50 atmospheres at 20.0 °C. Calculate Q (J), W (J), ΔU (J) and ΔH (J) for the process!

☐ **3.** An amount of heat of 5.00 kJ at constant temperature 30.0 °C is conducted away from 3 moles of ideal gas. State Q (J), W (J), ΔU (J) and ΔH (J) for the process!

☐ **4.** In (3.27), $W_{1.2}$ and $Q_{1.2}$ are given as functions $f(V_1, V_2)$ of V. Formulate a similar expression for calculation of $W_{1.2}$ and $Q_{1.2}$ as a function: $g(p_1, p_2)$ of p!

☐ **5.** It it possible to carry out an isothermal change of state with an ideal gas in an adiabatic system, if only volume work W is done on the system?

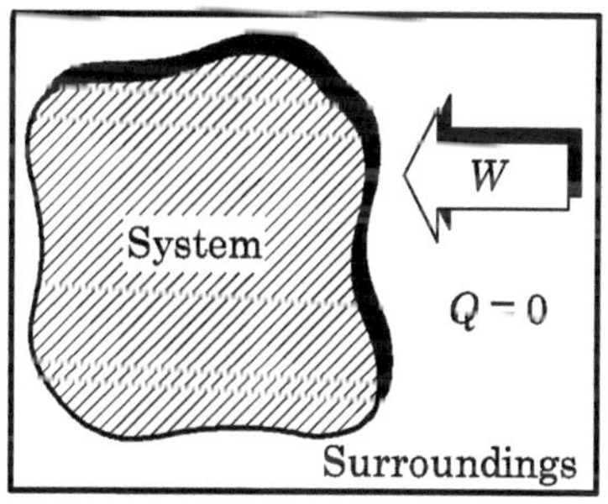

Figure 3.18. In an adiabatic process Q is equal to zero; therefore, the change in the internal energy of a system U in the adiabatic process is equal to the work W done on the system.

3.7 Adiabatic change of state

By an **adiabatic** change of state of an **ideal gas**, Q is equal to zero, i.e. exchange of heat between the system and its surroundings during the process is prevented. According to the first law (3.2), this implies that

$$dU = \delta Q + \delta W = 0 + \delta W = \delta W \tag{3.28}$$

Therefore, considering a **reversible**, **adiabatic** change of state, the differential dU of the internal energy per mole of gas is

$$dU = -pdV = -RT \cdot \frac{dV}{V} = -RT \cdot d\ln(V) \qquad (3.29)$$

The internal energy U of an ideal gas *solely* depends on the gas temperature T; thus, for a given temperature increment, dT, the increment in the molar internal energy, dU, is

$$dU = c_V \cdot dT \qquad \text{(ideal gas)} \qquad (3.30)$$

Using the identity between (3.29) and (3.30), and introducing the relation $R = c_p - c_V$, which was previously derived from (3.23), we get

$$\frac{dT}{T} = -\frac{R}{c_V} \cdot \frac{dV}{V} = -\left(\frac{c_p}{c_V} - 1\right) \cdot \frac{dV}{V} \qquad (3.31)$$

Adiabatic equations

The ratio (c_p/c_V) is called the **heat capacity ratio** γ. Introducing γ into (3.31) and integrating over a change of state from (1) to (2), we get

$$\int_{(1)}^{(2)} d\ln(T) = -(\gamma - 1) \cdot \int_{(1)}^{(2)} d\ln(V) \qquad (3.32)$$

$$\ln\left(\frac{T_2}{T_1}\right) = \ln\left(\left(\frac{V_1}{V_2}\right)^{\gamma-1}\right) \qquad (3.33)$$

$$T_1 \cdot V_1^{\gamma-1} = T_2 \cdot V_2^{\gamma-1} = \text{constant} \qquad (3.34)$$

Using the ideal gas law (1.8), the temperature T can be expressed as a function of the pressure p. Introducing $T = pV/R$, we get

$$p_1 \cdot V_1^{\gamma} = p_2 \cdot V_2^{\gamma} = \text{constant} \qquad (3.35)$$

Correspondingly, in the latter equation, V can be eliminated by introducing $V = RT/p$

$$p_1^{1-\gamma} \cdot T_1^{\gamma} = p_2^{1-\gamma} \cdot T_2^{\gamma} = \text{constant} \qquad (3.36)$$

Summarizing these expressions, we have the following equations for description of the adiabatic change of state

Adiabatic change of state, ideal gas (3.37)

For an **adiabatic** change of state, the following *general* expression applies

$$dU = \delta W; \qquad \delta Q = 0$$

For an **adiabatic** change of state, the **adiabatic equations** apply

$$T \cdot V^{\gamma-1} = \text{constant}; \quad p \cdot V^{\gamma} = \text{constant}; \quad p^{1-\gamma} \cdot T^{\gamma} = \text{constant}$$

where $\gamma = (c_p/c_V)$ is the **heat capacity ratio**; the three equations stated for (p, T, V) are called **adiabatic equations**.

In many cases, fast-developing changes of state in a gas phase will be adiabatic or near-adiabatic, so that an approximate description of the process can be given with the **adiabatic equations** (3.37). For example, this applies to changes of pressure in **sound waves**, combustion and explosion processes in gases, and changes of pressure in fast-working gas compressors.

Heat capacity ratio

Hydrogen H_2	1.410
Helium He	1.667
Oxygen O_2	1.401
Nitrogen N_2	1.404
Carbon dioxide CO_2	1.304
Atmospheric air	1.401

γ at 25 °C, 1 atmosphere

Sound wave in air

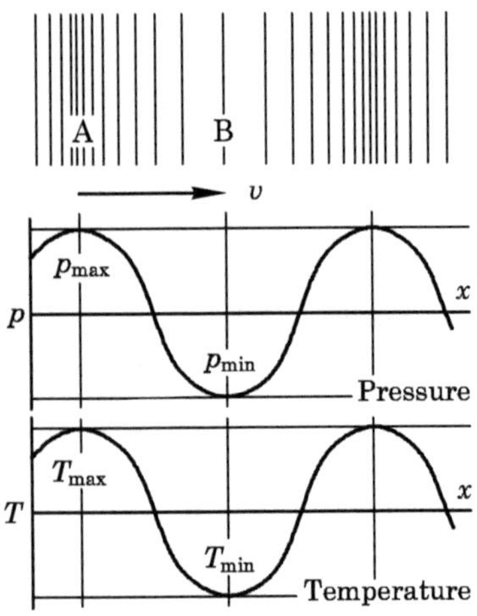

Figure 3.19. Sound propagates as pressure waves at a characteristic velocity v. In the process, the very fast, local pressure variation Δp produces a temperature variation ΔT in the mass of air, which is almost adiabatic.

Course of adiabats

Applying adiabatic equations in the following, it is useful to compare the **adiabatic** and the **isothermal** courses of a pV process of an ideal gas. In a pV diagram, isotherms are **hyperbolas** of the 1st degree

$$p = \frac{nRT}{V} = \text{constant} \cdot \left(\frac{1}{V}\right) \qquad \text{(isotherm)} \tag{3.38}$$

Illustrating an adiabatic change of state of an ideal gas in the pV diagram, the course according to the **adiabatic equations** (3.37) is

$$p = (\text{constant}) \cdot \left(\frac{1}{V}\right)^{\gamma} \qquad \text{(adiabat)} \tag{3.39}$$

Since $c_p > c_V$, the **heat capacity ratio** is $\gamma = (c_p/c_V) > 1$. This implies that the adiabat (3.39) is *steeper* than the isotherm (3.38) in a pV diagram. The physical explanation is that the gas temperature changes during the adiabatic process: when a gas is adiabatically compressed, the gas temperature T rises, and thus the pressure p is increased in relation to the isotherm.

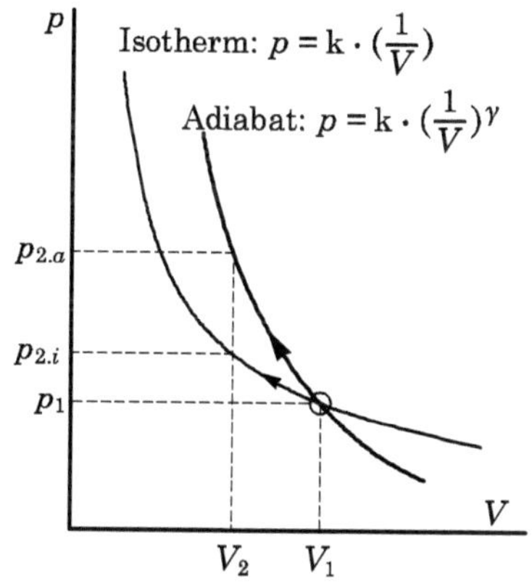

Figure 3.20. In a pV diagram the adiabats have a steeper course than isotherms; the reason is that the temperature T changes in the adiabatic process and thus influences the pressure p in the gas.

■ Assume a volume $V_1 = 1.000\,\text{m}^3$ of atmospheric air at 20.0 °C with $p_1 = 1.00\,\text{atm} = 101325\,\text{Pa}$ in its initial state. The air is assumed to be an ideal gas with $c_p = 29.0$ J/mol K. In a **reversible, adiabatic** compression, the volume is changed to $V_2 = 0.500\,\text{m}^3$. Calculate the pressure p_2 and the temperature θ_2 in the final state!

Solution. The **heat capacity ratio** $\gamma = c_p/c_V$ is determined from (3.23)

$$\gamma = \frac{c_p}{c_V} = \frac{c_p}{c_p - R} = \frac{29.0}{29.0 - 8.314} = 1.40$$

Through a simple rewriting of the **adiabatic equations** (3.37), we can then draw up the following expressions for determination of p_2 and T_2

$$p_2 = p_1 \cdot \left(\frac{V_1}{V_2}\right)^{\gamma} = 101325 \cdot \left(\frac{1.000}{0.500}\right)^{1.40} = 2.67 \cdot 10^5\,\text{Pa} \quad (2.64\,\text{atm})$$

$$T_2 = T_1 \cdot \left(\frac{V_1}{V_2}\right)^{\gamma-1} = 293.15 \cdot \left(\frac{1.000}{0.500}\right)^{0.40} = 386.8\,\text{K} \quad \Rightarrow \theta_2 = 114\,°\text{C}$$

In its final state, the gas pressure is **2.64 atm** and the gas temperature is **114 °C**. Compare these values with the results in the example, section (3.6), where the same change of volume has been calculated for an isothermal process.

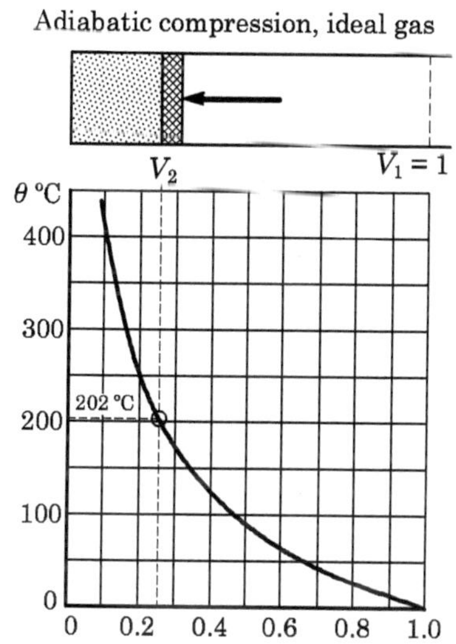

Figure 3.21. When a gas is compressed in an adiabatic process, the gas temperature θ is increased. The graph shows the temperature rise at adiabatic compression of atmospheric air; as an example, the air temperature increases to approximately 202 °C by compression to 25 % of the initial volume.

☐ **1.** Calculate the heat capacity ratio γ for an ideal gas, if: a) $c_p = 33.6$ J/mol K; b) $c_V = 12.5$ J/mol K and c) $c_p = 7.0$ cal/mol °C!

☐ **2.** Assume a reversible, adiabatic process with an ideal Ar gas. Given: (θ_1, p_1, V_1) are (0.0 °C, 1 atm, 1 m^3). After the process, $p_2 = 0.1$ atmospheres; calculate θ_2 (°C) and V_2 (m^3)!

☐ **3.** The heat capacity ratio of a gas is $\gamma = 1.35$ and $c_p = 24.9$ J/mol K. Determine C_V of unit (J/K) for a system containing 42.0 mol of the gas!

☐ **4.** The compression ratio of a high-speed diesel engine is $V_2{:}V_1 = 1{:}25$. Calculate θ_2 (°C) of the compacted intake air when $\theta_1 = 20$ °C and $\gamma = 1.40$!

☐ **5.** In a reversible, adiabatic process with an ideal gas with the initial temperature 25.0 °C, V is changed from 1.00 m^3 to 2.00 m^3. Calculate $W_{1,2}$ (J/mol) and ΔU (J/mol), when $c_p = 28.5$ J/mol K!

3.8 Thermochemical equation

Reaction enthalpy

In technical calculations comprising chemical reactions or phase transformations at atmospheric pressure, the **reaction enthalpy** ΔH is a central quantity. The reason for this was found in the relation (3.15), which was previously derived

> **Reaction enthalpy** (3.40)
>
> For a process developing at **constant pressure** p, and **temperature** T, the **reaction enthalpy** ΔH is *equal to* the supplied **heat** Q, if *only* volume work is done on the system.

Figure 3.22. The evaporation enthalpy ΔH for water denotes the "heat of evaporation" consumed to transform 1 mol of water to water vapour at constant pressure.

The reaction enthalpy ΔH is the precise thermodynamic notation for the "amount of heat" referenced colloquially. Thus, the amount of heat consumed to evaporate 1 kg of water (colloquially called the "heat of evaporation") is the reaction enthalpy ΔH (kJ/kg) for the phase transition $H_2O(\ell) \rightarrow H_2O(g)$ in the actual state. Thus, the concept of reaction enthalpy is neither *new* nor *exotic*, but because of the precise definition (3.11), the thermodynamic concept ΔH is an unambiguous calculation parameter.

Reaction equations

The enthalpy H is a **state function**; for a change of state in a system, ΔH solely depends on the **initial state** and the **final state** of the system. This important property is e.g. used in thermochemical calculations that describe changes of enthalpy during chemical reactions or phase transitions.

A **thermochemical equation** is a reaction equation for a chemical reaction or phase transition with specification of the reaction enthalpy ΔH for an isobaric process at a given temperature T. In thermochemical equations the following notation is used

> **Thermochemical equation** (3.41)
>
> $$\{\text{Reactants}\} \quad \rightarrow \quad \{\text{Products}\} \qquad \Delta H = \quad \text{kJ/mol}$$
> $$a\text{A} + b\text{B} \quad \rightarrow \quad c\text{C} + d\text{D} \qquad \Delta H = \quad \text{kJ/mol}$$
>
> Example: ($\theta = 25\,°\text{C}$)
>
> $$1 \cdot \text{CaO(s)} + 1 \cdot \text{H}_2\text{O}(\ell) \quad \rightarrow \quad 1 \cdot \text{Ca(OH)}_2\text{(s)} \qquad \Delta H = \quad -65.2\,\text{kJ/mol}$$

When thermochemical equations are specified, the **state** of the substances *must* be stated, if they can occur in different states. This is necessary because ΔH depends on the state of reactants and products. Thus, the example in (3.41) assumes the form

$$1 \cdot \text{CaO(s)} + 1 \cdot \text{H}_2\text{O(g)} \quad \rightarrow \quad 1 \cdot \text{Ca(OH)}_2\text{(s)} \qquad \Delta H = -109.2\,\text{kJ/mol}$$

if the reaction incorporates $H_2O(g)$ instead of $H_2O(\ell)$. For specifying the **state** of the substances, the following notation is used

Symbol	Meaning	Examples	
(s)	solid phase ("solid")	$H_2O(s)$	ice
(ℓ)	liquid phase ("liquid")	$H_2O(\ell)$	water
(g)	gas phase ("gas")	$H_2O(g)$	water vapour
(aq)	ion, aqueous solution ("aqua")	$OH^-(aq)$	hydroxide ion

Reaction heat

Chemical reactions are normally connected with a "heat exchange", which is manifested as absorption or release of heat during the reaction. The connection between heat exchange and reaction enthalpy ΔH is

Figure 3.23. The reaction enthalpy ΔH for dissolution of ammonium nitrate NH_4NO_3 in water is +25.7 kJ/mol, i.e. the process is heavily endothermic ("heat absorbing"). When ammonium nitrate is dissolved in water at room temperature, the temperature of the solution may drop below the freezing point.

> **Heat exchange by chemical reaction** (3.42)
>
> When a chemical reaction develops at constant (p, T) the following applies
>
> $\Delta H > 0$ for **endothermic** ("heat-absorbing") reactions
>
> $\Delta H < 0$ for **exothermic** ("heat-releasing") reactions

Thermochemical calculation

The procedure in thermochemical calculations is best illustrated by explanation of a simple example: determination of ΔH for **sublimation** of ice.

Assume a **closed** system consisting of 1 mol of H_2O at 0 °C. The pressure p is assumed to be constant at 1 atmosphere. In this state, water is found in three forms, namely as **ice**, as **water**, and as **water vapour**. At 0 °C, 1 mol of ice can be transformed into 1 mol of water if the system is *supplied* with $Q = 6026$ J; the "melting heat" of the ice is 6026 J/mol. The process can be described by the following **thermochemical equation**

(a) **Melting**	$H_2O(s)$	$\rightarrow$	$H_2O(\ell)$	$\Delta H_a =$	+6026 J/mol

According to (3.40), the enthalpy increase ΔH_a during an **isobaric** process is equal to the *supplied* heat Q. The melting is **endothermic**, and $\Delta H > 0$, cf. (3.42); this means that the enthalpy content of the system has been *increased* by 6026 J/mol during melting.

Similarly, 1 mol of water at 0 °C can be transformed into 1 mol of water vapour if the system is supplied with an amount of heat of $Q = 45050$ J/mol, the so-called "heat of evaporation". This transformation is described by a thermochemical equation of the form

(b) **Evaporation**	$H_2O(\ell)$	$\rightarrow$	$H_2O(g)$	$\Delta H_b =$	+45050 J/mol

According to (3.40) the enthalpy increase ΔH_b during an **isobaric** evaporation process is equal to the supplied amount of heat Q. When water is transformed into water vapour, the enthalpy of the system is increased by 45050 J/mol. According to (3.42), the evaporation process is therefore **endothermic**.

Figure 3.24. At temperatures below 0 °C, ice can evaporate and turn into water vapour in the same way as water can evaporate at temperatures above 0 °C; the ice sublimes. The phenomenon often occurs in the nature during winter.

Melting heat of ice and evaporation heat of water are simple, measurable physical quantities, and their numerical values can be found in most tables of reference. The same does *not* apply to the numerical value for the sublimation heat of ice ΔH_c at 0 °C, corresponding to the process

(c) **Sublimation**	$H_2O(s)$	$\rightarrow$	$H_2O(g)$	$\Delta H_c =$	? J/mol

As the enthalpy increase ΔH_c is *independent* of the process path, the enthalpy change of process (c) will correspond to the enthalpy change of process (a) + process (b), since the total of these two processes corresponds to the same change of state as does process (c). Thus, for phase transformation at 0 °C, we get

(a) **Melting**	$H_2O(s)$	$\rightarrow$	$H_2O(\ell)$	$\Delta H_a =$	+6026 J/mol
(b) **Evaporation**	$H_2O(\ell)$	$\rightarrow$	$H_2O(g)$	$\Delta H_b =$	+45050 J/mol

(c) **Sublimation** $H_2O(s) \rightarrow H_2O(g)$ $\Delta H_c = 6026 + 45050$ J/mol

the net reaction (c) is produced by *addition* of process (a) and process (b); similarly, the increase in enthalpy ΔH_c is produced by *addition* of ΔH_a and ΔH_b. Thus, we have determined the **sublimation enthalpy** ΔH_c of the ice to be +51076 J/mol.

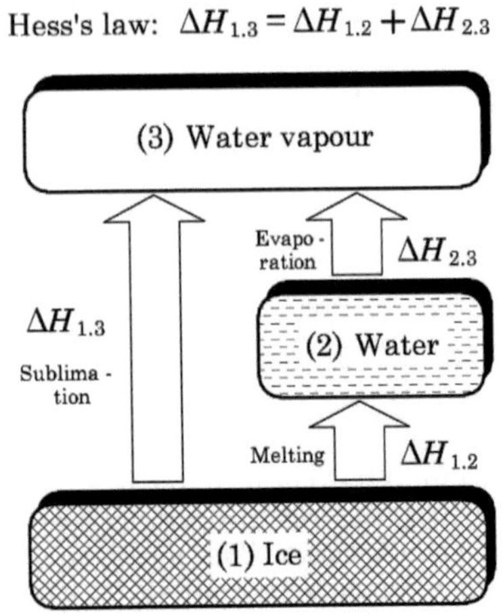

Figure 3.25. In thermochemical calculation it is utilized that the enthalpy H is a state function; therefore, for a process connecting states of equilibrium, ΔH is independent of the process path (Hess's law).

Calculation rules

Equations of the forms (a), (b), and (c) are called **thermochemical equations.** Generally, these kinds of equations can be *added, subtracted, reversed* and *multiplied* by constants. Thus, it is possible to obtain information of changes of enthalpy for processes for which no thermodynamic data are available, or of processes for which experimental investigations are difficult.

The procedure by *addition* and *subtraction* of thermochemical equations can be seen from the example. *Reversing* a thermochemical equation can be made as follows:

(a) $a \cdot A + b \cdot B \rightarrow c \cdot C + d \cdot D$ $\qquad \Delta H_a$

(b) $c \cdot C + d \cdot D \rightarrow a \cdot A + b \cdot B$ $\qquad \Delta H_b = -\Delta H_a$

i.e. by *interchanging* reactants and products and *at the same time* changing the sign of ΔH.

When multiplying a thermochemical equation by a constant, *both* the stoichiometric coefficients *and* ΔH are multiplied by the constant. Thus, for the above, formal reaction equation (a), multiplication by a constant k will often result in the equation

(c) $ka \cdot A + kb \cdot B \rightarrow kc \cdot C + kd \cdot D$ $\qquad k \cdot \Delta H_a$

The thermochemical calculation method is directly based on the fact that the enthalpy H is a **state function.** The calculation principle is often referred to as **Hess's law**, since this principle was originally formulated as an empirical rule by the Swiss physicist **G.H. Hess** (1840).

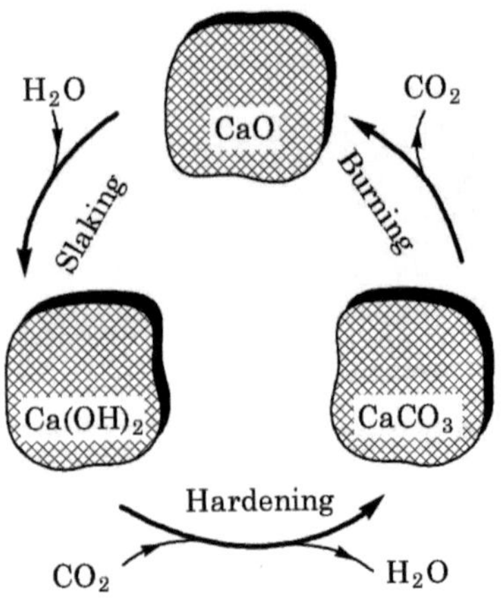

Figure 3.26. The figure shows a schematic outline of the cycle of the lime. Calcium hydroxide $Ca(OH)_2$ has binder properties, which are e.g. utilized in air-hardening lime mortar.

■ For brick-laying, a **lime mortar** is used. It hardens and achieves strength when subjected to the carbon dioxide of air $CO_2(g)$. The net reaction is

(a) **Hardening** $Ca(OH)_2(s) + CO_2(g) \rightarrow CaCO_3(s) + H_2O(\ell)$

During hardening, calcium hydroxide and carbon dioxide form the sparingly soluble compound calcium carbonate that has binder properties. The raw material $Ca(OH)_2$ form naturally occurring **calcite** $CaCO_3$ by the following two steps

(b) **Burning** $CaCO_3(s) \rightarrow CaO(s) + CO_2(g)$ $\qquad \Delta H_b = +177.9$ kJ/mol

(c) **Slaking** $CaO(s) + H_2O(\ell) \rightarrow Ca(OH)_2(s)$ $\qquad \Delta H_c = -\ 65.2$ kJ/mol

Calculate the heat developed in the mortar, when 1 kg of $Ca(OH)_2$ is transformed into $CaCO_3$ during hardening of the mortar; $M(Ca(OH)_2) = 74.10$ g/mol.

Solution. Adding the **endothermic** lime burning reaction (b) to the **exothermic** slaking reaction (c), the following net reaction is obtained

(d) $CaCO_3(s) + H_2O(\ell) \rightarrow Ca(OH)_2(s) + CO_2(g)$ $\Delta H_d = +112.6$ kJ/mol

The components in reaction (d) correspond to those searched for in (a), but the process goes in the opposite direction. Therefore, the hardening reaction (a) is produced by *reversing* the reaction (d)

(a) $Ca(OH)_2(s) + CO_2(g) \rightarrow CaCO_3(s) + H_2O(\ell)$ $\Delta H_a = -112.6$ kJ/mol

If ΔH_a is *negative*, the reaction during hardening of the mortar is **exothermic**, and 112.6 kJ/mol are released. The amount of substance n in 1 kg of $Ca(OH)_2$ is

$$n = \frac{m}{M} = \frac{1\,\mathrm{kg}}{74.10 \cdot 10^{-3}\,\mathrm{kg/mol}} = 13.5\,\mathrm{mol}$$

Thus, during hardening, $13.5 \cdot 112.6$ kJ/kg **=1520 kJ/kg**

☐ **1.** State whether ΔH is unambiguously determined by the following equations: a) $CO_2 + H_2O \rightarrow HCO_3^- + H^+$ b) $CaSO_4 \cdot \frac{1}{2}H_2O + 1\frac{1}{2}H_2O \rightarrow CaSO_4 \cdot 2H_2O$!

☐ **2.** State whether the following processes are *exothermic* or *endothermic*: a) freezing of water b) hardening of $Fe(\ell)$ c) evaporation of ice d) combustion of $H_2(g) + O_2(g)$!

☐ **3.** State the reaction enthalpy ΔH in (J/mol) at 0°C for the following processes: a) $H_2O(g) \rightarrow H_2O(\ell)$; b) $H_2O(\ell) \rightarrow H_2O(s)$; c) $H_2O(g) \rightarrow H_2O(s)$!

☐ **4.** Reverse the following thermochemical equation and multiply it by the constant $k = \frac{3}{2}$: $2Fe(s) + Al_2O_3(s) \rightarrow Fe_2O_3(s) + 2Al(s)$; $\Delta H = +851.1$ kJ/mol!

☐ **5.** In the above example, ΔH for hardening of lime mortar after reaction (a) has been determined as -112.6 kJ/mol, when $H_2O(\ell)$ is formed by the reaction. Will ΔH be larger or smaller than this value, if $H_2O(g)$ is formed by the reaction?

Figure 3.27. Shaft kiln for lime burning; the raw material is natural calcite $CaCO_3$ that will calcine and transform into burnt lime CaO when heated above 800°C.

3.9 Standard enthalpy

The absolute enthalpy content H in a thermodynamic system cannot be measured; by experiments, *only* an **enthalpy change** ΔH in a system can be determined as

$$\Delta H = H_2 - H_1 \qquad (3.43)$$

where H_1 and H_2 denote the total enthalpy content of the system in its initial and final state, respectively.

In thermochemical calculations, however, it is desirable that any **element** and any **compound** can be ascribed an **absolute**, molar enthalpy H. Then it will be possible to tabulate the thermodynamic properties of the substances in a form that is fit for technical calculations.

By introducing the concept of **standard enthalpy** $H^\ominus$, such absolute enthalpy scale for elements and compounds can be determined. This enthalpy scale has an **arbitrary zero** determined by the following definition

Standard enthalpy (3.44)

For **elements** and **compounds**, standard enthalpy $H^\ominus_{298}$ is defined by the standard state: $p^\ominus = 101325$ Pa; $c^\ominus = 1$ mol/ℓ

- **Elements:**

  $H^\ominus_{298} \stackrel{\text{def}}{=} 0$ for elements in their stable form at 298.15 K

- **Compounds:**

  $H^\ominus_{298} \stackrel{\text{def}}{=} \Delta H^\ominus_{298}$ for formation of 1 mol of the compound from elements in their stable form at 298.15 K.

Definition of standard enthalpy $H^\ominus_{298}$

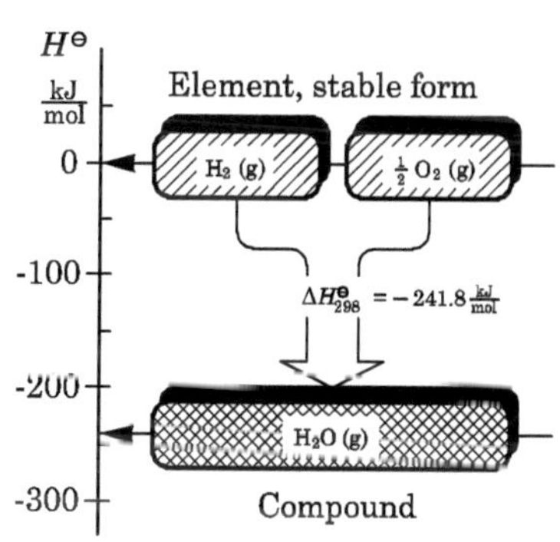

Figure 3.28. Definition of standard enthalpy: Arbitrarily, $H^\ominus_{298} = 0$ is ascribed to any element in its stable form at 25°C; compounds are ascribed to the standard enthalpy $H^\ominus_{298} = \Delta H^\ominus_{298}$ corresponding to ΔH of formation from their elements in their stable form at 25°C.

In the definition (3.44) of standard enthalpy $H^\ominus_{298}$, it is important to note the following: **the standard state** ($^\ominus$) means that the **pressure** is $p = p^\ominus = 101325$ (Pa), *and* that the **concentration** of dissolved substances is $c = c^\ominus = 1$ (mol/ℓ). As **reference temperature** for table data, $T = 298.15$ K ($\theta = 25$ °C) is used. Thus, the standard symbol ($^\ominus$) is *not* in itself connected with the temperature 298.15 K.

Further, it should be emphasized that the **standard state** ($p^\ominus, c^\ominus$) is a *computational* **state of reference**, which is not necessarily physically possible for all compounds. For example, standard enthalpy $H^{\ominus}_{298}$ for $H_2O(g)$ denotes the computational enthalpy $H^{\ominus}_{298}$ for water vapour at 25 °C for a partial pressure of $p^{\ominus} = 101325$ Pa. This state *cannot* be established experimentally, and therefore, a tabular value *only* serves as a computational reference value.

From table data of standard enthalpy $H^{\ominus}_{298}$ for elements and compounds, the enthalpy of the substances at other temperatures and pressures can be calculated.

Pressure-dependence of enthalpy

Generally the pressure-dependence of enthalpy can be disregarded unless major changes of pressure are dealt with. Previously, based on **Joule's law** (3.19) and (3.23), we have shown that for an **ideal gas**, enthalpy is *independent* of the pressure. For solids and liquids, the pressure only has very little influence on the enthalpy content; According to (3.11), for an adiabatic change of pressure dp in a condensed phase, dH is

$$dH = dU + d(pV) = 0 - pdV + pdV + Vdp = Vdp \qquad (3.45)$$

As the molar volume V for solid matters and liquids is typically of the order of magnitude $1 \cdot 10^{-4}$ to $1 \cdot 10^{-5}$ m^3/mol, the change in enthalpy is of the order of magnitude 0.01 to 0.001 (kJ/mol·atm). For technical calculations, therefore, it is a reasonable approximation to use

Pressure-dependence of enthalpy (3.46)

The enthalpy H of **solid matters**, **liquids** and **gases** is *approximately* independent of the pressure when no extreme changes of pressure are dealt with.

Temperature-dependence of enthalpy

The enthalpy of substances H_T at a temperature of $T \neq 298.15$ K can be determined from the expression (3.16)

$$H_T = H^{\ominus}_{298} + \int_{298.15}^{T} \left(\frac{\partial H}{\partial T}\right)_p dT = H^{\ominus}_{298} + \int_{298.15}^{T} c_p(T)dT \qquad (3.47)$$

Over moderate temperature intervals $(T - 298.15)$, c_p can, to a good approximation, be assumed to be constant. In this case, integration leads to the following standard procedure for temperature correction

Temperature-dependence of enthalpy (3.48)

$$H_T = H^{\ominus}_{298} + c_p \cdot (T - 298.15) \qquad \text{(kJ/mol)}$$

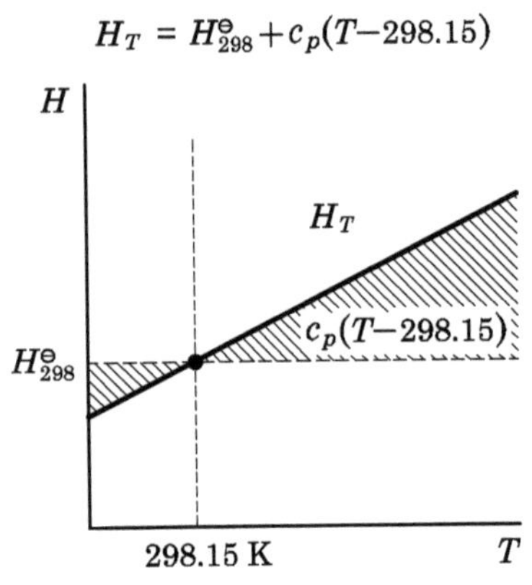

Figure 3.29. Schematic outline of the temperature dependence of enthalpy: H_T is calculated by adding c_p multiplied by the temperature increase to the standard enthalpy $H^{\ominus}_{298}$ at 298.15 K.

For temperature correction of standard enthalpy according to (3.47) or (3.48), tables of the standard enthalpy of substances $H^{\ominus}_{298}$ normally contain data for the molar heat capacity c_p of substances at the standard pressure $p^{\ominus}$.

■ In the handbook *Physical Chemistry* by P.W.Atkins, the following thermodynamic data are given for water $H_2O(\ell)$

$$H^{\ominus}_{298} = -285.83\,\text{kJ/mol}; \qquad c_p = 75.291\,\text{J/mol K}$$

Calculate the molar standard enthalpy of water at 0 °C and at 100 °C!

Solution. It is assumed that c_p is constant in the temperature range considered; the sought values are calculated by (3.48)

$$H_T = H^{\ominus}_{298} + c_p \cdot (T - 298.15) \qquad \text{(kJ/mol)}$$

$$H_{273} = -285.83 + 75.291 \cdot 10^{-3} \cdot (273.15 - 298.15)\ \text{kJ/mol} = \mathbf{-287.71\,kJ/mol}$$

$$H_{373} = -285.83 + 75.291 \cdot 10^{-3} \cdot (373.15 - 298.15) \text{ kJ/mol} = \mathbf{-280.18\,kJ/mol}$$

Note: the molar heat capacity c_p will normally be given in the unit **(J/mol K)**, while in most tables the standard enthalpy $H^{\ominus}_{298}$ is given in the unit **(kJ/mol)**. Use expression (3.48) to harmonize these units!

☐ **1.** Determine the standard enthalpy $H^{\ominus}_{T}$ of: a) $Fe_2O_3(s)$ at 25 °C; b) $CO_2(g)$ at 0 °C; c) $OH^-(aq)$ at 50 °C, and d) $Hg(\ell)$ at 60 °C!

☐ **2.** Determine the enthalpy H_T of: a) saturated water vapour at 40.0 °C; b) water vapour at 10.0 °C; RH = 50 %!

☐ **3.** State H_{298} for water at 10 atmospheres, if: a) the approximation (3.46) applies and b) in relation to the pressure correction (3.45), and c) state the difference in %!

☐ **4.** In *Handbook of Chemistry and Physics*, $H^{\ominus}_{298}$ is given for H(g) as 52.095 kcal/mol, and for $H_2(g)$ as 0 kcal/mol; discuss these numerical values!

☐ **5.** $H^{\ominus}_{298} = -285.8$ kJ/mol for $H_2O(\ell)$, and $H^{\ominus}_{298} = -241.8$ kJ/mol for $H_2O(g)$ are given; discuss the physical significance of the difference between these two numerical values.

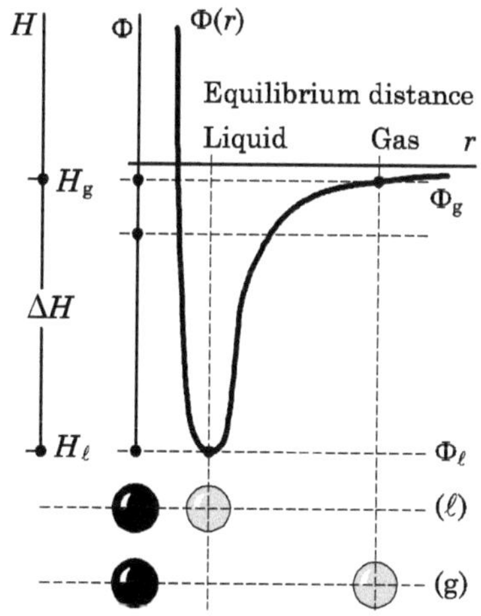

Figure 3.30. For the isothermal transformation of a liquid to a gas (evaporation), the reaction enthalpy ΔH is a measure of the increase in the potential binding energy $\Delta\Phi$ of the molecules.

3.10 Reaction enthalpy

We consider a simple chemical reaction or phase transformation, which can be described by the following formal reaction equation

$$aA + bB \quad \rightarrow \quad cC + dD \tag{3.49}$$

$$\text{(reactants)} \quad \rightarrow \quad \text{(products)}$$

where a, b, c, and d are non-dimensional **stoichiometric coefficients**. Assume that this reaction takes place in a system at constant pressure p and temperature T. The enthalpy increase created by reaction in the system is called the **reaction enthalpy** $\Delta_r H$. Since enthalpy is a state function, the reaction enthalpy $\Delta_r H$ can be expressed as the total enthalpy of the system in its final state minus the total enthalpy of the system in its initial state.

Reaction enthalpy (3.50)

The **reaction enthalpy** $\Delta_r H$ for a chemical reaction or transformation taking place at **constant pressure** p and **temperature** T is

$$\Delta_r H = \sum H(\text{products}) - \sum H(\text{reactants})$$

Reaction enthalpy specifies the total enthalpy in the **final state** minus the total enthalpy in the **initial state**.

We shall now express the reaction enthalpy $\Delta_r H_T$ for one mol of the specified reaction equation (3.49) by the standard enthalpy $H^{\ominus}_{298}$ and molar heat capacity c_p. If c_p is assumed to be constant in the considered temperature range from 298.15 K to T (K) we find from (3.48) and (3.49)

Reaction enthalpy

$\Delta_r H = \sum H_{\text{prod}} - \sum H_{\text{react}}$

$$\Delta_r H_T = (c \cdot H^{\ominus}_{298.C} + d \cdot H^{\ominus}_{298.D}) - (a \cdot H^{\ominus}_{298.A} + b \cdot H^{\ominus}_{298.B}) + ((c \cdot c_{p.C} + d \cdot c_{p.D}) - (a \cdot c_{p.A} + b \cdot c_{p.B})) \cdot (T - 298.15) \tag{3.51}$$

Using the following notation

$$\Delta_r H^{\ominus}_{298} = (c \cdot H^{\ominus}_{298.C} + d \cdot H^{\ominus}_{298.D}) - (a \cdot H^{\ominus}_{298.A} + b \cdot H^{\ominus}_{298.B}) \tag{3.52}$$

$$\Delta_r c_p = (c \cdot c_{p.C} + d \cdot c_{p.D}) - (a \cdot c_{p.A} + b \cdot c_{p.B}) \tag{3.53}$$

we can rewrite (3.51) to the more appropriate form

$$\Delta_r H_T = \Delta_r H^{\ominus}_{298} + \Delta_r c_p \cdot (T - 298.15) \tag{3.54}$$

Here it is important to note the following as regards units: the stoichiometric coefficients a, b, c and d are **non-dimensional** quantities. Therefore, the unit (J/mol) is used for the reaction enthalpy $\Delta_r H_T$, where the notation "mol" refers to the actual reaction equation (3.49). Expressing this verbally, the unit of the reaction enthalpy is "*Joule per mole of reaction equation*".

Calculation procedure

Summing up, we then have the following **procedure** for calculation of reaction enthalpy $\Delta_r H_T$ for a given chemical reaction or transformation

Reaction enthalpy (3.55)

For a reaction or transformation $a\text{A}+b\text{B} \rightarrow c\text{C}+d\text{D}$, the reaction enthalpy $\Delta_r H_T$ at the temperature T is calculated using the procedure:

(a) With tabular values for **molar standard enthalpy** $H^{\ominus}_{298}$, first calculate

$$\Delta_r H^{\ominus}_{298} = (c \cdot H^{\ominus}_{298.\text{C}} + d \cdot H^{\ominus}_{298.\text{D}}) - (a \cdot H^{\ominus}_{298.\text{A}} + b \cdot H^{\ominus}_{298.\text{B}})$$

(b) With tabular values for **molar heat capacity** c_p, then calculate

$$\Delta_r c_p = (c \cdot c_{p.\text{C}} + d \cdot c_{p.\text{D}}) - (a \cdot c_{p.\text{A}} + b \cdot c_{p.\text{B}})$$

(c) Then, the **reaction enthalpy** $\Delta_r H_T$ at the **temperature** T is

$$\Delta_r H_T = \Delta_r H^{\ominus}_{298} + \Delta_r c_p \cdot (T - 298.15)$$

With respect to calculation of reaction enthalpy $\Delta_r H_T$ according to (3.55), the following assumptions should be noted in the procedure:

- The molar heat capacity of the components c_p is assumed to be **constant** in the temperature range considered, i.e. 298.15 K to T (K)
- The temperature T of the system is assumed to be the *same* in the initial state and the final state; the temperature state *during* the process is of no importance.

The calculation procedure (3.55) provided can be adjusted to reactions with either more or fewer reactants and products. Furthermore, when a reaction takes place at a very high or a very low temperature, i.e. a temperature far removed from that of the standard state, the c_p values used can be introduced as temperature-dependent quantities by integration, cf. the expression (3.47).

■ **Steam heating** in sealed curing facilities is often used for heat curing of concrete elements and concrete products. The advantage is that a considerable amount of heat can be transferred to the concrete by **condensation** of water vapour on the surface of the concrete.

Calculate how many (kJ) are released by condensation of 1.000 kg of saturated water vapour at 50 °C, when the water vapour $H_2O(g)$ is assumed to be an ideal gas with molar mass 18.02 g/mol!

Solution. We consider the phase transformation where water vapour $H_2O(g)$ is transformed into water $H_2O(\ell)$ by the following process

$$H_2O(g) \quad \rightarrow \quad H_2O(\ell) \qquad T = 323.15\,\text{K} \;\; (50\,°\text{C})$$

The water vapour is assumed to be an ideal gas, i.e. the enthalpy is independent of the pressure p. The necessary thermochemical standard values can be found in reference tables.

$$H_2O(\ell): \quad H^{\ominus}_{298} = -285.8\,\text{kJ/mol}; \quad c_p = 75.3\,\text{J/mol K}$$

$$H_2O(g): \quad H^{\ominus}_{298} = -241.8\,\text{kJ/mol}; \quad c_p = 33.6\,\text{J/mol K}$$

Using the procedure (3.55), the calculations can be carried out as follows:

(a) $\Delta_r H^{\ominus}_{298} = (1 \cdot (-285.8) - (1 \cdot (-241.8))= -44.0\,\text{kJ/mol}$

(b) $\Delta_r c_p \quad = (1 \cdot 75.3 - 1 \cdot 33.6) = +41.7\,\mathrm{J/mol\,K}$

(c) $\Delta_r H_{323} = -44.0 + 41.7 \cdot 10^{-3} \cdot (323.15 - 298.15) = \mathbf{-43.0\,kJ/mol}$

The reaction enthalpy $\Delta_r H_T$ is negative, i.e. condensation of water vapour is an **exothermic** process. By condensation of 1.000 kg of water vapour, the following amount of heat is developed:

$$Q_{\text{developed}} = -n \cdot \Delta_r H_{323} = -\left(\frac{1000\,\mathrm{g} \cdot (-43.0)\ \mathrm{kJ/mol}}{18.02\,\mathrm{g/mol}}\right) = 2.39 \cdot 10^3\,\mathrm{kJ}$$

Thus, by condensation of 1 kg of water vapour, approximately 2500 kJ of heat is released!

☐ **1.** From table data of $H^\ominus_{298}$ and c_p calculate the magnitude of the evaporation heat of water at 5 °C, at 50 °C, and at 95 °C!

☐ **2.** From table data, calculate $\Delta_r H_{303}$ (kJ/mol) for hardening of plaster by the process: $CaSO_4 \cdot \frac{1}{2}H_2O(s) + 1\frac{1}{2}H_2O(\ell) \rightarrow CaSO_4 \cdot 2H_2O(s)$!

☐ **3.** From table data, calculate for c_p the change of the evaporation heat of water when the temperature is raised from 20 °C to 100 °C!

☐ **4.** Calculate the development of heat $Q_{\text{developed}}$ (kJ/mol) at 25 °C for the reaction of a mixture of oxygen and hydrogen: $H_2(g) + \frac{1}{2}O_2(g) \rightarrow H_2O(g)$!

☐ **5.** Investigate whether solution of ammonium nitrate $NH_4NO_3(s)$ in water is an exothermic or an endothermic dissolution process and specify $\Delta_r H^\ominus_{298}$ (kJ/mol) for the process!

List of key ideas

The following overview lists the key definitions, concepts and terms introduced in Chapter 3.

Energy .... **Energy** is a measure of the *capacity* to do **work** or the *capacity* to release **heat**.

Energy conservation .... **Energy** *cannot* be created or destroyed; the energy in an isolated thermodynamic system is **constant**.

First law .... $dU = \delta Q + \delta W$; the increment in the **internal energy** dU in a closed system is the sum of the *supplied* **heat** δQ and the **work** δW *done* on the system.

Internal energy U .... $\int_{(1)}^{(2)} dU = U_2 - U_1\,; \quad \oint dU = 0$ state function

Internal energy of a **system** (J)
Molar internal energy of a substance (J/mol)

Definition of enthalpy H .. $H \stackrel{\text{def}}{=} U + pV$

Enthalpy H .... $\int_{(1)}^{(2)} dH = H_2 - H_1\,; \quad \oint dH = 0$ state function

Enthalpy H of a **system** (J)
Molar enthalpy H of a substance (J/mol)

Heat capacity .... $c_V = \left(\frac{\partial U}{\partial T}\right)_V$ **isochoric** (J/mol)

$c_p = \left(\frac{\partial H}{\partial T}\right)_p$ **isobaric** (J/mol)

Joule's law .... $U = U(T)$ applies to **ideal gas**

Ideal gas, general .... $pV = nRT$ complies with the **ideal gas law**

$U = U(T)$ **internal energy** U *only* depends on T

$H = H(T)$ **enthalpy** H *only* depends on T

$c_p - c_V = R$ specific heat difference is **constant**

Isothermal process .... *Generally*, for an **isothermal** change of state with an **ideal gas**,

$\Delta U = Q_{1.2} + W_{1.2} = 0$

Specifically, for a **reversible, isothermal** change of state with an **ideal gas**

$$W_{1.2} = -nRT \cdot \ln\left(\frac{V_2}{V_1}\right); \quad Q_{1.2} = +nRT \cdot \ln\left(\frac{V_2}{V_1}\right)$$

The heat capacity ratio .... $\gamma = c_p/c_V$ (non-dimensional)

Adiabatic process .... *Generally*, for an **adiabatic** change of state with an **ideal gas**,

$\Delta U = W_{1.2}; \quad Q_{1.2} = 0$

Specifically, for a **reversible, adiabatic** change of state with an **ideal gas**, the **adiabatic equations** apply

$$T \cdot V^{\gamma-1} = \text{k}; \qquad p \cdot V^{\gamma} = \text{k}; \qquad p^{1-\gamma} \cdot T^{\gamma} = \text{k}.$$

Endothermic process ...... An **endothermic** process "absorbs" heat; the process has **positive reaction enthalpy**: $\Delta H > 0$.

Exothermic process .... An **exothermic** process "releases" heat; the process has **negative reaction enthalpy**: $\Delta H < 0$.

Standard enthalpy .... In the **standard state** $(p^{\ominus}, c^{\ominus})$ and at the reference temperature $T = 298.15$ K, an **absolute** standard enthalpy is defined by

$H^{\ominus}_{298} = 0$ for **elements** in their stable form

$H^{\ominus}_{298} = \Delta H^{\ominus}_{298}$ for **compounds**

Standard state .... The **standard state** $(p^{\ominus}, c^{\ominus})$ is defined by the pressure $p^{\ominus} = 101325$ Pa, and for *dissolved* substances it is defined by the concentration $c^{\ominus} = 1$ mol/ℓ.

Reaction enthalpy .... $\Delta H = \sum H(\text{products}) - \sum H(\text{reactants})$

Reaction enthalpy .... $\Delta_r H_T = \Delta_r H^{\ominus}_{298} + \Delta_r c_p \cdot (T - 298.15)$ (J/mol)

Examples

The following examples illustrate how subject matter explained in Chapter 3 can be applied in practical calculations.

Figure 3.31. Chamber kiln for burning gypsum from around 1700.

Example 3.1

■ Temperature rise in hardening plaster of Paris

Gypsum is a **calcium sulphate-dihydrate** of the composition $CaSO_4 \cdot 2H_2O$. In some countries, gypsum does not occur naturally, but is obtained in large quantities as a **by-product** from the chemical industry. In construction, considerable quantities of gypsum are used in the manufacture of, for example, prefabricated boards for lining walls and ceilings. Plasterboards are characterized as being extremely resistant to fire, because a considerable amount of heat is consumed for evaporation of the hydrate water in the gypsum during a fire.

Gypsum, $CaSO_4 \cdot 2H_2O$, in itself has no binder properties. Before application, the gypsum is burnt so that the hydrate water is partly or completely removed. Thus, *active* forms of **anhydrite** $CaSO_4$ or **hemihydrate** $CaSO_4 \cdot \frac{1}{2}H_2O$ are produced, which can react with water and harden.

The so-called **plaster of Paris** is a hemihydrate that hardens during re-formation of the dihydrate: the hardening is a result of the following reaction

$$CaSO_4 \cdot \tfrac{1}{2}H_2O(s) + 1\tfrac{1}{2}H_2O(\ell) \quad \rightarrow \quad CaSO_4 \cdot 2H_2O(s) \qquad (a)$$

Plaster of Paris reacts *quickly* when water is supplied and the hardening reaction is accompanied by a considerable **development of heat**. During production of plasterboard the temperature rise that will be produced by the heat of hydration must be taken into account.

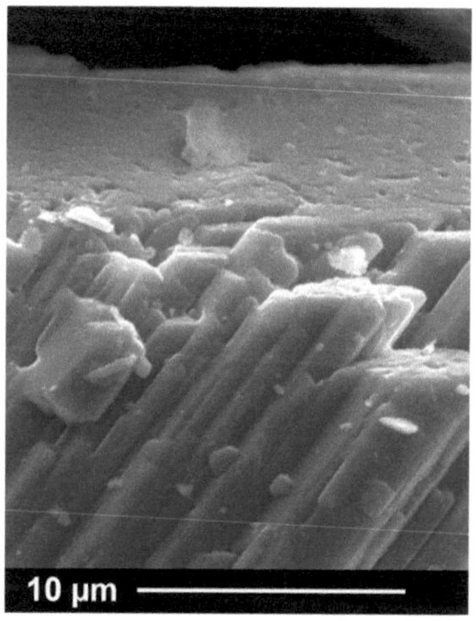

Figure 3.32. Electron microscope picture of gypsum which consists of flat, elongated laths of monoclinic crystals.

Problem. Calculate the adiabatic temperature rise in a mix of plaster of Paris and water in the weight proportion (1:0.5), when the mix is cast at 25 °C.

Conditions. In the calculations a complete transformation of the hemihydrate into dihydrate is assumed. Also, the heat capacity c_p of the components is assumed to be constant in the temperature range considered. The calculations are based on the following data taken from: P.W. Atkins: *Physical Chemistry* and H.A. Bent: *Entropy*.

Component	$H^{\ominus}_{298}$ (kJ/mol)	c_p (J/mol K)	M (g/mol)
$CaSO_4 \cdot 2H_2O(s)$	−2022.1	186.3	172.17
$CaSO_4 \cdot \frac{1}{2}H_2O(s)$	−1575.9	119.7	145.14
$H_2O(\ell)$	− 285.8	75.3	18.02

Solution. The reaction equation (a) is first converted into a form that reflects the actual weight proportion (1:0.5) between plaster of Paris and water. Considering that 1 mol of hemihydrate corresponds to 145.14 g, the following shall be added to 1 mol of reaction equation

$$n = 0.5 \cdot 145.14\text{g}/(18.02\text{ g/mol } H_2O) = 4.03\text{ mol } H_2O$$

During transformation of 1 mol of hemihydrate, 1.50 mol of water is consumed and the surplus water is, therefore, (4.03 − 1.50) = 2.53 mol. Thus, the computational reaction equation including surplus water is

$$CaSO_4 \cdot \tfrac{1}{2}H_2O(s) + 4.03 \cdot H_2O(\ell) \quad \rightarrow \quad CaSO_4 \cdot 2H_2O(s) + 2.53 \cdot H_2O(\ell) \qquad (b)$$

The reaction enthalpy at 25 °C is calculated for the reaction equation (a) using the procedure (3.55)

$$\Delta_r H_{298} = 1 \cdot (-2022.1) - (1\tfrac{1}{2} \cdot (-285.8) + 1 \cdot (-1575.9)) = -17.5\text{ kJ/mol}$$

According to (3.42), the negative reaction enthalpy shows that the process is **exothermic** ("heat-releasing"). For the *products* formed, we shall now determine the heat capacity per mole of reaction equation (b) that corresponds to the actual composition including surplus water

$$c_p(\text{products}) = 1 \cdot 186.3 + 2.53 \cdot 75.3\text{ J/mol K} = 376.8\text{ J/mol K}$$

The adiabatic temperature rise corresponds to an **isobaric** process with $\Delta H = 0$. The calculated *increase* in enthalpy during the reaction at 25 °C must therefore be removed from the system; this corresponds to an adiabatic temperature rise of ΔT_a and a final adiabatic temperature of T_a, determined by

$$\Delta T_a = (-\Delta_r H_{298}/c_p(\text{products})) = -(-17500)/376.8\text{ K} \simeq \mathbf{+46.4\text{ K}}$$

$$T_a = T_{mix} + \Delta T_a = 298.15\text{ K} + 46.4\text{ K} \simeq \mathbf{344.6\text{ K}}$$

Thus, an adiabatic temperature rise of approximately 46 °C can be expected; the final temperature developed as a result of adiabatic hardening is approximately 71 °C.

Figure 3.33. Plasterboards; the individual boards are supplied with a groove and tongue joint so that the boards can be assembled and fixed in place.

Discussion. Experimental determination of the adiabatic heat development can, for example, be performed with an **adiabatic calorimeter**, as shown in figure 2.29. For casting and hardening of small units under normal, i.e. non-adiabatic, conditions the temperature rise is limited by the heat loss during hardening.

Example 3.2

■ Computerized calculation of the evaporation heat of water

Concrete is a mixture of **cement**, **water**, and **sand and coarse aggregates**. The concrete hardens and gains strength when the cement reacts with the added water. During hardening, considerable heat is developed in the concrete. To attain a dense concrete with high strength, it is necessary to protect the concrete against desiccation during the first days after casting.

In industrial production of concrete elements, such as concrete pipes or concrete slabs, etc., **heat-curing** of the concrete is often used. After casting the units are conveyed through curing chambers, where the concrete temperature is raised to 40-60 °C, as a result of which, the hardening is strongly accelerated. However, heat curing increases the risk of early desiccation of the concrete, and may result in an impairment of the concrete quality. A contributory cause of this desiccation is the heat development taking place in the hardening concrete.

Process simulation by computer is used to an increasing extent for determining the dimensions of plants for heat-curing of concrete products. In this way it is possible to predict which temperature and moisture conditions will ensure the best hardening and thus the best concrete quality.

For calculation of heat and moisture transfer between concrete and its environments, the **evaporation heat** of water is an important quantity.

Figure 3.34. For major concrete work it is general practice to plan the hardening of the concrete; computer simulation makes it possible to choose the method that ensures the best hardening and thus the best concrete quality.

Problem. Prepare and test a computer function

$$hvap(\theta) \qquad 0\,^{\circ}\mathrm{C} \le \theta \le 80\,^{\circ}\mathrm{C} \qquad (a)$$

which by input of the temperature θ (°C) will return the evaporation heat of water with the unit (kJ/kg). The maximum allowable deviation between calculated and tabular evaporation heat is 0.5 % in the temperature range $0\,^{\circ}\mathrm{C} \le \theta \le 80\,^{\circ}\mathrm{C}$. There are no requirements for optimization of the calculation speed.

Conditions. In the calculations the following thermodynamic standard values for $H_2O(\ell)$ and $H_2O(g)$ are used

Component	M (g/mol)	$H^{\ominus}_{298}$ (kJ/mol)	c_p (J/mol K)
$H_2O(\ell)$	18.02	−285.8	75.3
$H_2O(g)$	18.02	−241.8	33.6

The water vapour is assumed to be an ideal gas and c_p values are assumed to be constant in the temperature range considered, $0\,^{\circ}\mathrm{C} \le \theta \le 80\,^{\circ}\mathrm{C}$.

Solution. At a given thermodynamic temperature T (K), the evaporation heat is equal to the reaction enthalpy $\Delta_r H_T$ for the transformation

$$H_2O(\ell) \quad \rightarrow \quad H_2O(g) \qquad (b)$$

With the subscript (ℓ) for water and (g) for water vapour, respectively, and using the expression for standard reaction enthalpy (3.55), we obtain

$$\Delta_r H_T = H^{\ominus}_{298.g} - H^{\ominus}_{298.\ell} + (c_{p.g} - c_{p.\ell}) \cdot (T - 298.15) \quad \text{(kJ/mol)} \qquad (c)$$

Converting from the unit (kJ/mol) to the unit (kJ/kg), the expression (c) must be multiplied by the following conversion factor k

$$k = M^{-1} = (18.02 \cdot 10^{-3}\,\text{kg/mol})^{-1} = 55.49\,\text{mol/kg}$$

The required computer function can now be outlined in the following pseudo code, which in a simple way can be converted into structured computer languages such as *Matlab* or *C*.

```
FUNCTION hvap(T:real):real;                          (* enthalpy function *)
const H_H2OL = −285.8; C_H2OL = 0.0753;              (* H2O(ℓ)-data *)
      H_H2OG = −241.8; C_H2OG = 0.0336;              (* H2O(g)-data *)
      M      = 0.01802;                              (* molar mass *)
var   DELTAH, DELTAC : real;
begin
  T:= T + 273.15;                                    (* °C → K *)
  if (T < 273.15) or (T > 353.15)                    (* range test *)
    then writeln('Out of range');                    (* Error note *)
  else
    begin                                            (* calculate hvap *)
      DELTAH:= H_H2OG − H_H2OL;
      DELTAC:= C_H2OG − C_H2OL;
      hvap := (DELTAH + DELTAC * (T − 298.15))/M;
    end;
end;                                                 (* return value *)
```

Evaporation enthalpy

$\Delta H = H(\mathrm{g}) - H(\ell)$

To test the setup function, compare with the values for the evaporation enthalpy of water tabulated in *Tables of physical and chemical constants* by G.W.C. Kaye and T.H. Laby. In the following table the table value and the calculated value are compared for 0, 20, 40, 60 and 80 °C.

Temperature θ	0	20	40	60	80	°C
Table value ΔH	2501	2454	2406	2358	2308	kJ/kg
$hvap(\theta)$	2500	2453	2407	2361	2314	kJ/kg
Deviation	0.04	0.04	0.04	0.10	0.26	%

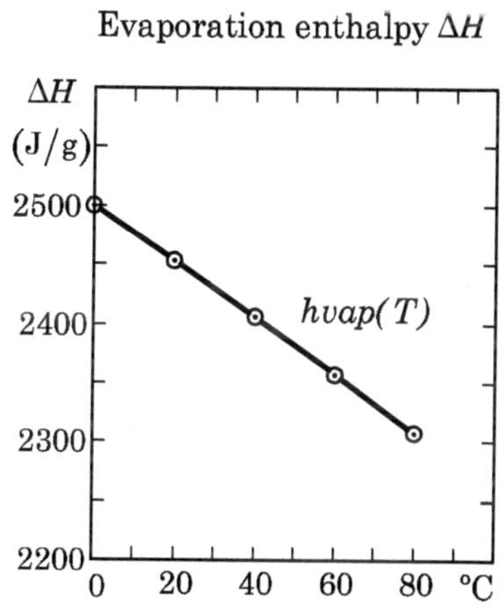

Figure 3.35. The evaporation enthalpy of water ΔH will decrease with rising temperatures. The full-drawn graph shows ΔH calculated with the function hvap(0); the table values are given as ⊙.

Thus, the computer function $hvap(\theta)$ shown above fulfils the requirement that the largest deviation between calculated and tabulated evaporation heat is $\leq 0.5\,\%$ in the temperature range $0\,°\mathrm{C} \leq \theta \leq 80\,°\mathrm{C}$.

Discussion. For the development of actual calculation programs, the thermodynamic standard values should be declared **global constants** in the main program so that these data are available for all **functions** and **procedures** of the program. For example, the same thermodynamic standard values for water and water vapour, which are declared constants in $hvap(\theta)$, are also included in a thermodynamic calculation of the vapour pressure curve of water.

Example 3.3

■ Heat development and hydration of clinker minerals

Portland cement can be produced by **burning** a homogenized mix of materials containing **clay** and **lime**. When partly melting at 1400-1500 °C, a number of **calcium silicates** and **calcium aluminates**, the so-called clinker minerals, are formed. The burnt product - the cement clinkers - are then ground with addition of a few per cent of **gypsum**, which results in the commercial product **Portland cement**.

The most important clinker minerals in Portland cement are

Name	Chemical composition	Cement-chemical name	Typical content
Tricalcium silicate	$3CaO \cdot SiO_2$	C_3S	55 %
Dicalcium silicate	$2CaO \cdot SiO_2$	C_2S	20 %
Tricalcium aluminate	$3CaO \cdot Al_2O_3$	C_3A	7 %
Tetracalcium aluminoferrite	$4CaO \cdot Al_2O_3 \cdot Fe_2O_3$	C_4AF	9 %

For the cement-chemical description, the following names or "abbreviations" are used for the oxides that form the clinker minerals

$$\text{'C'} = \mathrm{CaO}; \quad \text{'S'} = \mathrm{SiO_2}; \quad \text{'A'} = \mathrm{Al_2O_3}; \quad \text{'F'} = \mathrm{Fe_2O_3}$$

The four clinker minerals mentioned have characteristic properties with regard to rate of **heat development** and rate of **strength development** during hydration; subsequent concrete **durability**; and **colour**. Therefore, the useful qualities of Portland cement depend, at least in part, on the quantitative proportions of these clinker minerals in the cement.

Figure 3.36. Rotary kiln for burning of Portland cement; modern kilns can produce up to 5000 tons cement every 24 hours.

Problem. In the book *Thermodynamik der Silikate* by Mchedlov-Petrossyan it is stated that a potential chemical reaction of hydration of tricalcium silicate $3CaO \cdot SiO_2$ is the formation of **hillebrandite** $2CaO \cdot SiO_2 \cdot 1.17H_2O$ according to the following reaction equation

$$3CaO \cdot SiO_2(s) + 2.17\,H_2O(\ell) \rightarrow 2CaO \cdot SiO_2 \cdot 1.17H_2O(s) + Ca(OH)_2(s) \qquad \text{(a)}$$

Calculate the heat development with the unit (J/g) at total hydration of tricalcium silicate at 25 °C if the reaction follows reaction equation (a)!

Conditions. The following thermodynamic standard values are given by Mchedlov-Petrossyan in: *Thermodynamik der Silikate*

Substance	$3CaO \cdot SiO_2(s)$	$2CaO \cdot SiO_2 \cdot 1.17H_2O(s)$	$Ca(OH)_2(s)$	$H_2O(\ell)$
$H^{\ominus}_{298}$ kcal/mol	-688.1	-624.8	-235.8	-68.32

The molar mass M for tricalcium silicate is 228.26 g/mol; the unit 1 cal = 4.184 J.

Specific surface

$$S \stackrel{\text{def}}{=} \frac{\text{surface area}}{\text{mass}}$$

Solution. In the standard state $p^{\ominus} = 1$ atm and 25 °C, the heat development Q for the reaction (a) is identical with $-\Delta_r H^{\ominus}_{298}$. Using the procedure (3.55), we get

$$\Delta_r H^{\ominus}_{298} = (1 \cdot (-624.8) + 1 \cdot (-235.8)) - (1 \cdot (-688.1) + 2.17 \cdot (-68.32)) \;\; \text{kcal/mol}$$

$$\Delta_r H^{\ominus}_{298} = -24.25\,\text{kcal/mol}$$

$$\Delta_r H^{\ominus}_{298} = \frac{-\,24.25\,\text{kcal/mol} \cdot 1000\,\text{cal/kcal} \cdot 4.184\,\text{J/cal}}{228.26\,\text{g/mol}} = \mathbf{-445\,J/g}$$

$$\text{Heat development:} \quad Q = -\Delta_r H^{\ominus}_{298} \simeq \mathbf{+445\,J/g}$$

Discussion. The heat of hydration of approximately 445 J/g developed for the reaction (a) is *solely* due to the chemical transformations during the reaction.

The deposited hydration products consist of **colloidal** solid particles - the so-called "cement gel" - and therefore, they have an extremely large specific surface: 200-300 m^2 per gram of solid matter. During hydration of the cement, some of the water added is adsorbed on the surface by this cement gel; by the adsorption a certain adsorption heat is released. According to Powers,T.C. & Brownyard,T.L.: *The Thermodynamics of Adsorption of Water on Hardened Cement Paste*, J.of ACI, vol.18, No.5 (1947), this adsorption heat is of the order of magnitude of 28 cal per gram of cement. The total heat development during the hydration, therefore, will typically be $\sim$ 25 % higher than the calculated value for the chemical transformations according to (a).

Figure 3.37. The resistance of plaster boards to fire: After fire behind the light partition shown, the door was burnt through while the plaster board partition remained intact.

Example 3.4

■ Fire resistance of plaster

Pre-fabricated sheathings of plaster are extremely resistant to **fire**. This is especially based on two facts: 1) heat is consumed for evaporation of hydrate water when the plaster is dehydrated during heating, and 2) the heat-insulating properties of the dehydrated plaster are good.

Without getting further into purely fire technical mechanisms, the following example may serve as an illustration of the above-mentioned effect - the absorption of heat by **dehydration** of plaster during fire exposure.

Problem. In their book *Gips*, Schwiete & Knauf state that dehydration of plaster takes place in two steps at the following approximate transformation temperatures

$$CaSO_4 \cdot 2H_2O(s) \rightarrow CaSO_4 \cdot \tfrac{1}{2}H_2O(s) + 1\tfrac{1}{2}H_2O(g) \qquad 370\,\mathrm{K}\ (97\,^\circ\mathrm{C}) \qquad (a)$$

$$CaSO_4 \cdot \tfrac{1}{2}H_2O(s) \rightarrow CaSO_4(s) + \tfrac{1}{2}H_2O(g) \qquad 373\,\mathrm{K}\ (100\,^\circ\mathrm{C}) \qquad (b)$$

During the reaction (a), the **dihydrate** $CaSO_4 \cdot 2H_2O(s)$ liberates water and is transformed into **hemihydrate** $CaSO_4 \cdot \frac{1}{2}H_2O(s)$. During the reaction (b), the hemihydrate liberates more water and is transformed into **anhydrite** $CaSO_4(s)$. Anhydrite is stable at all temperatures up to 450 K (177 °C).

Calculate the amount of heat Q (kJ) to be added by reversible heating of 1 kg of dihydrate $CaSO_4 \cdot 2H_2O(s)$ and 1 kg of anhydrite $CaSO_4(s)$, respectively, from 25 °C to 125 °C!

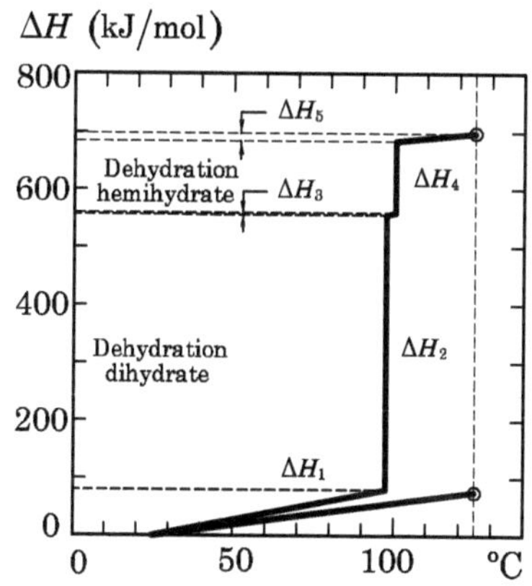

Figure 3.38. The graph shows ΔH for heating of dihydrate and anhydrite, as calculated in example 3.4. The vertical sections of the curve mark the dehydration reactions.

Conditions. According to table information in H.A. Bent: *The Entropy*, the following data are given for the components in the reaction equations (a) and (b)

Substance	M (g/mol)	$H^{\ominus}_{298}$ (kJ/mol)	c_p (J/mol K)
$CaSO_4 \cdot 2H_2O(s)$	172.17	−2022.1	186.3
$CaSO_4 \cdot \frac{1}{2}H_2O(s)$	145.14	−1575.9	119.7
$CaSO_4(s)$	136.14	−1433.4	99.6
$H_2O(g)$	18.02	−241.9	33.6

The molar heat capacity c_p is assumed to be constant in the considered temperature range.

Solution. The amount of substance of reactants in reaction equations (a) and (b) is calculated for the actual mass of 1 kg

$$n_a = \frac{m}{M} = \frac{1000\,\mathrm{g}}{172.17\,\mathrm{g/mol}} = 5.81\,\mathrm{mol} \qquad \text{(dihydrate)}$$

$$n_b = \frac{m}{M} = \frac{1000\,\mathrm{g}}{136.14\,\mathrm{g/mol}} = 7.35\,\mathrm{mol} \qquad \text{(anhydrite)}$$

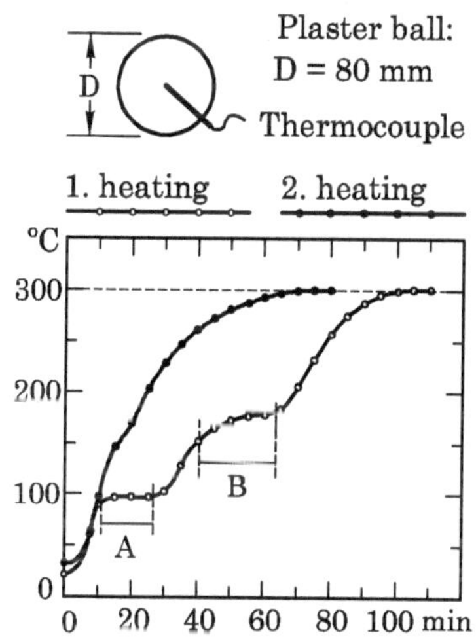

Figure 3.39. Demonstration of the special properties of plaster subjected to fire. A 80 mm-diameter plaster ball is placed in a kiln at 300 °C. During the first heating, the temperature rise is delayed due to dehydration reactions at A and B. At the second heating, the plaster has been transformed into anhydrite and then the temperature rises much faster.

Heating of dihydrate

- By reversible heating from 25 °C to 97 °C, and according to (3.48), the enthalpy increase ΔH_1 is

$$\Delta H_1 = n_a \cdot c_p \cdot \Delta T = 5.81 \cdot 0.1863 \cdot (370 - 298)\,\mathrm{K} = \mathbf{77.9\ kJ}$$

- By reversible transformation of dihydrate into hemihydrate at 97 °C, ΔH_2 is determined using the procedure (3.55)

$$\Delta_r H^{\ominus}_{298} = 1 \cdot (-1575.9) + 1\tfrac{1}{2} \cdot (-241.9) - 1 \cdot (-2022.1)\,\mathrm{kJ/mol} = 83.4\,\mathrm{kJ/mol}$$

$$\Delta_r c_p = 1 \cdot 119.7 + 1\tfrac{1}{2} \cdot 33.6 - 1 \cdot 186.3\,\mathrm{J/mol\,K} = -16.2\,\mathrm{J/mol\,K}$$

$$\Delta_r H_{370} = 83.4 - 16.2 \cdot 10^{-3} \cdot (370 - 298)\,\mathrm{kJ/mol} = 82.2\,\mathrm{kJ/mol}$$

Thus, for the considered amount of substance of 5.81 mol, the transformation enthalpy ΔH_2 is

$$\Delta H_2 = n_a \cdot \Delta_r H_{370} = 5.81\,\mathrm{mol} \cdot 82.2\,\mathrm{kJ/mol} = \mathbf{477.6\ kJ}$$

- By heating of hemihydrate from 97 °C to 100 °C, the enthalpy increase ΔH_3 according to (3.48) is

$$\Delta H_3 = n_a \cdot c_p \cdot \Delta T = 5.81\,\mathrm{mol} \cdot 0.1197\,\mathrm{kJ/mol\,K} \cdot (373 - 370)\,\mathrm{K} = \mathbf{2.1\ kJ}$$

• By reversible transformation of hemihydrate into anhydrite at 100 °C, the enthalpy increase ΔH_4 can be determined by the procedure (3.55)

$$\Delta_r H^{\ominus}_{298} = 1 \cdot (-1433.4) + \tfrac{1}{2} \cdot (-241.9) - 1 \cdot (-1575.9)\,\text{kJ/mol} = 21.6\,\text{kJ/mol}$$

$$\Delta_r c_p = (1 \cdot 99.6 + \tfrac{1}{2} \cdot 33.6) - 1 \cdot 119.7\,\text{J/mol K} = -3.3\,\text{J/mol K}$$

$$\Delta_r H_{373} = 21.6 - 3.3 \cdot 10^{-3} \cdot (373 - 298)\,\text{kJ/mol} = 21.4\,\text{kJ/mol}$$

Thus, for the considered amount of substance of 5.81 mol, the transformation enthalpy ΔH_4 is

$$\Delta H_4 = n_a \cdot \Delta_r H_{373} = 5.81\,\text{mol} \cdot 21.4\,\text{kJ/mol} = \mathbf{124.3\,kJ}$$

• Heating of anhydrite from 100 °C to 125 °C concludes the process; the enthalpy increase ΔH_5 is determined from (3.48)

$$\Delta H_5 = n_a \cdot c_p \cdot \Delta T = 5.81\,\text{mol} \cdot 0.0996\,\text{kJ/mol K} \cdot (398 - 373)\,\text{K} = \mathbf{14.5\,kJ}$$

The total enthalpy increase, and thus the total amount of heat consumed by the heating, can now be determined by addition of the calculated individual contributions, i.e.

$$\Delta H_{total} = 77.9 + 477.6 + 2.1 + 124.3 + 14.5\,\text{kJ} \simeq \mathbf{696.4\,kJ}$$

Heating of anhydrite

• Reversible heating of anhydrite from 25 °C to 125 °C does *not* involve any phase changes. Therefore, from (3.48) we can calculate the total enthalpy increase and thus the total amount of heat consumed by the heating

$$\Delta H_{total} = n_b \cdot c_p \cdot \Delta T = 7.35\,\text{mol} \cdot 0.0996\,\text{kJ/mol K} \cdot (398 - 298)\,\text{K} \simeq \mathbf{73.2\,kJ}$$

Discussion. By reversible heating of 1 kg of **dihydrate** $CaSO_4 \cdot 2H_2O(s)$ from 25 °C to 125 °C, the enthalpy increase, and thus the heat consumption, is approximately **696 kJ**. At a similar heating of 1 kg of **anhydrite** $CaSO_4(s)$, the enthalpy increase, and thus the heat consumption, is approximately **73 kJ**. Thus, for heating of the dihydrate, the heat consumption is almost 10 times that required for heating of anhydrite.

Plaster sheathings mainly consist of fiber-reinforced dihydrate; the fine fire resistance properties are strongly based on the heat-consuming liberation of hydrate water by heating as shown here.

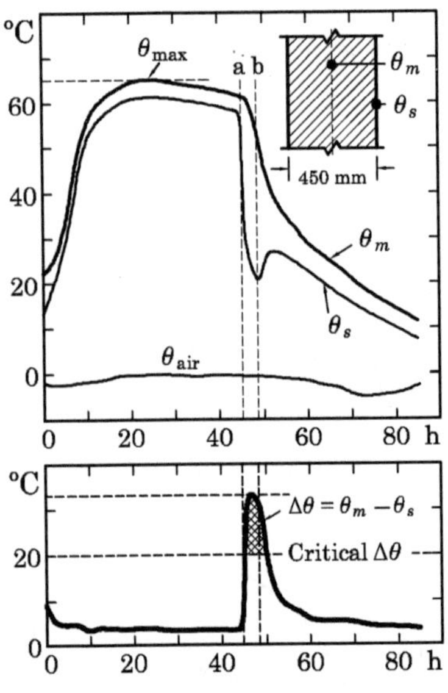

Figure 3.40. Example of temperature development in a hardening concrete wall of thickness 450 mm, which is cast in an insulated form. At (a) the form is removed and at (b) the wall is covered with a tarpaulin. After stripping at (a), the temperature difference $\Delta\theta$ between the middle and the surface is increased to approximately 34 °C, and cracks appear on the concrete surface (Jensen, B.C. & Westergaard, J.: Temperaturmålinger i betonkonstruktioner) (Temperature Measurements in Concrete Structures (in Danish)).

Example 3.5

■ Measurement of hydration heat with a solution calorimeter

The binder in concrete is a reacted mixture of **Portland cement** and **water**, the so-called **cement paste**. The cement reacts chemically with the water and generates hydrates; the cement paste sets by this depositing of hydrates and the newly formed solid then gains strength. The chemical reactions between cement and water are strongly **exothermic**, i.e. heat-releasing. In practice, this is seen as a temperature rise in the hardening concrete. For casting of large, solid concrete structural members with substantial cross-sections, a temperature increase of 50-70 °C in the hardening concrete is not uncommon. This self-heating may cause serious **cracking** in the concrete in case of temperature differences in the different parts of hardened cross-sections.

The temperature rise in the hardening concrete depends on, among other things, the **cement content**, and on the **chemical composition** and **fineness** of the cement. Thus, there are specific cement types produced with particularly low and slow development of heat - low-heat cements - which are used for casting of heavy concrete structures such as dams.

The heat development properties of cement is determined by calorimeter measurements. In practice, **solution** calorimetry, **isothermal** calorimetry and **adiabatic** calorimetry have all been used. The solution calorimeter is interesting in that it produces data that can be directly applied to **thermochemical** calculations.

Problem. In their paper *Studies of the Physical Properties of Hardened Portland Cement* (J.of ACI, Vol 18, No.3), T.C.Powers & T.L.Brownyard mention an investigation of the heat development properties of cement using solution calorimetry. The paper specifies, among other things, the solution heat of a cement paste that is fully dissolved in an acid.

- solution heat, fresh cement paste ($t = 0$ d) 623.0 cal/g cement
- solution heat, hardened cement paste ($t = 27$ d) 509.5 cal/g cement

From the above information, calculate: (a) the quantity Q (J/g) of the heat developed by the cement during 27 days of hardening, and (b) the adiabatic temperature rise $\Delta\theta_a$ after hardening for 27 days of a concrete that contains 375 kg/m^3 of the cement in question!

Figure 3.41. Simple solution calorimeter for measuring the reaction enthalpy ΔH; solution calorimetry is, for example, used for determination of the hydration heat of cement.

Conditions. In the calculations the mass-specific heat capacity c_p of the concrete is estimated to be 1.10 kJ/kg K; the density ϱ of the concrete is estimated to be 2350 kg/m^3.

Solution. For the calculations, the following notation is used: $\{PC\}$ denotes 1 g of unhydrated Portland cement; $\{nH_2O\}$ denotes the amount of water added to 1 g of cement; $\{PC \cdot nH_2O\}$ denotes hardened cement paste after 27 days; $\{A\}$ denotes the amount of acid added to the calorimeter; $\{SOL\}$ denotes the solution formed by cement and acid in the calorimeter. With this notation, this information can be written by the following thermochemical equations:

Solution of fresh cement paste in acid at the time $t = 0$ days

$$\{PC\} + \{nH_2O\} + \{A\} \rightarrow \{SOL\} \qquad \Delta H_a = -623.0\,\text{cal/g} \qquad \text{(a)}$$

Hydration of cement paste from $t = 0$ days to $t = 27$ days

$$\{PC\} + \{nH_2O\} \rightarrow \{PC \cdot nH_2O\} \qquad \Delta H_b = \ ? \qquad \text{(b)}$$

Solution of hardened cement paste in acid at the time $t = 27$ days

$$\{PC \cdot nH_2O\} + \{A\} \rightarrow \{SOL\} \qquad \Delta H_c = -509.5\,\text{cal/g} \qquad \text{(c)}$$

In these three reactions there is a change of state between equilibrium states; the reaction enthalpy ΔH is, therefore, dependent on the process path. Addition of equation (b) and equation (c) results in the net reaction

$$\{PC\} + \{nH_2O\} + \{A\} \rightarrow \{SOL\} \qquad \Delta H_d = \Delta H_b + \Delta H_c \qquad \text{(d)}$$

As the reaction (d) is identical with the reaction (a), the unknown reaction enthalpy ΔH_b for hydration of 1 g of Portland cement can be determined as

$$\Delta H_a = \Delta H_d = \Delta H_b + \Delta H_c$$

$$\Delta H_b = \Delta H_a - \Delta H_c = -623.0 - (-509.5)\,\text{cal/g} = \mathbf{-113.5\,cal/g}$$

The negative reaction enthalpy shows that the hydration is **exothermic**; the curing heat *developed* is: $Q = -\Delta H_b$. Thus, conversion into the unit (J/g) gives

$$Q = -(-113.5\,\text{cal/g}) \cdot 4.184\,\text{J/cal} = \mathbf{475\,J/g}$$

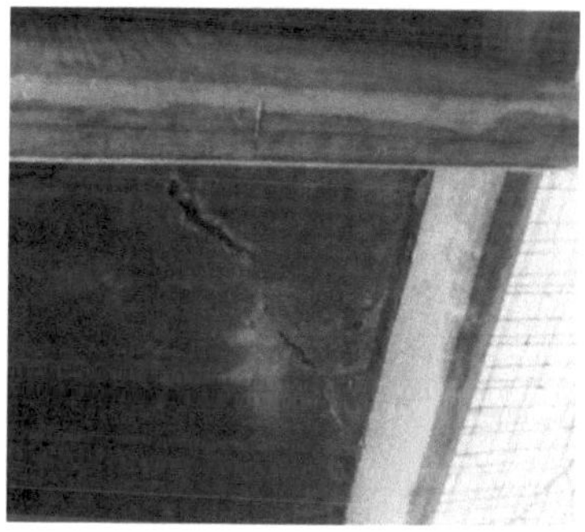

Figure 3.42. Example of cracks in a concrete floor caused by temperature differences in the hardening concrete; the photo shows a newly cast ribbed concrete floor seen from beneath.

Now, we assume a system consisting of 1 m^3 of the specified concrete with a cement content of 375 kg/m^3; the adiabatic temperature rise ΔT (K) can then be determined from

$$\Delta T = \frac{Q}{V \varrho c_p} = \frac{475\,\text{kJ/kg} \cdot 375\,\text{kg}}{1\text{m}^3 \cdot 2350\,\text{kg/m}^3 \cdot 1.1\,\text{kJ/kg K}} \simeq \mathbf{69\,K}$$

Discussion. From the results above, the heat development of the cement after 27 days of hardening is found to be approximately **475 J/g**. For adiabatic hardening of a concrete with 375 kg/m^3 of the cement concerned, a temperature rise of approximately **69 K** can be expected.

It shall be noted that the heat development determined is the sum of the chemical reaction heat from the hydration of the cement, and the adsorption heat released by adsorption of the water on the surface of the cement gel, which is formed (see the discussion of this phenomenon in example 3.3).

Example 3.6

■ Computerized enthalpy calculation

The energy quantities **enthalpy** H and **enthalpy change** ΔH are fundamental calculation parameters in technical calculations of materials. Nearly all calculations dealing with *equilibrium* in, or *transformations* of systems of matter, contain the quantities H and ΔH. Examples include:

- *Phase equilibria in systems of matter*: vapour pressure calculations, freezing and freezing pressure, shrinkage and swelling mechanisms, vapour pressure over capillary systems, phase diagrams, etc.
- *Chemical reactions*: heat of hydration for cement, carbonation of concrete, alkali-silicate reactions, oxidation of metals, etc.
- *Electrochemical processes*: electrochemical corrosion, electrometallurgy and electroforming, hydrogen brittleness, etc.

Many engineering calculations are commonly made on a PC; therefore, it can be desirable to program computer functions in order to utilize thermodynamic table data optimally. The following example shows how to develop a simple function to calculate the enthalpy of substances $H(T)$ from table data for standard enthalpy $H^{\ominus}_{298}$. In the calculation, the temperature dependence of the heat capacity is taken into account so that the best possible determination of the enthalpy of the substance H at the temperature T is achieved with the given table data set.

Vapour pressure curve: H_2O

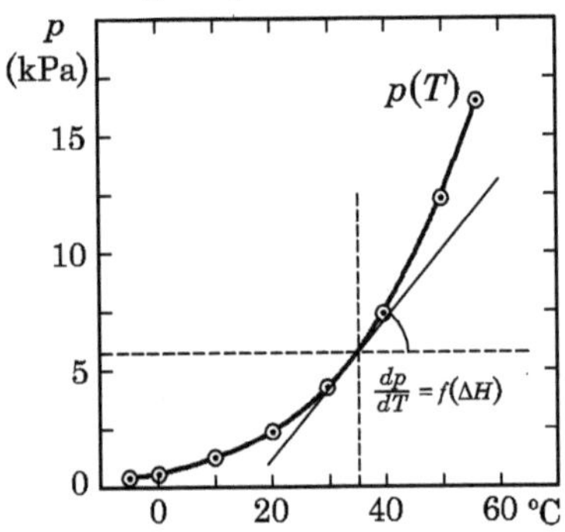

Figure 3.43. The reaction enthalpy ΔH is an important parameter for describing chemical reactions and phase transitions; for example, the slope of the vapour pressure curve is proportional to the evaporation enthalpy ΔH.

Problem. Normally, standard references of thermodynamic substance data specify **standard enthalpy** $H^{\ominus}_{298}$ (kJ/mol) and **molar heat capacity** c_p (J/mol K) at $T =$ 298.15 K (25 °C). Furthermore, in reference tables such as: *Handbook of Chemistry and Physics*, a set of constants (a, b, c, d) for determining the heat capacity c_p as a function of the temperature T are given in the form

$$c_p(T) = f(a, b, c, d) = a + b \cdot T + c \cdot T^2 + d \cdot T^{-2} \quad \text{(a)}$$

where one or more of the constants a, b, c and d are given. Work out and test a computer function

$$fnh(H^{\ominus}_{298},\ T,\ a,\ b,\ c,\ d) \quad \text{(b)}$$

which by input of the arguments: standard enthalpy $H^{\ominus}_{298}$ (kJ/mol), thermodynamic temperature T (K) and the constants a, b, c and d returns the substance enthalpy H_T (kJ/mol) at the given temperature T (K). No specific requirements for optimization of calculation speed are made.

Conditions. In the calculations it is assumed that there are *no* phase changes in the substance within the considered temperature range: 298.15 → T (K). Further, it is assumed that arguments in the function (b) are matching SI units (kJ/mol) and (J/mol K) for $H^{\ominus}_{298}$ and c_p, respectively.

Solution. For calculation of the enthalpy $H(T)$, the expression for the temperature dependence of the enthalpy (3.47) is used. Introducing c_p from (a) in (3.47), we get

$$H(T) = H^{\ominus}_{298} + \int_{298}^{T} (a + b \cdot T + c \cdot T^2 + d \cdot T^{-2})dT \quad \text{(c)}$$

Enthalpy calculation

$$H_T = H^{\ominus}_{298} + \int_{298}^{T} c_p(T)dT$$

By integration and introduction of limits, (c) can then be rewritten to a form which is suitable for programming

$$H(T) = H^{\ominus}_{298} + \left[a \cdot T + \tfrac{1}{2}b \cdot T^2 + \tfrac{1}{3}c \cdot T^3 - d \cdot T^{-1}\right]_{298}^{T} \quad \text{(d)}$$

$$H(T) = H^{\ominus}_{298} + \Big[T \cdot (a + T \cdot (b/2 + T \cdot (c/3))) - d/T\Big]_{298}^{T} \quad \text{(e)}$$

The targeted calculation function $fnh(H,\ T,\ a,\ b,\ c,\ d)$ can now be formulated in a pseudo code which in a simple way can be rewritten to, for example, *Matlab* or *C*.

```
FUNCTION fnh(H, T, a, b, c, d : real) : real;      (* enthalpy function *)
const R = 298.15;                                   (* reference temperature *)
begin                                               (* calculate fnh *)
  fnh:= +(T * (a + T * (b/2 + T * (c/3))) - d/T)
        - (R * (a + R * (b/2 + R * (c/3))) - d/R);
  fnh:=   H + fnh/1000;
end;                                                (* returns value *)
```

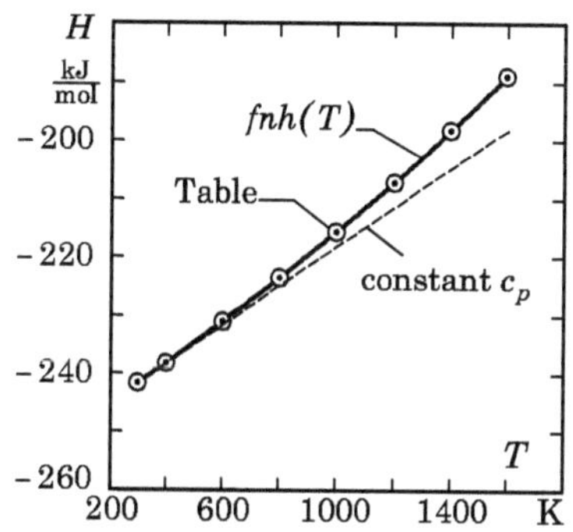

Figure 3.44. Enthalpy of water vapour, example 3.6; the graph shows table values (⊙) compared with a full-drawn curve, which is calculated using fnh(T). The broken line shows H_T calculated with the constant c_p.

The handbook: *Chemical Engineers' Handbook* by Perry & Chilton contains a comprehensive overview of the c_p constants a, b, c and d that are included in $fnh()$. To test the developed function $fnh()$, the following constants for water vapour $H_2O(g)$ from this work of reference are used

$$a = 34.4\,\text{J/mol K}; \quad b = 6.276 \cdot 10^{-4}\,\text{J/mol K}^2; \quad c = 5.609 \cdot 10^{-6}\,\text{J/mol K}^3; \quad d = 0$$

Using $fnh()$ we now calculate the enthalpy H (kJ/mol) for $H_2O(g)$ in the temperature range 298 – 1600 K. The result is compared with the reference table from the internationally recognized work: *Thermochemical Data of Pure Substances* by Barin. Further, the enthalpy H is calculated in the same temperature range with the constant $c_p = 33.6$ J/mol K according to the expression (3.48). The result of this test is shown in the following table.

T (K)	Ref.Table H (kJ/mol)	$fnh()$ (kJ/mol)	$\|\Delta\|$ (kJ/mol)	constant c_p H (kJ/mol)	$\|\Delta\|$ (kJ/mol)
298	−241.826	−241.826	0.000	−241.826	0.000
400	−238.375	−238.230	0.145	−238.404	0.029
600	−231.326	−231.003	0.323	−231.634	0.308
800	−223.825	−223.482	0.343	−224.964	1.139
1000	−215.827	−215.576	0.251	−218.244	2.417
1200	−207.321	−207.197	0.124	−211.524	4.203
1400	−198.334	−198.254	0.080	−204.804	6.470
1600	−188.919	−188.658	0.261	−198.084	9.165

Discussion. The function fnh was tested by calculation of the enthalpy $H(T)$ for water vapour $H_2O(g)$ in the temperature range 298 – 1600 K. In a comparison with recognized table values, satisfactory agreement in the entire temperature range was found; the largest absolute deviation in the range was approximately 0.3 kJ/mol, corresponding to about 0.1 %.

Calculation of the enthalpy $H(T)$ for water vapour assuming constant c_p shows satisfactory agreement up to approximately 600 K (300 °C). Above this value, the deviation between reference value and calculated value is drastically increased. At 1600 K the absolute deviation is approximately 9 kJ/mol, corresponding to about 5 %; this deviation is about 35 times larger than that achieved with $fnh()$. Thus, the test shows that over moderate temperature ranges 298 → T (K), (3.48) with constant c_p can be used for calculation of the enthalpy H_T. For high-temperature processes – i.e. combustion processes, etc. – the enthalpy H_T should be calculated from the expanded expression (3.47) taking into account the temperature dependence of the heat capacity $c_p = c_p(T)$.

Exercises

The following exercises can be used when learning the subjects addressed in Chapter 3. The times given are suggested guidelines for use in assessing the effectiveness of the teaching.

Figure 3.45. Cumulus clouds are created by adiabatic expansion, and thus cooling, of air in thermal upwind; when the air temperature has fallen to the dew point, the water vapour will condense into this well-known cloud formation.

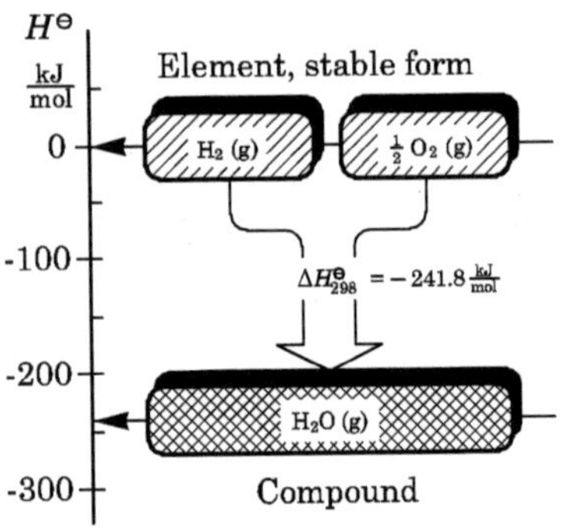

Figure 3.46. The standard enthalpy $H^{\ominus}_{298}$ for compounds is defined as formation enthalpy based on elements in their stable form at 298.15 K; the example shows this definition used for determination of standard enthalpy for water vapour.

Exercise 3.1

☐ Isobaric heating of ideal gas *(10 min.)*

Assume a system containing 8 mol of **ideal gas** at the pressure $p = 0.5$ atm. In the initial state (1), the gas temperature is $\theta = 25\,°C$. By **reversible**, **isobaric** heating the gas temperature is increased to 80 °C in state (2). The molar heat capacity of the gas at constant volume is $c_V = 12.5\,J/mol\,K$. Calculate the supplied heat $Q_{1.2}$ (J), the work done on the system $W_{1.2}$ (J), and the change in the internal energy of the system ΔU (J) during the change of state!

Exercise 3.2

☐ Temperature rise by adiabatic compression *(6 min.)*

In a diesel engine the air is so strongly heated during compression that the fuel is ignited when it is atomized and injected into the cylinder and is brought into contact with the air. Calculate the **temperature rise** for 20 °C atmospheric air by **adiabatic**, **reversible** compression 1:20 ($V_2 = V_1/20$); the air is assumed to be an ideal gas with constant heat capacity $c_p = 29\,J/mol\,K$.

Exercise 3.3

☐ Formation of a chemical compound from elements *(3 min.)*

According to table information, **water vapour** at $p^{\ominus} = 1$ atm and $T = 298.15\,K$ has the **standard enthalpy** $H^{\ominus}_{298} = -241.8\,kJ/mol$. At this temperature, the *stable* form of the elements **hydrogen** and **oxygen** are $H_2(g)$ and $O_2(g)$, respectively. Calculate, based on this information *only*, the heat-release during the reaction

$$3 \text{ mol } H_2(g) + 1\tfrac{1}{2} \text{ mol } O_2(g) \quad \rightarrow \quad 3 \text{ mol } H_2O(g)$$

if the reaction occurs at the temperature $T = 298.15\,K$ and the pressure $p = p^{\ominus}$ atm.

Exercise 3.4

☐ Enthalpy of compound system *(5 min.)*

A system consists of 1000 g of **gypsum** $CaSO_4 \cdot 2H_2O(s)$ and 200 g of **water** $H_2O(\ell)$. For the gypsum, the following table data are given: $H^{\ominus}_{298} = -2022.1\,kJ/mol$ and $M = 172.17\,g/mol$, and for water $H_2O(\ell)$, the following information is available: $H^{\ominus}_{298} = -285.8\,kJ/mol$ and $M = 18.02\,g/mol$. Calculate the total enthalpy content H (kJ) of the system at 25 °C!

Exercise 3.5

☐ Thermochemical equation for hydration of plaster of Paris *(10 min.)*

When **gypsum** is heated to form **hemihydrate** ("plaster of Paris"), the hydrate water is removed in the form of water vapour. At 25 °C, the dehydration can be described by the following thermochemical equation (a)

$$CaSO_4 \cdot 2H_2O(s) \quad \rightarrow \quad CaSO_4 \cdot \tfrac{1}{2}H_2O(s) + 1\tfrac{1}{2}H_2O(g) \qquad \text{(a)}$$

where $\Delta H_a = +83.35\,kJ/mol$. At the same temperature, transformation of water $H_2O(\ell)$ into water vapour $H_2O(g)$ can be described by the following thermochemical equation (b)

$$H_2O(\ell) \quad \rightarrow \quad H_2O(g) \qquad \Delta H_b = +43.9\,kJ/mol \qquad \text{(b)}$$

Based on these thermochemical equations (*Hess's law*), calculate the amount of the **heat developed** during the hydration of 1.000 kg of plaster of Paris according to the following reaction equation

$$CaSO_4 \cdot \tfrac{1}{2}H_2O(s) + 1\tfrac{1}{2}H_2O(\ell) \quad \rightarrow \quad CaSO_4 \cdot 2H_2O(s) \qquad \text{(c)}$$

Exercise 3.6

☐ Isobaric heating of ideal gas *(15 min.)*

A **closed** thermodynamic system contains 3 mol of **ideal gas** at the pressure $p = 1.000$ atm. During a **reversible**, **isobaric** heating, an amount of heat $Q = 482\,J$ is supplied to the system so that the gas temperature is increased from 21.5 °C to 28.8 °C. From this information, calculate: a) the increase in internal energy of the system ΔU (J); b) the increase in enthalpy of the system ΔH (J); c) the molar heat capacity of the gas c_p (J/mol K); d) the molar heat capacity of the gas c_V (J/mol K)!

Exercise 3.7

☐ Isochoric cooling of ideal gas *(12 min.)*

A closed thermodynamic system contains 7 mol of **ideal gas**. By **isochoric** cooling, the amount of heat $Q = -427$ J is supplied to the system, causing the gas temperature to drop from 32.3 °C to 27.5 °C. Calculate, based on this information: a) the increase in internal energy of the system ΔU (J); b) the molar heat capacity of the gas c_V (J/mol K); c) the molar heat capacity of the gas c_p (J/mol K)!

Thermodynamic process types

isobaric	$\Delta p = 0$
isochoric	$\Delta V = 0$
isothermal	$\Delta T = 0$
adiabatic	$Q = 0$

Exercise 3.8

☐ Isothermal expansion of ideal gas *(12 min.)*

In its initial state, a closed thermodynamic system contains $V_1 = 5.000\,\ell$ of **ideal gas** at the pressure $p_1 = 1.000$ atm; the gas temperature is $\theta_1 = 10.0$ °C. By a **reversible, isothermal** change of state the gas volume is increased to $V_2 = 6.000\,\ell$. Based on this information, calculate: a) the gas pressure p_2 (Pa) in the final state; b) the supplied heat $Q_{1.2}$ (J) during the process; c) the work $W_{1.2}$ (J) done on the system during the process!

Exercise 3.9

☐ Adiabatic compression of ideal gas *(12 min.)*

In its initial state, a closed system contains 2 mol of **ideal gas** at the pressure $p_1 = 101325$ Pa; in this state the gas temperature is $\theta_1 = 20.0$ °C. By a **reversible, adiabatic** compression the gas pressure is increased to $p_2 = 251050$ Pa. The molar heat capacity of the gas is $c_V = 12.8$ J/mol K. Based on this information, calculate: a) the gas temperature θ_2 (°C) in its final state; b) the gas volume V_2 (ℓ) in its final state; c) the increase in the internal energy ΔU (J) of the system at the described change of state!

Exercise 3.10

☐ Temperature dependence of enthalpy *(10 min.)*

Calculate from table data for **standard enthalpy** $H^{\ominus}_{298}$ and **molar heat capacity** c_p, the enthalpy H_T (kJ/mol) for the following substances at the given temperatures: a) water $H_2O(\ell)$ at 65 °C; b) copper sulphate trihydrate $CuSO_4 \cdot 3H_2O(s)$ at 0 °C; c) quartz $SiO_2(s)$ at 500 °C; d) white tin Sn(s) at 150 °C.

Exercise 3.11

☐ Calculation of standard reaction enthalpy *(15 min.)*

Calculate from table data for **standard enthalpy** $H^{\ominus}_{298}$, the value of the standard reaction enthalpy $\Delta_r H^{\ominus}_{298}$ (kJ/mol) for the following reactions and transformation processes

$$2Al(s) + 1\tfrac{1}{2}O_2(g) \rightarrow Al_2O_3(s) \quad (a)$$
$$CaO(s) + H_2O(\ell) \rightarrow Ca(OH)_2(s) \quad (b)$$
$$CO_2(g) + H_2O(\ell) \rightarrow H^+(aq) + HCO_3^-(aq) \quad (c)$$
$$CuSO_4 \cdot 3H_2O(s) \rightarrow Cu^{++}(aq) + SO_4^{--}(aq) + 3H_2O(\ell) \quad (d)$$

Reaction enthalpy

$$\Delta_r H^{\ominus}_{298} = \sum H^{\ominus}_{\text{prod}} - \sum H^{\ominus}_{\text{react}}$$

Exercise 3.12

☐ Temperature dependence of reaction enthalpy *(12 min.)*

Calculate from table data for **standard enthalpy** $H^{\ominus}_{298}$ and **heat capacity** c_p, the value of the reaction enthalpy $\Delta_r H_T$ (kJ/mol) at the given temperature T (K) for the following reactions and transformations

$$T = 1100\,\text{K}: \quad CaCO_3(s) \rightarrow CaO(s) + CO_2(g) \quad (a)$$
$$T = 664\,\text{K}: \quad Hg(\ell) \rightarrow Hg(g) \quad (b)$$
$$T = 800\,\text{K}: \quad Ca(OH)_2(s) \rightarrow CaO(s) + H_2O(g) \quad (c)$$

In the calculations, constant heat capacity c_p in the relevant temperature ranges is assumed.

Figure 3.47. Flame cutting of steel. A sharply delimited jet of pure oxygen burns (oxidizes) and melts away the steel in the cutting zone; the process is strongly exothermic so that the oxygen supply ensures continued heating of the steel in the cutting zone.

Exercise 3.13

☐ Deoxidation of steel with aluminium *(5 min.)*

In the manufacture of steel, **aluminium** Al is used as a powerful **deoxidation agent**. The deoxidizing effect is caused by the great affinity of aluminium for oxygen. Deoxidation of ferric oxide Fe_2O_3 takes place by the following transformation

$$Fe_2O_3(s) + 2Al(s) \quad \rightarrow \quad Al_2O_3(s) + 2Fe(s) \qquad \text{(a)}$$

The reaction (a) is strongly **exothermic**. Calculate $\Delta_r H^\ominus_{298}$ for the reaction (a), based on the following data for the standard enthalpy of the components

Substance:	$Fe_2O_3(s)$	$Al(s)$	$Al_2O_3(s)$	$Fe(s)$
$H^\ominus_{298}$ (kJ/mol)	−824.2	0.0	−1675.7	0.0

Exercise 3.14

☐ Flame cutting of steel *(8 min.)*

Flame cutting is often used for shaping of heavy specimens of steel. The principle of flame cutting is as follows: A special gas burner is used for preheating the steel with a flame of **acetylene** and **oxygen**; when the steel has reached a temperature that is slightly below the melting temperature, a sharp jet of pure oxygen is directed towards the preheated steel that now oxidizes ("burns") by the following reaction

$$2Fe(s) + 1\tfrac{1}{2}O_2(g) \quad \rightarrow \quad Fe_2O_3(s) \qquad \text{(a)}$$

The reaction (a) is strongly **exothermic** so that the developed heat ensures continued heating and burning of the steel. During cutting, the oxides formed are blown away from the fire zone in the form of melts. The following thermodynamic data are given $Fe_2O_3(s): H^\ominus_{298} = -824.2\,\text{kJ/mol}$; $Fe(s): c_p = 25.2\,\text{J/mol K}$. Based on this, calculate $\Delta_r H^\ominus_{298}$ for the reaction (a), and calculate for constant c_p the amount of heat (kJ) to be supplied with 2 mol of Fe to increase the temperature from 25 °C to the melting point of iron! – compare and discuss the results briefly!

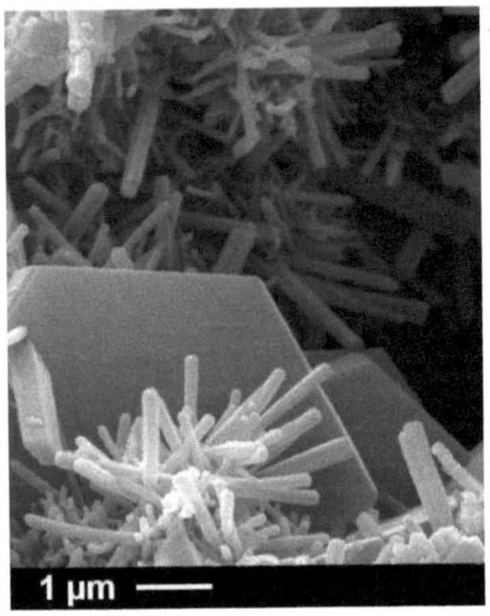

Figure 3.48. Needle-like ettringite is formed by reaction between the clinker mineral C_3A and gypsum, both are components of Portland cement.

Exercise 3.15

☐ Ettringite formation by hydration of tricalcium aluminate *(12 min.)*

Typically, Portland cement contains approximately 7 wt-% of **tricalcium aluminate** $3CaO{\cdot}Al_2O_3(s)$. *Tricalcium aluminate reacts very quickly when mixed with water; the mix sets and hardens within a few minutes. Therefore, in order to make Portland cement fit for manufacture of concrete, it is necessary to retard the reaction between tricalcium aluminate and water. This retardation – delay – is achieved by grinding 2 – 5 wt-% of* **gypsum** $CaSO_4 \cdot 2H_2O(s)$ *together with the cement. When water is added the gypsum reacts with tricalcium aluminate and forms the hydrate* **ettringite** $3CaO \cdot Al_2O_3 \cdot 3CaSO_4 \cdot 31H_2O$ *by the following reaction*

$$3CaO \cdot Al_2O_3(s) + 3(CaSO_4 \cdot 2H_2O(s)) + 25H_2O(\ell)$$
$$\rightarrow \quad 3CaO \cdot Al_2O_3 \cdot 3CaSO_4 \cdot 31H_2O(s) \qquad \text{(a)}$$

The reaction products precipitate on the surface of the reactive tricalcium aluminate so that the reaction is gradually slowed. In: Thermodynamik der Silikate, Mchedlov-Petrossyan specify the following values of the standard enthalpy of the components

Substance:	$3CaO \cdot Al_2O_3(s)$	$CaSO_4 \cdot 2H_2O(s)$	$H_2O(\ell)$	*"Ettringite"*
$H^\ominus_{298}$ (kJ/mol)	−3558.1	−2022.1	−285.9	−17208.1

Calculate the *developed* heat in (J) per gram of tricalcium aluminate by the reaction (a)! – The molar mass of tricalcium aluminate is 270.18 g/mol.

Exercise 3.16

☐ Calorific value of welding gases *(15 min.)*

In **gas welding** ("autogenous welding") of steel, the welding heat is produced by combustion of an inflammable gas with pure oxygen. Normally **acetylene** $C_2H_2(g)$ or **hydrogen** $H_2(g)$ is used as the fuel gas. For welding of load-bearing structures, gas welding has gradually been superseded by electric arc welding. One of the reasons for this is that gas welding produces an undesirably large diffusion of heat in the steel around the welding zone. On the other hand, gas burners are used extensively for **preheating** of metals for welding and in connection with bending of specimens.

Complete combustion of acetylene $C_2H_2(g)$ and hydrogen $H_2(g)$ in pure oxygen follow the reaction equations:

$$C_2H_2(g) + 1\tfrac{1}{2}O_2(g) \rightarrow 2CO_2(g) + H_2O(g) \quad \text{(a)}$$
$$H_2(g) + \tfrac{1}{2}O_2(g) \rightarrow H_2O(g) \quad \text{(b)}$$

- Calculate the theoretical calorific value $Q = -\Delta_r H^\ominus_{298}$ of acetylene from reaction equation (a) and of hydrogen from reaction equation (b); unit (kJ/kg)
- Calculate the calorific value per m^3 $C_2H_2(g)$ and per m^3 $H_2(g)$ at 20 °C and 1 atmosphere, when the gases are assumed to comply with the ideal gas law; unit (kJ/m^3).

Exercise 3.17

☐ Calculation of enthalpy with temperature-dependent heat capacity *(20 min.)*

For thermochemical calculations close to the reference temperature 298 K (25 °C), one can, to a good approximation, assume that the heat capacity c_p of any substance is constant. The enthalpy H_T is then determined from the expression (3.48)

$$H_T = H^\ominus_{298} + c_p \cdot (T - 298.15) \quad \text{(kJ/mol)} \quad \text{(a)}$$

If the temperature deviates much from the reference temperature 298 K, it can be necessary to take the temperature dependence of the heat capacity into account; in this case the enthalpy is determined by the expression (3.47)

$$H_T = H^\ominus_{298} + \int_{298.15}^{T} c_p(T)dT \quad \text{(kJ/mol)} \quad \text{(b)}$$

In *Thermodynamik der Silikate*, Mchedlov-Petrossyan lists the following thermodynamic standard values of **calcite** $CaCO_3(s)$: $H^\ominus_{298} = -1207.5$ kJ/mol; $c_p = 81.9$ J/mol K at 298 K. The temperature-dependence of the heat capacity for calcite is specified as

$$c_p(T) = 104.6 + 0.0219 \cdot T - 26 \cdot 10^{5} \cdot T^{-2} \quad \text{(J/mol K)} \quad \text{(c)}$$

Calculate according to (a) and (b), respectively, the enthalpy H_T of calcite at $\theta =$ 100 °C, 500 °C and 1000 °C, and determine the deviation between these results in per cent!

Exercise 3.18

☐ Stretching of a metal wire *(12 min.)*

A 15 m long 1 mm-diameter wire of hardened steel is by a **reversible, adiabatic** elongation subjected to a **tensile force** F, that increases from $F_1 = 0$ N in its initial state (1) to $F_2 = 600$ N in its final state (2). The **stress-strain relation** for the steel is given by the following relation between **stress** σ (MPa) and strain ε

$$\sigma = 0.20 \cdot 10^6 \cdot \varepsilon - 6.0 \cdot 10^6 \cdot \varepsilon^2 \quad \text{(MPa)} \quad \text{(a)}$$

Calculate, for the change of state: a) the amount of work done $W_{1.2}$ (J) on the wire, and b) the change of the internal energy ΔU (J) of the wire during the process!

Exercise 3.19

☐ Heat development by hydration of silica fume *(15 min.)*

Silica fume – a residual product from the manufacture of silicon or ferrosilicon – is used as an additive for concrete, in quantities up to 10 % of the cement weight. Addition of silica fume results in a significant change of the strength properties of the concrete, as well as a dense and durable concrete. Mainly, silica fume consists of amorphous silicon dioxide SiO_2(amorphous); in the strongly alkaline environment – pH $\simeq$ 13.5 – of the concrete, silica fume reacts with calcium hydroxide $Ca(OH)_2$ derived from the cement hydration, and thus forms calcium silicate hydrates with binder properties. The following reaction is an imperfect, but useful approximation of the reaction that occurs:

$$3Ca(OH)_2(s) + 2SiO_2(\text{amorphous}) \rightarrow 3CaO \cdot 2SiO_2 \cdot 3H_2O(s) \quad \text{(a)}$$

The hydrate $3CaO \cdot 2SiO_2 \cdot 3H_2O(s)$ that is formed is C-S-H gel and is similar to the C-S-H gel that is formed by the hydration of calcium silicate compounds in Portland cement. According to *Thermodynamik der Silikate* by Mchedlov-Petrossyan, $H^\ominus_{298} = -847.7$ kJ/mol for SiO_2(amorphous) and $H^\ominus_{298} = -4682.0$ kJ/mol for tobermorite $3CaO \cdot 2SiO_2 \cdot 3H_2O(s)$. (Note: Tobermorite is a crystalline form of calcium silicate somewhat similar to C-S-H gel).

a) Calculate $\Delta_r H^\ominus_{298}$ (kJ/mol) for the hydration reaction (a), assuming that the C-S-H gel produced has the same properties as tobermorite; and b) calculate the amount of heat Q (J) *developed* by complete transformation of 1 g of SiO_2(amorphous) according to reaction equation (a)!

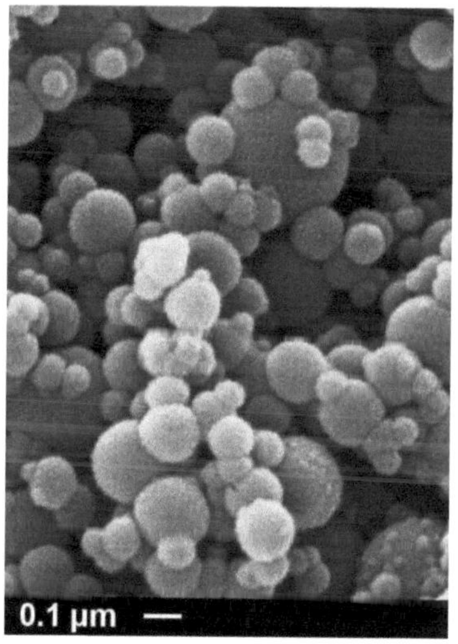

Figure 3.49. The particle dimension of silica fume particles is typically 0.1 µm. As the typical particle dimension of cement is 10 µm, silica fume can fill voids between cement particles and thus produce an extremely dense and high-strength binder phase.

Exercise 3.20

☐ **Calcination of calcium carbonate by lime burning** *(15 min.)*

In its initial state, a thermodynamic system consists of 2 moles of calcium carbonate $CaCO_3$ confined at 25 °C by a frictionless piston. The system is subjected to reversible heating. As a result, the calcium carbonate starts to decompose ("calcination") at 800 °C according to the following reaction equation

$$CaCO_3(s) \quad \rightarrow \quad CaO(s) + CO_2(g) \tag{a}$$

Volume work

$$W_{1.2} = -\int_{V_1}^{V_2} p(V)\,dV$$

During the decomposition, the external pressure on the piston is maintained constant at 1 atm. By a reversible work process, the system undergoes thermal expansion at 800 °C during the formation of $CO_2(g)$. The CO_2 gas formed is assumed to be ideal, and the volumes of the solid phases should be disregarded. Based on these assumptions, calculate a) the work W (J) done on the system by the surroundings when the calcination of $CaCO_3(s)$ is complete; b) the reaction enthalpy $\Delta_r H_T$ (kJ/mol) for the transformation (a) at 800 °C !

Literature

The following literature and supplementary readings are recommended for further reference on the subjects covered in Chapter 3.

References

In Chapter 3, a number of references to the literature have been made; the following is a complete list of citations of the references used.

- Atkins, P.W.: *Physical Chemistry*, Oxford University Press 1986.
- Mchedlov-Petrossyan: *Thermodynamik der Silikate*, VEB Verlag für Bauwesen, Berlin 1966.
- Bent, H.A.: *The Second Law; an Introduction to Classical and Statistical Thermodynamics*, Oxford University Press, New York 1965.
- Kaye, G.W.C. & Laby, T.H.: *Tables of Physical and Chemical Constants*, 5th ed., Longman, Harlow, 1992
- Powers, T.C. & Brownyard, T.L.: *The Thermodynamics of Adsorption of Water on Hardened Cement Paste*, Journal of ACI, Vol.18, No.5 1947.
- Schwiete, H.E. & Knauf, A.N.: *Gips – Alte und neue Erkentnisse in der Herstellung und Anwendung der Gipse*, Merziger Verlag (undated).
- Powers, T.C. & Brownyard, T.L.: *Studies of the Physical Properties of Hardened Portland Cement Paste*, Journal of ACI, Vol.18, No.3 1946.
- Jensen, B.C. & Westergaard, J.: *Temperaturmålinger i betonkonstruktioner* (Temperature Measurements in Concrete Structures)(in Danish), Nordisk Beton, 4 1979.
- Weast, R.C. (ed.): *CRC Handbook of Chemistry and Physics*, CRC Press, Inc., Florida 1983.
- Perry, R.H. & Chilton, C.H. (ed.): *Chemical Engineers' Handbook*, 5.th. ed., McGraw-Hill Kogakusha LTD. 1973.
- Barin, I.: *Thermochemical Data of Pure Substances*, Part I, VCH Verlagsgesellschaft, Weinheim 1989.

Supplementary literature

A comprehensive collection of data of elements and compounds is given in

- Barin, I. & Knacke, O.: *Thermochemical Properties of Inorganic Substances*, Springer-Verlag, Berlin 1973.

It shall be noted that these tables contain constants for calculation of the temperature dependence $c_p(T)$ of specific heat for a large number of substances. The tables

- Barin, I.: *Thermochemical Data of Pure Substances*, Part I, Part II, VCH Verlagsgesellschaft, Weinheim 1989.

contain an updated list of thermochemical data for approximately 2400 elements and compounds; it should be noted that these tables contain data for a large number of oxides and hydrates which are included in cement-chemical calculations. The same applies to the books

- Mchedlov-Petrossyan: *Thermodynamik der Silikate*, VEB Verlag für Bauwesen, Berlin 1966.
- Babushkin, Matveyev & Mchedlov-Petrossyan: *Thermodynamics of Silicates*, Springer-Verlag, Berlin 1985.

which specifically address the compounds of the cement system. Updated tables containing a comprehensive collection of physical and thermochemical data for water and water vapour can be found in

- Schmidt, E. & Grigull, U. (ed.): *Properties of Water and Steam in SI-Units*, Springer-Verlag, Berlin 1989.

Double-acting Watt steam engine from the late 18th century; within a few years of its introduction, Watt's steam engine ousted all existing Newcomen engines because of its far better thermal efficiency.

CHAPTER 4

Second law

The second law of thermodynamics (along with the concept of entropy) was a turning point in the scientific research of the 1800s. The second law proved to be one of the most fruitful and extensive axioms in physics.

The origins of the second law of thermodynamics were investigations by the French engineer **Sadi Carnot** (1824) into the ability of heat to do work. Carnot's original theory was that the work capacity of heat was solely based on the change from a higher to a lower temperature of an imperishable heat substance.

Around 1850, however, the German physicist **Rudolf Clausius** and the English physicist **William Thomson** (later Lord Kelvin) independently showed that the concept of energy conservation implied that the work capacity of heat included the actual conversion of heat into work. Clausius and Thomson each independently formulated the limits for energy conversion processes of the second law. In 1865, Clausius postulated the fundamental principle of the constant increase of entropy.

Like the first law of thermodynamics, the second law is a postulate, based on experience and *cannot* be proved. The second law formed the basis for treating the concept of equilibrium within physical chemistry. In this chapter the second law is introduced along with the use of the concept of equilibrium; their applications are illustrated by practical problems within the science of construction materials.

Rudolf Clausius (1822-1888)

German physicist, known for developing the second law of thermodynamics, including the introduction of the concept of entropy.

Contents

4.1 Introduction

The **first law** of thermodynamics can be seen as an expansion of the energy principles of mechanical physics. The first law expresses that the total energy – *all* forms considered – is **constant**. For a given process, therefore, the first law can ensure **energy conservation**; however, there is *no* guarantee that processes fulfilling the first law are **possible**.

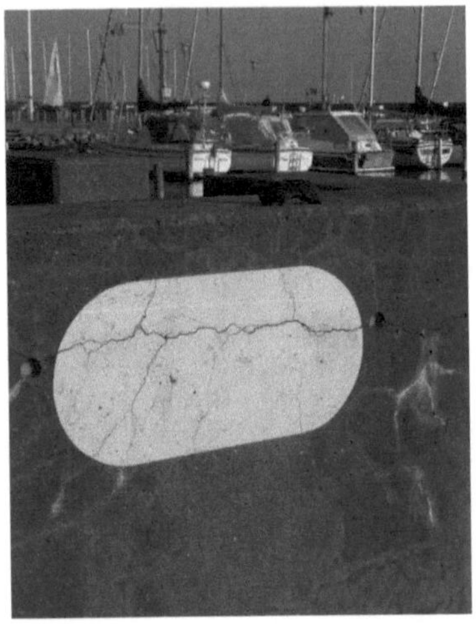

Figure 4.1. Decay of concrete wall due to alkali reactions in the aggregate used; improving the durability of concrete requires knowledge of the direction of chemical reactions.

Spontaneous processes

Some processes run **spontaneously** – others do not. In some cases, the development of a process can be predicted from experience. If, for example, a warm block of metal is placed in a refrigerator, the block will spontaneously release heat to its surroundings. The temperature of the block will decrease until the block has reached thermal equilibrium with its surroundings in the refrigerator. The opposite process – *that* the block will spontaneously absorb heat and obtain a higher temperature than its surroundings – will not occur, according to *experience*.

If the interior of an evacuated container is brought into contact with the atmosphere, we *know* that air will spontaneously flow into the container until equilibrium of pressure is obtained. The opposite process – *that* air will spontaneously flow out of the container and leave a vacuum – is inconsistent with *experience*.

In both of the above examples we *intuitively* know which process will be spontaneous and which process is impossible. However, for a number of chemical reactions and transformations that are important in science, we *cannot* lean on simple intuition or experience. For example, what are the conditions under which calcium hydroxide will absorb CO_2 from the air so that a mortar hardens? What is the condition that water will evaporate from a newly cast concrete surface and thus impair the concrete quality? And what conditions lead to transformation of concrete reinforcement into rust so that the load-carrying capacity of a structure is impaired?

For such questions, we cannot refer to the **first law** of thermodynamics. The first law does *not* distinguish between possible and impossible processes; the first law only describes the changes of energy in a thermodynamic system *if* a specified process develops. The first law is the **energy account** of thermodynamics, i.e. the bookkeeping that monitors the conservation of energy.

Viewed in this light, the **second law** of thermodynamics has a special meaning because it tells us whether a system is in **equilibrium**, i.e. whether it can change **spontaneously**, or whether spontaneous change is **impossible**.

In applying the first law, two important state functions have been introduced and defined, namely the **internal energy** U and the **enthalpy** H. The second law introduces and defines a new fundamental state function: **entropy** S. Before introducing this new state function, however, it is necessary to define some previously mentioned concepts.

Figure 4.2. Corrosion of reinforcement in a balcony parapet makes the concrete spall; prevention of corrosion damage in structures requires knowledge of the direction of chemical reactions.

Thermodynamic equilibrium

Classical thermodynamics describes **states of equilibrium** and **processes** that connect states of equilibrium. **State functions** such as **internal energy** U and **enthalpy** H, which are independent of the process path in a change of state, are used.

The macroscopic quantities used for describing the state of internal energy of a system are called **state variables**. For an ideal gas, these may, for example, be **volume** V, **pressure** p, **temperature** T, and **amount of substance** n. Added to these are the mechanical quantities required to determine the macroscopic potential and kinetic energy, i.e. quantities such as the **velocity** v and **position** (x, y, z) of the system. In the following, reference will mainly be made to the internal energy of systems.

It is characteristic of a thermodynamic state of equilibrium that the considered system is in **thermal** equilibrium, in **mechanical** equilibrium, and in **chemical** equilibrium at the same time.

Thermodynamic equilibrium

= *Thermal equilibrium*
+ *Mechanical equilibrium*
+ *Chemical equilibrium*

Thermodynamic equilibrium (4.1)

A system is in thermodynamic equilibrium if, at the same time, it is in

Thermal equilibrium: the *same* temperature T in the entire system.

Mechanical equilibrium: no *unbalanced* forces occur.

Chemical equilibrium: no *spontaneous* transformations can occur in the system.

A system in thermodynamic equilibrium cannot change state spontaneously.

Thermodynamic process

The processes connecting two different thermodynamic states of equilibrium can be divided into two types: **reversible** processes and **irreversible** processes.

The **reversible** process is a theoretical, idealized process. During a reversible process, the system undergoes states that deviate only *infinitesimally* from equilibrium. By a reversible heat exchange, the heat flow to the system is assumed to be caused by an infinitesimal temperature difference dT between a system and its surroundings. Correspondingly, by a reversible compression or expansion of a gas, it is assumed that only infinitesimal difference in pressure dp occurs between a system and its surroundings.

Thus, during the reversible process, the system undergoes a sequence of states, all of which are infinitely close to true thermodynamic equilibrium. Therefore, the reversible process can be unambiguously depicted in a diagram, for example, as a curve showing a reversible, isothermal change of state for a gas in a pV diagram.

As the designation indicates, a reversible process can be *reversed* by an infinitesimal change of the influence exerted on the system by the surroundings. In a reversible *cyclic process* returning a system to its initial state, no permanent change of the thermodynamic universe will occur.

All processes that are *not* reversible are called **irreversible** processes. The following examples can illustrate the nature of the irreversible process. Work W is done on a system; the work is wholly or partly transformed into heat by friction within the system – the process *cannot* be reversed, it is irreversible. A gas expands into a vacuum; during the expansion, pressure and temperature vary from one location to another inside the gas and thus, the gas is not in equilibrium during the expansion – the process *cannot* be reversed; it is irreversible. A system at low temperature absorbs heat from surroundings with a high temperature; during the heat exchange, the system is *not* in thermodynamic equilibrium – the process *cannot* be reversed; it is irreversible.

Always, if the initial state of the system is re-created after an irreversible cyclic process, a permanent change of the thermodynamic universe will have occurred. This change – an increase of the **entropy** of the universe – forms the basis of the **second law** of thermodynamics.

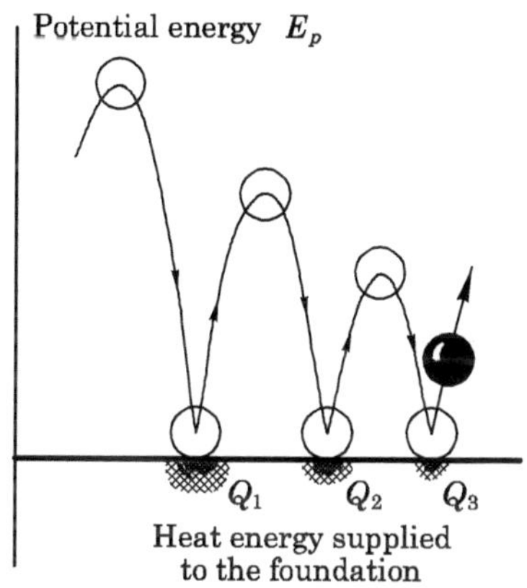

Figure 4.3. A ball changing from jumping to a state of rest is an example of an irreversible process. The ball's organized movement is spontaneously transformed into heat, i.e. disorganized, thermal molecular movements in the ball and its foundation. The opposite process, that disorganized molecular movements are spontaneously transformed into an organized, macroscopic movement, is not known.

■ An insulated, closed glass flask contains 100 mℓ of water $H_2O(\ell)$ and 100 mℓ of water vapour $H_2O(g)$. The overall temperature in the system is 35 °C. The partial pressure of water vapour in the flask is $p = 4500$ Pa, and the system is at rest. Find out, whether the system is in thermodynamic equilibrium!

Solution. We analyse for thermodynamic equilibrium using (4.1). The given information shows that there is **thermal equilibrium,** since the temperature is 35 °C all over the system. The system is at rest, so there is nothing to indicate the presence of *unbalanced* forces in the system, i.e. the system is in **mechanical equilibrium**. The partial pressure of the water vapour in the system is $p = 4500\,\text{Pa}$. At $35\,^\circ\text{C}$, the partial pressure of **saturated water vapour**, cf. the table on water and water vapour in Appendix B, is $5626.7\,\text{Pa}$. As $p < p_s$, a *spontaneous* evaporation of water will take place until the partial pressure is equal to p_s; during this process, the temperature of the system will decrease a little due to the heat consumed by evaporation. Therefore, the system is *not* in **chemical equilibrium**. Conclusion: the described system is **not** in thermodynamic equilibrium!

☐ **1.** A flask contains 100 mℓ of water mixed with 1 g of NaCl crystals. Left at rest, NaCl is slowly dissolved into the water. Is the process reversible or irreversible?

First law

$dU = \delta Q + \delta W$

☐ **2.** An isolated thermodynamic system is in thermodynamic equilibrium. Will it be possible to create a reversible or an irreversible process in this system ?

☐ **3.** Explain why the first law cannot be used to decide whether a process in a thermodynamic system can run spontaneously?

☐ **4.** For a system to be in thermodynamic equilibrium, explain whether or not the pressure p shall be uniform all over the system?

☐ **5.** A closed glass flask is partly filled with water at $40\,^\circ\text{C}$. State at which partial pressures p of $H_2O(g)$ evaporation of H_2O will be a) reversible b) irreversible c) impossible!

4.2 The Carnot cycle

The **second law** of thermodynamics stems from the studies in the 1800's of heat engines, and in the this period's theories on the **motive power of heat**. Therefore, the second law is introduced, along with the concept of entropy, through the **Carnot cycle** for a heat engine operating with an ideal gas. The energy considerations used in the Carnot process are universal and thus they lead to general conditions of equilibrium for thermodynamic systems.

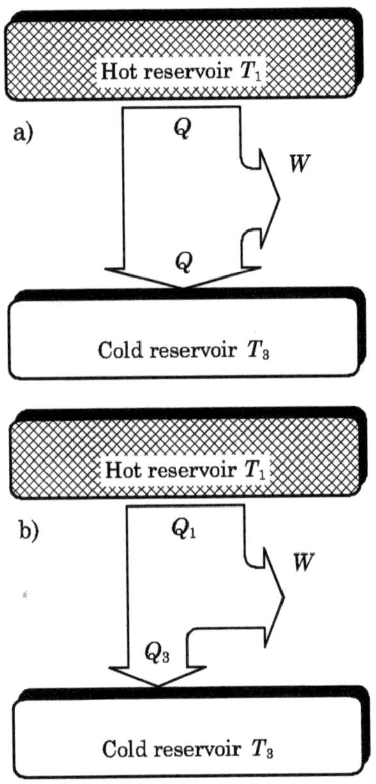

Figure 4.4. Transformation of work into heat. Sadi Carnot's original theory was that the work capacity was based on the change of an imperishable "heat substance" from a high to a lower temperature (a). Later, Rudolf Clausius and William Thomson proved that the work capacity, in accordance with the theorem on energy conservation, involved real transformation of heat into work (b).

Thermal efficiency

According to the first law of thermodynamics, **heat** Q and **work** W are equivalent forms of energy exchange between a system and its surroundings. According to the first law, it is, therefore, possible

- *to* supply energy to a system in the form of **mechanical work** W and then remove the *same* amount of energy in the form of **heat** Q.
- *to* supply energy to a system in the form of **heat** Q and then remove the same amount of energy in the form of **mechanical work** W.

However, experience shows that there is a fundamental *qualitative* difference between **work** W and **heat** Q: work that is done can always be *completely* transformed into heat, and heat can only partially be transformed into work.

With the technical development of the steam engine, this limitation of the motive power of heat was an urgent question in the beginning of the 1800s, i.e. before the formulation of the first law. In 1824, the French engineer **Sadi Carnot** presented his theory on the motive power of heat – this theory later formed the basis for the second law of thermodynamics, and for the definition of the concept of entropy.

Carnot's work aimed at clarifying, among other things, how large a proportion of the supplied heat could be transformed into mechanical useful work in an ideal heat engine. This can be expressed by the **thermal efficiency** of the machine, defined in the following way

> **Thermal efficiency** (4.2)
>
> The thermal efficiency η of a periodically working heat engine, which by a cyclic process is supplied with the **heat** Q_1 at the **temperature** T_1 and thus does the **work** $-W_{cycle}$ on its surroundings, is defined by
>
> $$\eta = \frac{-\text{Work performed on surroundings}}{\text{Heat absorbed from surroundings}} = \frac{-W_{cycle}}{Q_1}$$

In the following derivation we shall see how the reply to this technical question reveals the existence of one of the most fundamental state functions, the **entropy** S.

The Carnot cycle

The Carnot cycle is a **reversible cyclic process** during which a thermodynamic system cycles between two heat reservoirs of different temperatures. A Carnot process with an **ideal gas** will be described here as an illustration, but as will be shown later, there is *no* restriction on the nature of the system.

Assume a **closed system** of 1 mol of an **ideal gas**, which is confined in a cylinder below a non-frictional piston. In its surroundings there are two heat reservoirs, one with constant, **high temperature** T_1, and one with constant, **low temperature** T_3. In its initial state (a), the system is in contact with the heat reservoir; the gas temperature is T_1 and the volume is V_a. Now the following **reversible** cyclic process is performed

process 1 Reversible **isothermal expansion**: $V_a \rightarrow V_b;\quad T = T_1$
The system performs work *on* its surroundings and is supplied with heat from reservoir 1 with high temperature T_1.

process 2 Reversible **adiabatic expansion**: $V_b \rightarrow V_c;\quad T_1 \rightarrow T_3$
The system performs work *on* its surroundings. No heat is exchanged during the process; therefore, the system temperature drops to T_3 by the adiabatic expansion.

process 3 Reversible **isothermal compression**: $V_c \rightarrow V_d;\quad T = T_3$
Work is performed by the surroundings *on* the system which at the same time releases heat to the reservoir with low temperature T_3.

process 4 Reversible **adiabatic compression**: $V_d \rightarrow V_a;\quad T_3 \rightarrow T_1$
Work is performed by the surroundings *on* the system. No heat is exchanged during the process; therefore, the system temperature is increased to T_1 by the adiabatic compression.

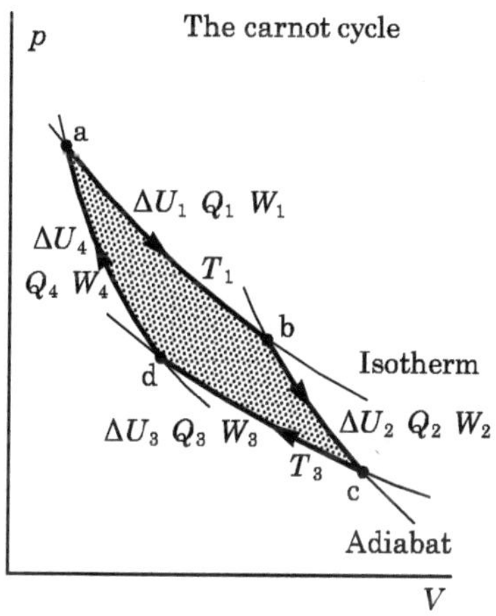

Figure 4.5. Reversible Carnot cycle abcda with an ideal gas; the work performed on the system during the cyclic process corresponds to the shaded area in the pV diagram.

After this sequence of processes, the initial state (a) is restored; therefore, taken as a whole, the processes constitute a **reversible cyclic process**. From the **first law**, it is now possible to set up a total energy account for the cyclic process

Process	ΔU	W	Q	Comments
1	0	W_1	Q_1	$\Delta U = 0$ cf. (3.27)
2	ΔU_2	W_2	0	$Q = 0$ cf. (2.27)
3	0	W_3	Q_3	$\Delta U = 0$ cf. (3.27)
4	ΔU_4	W_4	0	$Q = 0$ cf. (2.27)

The internal energy U is a **state function**; after the cyclic process, therefore, U is unchanged. By addition of all ΔU contributions for the cyclic process, the following condition is fulfilled

$$\sum \Delta U = \Delta U_2 + \Delta U_4 = W_2 + W_4 = 0 \tag{4.3}$$

The total work performed *on* the system during the cyclic process, W_{cycle}, can then be expressed by

$$W_{cycle} = (W_2 + W_4) + (W_1 + W_3) = 0 + (W_1 + W_3) = W_1 + W_3 \tag{4.4}$$

For process 1 and process 3, $W = -Q$ according to (3.27), and thus

$$W_{cycle} = W_1 + W_3 = -(Q_1 + Q_3) \tag{4.5}$$

Figure 4.6. One of James Watt's first steam engines; James Watt established the systematic development of heat engines.

By introduction into (4.2), the **thermal efficiency** of the cyclic process can now be expressed

$$\text{Thermal efficiency}: \quad \eta = \frac{-W_{cycle}}{Q_1} = \frac{Q_1 + Q_3}{Q_1} = 1 + \frac{Q_3}{Q_1} \tag{4.6}$$

The ratio (Q_3/Q_1) is solely determined by the ratio between the temperatures T_3 and T_1 in the two heat reservoirs; this can be seen by introduction of the previously derived expression for Q by an isothermal change of state with an ideal gas (3.27)

$$1 + \frac{Q_3}{Q_1} = 1 + \frac{RT_3 \cdot \ln(V_d/V_c)}{RT_1 \cdot \ln(V_b/V_a)} = 1 + \frac{T_3 \cdot \ln(V_d/V_c)}{T_1 \cdot \ln(V_b/V_a)} \tag{4.7}$$

According to the adiabatic equation (3.37), the volume conditions fulfil the following condition

$$T_1 \cdot V_b^{\gamma-1} = T_3 \cdot V_c^{\gamma-1}; \quad \text{and} \quad T_1 \cdot V_a^{\gamma-1} = T_3 \cdot V_d^{\gamma-1} \tag{4.8}$$

Dividing the second equation by the first, it is seen that: $V_a/V_b = V_d/V_c$. By insertion into (4.7), we get the following expression for the thermal efficiency η as a function of the temperatures T_1 and T_3 in the reservoirs used

Reversible cyclic process

$$\Sigma\left(\frac{Q}{T}\right)_{rev} = 0$$

Thermal efficiency η (4.9)

$$\eta = 1 + \frac{Q_3}{Q_1} = 1 - \frac{T_3}{T_1}$$

– the thermal efficiency of a **reversible cyclic process** between the temperatures T_1 and T_3, where $T_1 > T_3$.

By subtraction of the two expressions for η in (4.9) and after rearranging the terms, we get the following important relation with the key to a new state function, the entropy S

Heat transfer by a reversible cyclic process (4.10)

In a **reversible** Carnot process with an **ideal gas**, the sum of the supplied quantities of heat Q divided by the thermodynamic temperature T by which the heat is supplied is equal to zero

$$\sum\left(\frac{Q}{T}\right)_{rev} = \frac{Q_1}{T_1} + \frac{Q_3}{T_3} = 0$$

Figure 4.7. Eskimos make fire. Work can always be fully transformed into heat. Transformation of heat into work, however, is subject to clearly defined limitations as expressed in the second law of thermodynamics.

In accordance with (4.9), the thermal efficiency η of a reversible **Carnot cycle** (4.9) *solely* depends on the ratio between the thermodynamic temperature in in the two heat reservoirs, in this case T_1 and T_3. This property of the Carnot process has made possible the definition of an absolute temperature scale, which is *independent* of the thermometer substances used (Lord **Kelvin**, 1854).

The derived expressions (4.9) and (4.10) can be shown to be *general* for reversible cyclic processes. Further, it can be shown that the reversible Carnot cycle with an ideal gas is the most efficient process possible for transforming heat into work. The thermal efficiency η expressed in (4.9), therefore, is the **theoretical maximum efficiency**.

■ Calculate the theoretical, maximum thermal efficiency of a heat engine operating with steam under pressure at 350 °C and a condenser temperature of 100 °C. Determine the amount of heat Q_3, which is transmitted to the condenser as loss when $Q_1 = 1\,\text{MJ}$ at 350 °C is supplied under these conditions !

Solution. Conversion into thermodynamic temperatures according to (2.8) gives $T_1 \simeq 623\,\text{K}$ and $T_3 \simeq 373\,\text{K}$; the theoretical maximum efficiency η is determined by (4.9)

$$\eta_{\max} = 1 - \frac{T_3}{T_1} = \left(1 - \frac{373}{623}\right) \cdot 100\,\% = \mathbf{40\,\%}$$

The *supplied* heat Q_3 at the condenser temperature T_3 is determined by introduction into (4.10)

$$Q_3 = -Q_1 \cdot \frac{T_3}{T_1} = -1\,\text{MJ} \cdot \frac{373}{623} = \mathbf{-600\,kJ}$$

The negative sign indicates that 600 kJ is *released* from the engine to the condenser.

☐ **1.** What value should T_1 and T_3 approach to obtain the efficiency $\eta \to 1$ for a reversible Carnot cycle with an ideal gas ?

☐ **2.** Determine the maximum thermal efficiency η of a reversible heat engine which is supplied with heat at 650 °C and releases heat at 10 °C!

☐ **3.** Which of the following will increase the efficiency η of a heat engine more, increasing the high temperature by 10 °C, or reducing the low temperature by 10 °C?

☐ **4.** Will an ideal heat engine perform work, when it receives heat from and releases heat to the same heat reservoir? - explain the answer!

☐ **5.** An ideal heat engine operates with the efficiency $\eta = 55\,\%$. Determine the temperature T_1 at which the engine operates, when the condenser temperature is 50 °C!

Definition of entropy S

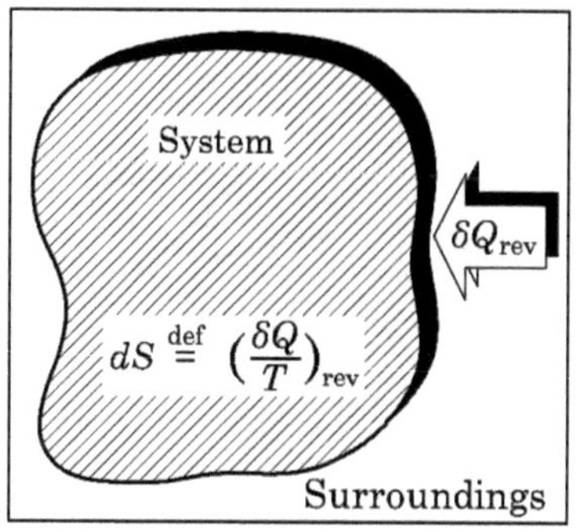

Figure 4.8. Entropy S is a state function; the entropy increase dS in a system is defined as the reversibly supplied heat δQ divided by the system temperature T.

4.3 Second law

The **second law** of thermodynamics is a *postulate*; i.e. its statement *cannot* be proven. However, both the first and the second law theorems are based on comprehensive experience to such degree that they can be taken as general natural laws.

The first law of thermodynamics was connected with the definition of a state function: **internal energy** U. Similarly, the second law leads to the definition of a state function: **entropy** S.

> **The second law of thermodynamics** (4.11)
>
> It is impossible for a system performing a **cyclic process** to perform work on its surroundings, if heat is supplied from one reservoir only!

"The Calusius inequality"

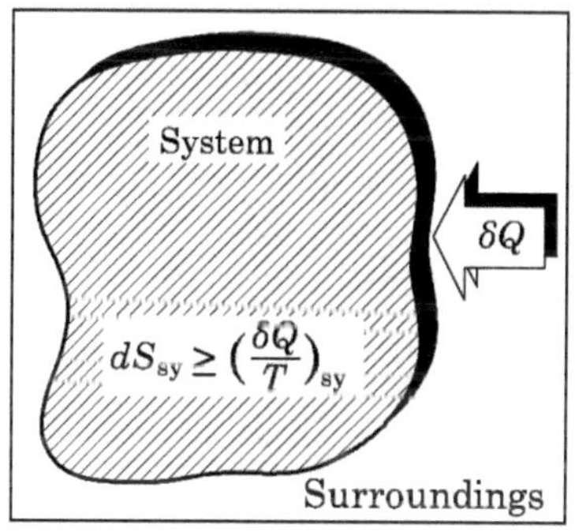

Figure 4.9. According to the Clausius inequality, the increase in system entropy dS_{sy} by a process is greater than or equal to $(\delta Q/T)_{sy}$; for an adiabatic system, where $\delta Q = 0$, dS_{sy} is greater than or equal to 0.

In this formulation, the second law can be seen as a generalization of the expression (4.9) of the thermal efficiency for a Carnot cycle. Heat absorbed from a reservoir at one temperature can *only* be transformed into work if at the same time, heat is released to another reservoir at a *lower* temperature.

Entropy S

The second law contains the fundamental principles for description of thermodynamic **equilibrium** and for determining the **direction** of irreversible processes.

The mathematical formulation of the principle of equilibrium that can be derived from the second law is

The second law (4.12)

- $dS \overset{\text{def}}{=} \dfrac{\delta Q_{rev}}{T}$ (definition)
- S is a **state function** (postulate)
- $dS_{system} \geq \left(\dfrac{\delta Q}{T}\right)_{system}$ ("the Clausius inequality")

Entropy of an **adiabatic** system ($\delta Q = 0$) can *never* decrease; the entropy *increases* at **irreversible** processes and remains *constant* at **reversible** processes.

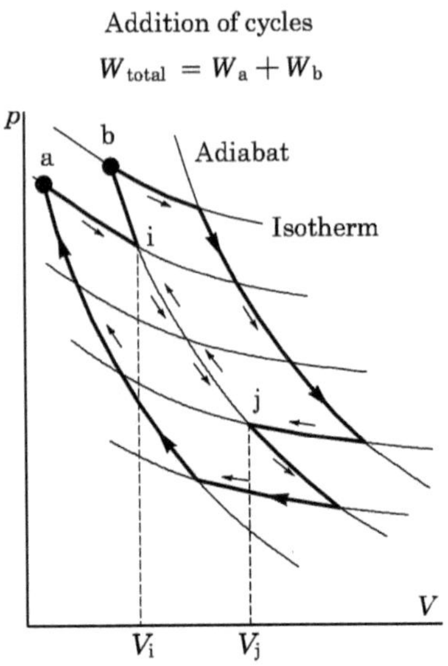

Figure 4.10. Addition of two, reversible Carnot cycles with partially identical adiabat. The work contributions W_{ij} and W_{ji} have opposite signs and neutralize each other; therefore, the composite process corresponds to a cyclic process following an outer contour line.

The definition of entropy S can be seen as a natural generalization of the previously derived condition (4.10) for a reversible Carnot cycle

$$\frac{Q_1}{T_1} + \frac{Q_3}{T_3} = 0 \qquad \text{(reversible Carnot cycle)}$$

The expression shows that in a simple reversible cyclic process, the quantity $\sum Q_{rev}/T = 0$. Since any reversible cyclic process in principle can be decomposed into an arbitrary number of sub-processes fulfilling this term, it can be shown that

$$\sum \left(\frac{Q}{T}\right)_{rev} \quad \rightarrow \quad \oint \frac{\delta Q_{rev}}{T} = \oint dS = 0 \tag{4.13}$$

It follows that $dS = \delta Q_{rev}/T$ is an **exact** differential and thus that the entropy S is a **state function**. The change ΔS in entropy at a change of state is calculated from (4.12) as

ΔS at a change of state (4.14)

$$\Delta S = \int_{S_1}^{S_2} dS = \int_{(1)}^{(2)} \frac{\delta Q_{rev}}{T}$$

When a system is changed from state (1) to state (2), the system **increase** in entropy ΔS is calculated as the integral of $\delta Q_{rev}/T$ for an *arbitrary* **reversible** process from (1) to (2).

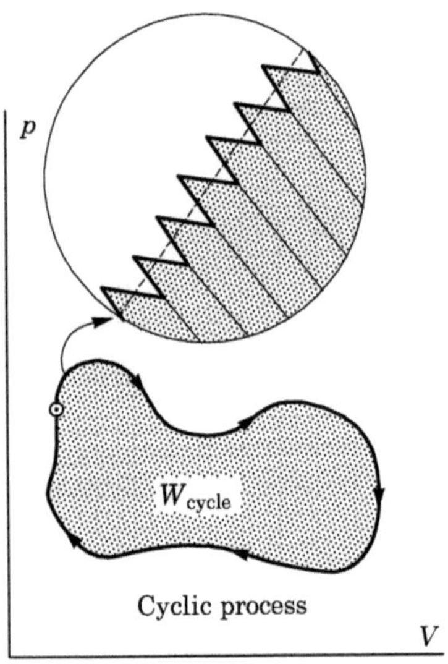

Figure 4.11. Any reversible cyclic process can be described as the sum of reversible Carnot cycles; an arbitrarily good approximation of the process can be obtained by the choice of division of the process.

It is important to note the following: A change of entropy ΔS is *always* calculated for a **reversible** change of state (1) $\rightarrow$ (2). As S is a state function, ΔS is *independent* of the process path; the calculated change of entropy ΔS in the reversible process, therefore, is identical with ΔS for an *arbitrary*, reversible or irreversible process connecting the same states of equilibrium (1) and (2).

■ A cylinder contains 2 mol of **ideal gas**, confined under a non-frictional piston. In its initial state (1), the gas temperature is $\theta = 200^{\circ}$C. By a **reversible, isothermal** process (1) $\rightarrow$ (2), an amount of heat $Q_{1.2} = 6000$ J is supplied to the gas. Calculate for the system: ΔU (J), $W_{1.2}$ (J) and the system increase in entropy ΔS (J) by the corresponding change of state!

Solution. The change of state is isothermal; then, according to *Joule's law* (3.19) and (3.27), we have for an ideal gas

$$\Delta U = Q_{1.2} + W_{1.2} = \mathbf{0\,J}$$

It follows directly that the work $W_{1.2}$ performed *on* the gas is

$$W_{1.2} = -Q_{1.2} = -6000\,\text{J} = \mathbf{-6000\,J}$$

The system is supplied with the amount of heat $Q_{1.2} = 6000\,\text{J}$ in a **reversible** process; then, according to the second law (4.12), we have

$$\Delta S = \int_{(1)}^{(2)} dS = \int_{(1)}^{(2)} \frac{\delta Q_{rev}}{T} = \frac{1}{T}\int_{(1)}^{(2)} \delta Q_{rev} = \frac{6000}{473.15} = \mathbf{12.7\,J/K}$$

☐ **1.** A system contains ideal gas. By a reversible, isothermal process, the system is supplied with $Q_{1.2} = 3000\,\text{J}$; during the process, $\Delta S = 10\,\text{J/K}$. Determine the gas temperature T (K)!

☐ **2.** Explain, based on the second law (4.12), *why* the entropy S in an adiabatic, thermodynamic system can never decrease!

☐ **3.** A process runs in an isolated thermodynamic system. Estimate, based on the second law (4.12), whether the process is reversible or irreversible!

☐ **4.** In a reversible, isothermal process, the work $W_{1.2} = 4000\,\text{J}$ is performed on an ideal gas at 25 °C. Calculate the entropy increase ΔS of the system during the process!

☐ **5.** In a chart with abscissa T and ordinate S, a reversible Carnot cycle with an ideal gas is plotted. Show that the graph will always form a rectangle whose sides are parallel to the axes of the TS coordinate system!

4.4 Temperature dependence of entropy

In the following, the change of entropy ΔS in a system of matter subjected to *heating* or *cooling* will be determined. Once more, it is important to bear in mind that the change ΔS shall be calculated for a **reversible** change of state that connects the two states of equilibrium. The change of entropy ΔS will then apply to *both* **reversible** and **irreversible** processes connecting these terminal states.

In a reversible heating, an infinitesimal temperature difference dT between the system and its surroundings is assumed during the process. Therefore, in each step of the process, the system will be infinitely close to **thermal equilibrium**. For a reversible temperature change developing at **constant pressure** p, we obtain according to (3.15) and (3.16)

$$(dH_{rev} = \delta Q_{rev})_p; \qquad (dH = nc_p \cdot dT)_p \tag{4.15}$$

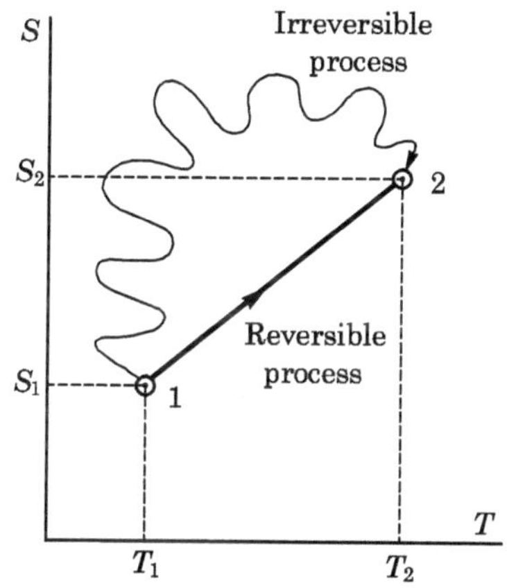

Figure 4.12. Entropy S is a state function; therefore, ΔS calculated for a given, reversible process 1 → 2 is equal to ΔS for any other reversible or irreversible process connecting the states (1) and (2).

The entropy increase $\Delta S_{1.2}$ by a reversible temperature change from T_1 to T_2 is now calculated from (4.14)

$$\Delta S_{1.2} = \int_{S_1}^{S_2} dS = \int_{(1)}^{(2)} \frac{\delta Q_{rev}}{T} = \int_{(1)}^{(2)} \frac{dH_{rev}}{T} = \int_{T_1}^{T_2} nc_p \cdot d\ln(T) \tag{4.16}$$

For **constant heat capacity** c_p in the temperature range concerned, integration gives

$$\Delta S_{1.2} = nc_p \int_{T_1}^{T_2} d\ln(T) = nc_p \cdot (\ln(T_2) - \ln(T_1)) = nc_p \cdot \ln\left(\frac{T_2}{T_1}\right) \tag{4.17}$$

A similar deduction can be made for a system of matter heated at **constant volume** V; in this case, the increase in internal energy ΔU according to (3.9) is equal to the supplied heat $Q_{1.2}$, and in accordance with (3.10), c_V is introduced into the expression for dU. This results in the following expression for calculation of the entropy change due to heating or cooling of a system of matter.

Temperature dependence of entropy (4.18)

A temperature change from T_1 to T_2 changes the entropy of a system of matter with

$$\Delta S_{1.2} = nc_p \cdot \ln\left(\frac{T_2}{T_1}\right) \qquad \text{(constant } p\text{)}$$

$$\Delta S_{1.2} = nc_V \cdot \ln\left(\frac{T_2}{T_1}\right) \qquad \text{(constant } V\text{)}$$

where nc_p and nc_V denote the thermal capacity of the system of matter at constant **pressure** p and constant **volume** V, respectively.

Temperature change

$\Delta S_{1.2} = nc_p \cdot \ln(\frac{T_2}{T_1})$

$\Delta S_{1.2} = nc_V \cdot \ln(\frac{T_2}{T_1})$

Again, it shall be noted that the deduction assumes a **reversible** process development; the expression found for ΔS, however, applies to any reversible or irreversible process connecting the actual initial and final states.

■ Assume two adiabatic systems (a) and (b) each consisting of a steel block with the mass m = 1.000 kg. The mass-specific heat capacity of the steel is c_p = 0.481 kJ/kg K. In the initial state, the temperatures of the two blocks are $\theta_a = 0\,°\mathrm{C}$ and $\theta_b = 100\,°\mathrm{C}$, respectively. At the time t_1, the two steel blocks are brought into mutual thermal contact. At the time t_2, the two blocks have reached thermal equilibrium at $\theta_{ab} = 50°\mathrm{C}$. Calculate for the described process (1) → (2), the increase in entropy for the two subsystems ΔS_a and ΔS_b, respectively, as well as the increase in entropy ΔS_{ab} of the total system!

Solution. The sought quantities ΔS_a, ΔS_b, and ΔS_{ab} are determined from the relation (4.18). In the calculation the thermodynamic temperatures T_a, T_b, T_{ab} are inserted corresponding to the specified Celsius values

$$\Delta S_a = mc_p \cdot \ln\left(\frac{T_{ab}}{T_a}\right) = 1.000 \cdot 481 \cdot \ln\left(\frac{323.15}{273.15}\right) = \mathbf{+80.9\ J/K}$$

$$\Delta S_b = mc_p \cdot \ln\left(\frac{T_{ab}}{T_b}\right) = 1.000 \cdot 481 \cdot \ln\left(\frac{323.15}{373.15}\right) = \mathbf{-69.2\ J/K}$$

$$\Delta S_{ab} = \Delta S_a + \Delta S_b = +80.9 - 69.2\,\mathrm{J/K} = \mathbf{+11.7\,J/K}$$

During the adiabatic process, the two systems did not exchange heat with their surroundings. In the process, the system entropy has been increased by 11.7 J/K; therefore, according to **the Clausius inequality**(4.12), this is an irreversible process.

Reversible isothermal expansion

$\Delta S_{1.2} = nR \cdot \ln(V_2/V_1)$

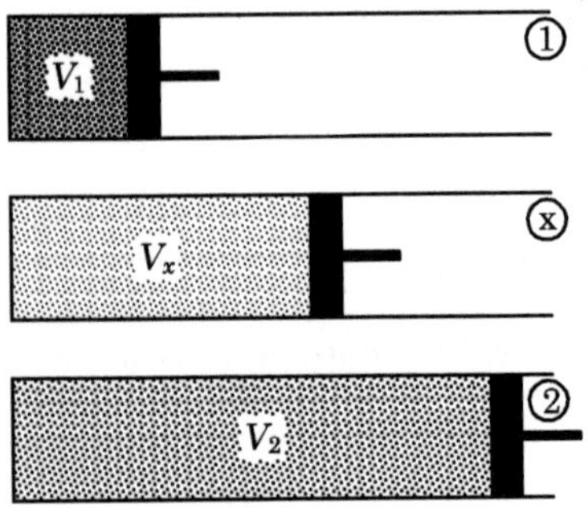

$\Delta U = 0 \quad Q > 0 \quad W = -Q$

Figure 4.13. In a reversible, isothermal change of state with an ideal gas the internal energy U is constant; the supplied heat Q corresponds to the work performed by the gas on its surroundings. In this case, ΔS can be calculated from the equation of definition.

☐ **1.** 5 mol of ideal gas with $c_p = 21.0\,\mathrm{J/mol\,K}$ are cooled at constant pressure p from 25 °C to −12 °C. Calculate the increase ΔS (J/K) in the system entropy!

☐ **2.** A system contains 3.000 g of $H_2O(\ell)$ at 0 °C. Calculate the increase in the molar entropy of the water ΔS (J/mol K), when the system is supplied with Q = 80.0 kJ!

☐ **3.** Assume that quartz SiO_2(s) has constant c_p at all temperatures. Calculate ΔS for heating of 1 mol of quartz: a) from 5 K to 6 K; and b) from 5 °C to 6 °C!

☐ **4.** 2 mol of ideal gas with $c_p = 21\,\mathrm{J/mol\,K}$ are cooled at constant volume V from 25 °C to −12 °C. Calculate the increase ΔS (J/K) in the system entropy!

☐ **5.** A system consists of 100 g of water $H_2O(\ell)$ at the temperature θ_1 (°C). The system is supplied with an amount of heat $Q = -5000$ J, whereby the molar entropy of the water is increased by $\Delta S = -3.01$ J/mol K. Calculate the initial temperature of the water θ_1!

4.5 Entropy change, ideal gas

Within the science of construction materials, calculations of equilibrium often include substances in the gas phase. As examples, **sorption equilibrium** for water vapour, **oxygen equilibrium** at electrochemical corrosion, and **evaporation**

equilibrium for moist, hardening concrete can be mentioned. In most cases, the considered gases will, to a good approximation, comply with the ideal gas law. The thermodynamic properties of ideal gases, therefore, are of considerable practical interest. In the following, expressions for calculation of entropy change by change of state of ideal gases are given.

Assume a system containing n mol of an ideal gas at the temperature T (K). By a **reversible, isothermal** change of state of the gas, the reversibly supplied heat Q_{rev} according to (3.27) will be determined by

$$Q_{rev} = nRT \cdot \ln\left(\frac{V_2}{V_1}\right)$$

where V_1 and V_2 are the system volume in the initial state and the final state, respectively. As the process is isothermal, the increase in entropy $\Delta S_{1.2}$ can be calculated from (4.14)

$$\Delta S_{1.2} = \frac{1}{T}\int_{(1)}^{(2)} \delta Q_{rev} = \frac{Q_{rev}}{T} = nR \cdot \ln\left(\frac{V_2}{V_1}\right) \tag{4.19}$$

This is an ideal gas; therefore, the volume ratio can be rewritten to a simple ratio between the pressures p_1 and p_2

$$\frac{V_2}{V_1} = \left(\frac{nRT}{p_2}\right) / \left(\frac{nRT}{p_1}\right) = \frac{p_1}{p_2} \tag{4.20}$$

Introducing this ratio into (4.19) and inverting the argument of the logarithm give

$$\Delta S_{1.2} = -nR \cdot \ln\left(\frac{p_2}{p_1}\right) \tag{4.21}$$

Thus, in conclusion, we have the following expression for calculation of the entropy change ΔS for isothermal changes of state with ideal gases

Irreversible isothermal expansion

$$\Delta S_{1.2} = nR \cdot \ln(V_2/V_1)$$

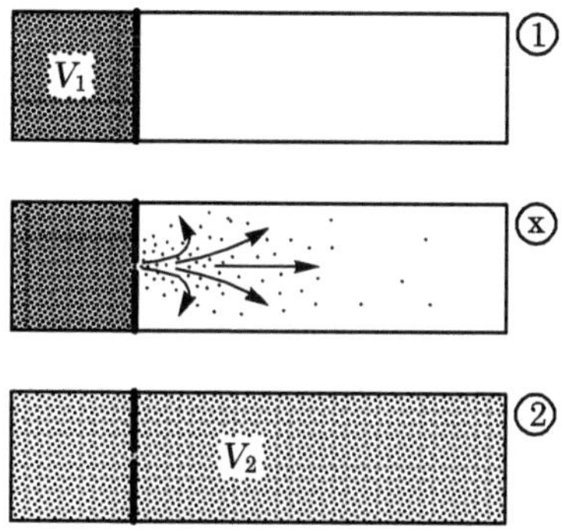

$\Delta U = 0 \quad Q = 0 \quad W = 0$

Figure 4.14. An irreversible expansion of an ideal gas into a vacuum is isothermal (Joule's law). Therefore, for this process, ΔS is the same as for the reversible, isothermal expansion described in figure 4.13.

ΔS for isothermal change of state, ideal gas (4.22)

For an **isothermal** change of state of n mol of **ideal gas** from (p_1, V_1) to (p_2, V_2), the *increase* in the gas **entropy** is

$$\Delta S_{1.2} = nR \cdot \ln\left(\frac{V_2}{V_1}\right); \qquad \Delta S_{1.2} = -nR \cdot \ln\left(\frac{p_2}{p_1}\right)$$

Once more it shall be noted that although ΔS has been determined for a reversible, isothermal change of state, (4.22), it also applies to an *arbitrary* reversible or irreversible change of state between the specified states of equilibrium (1) and (2); this is a consequence of the fact that the entropy S is a state function.

Generally, the entropy S of gases *increases* with increasing dilution, i.e. when the volume V is increased or when the pressure p is reduced. This is because the entropy S is a measure of the "degree of disorder" at molecular level. As will be seen in section 4.10, the molecular disorder, and thus the entropy, increases when crystalline substances melt and when liquids evaporate – i.e. by phase transformations that lead to increased disorder at molecular level.

Isothermal change of state ideal gas

$$\Delta S_{1.2} = nR \cdot \ln\left(\frac{V_2}{V_1}\right)$$

$$\Delta S_{1.2} = -nR \cdot \ln\left(\frac{p_2}{p_1}\right)$$

■ Saturated air at 20 °C and atmospheric pressure contains 17.3 g of water vapour per m^3. The water vapour and the atmospheric air are assumed to be ideal gases. Assume a system consisting of $V_1 = 1\,\text{m}^3$ of saturated atmospheric air at $\theta_1 = 20\,°\text{C}$ and $p_1 = 101325\,\text{Pa}$. Calculate the entropy change in the contained water vapour if the system volume is changed to $V_2 = 2\,\text{m}^3$ by an isothermal process!

Solution. In the initial state, the saturated air contains 17.3 g of water vapour per m^3; this corresponds to an **amount of substance**

$$n = \frac{m}{M} = \frac{17.3\,\text{g}}{18.02\,\text{g/mol}} = 0.96\,\text{mol}$$

The entropy increase of the water vapour $\Delta S_{1.2}$ during the process can now be determined from the expression (4.22)

$$\Delta S_{1.2} = nR \cdot \ln(V_2/V_1) = 0.96\,\text{mol} \cdot 8.314\,\text{J/mol K} \cdot \ln(\tfrac{2}{1}) = \mathbf{5.53\,J/K}$$

☐ **1.** By an isothermal expansion, the pressure p in 100 g of ideal $N_2(g)$ is changed from 101325 Pa to 8000 Pa. Calculate the increase in the system entropy ΔS (J/K)!

☐ **2.** By an isothermal process, the pressure p in 5 mol of ideal gas is changed from 0.500 atm to 0.125 atm. Calculate the increase in the molar entropy ΔS (J/mol K) of the gas!

☐ **3.** A 410 ℓ receptacle contains 476 g of ideal $N_2(g)$ at 0 °C. Calculate the entropy increase ΔS (J/K) by an isothermal process that changes the gas pressure to 10000 Pa!

☐ **4.** By an isothermal compression, the volume of 3 mol of ideal gas is halved. Calculate the increase in the molar entropy of the gas ΔS (J/mol K)!

☐ **5.** By an isothermal process with an ideal gas, the increment in the molar entropy of the gas is $\Delta S = -15.5$ J/mol K. Explain whether the gas pressure has been increased or reduced by the process!

4.6 Entropy change by phase transformation

Phase transformations play an important role within the science of construction materials. Different phenomena such as **hardening** of steel, **freezing damage** of concrete, **moisture absorption** in wood, and **chemical shrinkage** of hardening cement paste, are all connected with phase transformations in systems of matter. Table 4.1 contains an overview of the most important phase transformations between substances in the **solid**, **liquid**, and **gaseous** states.

Figure 4.15. Freezing damage of concrete is caused by expansion of the pore water due to freezing, i.e. by the phase transformation $H_2O(\ell) \rightarrow H_2O(s)$. The photo shows freezing damage on a concrete slab.

Transformation	Designation	$\Delta_r H$	Designation
$(s_1) \rightarrow (s_2)$	Transformation	$H(s_2) - H(s_1)$	Heat of transformation
$(s) \rightarrow (\ell)$	Melting	$H(\ell) - H(s)$	Heat of fusion
$(\ell) \rightarrow (s)$	Solidification, freezing	$H(s) - H(\ell)$	Heat of solidification
$(\ell) \rightarrow (g)$	Evaporation	$H(g) - H(\ell)$	Heat of evaporation
$(g) \rightarrow (s)$	Sublimation	$H(s) - H(g)$	Heat of sublimation
$(s) \rightarrow (g)$	Sublimation (deposition)	$H(g) - H(s)$	Heat of sublimation (deposition)
$(g) \rightarrow (\ell)$	Condensation	$H(\ell) - H(g)$	Heat of condensation

Table 4.1. Overview of designations of the major, simple phase transformations. The following abbreviations of the states have been used: (s) solid, (ℓ) liquid and (g) gaseous.

For describing phase equilibrium, the entropy change ΔS in phase transformations is an important parameter. Therefore, we shall set up an expression to calculate ΔS for **isothermal** phase transformations in systems of matter.

Normally, a phase transformation will be connected with absorption or release of heat; this is known, for example, in the form of the **heat of fusion** of ice and the **heat of evaporation** of water. This enthalpy change – the transformation enthalpy $\Delta_r H_T$ – is a direct measure of the entropy change by the phase transformation concerned.

For a reversible phase transformation at *constant* **pressure** p and **temperature** T, the enthalpy increase $\Delta_r H_T$ for the transformation according to (3.15) is equal to the reversible amount of heat Q_{rev} supplied to the system. Therefore, the entropy increase ΔS per mol of transformed substance can be determined by the expression (4.14).

Entropy change by phase transformation (4.23)

The entropy increase ΔS by an **isothermal** phase transformation at **constant pressure** p is

$$\Delta S = \frac{\Delta_r H_T}{T} \qquad (\text{J/mol K})$$

where T (K) is the transformation temperature and $\Delta_r H_T$ (J/mol) is the transformation enthalpy at T.

The calculated entropy increase ΔS according to (4.23) is independent of the process path, i.e. of the transformation course; it is *only* assumed that the system is in **equilibrium** in its initial and its final state at the transformation temperature T (K).

Reversible phase transformation

$$\Delta S = \frac{\Delta_r H}{T} \qquad (p, T \text{ constant})$$

■ A closed system contains 1 kg of ice $H_2O(s)$ at 0 °C and atmospheric pressure. The system is supplied with exactly such an amount of heat that the ice melts and is transformed into 1 kg of water $H_2O(\ell)$ at 0 °C. Calculate the increase in the system entropy ΔS (J/K) and the increase in the molar entropy ΔS (J/mol K) of H_2O by the transformation. The heat of fusion of the ice is 6008 J/mol at the transformation temperature.

Solution. First, we determine the amount of substance n of H_2O in the system

$$n = \frac{m}{M} = \frac{1000\,\text{g}}{18.02\,\text{g/mol}} = 55.5\,\text{mol}$$

The enthalpy increase by the phase transformation: $H_2O(s) \rightarrow H_2O(\ell)$ is exactly equal to the heat of fusion Q_{rev} supplied during a reversible transformation; the increase in the system entropy ΔS is then found using (4.23)

$$\Delta S = \frac{n \cdot \Delta_r H_{273}}{T} = \frac{55.5\,\text{mol} \cdot 6008\,\text{J/mol}}{273.15\,\text{K}} = \mathbf{1221\,J/K}$$

Finally, the increase of molar entropy is determined by

$$\Delta S_{mol} = \frac{1}{n} \cdot \Delta S_{sys} = \frac{1}{55.5} \cdot 1221\ \text{J/mol K} = \mathbf{22.0\,J/mol\,K}$$

Thus, when ice is transformed into water, the entropy is increased by 22 J/mol K; this is an indication of increased disorder in the molecular structure of the substance.

Figure 4.16. Snow crystal with a hexagonal structure. Ice has an organized crystalline structure with low entropy; by melting, a disorganized liquid state of water with a higher entropy content is formed. When the ice melts, the entropy increase ΔS is $\Delta H/T$.

□ **1.** The heat of evaporation of water is 2442 J/g at 25 °C. Calculate the increase in molar entropy ΔS (J/mol K) by evaporation of water at 25 °C!

□ **2.** The heat of fusion of ice at 0 °C is 6008 J/mol and the heat of evaporation of water is 45050 J/mol. Calculate ΔS for the sublimation: $H_2O(s) \rightarrow H_2O(g)$ at 0 °C!

□ **3.** For reversible evaporation of water at 100 °C, $\Delta S = 109.0$ J/mol K. Determine the heat of evaporation of water in (J/g) at 100 °C!

□ **4.** At approximately 11 °C, white tin Sn(s) can spontaneously be transformed into grey tin ("tin pest"); for the transformation, $\Delta_r H = -2.1$ kJ/mol. Determine ΔS for the transformation!

□ **5.** For a given phase transformation, the entropy increase is $\Delta S < 0$; what does this value indicate about the change of the molecular arrangement in the substance ?

4.7 Standard entropy

In technical calculations it is an advantage to assign an absolute **standard entropy** $S^{\ominus}_{298}$ to **elements** and **compounds** similarly to the introduction of the standard enthalpy $H^{\ominus}_{298}$ in section 3.9. Determination of a zero point for this absolute entropy scale can be made on the basis of the third law of thermodynamics.

The **third law** of thermodynamics postulates that the entropy of **chemically pure, crystalline** substances at absolute zero 0 K is zero. Originally, this postulate was put forward as **the Nernst heat theorem** (1906) and later proved by quantum mechanics (**Planck** 1912). From this absolute zero, the standard entropy $S(T)$ of substances can formally be determined at an arbitrary temperature T, by adding the entropy increases ΔS from heating and phase transformations to $S(0)$.

$$S(T) = S(0) + \int_0^T \frac{c_p(T)}{T} dT + \sum \Big(\frac{\Delta H}{T}\Big)_{trans} \tag{4.24}$$

The specific heat capacity $c_p(T)$ and transformation enthalpy ΔH_{trans} of the substances at phase transformations are determined experimentally.

Today, the standard entropy $S^{\ominus}_{298}$ is mapped and tabulated for a considerable number of elements and compounds. Conventionally, as reference temperature and reference pressure, 298.15 K (25 °C) and 101325 Pa, respectively, are used.

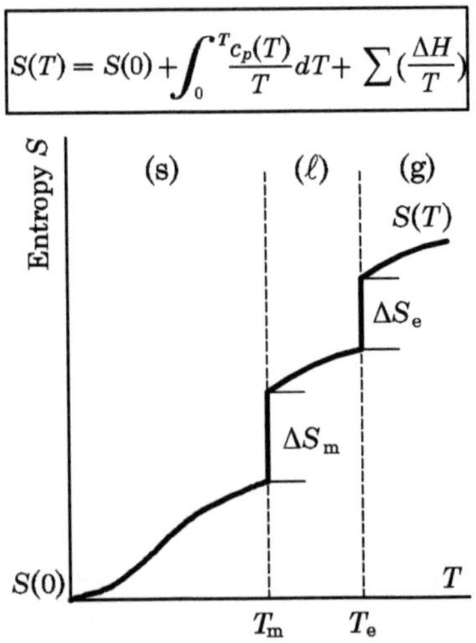

Figure 4.17. Calculation of absolute entropy $S(T)$, schematically. For a reversible heating from the thermodynamic zero point, $S(T)$ is calculated by integration of $dS = \delta Q/T$; in phase transformations, $\Delta S = \Delta H/T$ is added as shown for melting and evaporation.

Standard entropy (4.25)

At the **standard pressure** $p^{\ominus} = 101325\,\text{Pa}$ and for dissolved substances at the **standard concentration** $c^{\ominus} = 1\,\text{mol}/\ell$, the molar standard entropy of a substance is

$$S^{\ominus}_{298} \qquad (\text{J/mol K})$$

From tabulated values of the standard entropy of the substances $S^{\ominus}_{298}$, the entropy at other pressures and temperatures can be calculated using the expressions (4.18), (4.22), and (4.23) derived above. The following guidelines apply to such calculations.

For **solid** and **liquid** substances, the influence of pressure on the entropy can be disregarded if there are no extreme changes in pressure. The temperature dependence of entropy can normally be determined from (4.18) assuming constant heat capacity c_p in the considered temperature range. If the temperature deviates by more than 100-200 K from the reference temperature 298.15 K, the temperature dependence of the heat capacity should be taken into consideration for accurate calculations. In such a case, the calculation includes integration of the expression (4.16) with $c_p = c_p(T)$.

For **gases**, correction for the pressure dependence of the entropy must *always* be made if the pressure deviates from the standard value $p^{\ominus} = 101325\,\text{Pa}$. Normally the correction can be made by using the expression (4.22) assuming ideal gas conditions. The temperature correction is performed as described for solid and liquid substances.

Thus, in conclusion, we have the following **procedure** for calculation of the entropy $S(T, p)$ for solid, liquid and gaseous substances as a function of pressure and temperature

Calculation of entropy (4.26)

The **standard entropy** $S^{\ominus}_{298}$ and the **heat capacity** c_p of the substance are tabulated data; the entropy at the **temperature** T (K) and the **pressure** p (Pa) is determined by

- **Solid and liquid substances:**

$$S(T,p) = S^{\ominus}_{298} + c_p \cdot \ln\Big(\frac{T}{298.15}\Big) \qquad (\text{J/mol K})$$

- **Gaseous substances:**

$$S(T,p) = S^{\ominus}_{298} + c_p \cdot \ln\Big(\frac{T}{298.15}\Big) - R \cdot \ln\Big(\frac{p}{101325}\Big) \qquad (\text{J/mol K})$$

It is assumed in the calculation that c_p is **constant** in the considered temperature range and that there are *no* **phase transformations** or **chemical reactions** in the substance. As seen in the following example, the calculation work when using the procedure (4.26) is not very comprehensive (see Mathematical appendix, subchapter A8 *Maxwell's relations*).

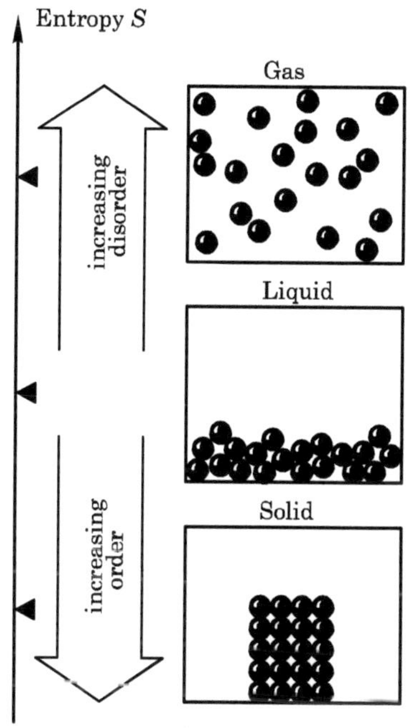

Figure 4.18. The entropy content of substances increases with the degree of molecular disorder. Transformation from a solid state to a liquid state (melting) increases the degree of disorder and thus also the entropy content; transformation from a liquid state to a gaseous state (evaporation) increases the degree of disorder and thus the entropy content.

■ According to Atkins: *Physical Chemistry*, water and water vapour have the following thermodynamic data in their standard state

Water	$H_2O(\ell)$	$S^{\ominus}_{298} = 69.91\,\text{J/mol K}$;	$c_p = 75.29\,\text{J/mol K}$
Water vapour	$H_2O(g)$	$S^{\ominus}_{298} = 188.83\,\text{J/mol K}$;	$c_p = 33.58\,\text{J/mol K}$

At 60 °C, the partial pressure of a saturated water vapour is $p_s = 19932\,\text{Pa}$. Calculate the entropy of water $H_2O(\ell)$ and water vapour $H_2O(g)$ at 60 °C!

Solution. The entropy of water at $T = 333.15\,\text{K}$ (60 °C) is determined by the procedure (4.26); constant c_p is assumed, and correction is *not* made for the influence of the pressure on the standard entropy since we are dealing with a **liquid** phase

$$S_\ell = 69.9 + 75.29 \cdot \ln\left(\frac{333.15}{298.15}\right) = \mathbf{78.3\,J/mol\,K}$$

The entropy of saturated water vapour at $T = 333.15\,\text{K}$ (60 °C) and $p = p_s = 19932\,\text{Pa}$ is also determined from (4.26); constant c_p is assumed and correction for the deviation of the pressure from $p^{\ominus}$ is made since we are dealing with a **gaseous** phase

$$S_g = 188.83 + 33.58 \cdot \ln\left(\frac{333.15}{298.15}\right) - 8.314 \cdot \ln\left(\frac{19932}{101325}\right) = \mathbf{206.1\,J/mol\,K}$$

☐ **1.** Determine the standard entropy of calcium hydroxide $Ca(OH)_2(s)$ and calcium oxide $CaO(s)$ at the following temperatures: a) 65 °C and b) −12 °C!

☐ **2.** Determine the entropy of oxygen $O_2(g)$ at 120 °C, if the gas pressure p is a) 101325 Pa and b) 200 Pa, respectively!

☐ **3.** A 40 ℓ reservoir contains 20 g of argon gas Ar(g) at 85 °C. Calculate S for the Ar gas when $S^{\ominus}_{298} = 154.84\,\text{J/mol K}$ and $c_p = 20.786\,\text{J/mol K}$!

☐ **4.** At 30 °C, atmospheric air contains 30.4 g of water vapour $H_2O(g)$ per m^3. Calculate the entropy of the water vapour in this state!

☐ **5.** Calculate the increase in entropy ΔS (J/K), when 1 mol of liquid mercury $Hg(\ell)$ is transformed into mercury vapour Hg(g) at 357 °C, the boiling point of mercury!

4.8 Reaction entropy

The **reaction entropy** $\Delta_r S$ is an important quantity of calculation for the analysis of reaction equilibrium. As we can now express the entropy of the substances at a given state $S(T,p)$, the standard reaction entropy $\Delta_r S^{\ominus}_T$ can be determined from the same principle as used for calculation of standard enthalpy $\Delta_r H^{\ominus}_T$.

Assume a simple **chemical reaction** or **phase transformation** which can be described by the following, formal reaction equation

$$\begin{array}{ccc} aA + bB & \rightarrow & cC + dD \\ \text{(reactants)} & \rightarrow & \text{(products)} \end{array} \qquad (4.27)$$

where a, b, c and d are non-dimensional, **stoichiometric coefficients**. Assume that this process occurs in a system at constant pressure p and temperature T. The entropy increase made by the reaction in the system is called the **reaction entropy** $\Delta_r S$. Since the entropy is a state function, the reaction entropy $\Delta_r S$ can be expressed as the total entropy of the system in its final state minus the total entropy of the system in its initial state.

Reaction entropy (4.28)

The **reaction entropy** $\Delta_r S$ of a chemical reaction or transformation occurring at **constant pressure** p and **temperature** T is

$$\Delta_r S = \sum S(\text{products}) - \sum S(\text{reactants})$$

– the reaction entropy denotes the total entropy of the system in its **final state** minus the total entropy of the system in its **initial state**.

Reaction entropy

$\Delta_r S = \sum S_{\text{prod}} - \sum S_{\text{react}}$

If we consider one mol of the reaction equation (4.27), the reaction entropy $\Delta_r S$ can be expressed in the following formal form

$$\Delta_r S = (c \cdot S_\text{C} + d \cdot S_\text{D}) - (a \cdot S_\text{A} + b \cdot S_\text{B}) \qquad (\text{J/mol K}) \qquad (4.29)$$

Here, it is important to note the following with regard to units: the stoichiometric coefficients a, b, c and d are **non-dimensional** quantities. Therefore, the unit for the reaction entropy $\Delta_r S$ is (J/mol K) where mol refers to the specific reaction equation. In verbal terms: "*Joule per mol of reaction equation per kelvin*".

Calculation procedure

Summing up, we thus have the following **procedure** for calculation of reaction entropy $\Delta_r S_T$ for a given chemical reaction or phase transformation

Reaction entropy (4.30)

For a reaction or transformation $a\text{A} + b\text{B} \rightarrow c\text{C} + d\text{D}$ the reaction entropy $\Delta_r S_T$ at the temperature T and the pressure p is calculated by the procedure:

a) Find the **thermodynamic standard values** $S^\ominus_{298}$ and c_p for the substances A,B,C and D, which are included in the reaction.

b) Calculate the **entropy** $S(T, p)$ for **reactants** and **products** at the temperature T and the pressure p using (4.26).

c) the **reaction entropy** $\Delta_r S_T$ for the reaction is determined by

$$\Delta_r S_T = (c \cdot S_\text{C} + d \cdot S_\text{D}) - (a \cdot S_\text{A} + b \cdot S_\text{B}) \qquad (\text{J/mol K})$$

For calculation of reaction entropy $\Delta_r S_T$ in accordance with (4.30), the system temperature T and pressure p are assumed to be identical in the initial state and in the final state. Other circumstances in the development of the process are not important.

The specified calculation procedure (4.30) can be directly adjusted to reactions with either more or fewer reactants and products. Furthermore, if a reaction or transformation occurs at very high or very low temperature, the temperature dependence of the heat capacity should be taken into account in accurate calculations. For such calculations, $c_p = c_p(T)$ should be integrated into (4.16).

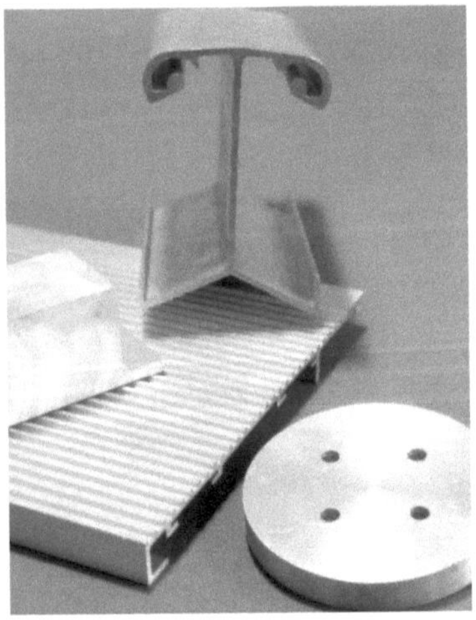

Figure 4.19. The metal aluminium forms a hard and wear-resisting surface layer of aluminium oxide Al_2O_3 "corundum" when exposed to the oxygen of the air. This oxide layer is crucial for the corrosion resistance of aluminium.

■ Metallic aluminium Al(s) forms a hard and wear-resistant surface layer of aluminium oxide Al_2O_3(s) ("corundum") when exposed to the oxygen of the air O_2(g). Oxidation occurs according to the following reaction equation

$$2\text{Al(s)} + \tfrac{3}{2}\text{O}_2\text{(g)} \quad \rightarrow \quad \text{Al}_2\text{O}_3\text{(s)} \qquad (a)$$

Determine the **reaction entropy** $\Delta_r S_T$ for the reaction (a) at 100 °C, when O_2(g) is absorbed from atmospheric air!

Solution. According to *table 1.4*, atmospheric air typically contains 20.95 vol-% $O_2(g)$; if the air is assumed to be an ideal gas mixture, the partial pressure of oxygen is

$$p(O_2(g)) = 0.2095 \cdot 101325\,\text{Pa} = 2.12 \cdot 10^4\,\text{Pa}$$

We now follow the procedure (4.30); according to item a), standard values are tabulated data

		Al(s)	$O_2(g)$	$Al_2O_3(s)$
$S^\ominus_{298}$	(J/mol K)	28.3	205.1	50.9
c_p	(J/mol K)	24.3	29.4	79.0

The entropy $S(T,p)$ of reactants and products is determined in accordance with item b) in the procedure (4.30). At $T = 373.15$ K (100 °C), the following figures can be calculated at the actual oxygen pressure

$$S(\text{Al}) = 28.3 + 24.3 \cdot \ln(\frac{373.15}{298.15}) = 33.8\,\text{J/mol K}$$

$$S(O_2) \quad = 205.1 + 29.4 \cdot \ln(\frac{373.15}{298.15}) - 8.314 \cdot \ln(\frac{2.12 \cdot 10^4}{101325}) = 224.7\,\text{J/mol K}$$

$$S(Al_2O_3) = 50.9 + 79.0 \cdot \ln(\frac{373.15}{298.15}) = 68.6\,\text{J/mol K}$$

The reaction entropy at (T,p) is determined in accordance with item c) in the procedure (4.30)

$$\Delta_r S_{373} = 1 \cdot 68.6 - (2 \cdot 33.8 + \tfrac{3}{2} \cdot 224.7) = \mathbf{-336.1\,J/mol\,K}$$

The high negative value for $\Delta_r S$ shows that the reaction has resulted in more regular (less random) molecular array in the system because the gas phase $O_2(g)$ is included in a more regular solid phase $Al_2O_3(s)$ after the reaction.

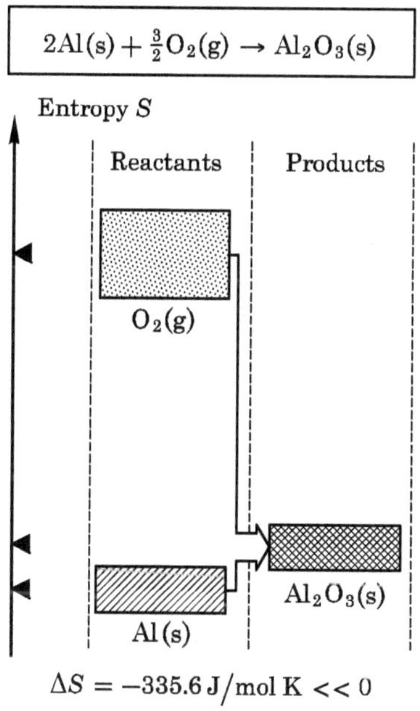

Figure 4.20. When $Al_2O_3(s)$ is formed from Al(s) and $O_2(g)$, ΔS is strongly negative; this is typical for processes where a gas phase with high entropy content is included as a reactant and where the resulting product is a condensed phase with low entropy content.

☐ **1.** From the reaction $Ca(OH)_2(s) + CO_2(g) \rightarrow CaCO_3(s) + H_2O(\ell)$, calculate the reaction entropy $\Delta_r S_T$ at 28 °C if $p(CO_2(g)) = 33$ Pa!

☐ **2.** From the reaction $H_2(g)+\frac{1}{2}O_2(g) \rightarrow H_2O(\ell)$, calculate the reaction entropy $\Delta_r S_T$ at 25 °C when $p(H_2) = 0.5$ atm, $p(O_2) = 0.5$ atm, and c_p is constant!

☐ **3.** Given the reaction $CaCO_3(s) \rightarrow CaO(s) + CO_2(g)$ and given the partial pressure at 1 atm $CO_2(g)$, calculate the standard reaction entropy at 800 °C at constant c_p!

☐ **4.** Specify whether the standard reaction entropy $\Delta_r S^\ominus_T$ will be increased or reduced, if $p(CO_2)$ in the above question 3 is reduced !

☐ **5.** For a given reaction, $\Delta_r S_T$ has a very high positive value; specify which change this will reflect in the molecular composition of the system !

4.9 Chemical equilibrium

We have now set up directions for calculating the entropy increase ΔS connected with **changes of state**, **chemical reactions**, and simple **phase transformations**. Next we shall describe how to combine these directions with the **second law** (4.12) into a calculation tool which makes it possible to address practical equilibrium problems.

The constant growth of entropy

The basis for setting up a thermodynamic equilibrium condition is the **Clausius inequality** in (4.12)

$$dS_{system} \geq \left(\frac{\delta Q}{T}\right)_{system} \qquad \text{"the Clausius inequality"} \qquad (4.31)$$

Assume a thermodynamic system that includes the entire thermodynamic universe. This – somewhat special – system constitutes, according to (2.2), an **isolated** thermodynamic system that can neither exchange matter nor energy. Thus, as a system, the thermodynamic universe is **adiabatic**, i.e. $\delta Q_{univ} = 0$. Then, by the Clausius inequality (4.31), $dS_{univ} \geq 0$ will apply. According to (4.12), this means that

$$dS_{univ} : \begin{cases} > 0 & \text{irreversible process (spontaneous)} \\ = 0 & \text{reversible process (equilibrium)} \\ < 0 & \text{impossible process} \end{cases} \quad (4.32)$$

The principle of the *constant increase of entropy* (Clausius 1865) is that any **irreversible** process leads to an increase of entropy in the universe; at **reversible** processes, the entropy in the universe remains unchanged.

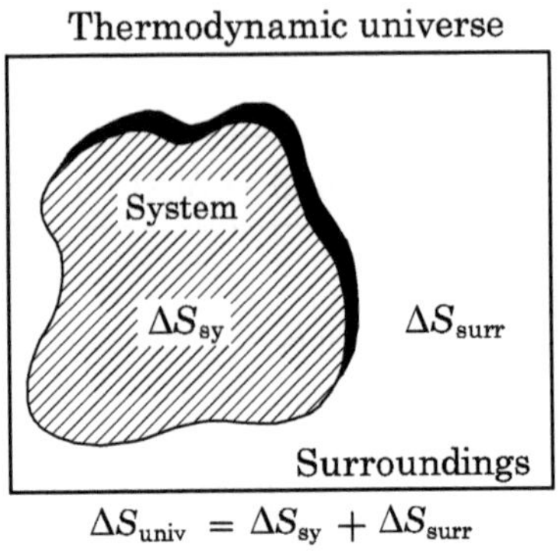

Figure 4.21. The increase in entropy in the thermodynamic universe is the sum of the increase in entropy of a system and its surroundings. Irreversible processes will always lead to an increase in the total entropy of the universe.

Thermodynamic equilibrium condition

Collectively, a **system of matter** and its **surroundings** form the thermodynamic universe; therefore, we can use the condition (4.32) for the system and its surroundings collectively.

It is now assumed that a process occurs at constant pressure p in the system of matter. By the process, the system is brought from the state of equilibrium (1) at (T, p) to the state of equilibrium (2) at (T, p). During the process the surroundings are assumed to be in thermodynamic equilibrium at the temperature T.

At constant pressure, the increase in the system enthalpy, $\Delta_r H_T$, p is equal to the supplied heat $Q_{1.2}$ (see 3.15). During the process, the heat is absorbed *from* the surroundings which are thus *supplied* reversibly with the amount of heat $Q_{surr} = -\Delta_r H_T$. If the unknown increase in entropy is denoted ΔS_{system}, introduction into (4.32) renders the following fundamental equilibrium condition

Reversible process

$$\Delta S_{\text{system}} - \frac{\Delta H}{T} = 0$$

Thermodynamic equilibrium condition I (4.33)

During a process at constant **pressure** p and **temperature** T, where *only* **volume work** is performed on a system, the increase in entropy is

$$\Delta S_{univ} = \Delta S_{system} - \frac{\Delta_r H_T}{T} : \begin{cases} > 0 & \text{irreversible process} \\ = 0 & \text{reversible process} \\ < 0 & \text{impossible process} \end{cases}$$

where $\Delta_r H_T$ denotes the **reaction enthalpy** for the process and ΔS_{system} denotes the increase in the system entropy.

Irreversible process

$$\Delta S_{\text{system}} - \frac{\Delta H}{T} > 0$$

In other words, any spontaneous process is producing entropy in the thermodynamic universe. A system has obtained equilibrium when the entropy in the system + surroundings reaches a maximum value $S_{\max}$. If we calculate the total increase in entropy (4.33) for a given process, we can, therefore, decide whether we are dealing with a spontaneous **irreversible process** or a **reversible equilibrium process**, or whether the process is **impossible**.

The following example shows how to calculate the partial pressure for saturated water vapour for the equilibrium condition (4.33). However, an equilibrium condition as in (4.33) is not very practical. Therefore, in chapter 5, we shall develop a more efficient calculation method by introducing another state function: the **Gibbs free energy** G.

■ Calculate by (4.33) the partial pressure p_s of saturated water vapour at 25 °C and compare the result with table data!

Solution. Assume a closed receptacle containing water $H_2O(\ell)$ and water vapour $H_2O(g)$ in equilibrium at 25 °C. In this system we consider the phase transition

$$H_2O(\ell) \rightarrow H_2O(g) \qquad T = 298.15\,\text{K} \quad \text{(a)}$$

If the system is in equilibrium and the partial pressure of the water vapour is $p = p_s$, the transformation (a) will be a **reversible process of equilibrium**; therefore, according to (4.33), the increase in entropy ΔS_{univ} will be zero. The thermodynamic standard values $H^{\ominus}_{298}$ and $S^{\ominus}_{298}$ for water and water vapour are tabulated data

$$\begin{array}{lll} H_2O(g): & H^{\ominus}_{298} = -241.8\,\text{kJ/mol}; & S^{\ominus}_{298} = 188.7\ \text{J/mol K} \\ H_2O(\ell): & -285.8\,\text{kJ/mol}; & 69.9\ \text{J/mol K} \end{array}$$

The system increase in entropy $\Delta_r S_{298}$ is calculated by the procedure (4.30);

$$\Delta_r S_{298} = (188.7 - R\cdot\ln(\frac{p_s}{101325})) - 69.9 \quad = 118.8 - R\cdot\ln(\frac{p_s}{101325})\ \text{J/mol K}$$

The reaction enthalpy $\Delta_r H_{298}$ of the system is calculated by the procedure (3.55); as the process occurs at the reference temperature 298.15 K, $\Delta_r c_p$ is not included in the calculation and we have

$$\Delta_r H_{298} = H^{\ominus}_{298.g} - H^{\ominus}_{298.\ell} = -241.8 - (-285.8) = 44.0\,\text{kJ/mol}$$

Now, by (4.33), the condition for equilibrium and thus for a reversible transformation of water into water vapour is

$$\Delta_r S_{298} - \frac{\Delta_r H_{298}}{298.15} = 118.8 - R\cdot\ln(\frac{p_s}{101325}) - \frac{44000}{298.15} = 0 \qquad \text{(b)}$$

By reduction, this equation determines the partial pressure of saturated water vapour p_s with the unit (Pa)

Partial pressure p_s of saturated water vapour calculated by (4.33) = **3180.7** Pa
Tabulated partial pressure of water vapour = 3169.1 Pa

The calculated partial pressure p_s of saturated water vapour deviates by approximately **0.4 %** from the specified table value.

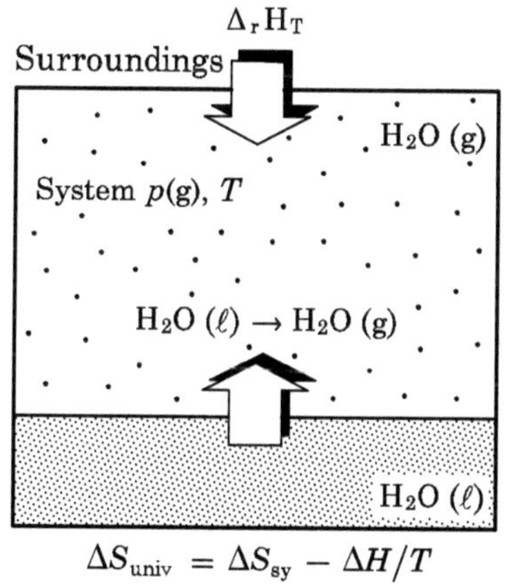

Figure 4.22. $\Delta S_{sy} - \Delta H/T = 0$ for a reversible phase transformation of water into water vapour; at constant p, T, this condition determines the partial pressure of saturated water vapour at the given temperature T.

☐ **1.** By a reversible process, 2 kg of ice are transformed into 2 kg of water at 0 °C; the heat of fusion of the ice is 6008 J/mol. Calculate ΔS_{univ}, ΔS_{system}, and ΔS_{surr} for the process!

☐ **2.** An adiabatic system contains 1 kg of water at $\theta = 20\,°C$. The work W on the system is performed by stirring which increases θ by 1 °C. Determine ΔS_{univ}, ΔS_{system}, and ΔS_{surr}!

☐ **3.** The reaction $CaSO_4(s) + 2H_2O(\ell) \rightarrow CaSO_4 \cdot 2H_2O(s)$ is given. Determine by (4.33) whether the reaction can occur spontaneously at 25 °C!

☐ **4.** Tin Sn(s) exists in two allotropic forms: White Sn and grey Sn. Find by (4.33) which state is stable at 25 °C!

☐ **5.** The reaction $Hg(\ell) \rightarrow Hg(g)$ is given. Calculate the partial pressure of saturated mercury vapour p_s (Pa) at 25 °C using the equilibrium condition (4.33)!

4.10 The concept of entropy

Classical thermodynamics *only* addresses the **macroscopic** properties of thermodynamic systems such as temperature, pressure, volume, internal energy, and entropy. The underlying molecular explanation of the properties of substances and changes hereof is treated within the special discipline called **statistical thermodynamics**.

In the **second law** of thermodynamics, the existence of a state function, the **entropy** S, is ascertained. This state function is then introduced without further explanation as a suitable calculation quantity. In the following, we shall briefly look at the the statistic nature of entropy and thus try to obtain a physical understanding – though feeble – of this difficult quantity.

Micro states

The macroscopic description of the state of a thermodynamic system does *not* address the spatial coordinates and velocities of the individual atoms. However, it is useful to realize the following ambivalence:

- *any* observed **macro state** of a system, e.g. an ideal gas, reflects one instantaneous **micro state** in the system determined by the energy state, location and velocity of the individual particles,

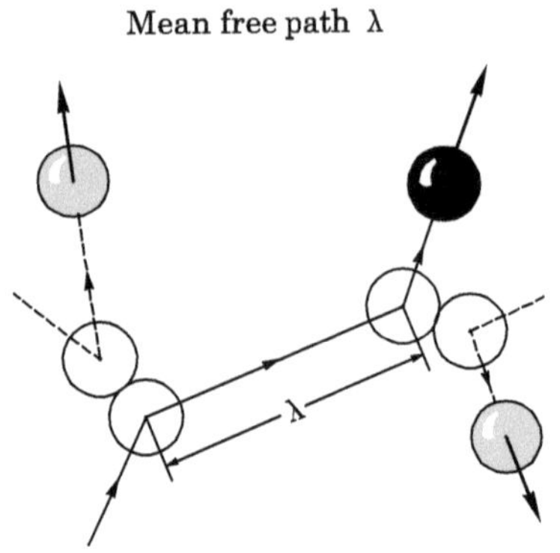

Figure 4.23. The mean free path specifies the average distance λ travelled by a gas particle between two collisions; at atmospheric pressure, $\lambda \simeq 10^{-7}$ m. This means that typically the individual gas particle changes its velocity and direction approximately 10 billion times per second at room temperature.

- a *certain* **macro state**, corresponds to a large number Ω of *different* **micro states** in the system.

In other words, we *cannot* tell the micro state of the system although the macro state of the system is known. We can only draw the conclusion that the micro state is one of the Ω states that correspond to the *same* actual macro state.

Now, assume that at any time we know the location and velocity of each individual atom or molecule – in the form of a table, for example – in one mol of gas in the standard state. This table of the micro state of the gas would then change at a chaotic velocity. In a gas, the individual atom or molecule typically moves at a velocity v of 1000 m/s (3600 km/h), see for example figure 1.13. The **mean free path** λ – i.e. the average distance travelled by a gas particle between two collisions – is approximately 10^{-7} m (0.1 μm) at atmospheric pressure. The individual atom or molecule in a gas at the standard state, therefore, changes velocity and direction about $v/\lambda = 10\,000\,000\,000$ times per second!

The number of possible **micro states** in a system with $6.02 \cdot 10^{23}$ particles is extremely large. On the other hand, there is extremely little probability that the particles at two different times can be retrieved in exactly the *same* micro state. Therefore, the probability that *all* $6.02 \cdot 10^{23}$ particles in a mol of gas are located at the same time in half of the available volume and here exert double pressure is extremely low. It is important to note the following: For an ideal gas, *all* micro states are in principle equally probable, including the micro states where the gas gathers in half of the volume and exerts double pressure.

What is then the reason why a gas usually spreads over an available volume and apparently conforms exactly to the macroscopic description of state of the ideal gas law?

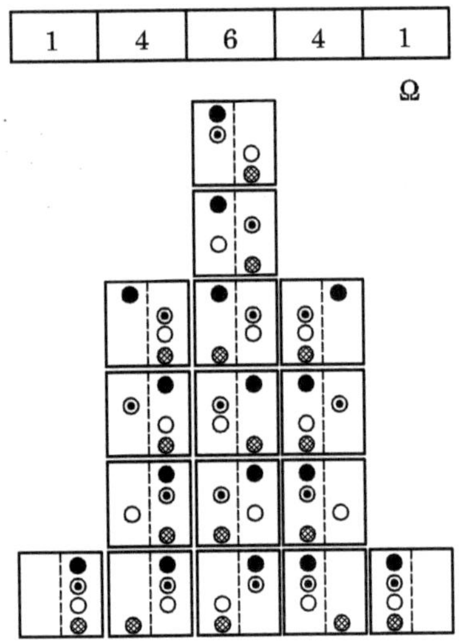

Figure 4.24. The probable disorder is shown schematically for a receptacle with four mutually independent gas particles. At a given time, the probability that the particles are evenly distributed with two in each half of the receptacle is (6/16); the probability that all particles are, for example, in the left half of the receptacle is only (1/16).

The probable disorder

That this is the case can best be illustrated by simple analogy. Assume a deck of cards sorted in a systematic order within each colour: *Spades 1,2,$\cdots$, Hearts 1,2,$\cdots$* etc. This arrangement of the cards mirrors *one* possible micro state of the deck of cards. If we shuffle the cards carefully there is very little probability that we retrieve exactly the same micro state with a systematic, sorted sequence of cards. Again, the probability that the cards are in the same, sorted order after shuffling as before shuffling is the *same* as the probability that any other distribution of the cards given *beforehand* is retrieved.

However, since a deck of cards – with its 52 "particles" – has an extremely large number of different micro states corresponding to *shuffled*, and only one micro state corresponding to *ordered*, the latter result of a shuffling process is regarded as an "impossible" event. Cutting the deck of cards results in a **spontaneous** change from the *sorted* state to a *shuffled* state with a higher "degree of disorder".

The Boltzmann relation

The constant growth of entropy postulated in the second law is closely related to the following principle: By the **irreversible** process the system is spontaneously directed towards the most frequently occurring set of micro states which corresponds to the same macro state of the system. The macroscopic state of equilibrium – where entropy is at its maximum – is the state which is linked to the largest number of different micro states. This basic concept is the basis of the **Boltzmann relation** that defines the entropy function S in statistical thermodynamics.

> **The Boltzmann relation** (4.34)
>
> The **entropy** S of a thermodynamic system is defined by the relation
>
> $S = k \cdot \ln(\Omega)$ (J/K)
>
> where Ω denotes the number of **micro states** that correspond to the *same*, actual **macro state**, and $k = R/\mathcal{N}$ is the **Boltzmann constant**.

The Boltzmann constant

$$k = \frac{R}{\mathcal{N}} = 1.3806 \cdot 10^{-23}\ \mathrm{J/K}$$

When interpreted statistically, the entropy S can be understood as a measure of "the degree of disorder" in a thermodynamic system. Growing entropy corresponds to a higher degree of molecular disorder. To calculate the quantity Ω of a system, one shall take account of the quantized distribution of energy on the elementary particles of the system as well as the distribution of the particles in the available volume.

Can entropy decrease?

According to **classical thermodynamics**, any process occurring in an isolated system is connected with an *increase* in entropy; according to the **Clausius inequality**, a spontaneous process reducing the entropy in an isolated system is an "impossible" process.

Within **statistical thermodynamics** the relationship is somewhat different: A process that spontaneously reduces the entropy of an isolated thermodynamic system is here possible – these kinds of processes are, however, unlikely.

To understand this apparently contradictory notion of the concept "impossible process", it can be useful – even if only roughly – to look into a specific example of the **Boltzmann relation** (4.34). Assume an isolated system containing 1 mol of ideal gas in thermodynamic equilibrium in state (1). What is the probability that a spontaneous change of state will reduce the entropy content of the system by $\Delta S = 1$ J/K?

We assume that any configuration of gas particles in the system is equally possible; the number of micro states Ω_1 and Ω_2 corresponding to the system in states (1) and (2), respectively, according to the Boltzmann relation (4.34) are

$$\Omega_1 = \exp(\frac{S_1}{k}); \qquad \Omega_2 = \exp(\frac{S_2}{k}) \qquad (4.35)$$

The relation Φ between the number of micro states Ω_2, which correspond to state (2), and the number of micro states Ω_1, which correspond to state (1), is then a measure of the probability of the spontaneous event (1)→(2). We therefore have

$$\Phi \sim \frac{\Omega_2}{\Omega_1} \sim \exp(\frac{S_2 - S_1}{k}) = \exp(\frac{-1}{k}) \sim 10^{-3.1 \cdot 10^{22}} \qquad (4.36)$$

Thus, the probability of a spontaneous reduction of the system entropy by 1 J/K is extremely little bordering the "impossible". If we tried to write Φ as a decimal number 0.00 ··· 001, the number would be approximately 4000 light years long after the dot - with the font used here!

$P = (\frac{1}{2})^n$

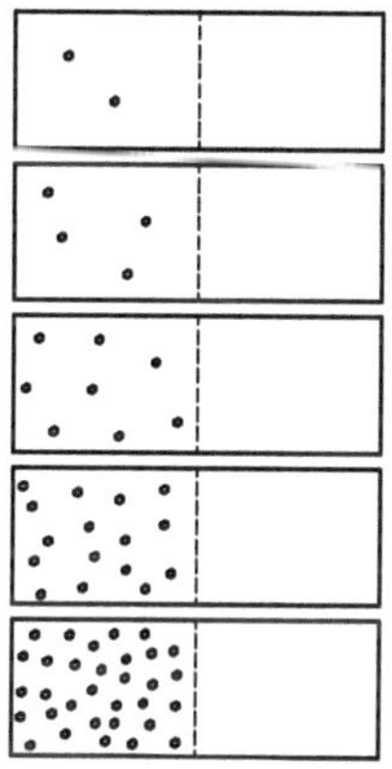

n	P
2	0.25
4	0.0625
8	$4 \cdot 10^{-3}$
16	$1 \cdot 10^{-5}$
32	$2 \cdot 10^{-10}$

Figure 4.25. The probability that n mutually independent gas particles are located at the same time in one half of an available volume decreases drastically with growing n. Such degenerate states are therefore not known in practical gas systems where n corresponds to the order of magnitude (mol), i.e. 10^{23}.

Phase transformation and disorder

At a given temperature, the number of possible micro states are smaller in a **crystal** than in a **liquid**; melting of substances, therefore, is *always* connected with an increase in entropy. When a substance is transformed from a liquid into a **gas**, the degree of disorder is increased; therefore, an evaporation process is *always* connected with an increase in entropy. For phase transition in systems of matter, therefore, it generally applies that

$$S_{solid} < S_{\ell iquid} < S_{gas} \qquad (4.37)$$

Through the Boltzmann relation, the macroscopic entropy concept of classical thermodynamics is connected with the molecular statistic quantity Ω that specifies the number of micro states corresponding to the same macro state. In many cases, this relationship can be used for simple, qualitative estimates of the properties and behaviour of the substances.

■ Assume n mol of **ideal gas** in thermodynamic equilibrium at the temperature T (K). The gas volume is V (m^3) and the molar mass of the gas is M (kg/mol). By statistical considerations it is shown that the number of micro states Ω that correspond to the state of equilibrium are determined by

$$\Omega = f(T, M, n) \cdot V^{n\mathcal{N}} \tag{a}$$

where $\mathcal{N}$ is the Avogadro constant. Derive from (a) an expression for the increase in entropy $\Delta S_{1.2}$ by an isothermal expansion of the gas from V_1 to V_2!

Solution. We express $\Delta S_{1.2} = S_2 - S_1$ using the Boltzmann relation (4.34); since the process is isothermal, $f(T, M, n) = \text{constant}$

$$\Delta S = S_2 - S_1 = k \cdot \ln(\frac{\Omega_2}{\Omega_1}) = k \cdot \ln(\frac{V_2}{V_1})^{n\mathcal{N}} \tag{b}$$

$$\Delta S = n \cdot \mathcal{N} \cdot k \cdot \ln(\frac{V_2}{V_1}) = n \cdot \mathcal{N} \cdot \frac{R}{\mathcal{N}} \cdot \ln(\frac{V_2}{V_1}) = nR \cdot \ln(\frac{V_2}{V_1}) \tag{c}$$

This statistically derived result is in conformity with the classical expression (4.22) found previously.

Phase transformation

$S_{solid} < S_{liquid} < S_{gas}$

☐ **1.** What is the probability that the cards in a well-shuffled deck of cards are placed in one particular order given beforehand?

☐ **2.** Calculate the value of the Boltzmann constant k using the unit (cal/K), which has previously been used frequently !

☐ **3.** Determine the number of micro states Ω in a) 1 mol of graphite C(s) and b) in 1 mol of diamond C(s); discuss the difference in entropy in the standard state!

☐ **4.** Find from the table of thermodynamic standard values minimum three examples showing that the inequality (4.37) is fulfilled!

☐ **5.** Determine the probability of a spontaneous drop in the gas entropy of 0.001 J/K in a mole of ideal gas!

List of key ideas

The following overview lists the key definitions, concepts and terms introduced in Chapter 4.

Thermodynamic equilibrium A system is in thermodynamic equilibrium if, at the same time, it is in **Thermal equilibrium**, **Mechanical equilibrium** and **Chemical equilibrium**.

Thermal equilibrium ...... A system is in thermal equilibrium when it has the same temperature T in the *entire* system and this temperature is the same as in the surroundings, or the system is adiabatic.

Mechanical equilibrium .... A system is in mechanical equilibrium when there are *no* **unbalanced forces** in the system.

Chemical equilibrium ...... A system is in chemical equilibrium when *no* **chemical reactions** or **phase transformations** are possible in the system.

Reversible process .... A process that leads a system through states with only **infinitesimal deviation** from **equilibrium**; in reversible processes, $\Delta S = 0$ in the thermodynamic universe.

Irreversible processes ..... Any process which is *not* reversible is called **irreversible**; in an irreversible process, $\Delta S > 0$ in the thermodynamic universe.

Thermal efficiency .... $\eta = \dfrac{\text{Work performed on surroundings}}{\text{Heat absorbed from surroundings}} = \dfrac{-W_{cycle}}{Q_1}$

For a **reversible** Carnot cycle,

$$\eta = 1 + \frac{Q_3}{Q_1} = 1 - \frac{T_3}{T_1} \quad \text{and} \quad \sum \left(\frac{Q}{T}\right)_{rev} = 0$$

Carnot cycle .... A cyclic process during which a thermodynamic system cycles between two heat reservoirs with different temperatures.

Second law .... It is *impossible* for a system performing a **cyclic process** to perform work on its surroundings, if heat is supplied from one reservoir only.

The third law .... The entropy of **chemically pure** crystals at the thermodynamic zero point 0 K is **zero**.

Entropy S .... $dS \stackrel{\text{def}}{=} \dfrac{\delta Q_{rev}}{T}$ (definition)

S is a **state function** (postulate)

$dS_{system} \geq \left(\dfrac{\delta Q}{T}\right)_{system}$ (Clausius' inequality)

The Clausius inequality ... $dS_{system} \geq \left(\dfrac{\delta Q}{T}\right)_{system}$ (general)

$dS_{system} \geq 0$ (adiabatic system)

ΔS, change of state .... ΔS is calculated for a **reversible** change of state; the result applies to *any* process connecting the **states of equilibrium** (1) and (2)

$$\Delta S = \int_{S_1}^{S_2} dS = \int_1^2 \left(\frac{\delta Q}{T}\right)_{rev}$$

ΔS, temperature change .. ΔS at a temperature change connecting the **states of equilibrium** (1) and (2)

$\Delta S_{1.2} = nc_p \cdot \ln\left(\dfrac{T_2}{T_1}\right)$ (p constant)

$\Delta S_{1.2} = nc_V \cdot \ln\left(\dfrac{T_2}{T_1}\right)$ (V constant)

ΔS, ideal gas .... For an **isothermal** change of state with an **ideal** gas connecting the **states of equilibrium** (1) and (2), the increase in the system entropy is

$$\Delta S_{1.2} = nR \cdot \ln\left(\frac{V_2}{V_1}\right) = -nR \cdot \ln\left(\frac{p_2}{p_1}\right)$$

ΔS, phase transformation For an **isothermal** phase transformation at **constant pressure** p the increase in entropy is

$\Delta S = \dfrac{\Delta_r H_T}{T}$ (T, p constant)

Entropy $S^\ominus_{298}$ .... The standard entropy $S^\ominus_{298}$ denotes **molar entropy** at the **standard pressure** $p^\ominus = 101325\,\text{Pa}$, and for dissolved substances at the **standard concentration** $c^\ominus = 1\,\text{mol}/\ell$ when the temperature is $T = 298.15\,\text{K}$. Can be found in **tables**.

Standard state .... The **standard state** ($p^\ominus, c^\ominus$) is defined by the pressure $p^\ominus = 101325$ Pa and for dissolved substances by the concentration $c^\ominus = 1\,\text{mol}/\ell$.

Entropy $S(T,p)$ .... For **solid** and **liquid** substances, $S(T,p)$ is determined by

$$S(T,p) = S^\ominus_{298} + c_p \cdot \ln\left(\frac{T}{298.15}\right)$$

For **ideal gases**, correction for the influence of pressure shall be made, and $S(T,p)$ is determined by

$$S(T,p) = S^\ominus_{298} + c_p \cdot \ln\left(\frac{T}{298.15}\right) - R \cdot \ln\left(\frac{p}{101325}\right)$$

Reaction entropy .... The reaction entropy $\Delta_r S_T$ for a **chemical reaction** or **phase transformation** denotes the total entropy of the system in its **final state** minus the total entropy of the system in its **initial state**

$$\Delta_r S = \sum S(\text{products}) - \sum S(\text{reactants})$$

Equilibrium condition I .... During a process at constant **pressure** p and **temperature** T, during which *only* **volume work** is performed on a system, the following condition is fulfilled

$$\Delta S_{system} - \frac{\Delta_r H_T}{T} : \begin{cases} > 0 & \textbf{irreversible}\text{ process} \\ = 0 & \textbf{reversible}\text{ process} \\ < 0 & \textbf{impossible}\text{ process} \end{cases}$$

The Boltzmann relation ... $S = k \cdot \ln(\Omega)$ (J/K)

where Ω denotes the number of **micro states** corresponding to the same, actual **macro state** of the system, and $k = R/\mathcal{N} = 1.3806 \cdot 10^{-23}$ J/K is the **Boltzmann constant**.

Phase transformations ..... Generally, for phase transformations in systems of matter, **melting** and **evaporation** bring along an increase in entropy, i.e.

$$S_{crystal} < S_{liquid} < S_{gas} \qquad \text{(general)}$$

Examples

The following examples illustrate how subject matter explained in Chapter 4 can be applied in practical calculations.

Example 4.1

■ Transformation of tin at a low temperature – "tin pest"

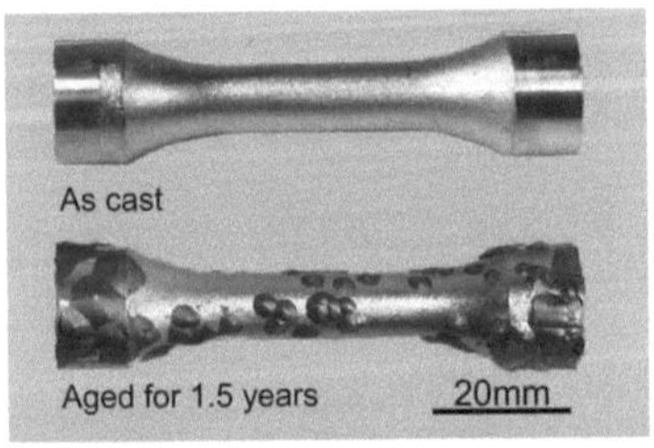

Figure 4.26. Tin pest attack on a specimen of white β-Sn; the disintegrating areas consist of grey α-Sn, which is the thermodynamically stable form of tin at temperatures below approximately 14°C.

The pure metal tin (*stannum* Sn) exists in two different allotropic forms. At higher temperatures, the stable form is a **silver-white metal** of poor hardness. This modification, so-called "white tin" or β-Sn, deposits in a **tetragonal** crystal lattice with density $7285\,\text{kg/m}^3$. At lower temperatures, the stable modification is a grey **metalloid** form. This low-temperature modification, so-called "grey tin" or α-Sn, deposits in a **cubic** crystal lattice with density $5769\,\text{kg/m}^3$.

Normally, the transformation from white tin into grey tin at low temperatures occurs very slowly; thus, below the transformation temperature, white thin can be *durable* for a long time until **nuclei** of grey tin are formed or supplied. When first the transformation has started, it will spread slowly. Since the density of grey tin is less than that of white tin, the transformation shows in the form of abscess-like growths on the surface of the white tin. This transformation into grey tin at low temperatures is therefore colloquially called "tin pest".

Problem. Calculate, based on thermodynamic standard values of α-Sn and β-Sn, the temperature T_{trans} at which white tin and grey tin are in thermodynamic equilibrium and specify in which temperature range there is a risk of "tin pest"!

Conditions. According to data from Barin: *Thermochemical Data of Pure Substances*, thermodynamic standard values of α-Sn and β-Sn are:

Metal	Chem. desig.	$H^{\ominus}_{298}$ (kJ/mol)	$S^{\ominus}_{298}$ (J/mol K)	c_p (J/mol K)
Grey tin	α-Sn(s)	-1.98	44.3	25.8
White tin	β-Sn(s)	0.00	51.2	27.0

For calculation of the transformation temperature T_{trans}, thermodynamic equilibrium in the system is assumed. Delayed transformation due to lack of nucleation formation is disregarded.

Solution. The transformation temperature T_{trans} denotes the temperature at which the two tin modifications are in thermodynamic equilibrium; therefore, we try to determine the temperature T, at which the transformation from white tin into grey tin is reversible (equilibrium).

$$\beta\text{-Sn(s)} \quad \rightarrow \quad \alpha\text{-Sn(s)} \qquad (T, p \text{ constant}) \qquad \text{(a)}$$

The reaction enthalpy $\Delta_r H_T$ for the transformation (a) can be expressed as a function of the temperature T; using the procedure (3.55), we get

$$\Delta_r H_T = \Delta_r H^{\ominus}_{298} + \Delta_r c_p \cdot (T - 298.15) \qquad \text{(J/mol)}$$

$$\Delta_r H_T = (-1980 - 0.0) + (25.8 - 27.0) \cdot (T - 298.15) = -1622 - 1.2 \cdot T \qquad \text{(J/mol)}$$

The reaction entropy $\Delta_r S_T$ is determined by (4.30) introducing the temperature T as a variable

$$\Delta_r S_T = \Delta_r S^{\ominus}_{298} + \Delta_r c_p \cdot \ln(\frac{T}{298.15}) \qquad \text{(J/mol K)}$$

$$\Delta_r S_T = (44.3 - 51.2) + (25.8 - 27.0) \cdot \ln(\frac{T}{298.15}) \; = -0.06 - 1.2 \cdot \ln(T) \; \text{(J/mol K)}$$

We can now use the equilibrium condition (4.33). In a reversible transformation, the total entropy increase in the system + surroundings is zero; when this condition is expressed by introduction of $\Delta_r H_T$ and $\Delta_r S_T$ we get

$$\Delta S_{univ} = \Delta S_{system} + \Delta S_{surr} = (-0.06 - 1.2 \cdot \ln(T)) - \frac{-1622 - 1.2 \cdot T}{T} = 0 \quad \text{(b)}$$

$$\Delta S_{univ} = 1.14 - 1.2 \cdot \ln(T) + 1622 \cdot T^{-1} = 0 \qquad \text{(c)}$$

The equilibrium temperature T can be determined iteratively from this equation, e.g. by using the **Newton-Raphson** method (see *Mathematical appendix, subchapter A.3*). The solution to (c) is $T_{trans} \simeq 287\,\text{K}$ (14 °C). Thus, it can be expected that there is a potential risk of tin pest at temperatures $\theta <$ approximately 14 °C.

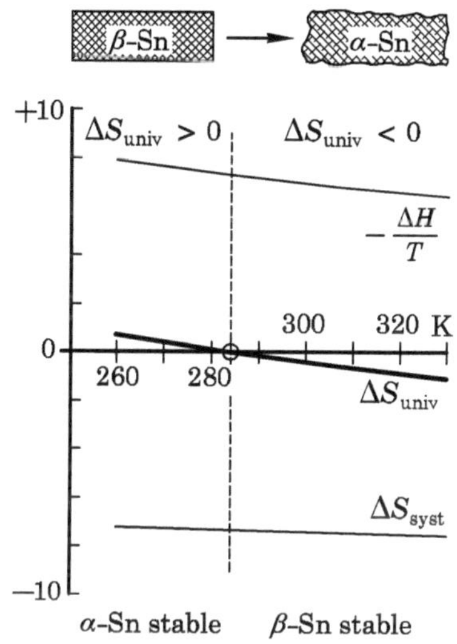

Figure 4.27. Determination of tin stability. For $T < 287\,K$, α-Sn is stable since the transformation of white tin into grey tin occurs spontaneously in this temperature range ($\Delta S_{univ} > 0$).

Discussion. According to Holleman–Wiberg: *Lehrbuch der Anorganischen Chemie*, the metallic, white tin can be transformed at temperatures below 13.2 °C; a destructive transformation, however, only occurs if nuclei of the grey α-Sn are formed at lower temperatures. Holleman–Wiberg state that the maximum transformation rate occurs at approximately −45 °C.

Tin pest has caused destruction of many tin objects in cold areas; the phenomenon is also relevant, for example, for storing objects of art in unheated buildings.

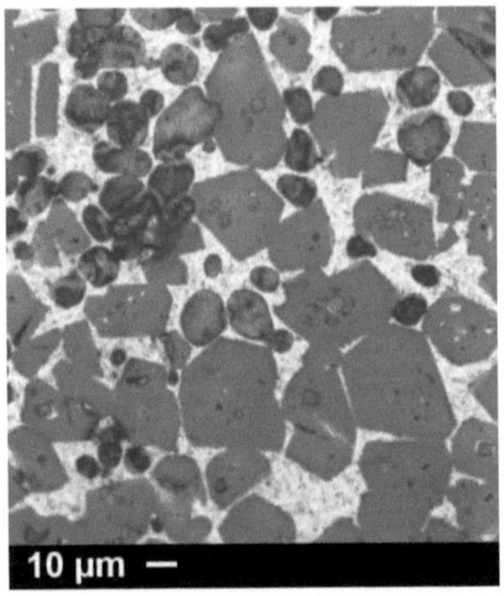

Figure 4.28. On plane-polished and etched cement clinkers, the individual clinker minerals can be seen in a microscope. The light groundmass consists of intermingled fine C_3A and C_4AF crystals. Darker crystals of C_3S and C_2S are embedded.

Figure 4.29. Modern tube mill for grinding of Portland cement; note the dimensions compared to the person standing on the floor. Such mills can grind 100-200 tons of cement clinkers per hour.

Example 4.2

■ Dehydration of gypsum when grinding Portland cement

Portland cement is produced by burning clay and lime-containing materials. During the burning at 1400-1500 °C, the raw materials **sinter** into sphere-shaped **cement clinkers**. During sintering the **calcium silicates** and **calcium aluminates** that form the binder phase in Portland cement are formed. The four most important of these so-called **clinker minerals** with binder properties are

Designation	Chemical composition	Cement-chemical name	Typical content
Tricalcium silicate	$3CaO \cdot SiO_2$	C_3S	55 %
Dicalcium silicate	$2CaO \cdot SiO_2$	C_2S	20 %
Tricalcium silicate	$3CaO \cdot Al_2O_3$	C_3A	7 %
Tetracalcium aluminoferrite	$4CaO \cdot Al_2O_3 \cdot Fe_2O_3$	C_4AF	9 %

In the frequently used cement-chemical designations, the following "abbreviations" for oxides in the clinker minerals are used

$$\text{'C'} = CaO; \quad \text{'S'} = SiO_2; \quad \text{'A'} = Al_2O_3; \quad \text{'F'} = Fe_2O_3;$$

After burning, the cement clinker is ground to a particle size distribution of mean size of about 10 μm. During grinding, 3-5 wt-% of **gypsum** $CaSO_4 \cdot 2H_2O$ is added. Addition of gypsum is necessary to give the cement the required useful qualities. The clinker mineral C_3A reacts very rapidly with water; therefore, a concrete made from cement *without* addition of gypsum may set and stiffen within a few minutes. This rapid setting by C_3A reaction is called "flash setting". Flash setting makes a concrete mix useless, since the concrete cannot be properly placed in forms. By adding gypsum, flash setting is prevented so that the concrete can be cast and be workable for several hours after mixing.

Grinding of cement clinkers is performed in large **ball mills**. During the grinding, the temperature in the ball mill can rise to more than 100 °C due to the violent mechanical processing. High temperature during grinding can cause total or partial **dehydration** of the added gypsum by the following transformation

$$CaSO_4 \cdot 2H_2O(s) \quad \rightarrow \quad CaSO_4 \cdot \tfrac{1}{2}H_2O(s) + 1\tfrac{1}{2}H_2O(g) \tag{a}$$

The **hemihydrate** $CaSO_4 \cdot \frac{1}{2}H_2O$ corresponds to Plaster of Paris and is formed by dehydration. A fresh concrete mix containing a large proportion of hemihydrate may exhibit "false setting" shortly after being mixed with water. The phenomenon is caused by the fact that during mixing the hemihydrate recrystallizes and produces large lath-shaped crystals of the dihydrate $CaSO_4 \cdot 2H_2O$. False setting in a concrete can make control of the concrete's consistency and workability difficult.

Problem. Provide a calculation of the maximum temperature θ (°C) to which the gypsum can be exposed during grinding if dehydration according to reaction equation (a) is to be *avoided*. Assume in the calculation that the partial pressure of the water vapour is 1 atm at any prospective transition temperature above 100 °C.

Conditions. In the calculation we assume the condition that the dehydration reaction (a) is reversible. This calculation gives **no** information of the possible rate of dehydration! Water vapour $H_2O(g)$ is assumed to be an ideal gas.

The following thermodynamic standard values are used for reactants and products in reaction equation (a)

Component	$H^{\ominus}_{298}$ (kJ/mol)	$S^{\ominus}_{298}$ (J/mol K)	c_p (J/mol K)
$CaSO_4 \cdot 2H_2O(s)$	−2021.1	194.0	186.2
$CaSO_4 \cdot \frac{1}{2}H_2O(s)$	−1575.2	130.5	119.7
$H_2O(g)$	−241.8	188.7	33.6

Solution. The reaction enthalpy $\Delta_r H_T$ for the dehydration reaction (a) is calculated by the procedure (3.55); thus, the targeted equilibrium temperature T is introduced as an unknown variable

$$\Delta_r H^{\ominus}_{298} = -1575.2 - 1\tfrac{1}{2} \cdot 241.8 - (-2021.1) = 83.2 \text{ kJ/mol}$$

$$\Delta_r c_p = 119.7 + 1\tfrac{1}{2} \cdot 33.6 - 186.2 = -16.1 \text{ J/mol K}$$

$$\Delta_r H_T = 83200 + (-16.1) \cdot (T - 298.15) = 88000 - 16.1 \cdot T \text{ (J/mol)}$$

Similarly, the reaction entropy $\Delta_r S_T$ is determined by the procedure (4.30); the temperature T is introduced as an unknown variable, and the partial pressure of the water vapour is assumed to be $p = 101325$ Pa. The quantity $\Delta_r c_p$ from the above calculation of $\Delta_r H_T$ is used.

$$\Delta_r S^{\ominus}_{298} = (130.5 + 1\tfrac{1}{2} \cdot 188.7) - 194.0 = 219.6 \text{ J/mol K}$$

$$\Delta_r S_T = 219.6 + (-16.1) \cdot \ln(\frac{T}{298.15}) = 311.3 - 16.1 \cdot \ln(T) \text{(J/mol K)}$$

We can now determine the temperature T_a, at which the dehydration reaction (a) is reversible, i.e. like an equilibrium reaction. By introduction into the thermodynamic equilibrium condition (4.33), we get the condition

$$\Delta_r S_T - \frac{\Delta_r H_T}{T} = (311.3 - 16.1 \cdot \ln(T)) - \frac{88000 - 16.1 \cdot T}{T} = 0 \qquad \text{(b)}$$

$$327.4 - 16.1 \cdot \ln(T) - 88000 \cdot T^{-1} = 0 \qquad \text{(c)}$$

Solving this equation in T by iteration (see *Mathematical appendix, subchapter A3*), we get the targeted transformation temperature: $T_a \simeq 380\,K$ (107 °C). Based on the given data and assumptions, therefore, it is calculated that **initial dehydration** of the gypsum can occur at temperatures of approximately 107 °C, i.e. that 107 °C is the maximum temperature at which gypsum will not dehydrate. As stated previously, the rate of dehydration *cannot* be assessed on the basis of the above.

Figure 4.30. Casting, compacting and striking off of a rolled concrete sheet. For such concrete products, the consistency and workability of the fresh concrete is decisive for the quality achieved; even moderate variations in the setting properties of the concrete may cause difficulties during casting.

Discussion. In the evaluation of the result it is important to note the following. The dehydration temperature T_a for the dihydrate depends on the partial pressure of the water vapour $p(H_2O)$ in the ambient atmosphere. In the example, $p(H_2O) \simeq 101325$ Pa was assumed. The reason for this assumption is that in practice the charge of clinker and gypsum being ground in cement mills is sprayed with a small amount of water spray during grinding to counteract dehydration of the gypsum.

If the partial pressure of water vapour in the ambient atmosphere is decreased, T_a is also reduced, i.e. the dehydration occurs at a lower temperature. At very low humidity levels, the dehydration may thus occur even at room temperature. It may be useful to compare the calculations with example 3.4 and example 4.4 both of which address the same problem.

Example 4.3

■ Computerized calculation of the partial pressure of saturated water vapour

Concrete is a mixture of **cement**, **water** and fine and coarse **aggregates**. The active binder phase is composed of the mixture of cement and water, the so-called **cement paste**.

During the first hours after casting, a concrete is plastic and workable. Setting, i.e. the time where the concrete stiffens and becomes dimensionally stable, will normally occur 3-5 hours after mixing. In the period immediately after setting, the strength of the concrete is very low. If the concrete is exposed to severe drying during this period, detrimental cracking can occur at the surface of the cast items. Such cracks occur due to so-called "plastic shrinkage" caused by capillary tensile stresses developed in the hardening cement paste. In particular, plastic shrinkage cracks cause problems when the concrete is cast at **low humidity** and under conditions of **high wind velocity** since the concrete surface rapidly dries out under such circumstances.

Figure 4.31. Plastic shrinkage with crack formation is a well-known phenomenon in nature and can be seen, for example, in summer where clayey or silty deposits dry out. The picture shows cracks due to plastic shrinkage in a dried-out lake bed.

To counteract crack damages due to plastic shrinkage, it is usually required that the newly cast concrete be protected against desiccation. This applies to **in-situ** cast concrete (cast on site) as well as concrete used in the **industrial production** of structural elements, pipes, slabs, etc. In this connection, a calculated estimate of the evaporation rate from concrete surfaces under different climatic conditions is often required. For such estimates, the **vapour pressure curve** of water is an important parameter. Therefore, in computer calculations it is preferable that the vapour pressure curve is expressed in analytic form: $p = p(T)$.

Figure 4.32. Plastic shrinkage on the surface of a bridge deck; the cracks are caused by desiccation during the period where the concrete is still plastic. Cracks due to shrinkage reduce the durability of the concrete when it is exposed to moisture, salt, or repeated freezing.

Problem. From thermodynamic considerations, find an analytic expression for the partial pressure p (Pa) of saturated water vapour as a function of the temperature θ (°C) in the temperature range 0 – 100 °C

$$p = p(\theta) \qquad \text{(Pa)}; \qquad 0 \leq \theta \leq 100\,^\circ\text{C} \tag{a}$$

and formulate a computer function *pvap*(T), which for the input argument: θ (°C) returns the partial pressure p of saturated water vapour at the temperature concerned; compare the calculated values from *pvap*(T) with table values for the partial pressure of saturated water vapour in the temperature range 0 – 100 °C!

Conditions. Water vapour $H_2O(g)$ is assumed to be an **ideal gas** in the temperature range 0 – 100 °C considered. In the calculations, the following thermodynamic standard values for H, S and c_p taken from Atkins: *Physical Chemistry*, are used.

Component		$H^\ominus_{298}$ (kJ/mol)	$S^\ominus_{298}$ (J/mol K)	c_p (J/mol K)
Water	$H_2O(\ell)$	−285.83	69.91	75.29
Water vapour	$H_2O(g)$	−241.82	188.83	33.58

Solution. We investigate an evaporation process where water $H_2O(\ell)$ is transformed into water vapour $H_2O(g)$ at constant temperature T and pressure p

$$H_2O(\ell) \quad \rightarrow \quad H_2O(g) \qquad (T, p \text{ constant}) \tag{b}$$

The water vapour is assumed to be an **ideal gas**. The reaction enthalpy $\Delta_r H_T$ is determined by the procedure (3.55), in which the temperature T is introduced as a variable. The simple reaction equation (b) makes it possible to write $\Delta_r H_T$ directly

$$\Delta_r H_T = \Delta_r H^\ominus_{298} + \Delta_r c_p \cdot (T - 298.15) \qquad \text{(J/mol)}$$

$$\Delta_r H_T = (-241820 - (-285830)) + (33.58 - 75.29) \cdot (T - 298.15) \qquad \text{(J/mol)}$$

$$\Delta_r H_T = 56446 - 41.71 \cdot T \qquad \text{(J/mol)}$$

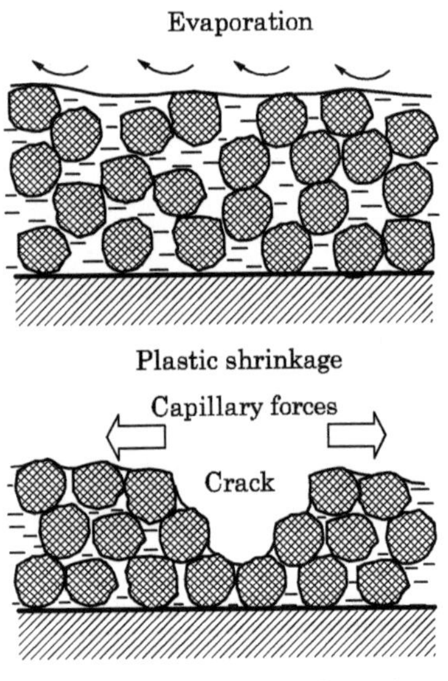

Figure 4.33. Formation of shrinkage crack shown schematically. If a water saturated particle system desiccates, a capillary underpressure arises in the pore water due to the surface tension of the water. The particles are thus influenced by capillary forces that may cause shrinkage and crack formation.

Next, the reaction entropy $\Delta_r S_T$ is determined by the procedure (4.30); here we introduce *both* the temperature T and the partial pressure of water vapour p as variables. In the expression, the symbol $p^\ominus$ is used for the standard pressure 101325 Pa

$$\Delta_r S_T = \Delta_r S^\ominus_{298} + \Delta_r c_p \cdot \ln(\frac{T}{298.15}) - R \cdot \ln(\frac{p}{p^\ominus})_g \qquad \text{(J/mol K)}$$

$$\Delta_r S_T = (188.83 - 69.91) + (33.58 - 75.29) \cdot \ln(\frac{T}{298.15}) - R \cdot \ln(\frac{p}{p^\ominus})_g \qquad \text{(J/mol K)}$$

$$\Delta_r S_T = 356.57 - 41.71 \cdot \ln(T) - R \cdot \ln(\frac{p}{p^\ominus})_g \qquad \text{(J/mol K)}$$

From the equilibrium relation (4.33) we can now set up the condition that the evaporation process (a) is reversible. This corresponds to phase equilibrium between water and water vapour.

$$\Delta S_{univ} = \Delta S_{system} + \Delta S_{surr} = \Delta S_{system} - \frac{\Delta_r H_T}{T} = 0 \tag{c}$$

$$\Delta S_{univ} = 356.557 - 41.71 \cdot \ln(T) - R \cdot \ln(\frac{p}{p^\ominus}) - \frac{56446 - 41.71 \cdot T}{T} = 0 \tag{d}$$

After reduction and isolation of the partial pressure of water vapour p, we get the relation (a) between p and T

$$p = p^\ominus \cdot \exp(47.905 - 5.017 \cdot \ln(T) - 6789.3 \cdot T^{-1}) \quad \text{(Pa)} \tag{e}$$

where $p^\ominus = 101325$ Pa. We can then formulate a computer function *pvap*(T) in pseudo code, which in a simple manner can be transcribed into, for example, *Matlab* or *C*.

```
FUNCTION pvap(T:real):real;                          (* vapour pressure function *)
const PSTAN = 101325;                                (* p⊖ value *)
      KA = 47.905; KB = -5.017; KC = -6789.3;        (* constants *)
begin
  T:= T + 273.15;                                    (* °C → K *)
  if (T < 273.15) or (T > 373.15)                    (* range test *)
    then writeln('Out of range');                    (* error message *)
  else
    pvap:= PSTAN*exp(KA + KB*ln(T) + KC/T);          (* calculate value *)
end;                                                 (* return value *)
```

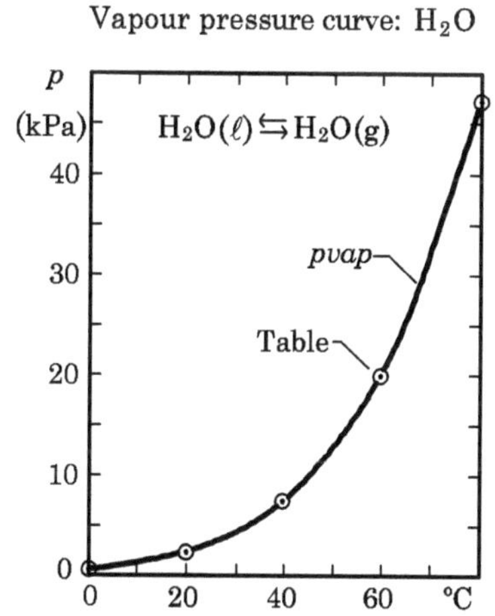

Figure 4.34. The vapour pressure curve of water calculated by the function pvap(), derived in example 4.3; the table values ⊙ are shown for comparison.

The developed function *pvap*(T) can then be tested against table data. Comparing the calculated values of the partial pressures of saturated water vapour to the given table values, we find the following correspondence in the temperature range 0 – 100 °C

Temperature	0	20	40	60	80	100	°C
p (table)	611	2339	7381	19932	47373	101320	Pa
p (calculated)	620	2372	7475	20136	47663	101311	Pa
$\|\Delta\|$	1.5	1.4	1.3	1.0	0.6	0.0	%

When calculating the partial pressure of saturated water vapour p using the developed function *pvap*(T), we find the largest deviation $|\Delta|$ of approximately 1.5 % in the temperature range 0 – 100 °C.

Discussion. In the calculation we have used "raw" data for thermodynamic standard values and we have not corrected for the temperature dependence of specific heat. In example 5.4, we shall see how a more refined calculation method can be used to express a high-precision vapour pressure curve.

Example 4.4

■ Differential thermal analysis of cement paste – DTA

Differential thermal analysis has often been used for laboratory investigations of materials. This measurement can be used, for example, to identify the composition of materials and also to test their durability at increased temperatures.

Differential thermal analysis, normally abbreviated "DTA", is based on the following measuring principle: In a furnace, two small identical crucibles are placed. In each of the crucibles, a **thermocouple** is mounted so that the crucible temperature can be registered. In one of the crucibles, a sample of the material to be tested is placed; in the other crucible, a corresponding amount of an **inactive substance**, for example rutile TiO_2, is placed. Typically, the quantity to be tested is a few milligrams.

The furnace is then heated at a **constant rate**, normally from 2 to 10 °C per minute. During heating, the temperature difference $\Delta\theta$ between the sample and the inactive substance is measured.

Exothermic or **endothermic** reactions in the material sample that take place during heating result in a transient **temperature difference** $\Delta\theta$ between the two crucibles. For example, an endothermic heat-absorbing phase transformation in the material sample will cause the temperature in its crucible to be a bit lower than in the crucible with the inactive reference sample. On the other hand, an exothermic heat-releasing reaction in the material sample will cause the temperature measured in its crucible to be a bit higher for a short period than that of the inactive reference substance. When a transformation is completed, heat flow between the two crucibles will rapidly eliminate the temperature difference, and the two crucibles will again have identical temperatures during further heating.

A graphical illustration of the differential temperature $\Delta\theta$ as a function of the crucible temperature is called a **DTA chart**. With typical "peaks", this chart shows the **phase transformations** and **reactions** occurring during heating of the material sample. When measuring the area under the "peaks" of the DTA curve, and after calibration, the **reaction enthalpy** $\Delta_r H_T$ for the individual transformations or reactions can be determined.

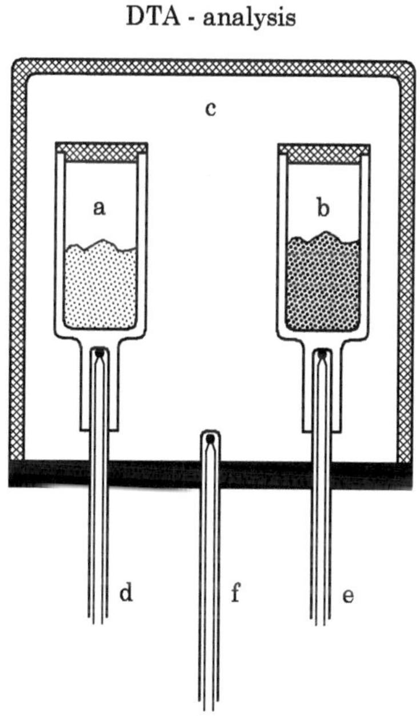

Figure 4.35. Furnace (c) with crucibles (a) and (b) for differential thermal analysis DTA. The crucible (a) contains an inactive substance and (b) contains the sample of the material. Due to a reaction or phase transformation in the sample (b), a temperature difference $\Delta\theta$ arises between the crucibles; $\Delta\theta$ is registered by the thermocouples (d) and (e).

DTA analysis of a finely ground sample of hardened cement paste or concrete makes it possible to estimate the degree of carbonation, the calcium hydroxide content, the content of silicate hydrates and aluminate hydrates, etc. Therefore, the DTA analysis is often used for investigation of the binder phase in hardened structural concrete, e.g. in connection with durability investigations.

Figure 4.36. Example of equipment for differential-thermal analysis DTA; the photo shows a furnace setup with crucibles and equipment for automatic registration of T and ΔT.

Problem. A finely ground sample of hardened concrete is to be investigated by DTA analysis. The purpose of the analysis is, among others, to determine the degree of carbonation of the concrete sample (see example 1.4). The content of **calcium hydroxide** $Ca(OH)_2$ in the sample shows as an **endothermic** peak starting around 500 °C. This peak occurs because heat is consumed for the dehydration of $Ca(OH)_2$ by the reaction

$$Ca(OH)_2(s) \quad \rightarrow \quad CaO(s) + H_2O(g) \qquad \text{(a)}$$

By carbonation of concrete, $Ca(OH)_2$ is slowly transformed into $CaCO_3$ when absorbing the **carbon dioxide** of the air CO_2. In a carbonated concrete, the DTA chart will therefore show a characteristic peak at the temperature where the **calcium carbonate** $CaCO_3$ decomposes and liberates CO_2

$$CaCO_3(s) \quad \rightarrow \quad CaO(s) + CO_2(g) \qquad \text{(b)}$$

Calculate the approximate temperature θ °C, where a peak on the DTA curve can be expected due to possible contents of calcium carbonate $CaCO_3$ in the sample!

Conditions. The reaction equation (b) shows that the equilibrium when $CaCO_3$ decomposes must depend on the partial pressure of the CO_2 gas in the crucible. Atmospheric air has a natural content of CO_2 of approximately 0.03 vol-% (see table 1.4). This corresponds to the CO_2 partial pressure

$$p(CO_2) \simeq 0.03 \cdot 10^{-2} \cdot 101325\,\text{Pa} \qquad = 30\,\text{Pa}$$

When $CaCO_3$ decomposes, CO_2 in the ground sample is liberated; thus, the partial pressure of CO_2 in the crucible is increased so that the equilibrium shifts to the left in the reaction equation (b). Unimpeded transformation of $CaCO_3$ in the partially closed crucible cannot take place until a partial pressure of $p(CO_2) \simeq 1$ atm is reached. Therefore, the peak of the DTA curve for decomposition of $CaCO_3$ can be expected to start at the temperature θ at which the reaction (b) is in equilibrium at a CO_2 partial pressure of 101325 Pa.

The calculations are performed with the following table data for the thermodynamic standard values of the components

Component	$H^\ominus_{298}$ (kJ/mol)	$S^\ominus_{298}$ (J/mol K)	c_p (J/mol K)
$CO_2(g)$	− 393.5	213.6	37.1
$CaO(s)$	− 635.5	39.7	42.8
$CaCO_3(s)$	−1206.9	92.9	81.9

In calculations, the CO_2 gas is assumed to be ideal; pure CO_2 atmosphere is assumed to build up in the crucible during decomposing of $CaCO_3$, i.e. $p(CO_2) \simeq 101325$ Pa. Constant heat capacity c_p in the actual temperature range is assumed.

Figure 4.37. Thermocouples for measurement of temperature difference between the active and inactive sample in the DTA analysis.

Solution. The temperature T at which decomposing of $CaCO_3$ due to reaction equation (b) is reversible (equilibrium) is to be determined. The reaction enthalpy $\Delta_r H_T$ is determined by the procedure (3.55); thus, the temperature T is introduced as an unknown variable

$$\Delta_r H^\ominus_{298} = (-393500 - 635500) - (-1206900) = 177900 \text{ J/mol}$$

$$\Delta_r c_p = (37.1 + 42.8) - 81.9 = -2.0 \text{ J/mol K}$$

$$\Delta_r H_T = 177900 - 2.0 \cdot (T - 298.15) = 178496 - 2.0 \cdot T \qquad \text{(J/mol)}$$

Correspondingly, the reaction entropy $\Delta_r S_T$ is determined by the procedure (4.30); $p(CO_2)$ is taken as 101325 Pa and the contribution then is $-R \cdot \ln(p/p^\ominus) = 0$. In the calculation, the value of $\Delta_r c_p$ as determined above is used.

$$\Delta_r S^\ominus_{298} = (213.6 + 39.7) - 92.9 = 160.4 \text{ J/mol K}$$

$$\Delta_r S_T = 160.4 - 2.0 \cdot \ln(\frac{T}{298.15}) = 171.8 - 2.0 \cdot \ln(T) \qquad \text{(J/mol K)}$$

According to (4.33), equilibrium for the reaction (b) means that the entropy increase in system + surroundings is zero. Thus we have the condition

$$\Delta S_{univ} = \Delta S_{system} + \Delta S_{surr} = \Delta_r S_T - \frac{\Delta_r H_T}{T} = 0 \qquad \text{(c)}$$

$$\Delta S_{univ} = 171.8 - 2.0 \cdot \ln(T) - \frac{178496 - 2.0 \cdot T}{T} = 0 \qquad \text{(d)}$$

$$\Delta S_{univ} = 173.8 - 2.0 \cdot \ln(T) - 178496 \cdot T^{-1} = 0 \qquad \text{(e)}$$

This equation in T can, for example. be solved by **Newton-Raphson** iteration (see *Mathematical appendix, subchapter A3*); thus, the equilibrium temperature $T \simeq$ 1117 K (844 °C) is found.

As we find $\Delta_r H_T > 0$ for the reaction, decomposition of $CaCO_3$ can be expected to be seen as an **endothermic** peak on the DTA curve in the range from approximately 850 °C and above. The appearance of this peak will depend on the measuring equipment used, on the rate of heating, and on the size of the sample.

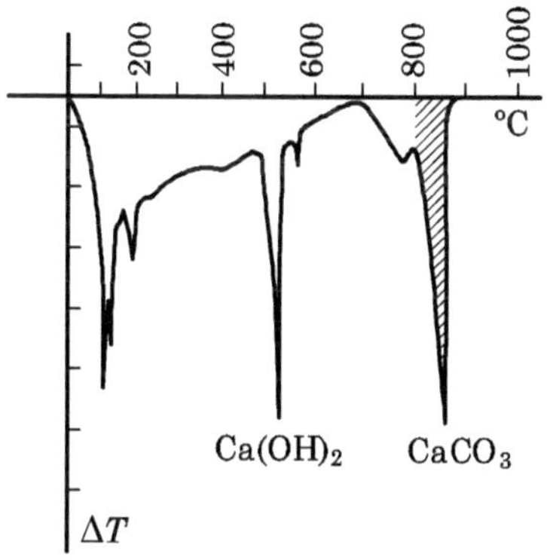

Figure 4.38. DTA analysis of hardened, partly carbonated cement paste. The peak above 800 °C indicated by the shaded area shows decomposition of $CaCO_3$; decomposition of $Ca(OH)_2$ is seen as a peak at approximately 500 °C.

Discussion. The calculated temperature for decomposition of calcium carbonate $CaCO_3$ thus found is in accordance with the experimentally determined values found in DTA analysis. For example, in *Thermoanalytical Methods of Investigation* by P.D. Garn, the transformation temperature 875 °C is given for the process. In figure 4.38, a DTA chart for partly carbonated cement paste is given. In this figure, the measured peak is found at approximately 850 °C.

Example 4.5

■ Vapour pressure of mercury – occupational exposure limit

Organic and inorganic substances used on a daily basis in laboratories and production may be **unhealthy**. Many poisonous substances have a characteristic smell and can be directly detected by the senses if they are present in a dangerous concentration. Other substances are completely odourless even in life-threatening concentrations. An example of this is **mercury** Hg.

Mercury has been extensively used in instruments for laboratory tests. Carelessness with mercury may cause inhalation of **mercury vapour** in a concentration that may be a serious health risk. Mercury is the only metallic element that is liquid at room temperature; at room temperature, the vapour pressure of the metal is so high that the air may contain hazardous concentrations of Hg.

Problem. Based on thermodynamic standard values, calculate the content of Hg(g) in mg per m^3 of atmospheric air that is *saturated* with Hg vapour at 20 °C and at 100 °C; compare the result with the recommended **occupational exposure limit** for mercury set by the American Conference of Governmental Industrial Hygienists as 0.025 mg/m^3!

Figure 4.39. Mercury Hg was often used as measuring fluid in laboratory thermometers.

Conditions. According to Atkins. *Physical Chemistry*, mercury has the following thermodynamic standard values:

State	$H^{\ominus}_{298}$ (kJ/mol)	$S^{\ominus}_{298}$ (J/mol K)	c_p (J/mol K)
Hg(g)	61.32	174.96	20.786
Hg(ℓ)	0.00	76.02	27.983

The molar mass of mercury Hg is M = 200.59 g/mol. In the calculation, equilibrium with a free, plane surface of mercury is assumed, and mercury vapour Hg(g) is assumed to be an ideal gas.

Solution. We investigate the following phase transformation in a closed system

$$\text{Hg}(\ell) \quad \rightarrow \quad \text{Hg(g)} \qquad (T, p \text{ constant}) \qquad \text{(a)}$$

and assume reversible transformation (equilibrium). The reaction enthalpy $\Delta_r H_T$ is determined by the procedure (3.55); thus, the temperature T is introduced as an unknown variable

$$\Delta_r H_T = \Delta_r H^{\ominus}_{298} + \Delta_r c_p \cdot (T - 298) \qquad \text{(J/mol)}$$

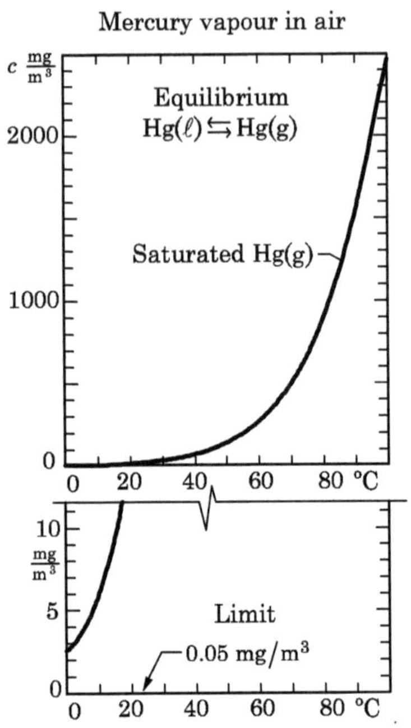

Figure 4.40. The concentration of Hg in atmospheric air that is saturated with mercury vapour increases strongly with the temperature; at room temperature, the saturation concentration is already approximately 600 times the occupational exposure limit value of 0.025 mg/m^3.

$$\Delta_r H_T = (61320 - 0.0) + (20.786 - 27.983) \cdot (T - 298.15) \qquad \text{(J/mol)}$$

$$\Delta_r H_T = 63466 - 7.197 \cdot T \qquad \text{(J/mol)}$$

The reaction entropy $\Delta_r S_T$ is then determined by the procedure (4.30); thus, the temperature T and the partial pressure p of Hg(g) are introduced as unknown variables

$$\Delta_r S_T = \Delta_r S^{\ominus}_{298} + \Delta_r c_p \cdot \ln(\frac{T}{298.15}) - R \cdot \ln(\frac{p}{p^{\ominus}})_g \qquad \text{(J/mol K)}$$

$$\Delta_r S_T = (174.96 - 76.02) - 7.197 \cdot \ln(\frac{T}{298.15}) - R \cdot \ln(\frac{p}{p^{\ominus}})_g \qquad \text{(J/mol K)}$$

$$\Delta_r S_T = 139.95 - 7.197 \cdot \ln(T) - R \cdot \ln(\frac{p}{p^{\ominus}})_g \qquad \text{(J/mol K)}$$

According to (4.33), equilibrium between Hg(ℓ) and Hg(g) means that the total entropy increase by the transformation (a) is zero. Thus, by introduction in the equation, we have the equilibrium condition

$$\Delta S_{univ} = \Delta S_{system} + \Delta S_{surr} = \Delta_r S_T - \frac{\Delta_r H_T}{T} = 0 \qquad \text{(b)}$$

$$139.95 - 7.197 \cdot \ln(T) - R \cdot \ln(\frac{p}{p^{\ominus}})_g - (63466 - 7.197 \cdot T) \cdot T^{-1} = 0 \qquad \text{(c)}$$

$$p = p^{\ominus} \cdot \exp(17.699 - 0.866 \cdot \ln(T) - 7633.6 \cdot T^{-1}) \quad \text{(Pa)} \qquad \text{(d)}$$

where the standard pressure is $p^{\ominus} = 101325\,\text{Pa}$. The partial pressure of Hg(g) is thus determined by introduction of $T = 293.15\,\text{K}$ (20 °C) and $T = 373.15\,\text{K}$ (100 °C), respectively, in (d)

$$p_{20} = 101325 \cdot \exp(17.699 - 0.866 \cdot \ln(293.15) - 7633.6 \cdot 293.15^{-1}) = \mathbf{0.18\,Pa}$$

$$p_{100} = 101325 \cdot \exp(17.699 - 0.866 \cdot \ln(373.15) - 7633.6 \cdot 373.15^{-1}) = \mathbf{38.1\,Pa}$$

Based on the ideal gas law (1.8), the sought concentrations of mercury vapour can be determined in (mg/m^3); to obtain the requested unit, the molar mass is introduced as $M = 200590$ (mg/mol) in the calculation.

$$c_{20} = \frac{m}{V} = \frac{pM}{RT} = \frac{0.18 \cdot 200590}{8.314 \cdot 293.15} = \mathbf{14.8\,mg/m^3}$$

$$c_{100} = \frac{m}{V} = \frac{pM}{RT} = \frac{38.1 \cdot 200590}{8.314 \cdot 373.15} = \mathbf{2463\,mg/m^3}$$

Saturated mercury vapour contains approximately 15 mg Hg(g) per m^3 at 20 °C, and approximately 2500 mg Hg(g) per m^3 at 100 °C. The occupational exposure limit value for mercury vapour in air set by the American Conference of Governmental Industrial Hygienists is 0.025 mg/m^3. Thus, saturated mercury vapour at 20 °C contains approximately 600 times the permitted value, and at 100 °C, it contains approximately 100 000 times the permitted value!

Discussion. Spilled mercury in the laboratory *may* constitute a considerable health risk, especially in closed or poorly ventilated rooms.

Example 4.6

■ Control of relative humidity RH by salt hydrates

Figure 4.41. Desiccator for conditioning of moisture of wood specimens; a saturated salt solution at the bottom of the chamber maintains constant, relative humidity RH around the test specimens shown.

In practical measurements in the laboratory, it is often necessary to maintain a known, constant relative humidity *RH* in a closed test chamber. For example, in connection with measurement of sorption isotherms, test specimens of wood or cement paste may need equilibration to moisture in a desiccator. In the absence of real, climate-controlled test rooms, constant relative humidity *RH* can be maintained in laboratory tests in small, closed test chambers by using

- *saturated salt solutions*: by choosing appropriate salts, *RH* can be adjusted to the required value
- *homogenous solutions*: aqueous solutions of strongly hygroscopic substances such as sulphuric acid H_2SO_4 and glycerol $C_3H_5(OH)_3$ can maintain constant relative humidity *RH* when adjusted to a certain concentration

- *salt hydrates*: many salts can form a number of hydrates that contain an increasing amount of hydrate water; when paired, these salts can form an equilibrium that keeps up constant humidity in a closed test chamber.

As an example of the latter method for controlling relative humidity, mention can be made of the salt pair **natrium sulphate anhydrite** Na_2SO_4(s) and **natrium sulphate decahydrate** $Na_2SO_4 \cdot 10H_2O$(s). At constant temperature *below* 32.4 °C, these salt pairs form a mutual equilibrium in a closed room; the equilibrium reaction is

$$Na_2SO_4 \cdot 10H_2O(s) \leftrightarrows Na_2SO_4(s) + 10H_2O(g) \qquad (T \text{ constant}) \qquad \text{(a)}$$

By this reaction, the partial pressure $p(H_2O)$ – and thus the relative humidity RH – is maintained in the ambient air through the following mechanism:

Relative humidity

$$RH = \frac{\text{actual partial pressure}}{\text{saturated vapour pressure}}$$

– if the air in the room is supplied with water vapour, e.g. from a moist wood specimen, the equilibrium (a) is shifted to the left; thus, decahydrate $Na_2SO_4 \cdot 10H_2O$ is formed until the partial pressure has again decreased to the equilibrium pressure at the given temperature,

– if water vapour is removed from the air, e.g. absorbed by a dry specimen of wood, the equilibrium (a) is shifted to the right; now, decahydrate is transformed into anhydrite Na_2SO_4 and liberates water vapour to the air during the process until equilibrium has been restored anew.

Problem. In a laboratory test, constant relative humidity RH needs to be maintained around some specimens of wood. For practical reasons, "dry" salt pairs shall be used for this purpose. In this connection, an informative calculation shall be made to find out what relative humidity RH can be maintained in a desiccator at the equilibrium reaction (a) if the temperature is constant, $\theta = 15\,°C$.

Conditions. The following thermodynamic standard values of the relevant substances in reaction equation (a) can be found from a table of standard values

Component	$H^\ominus_{298}$ (kJ/mol)	$S^\ominus_{298}$ (J/mol K)	c_p (J/mol K)
$Na_2SO_4 \cdot 10H_2O$(s)	−4326.1	593.2	587.7
Na_2SO_4(s)	−1385.1	149.6	127.7
H_2O(g)	− 241.8	188.7	33.6

It is assumed that the vater vapour is an ideal gas, and all hysteresis effects in connection with desorption and absorption are disregarded.

Solution. We investigate the transformation of sodium sulphate decahydrate $Na_2SO_4 \cdot 10H_2O$ into sodium sulphate anhydrite Na_2SO_4 by the reaction

$$Na_2SO_4 \cdot 10H_2O(s) \rightarrow Na_2SO_4(s) + 10H_2O(g) \qquad (T \text{ constant}) \qquad \text{(b)}$$

assuming that the transformation is reversible (equilibrium). For the reaction, the reaction enthalpy $\Delta_r H_T$ at 15 °C is determined by the procedure (3.55)

$$\Delta_r H^\ominus_{298} = (-1385.1 + 10 \cdot (-241.8)) - (-4326.1) = 523.0 \text{ kJ/mol}$$

$$\Delta_r c_p = (127.7 + 10 \cdot 33.6) - 587.7 = -124.0 \text{ J/mol K}$$

$$\Delta_r H_T = 523\,000 + (-124.0) \cdot (288.15 - 298.15) = 524\,240 \text{ J/mol}$$

Correspondingly, the standard reaction entropy $\Delta_r S^\ominus_T$ is determined by the procedure (4.30); the partial pressure p of water vapour is thereby introduced as an unknown variable. In particular, the stoichiometric coefficient 10 for H_2O(g) must be noted in the logarithmic correction term that adjusts the entropy of the water vapour to the actual pressure p.

$$\Delta_r S^\ominus_{298} = (149.6 + 10 \cdot 188.7) - 593.2 = 1443.4 \text{ J/mol K}$$

$$\Delta_r S_T = 1443.4 + (-124.0) \cdot \ln(\frac{288.15}{298.15}) - 10R \cdot \ln(\frac{p}{p^\ominus}) \qquad \text{(J/mol K)}$$

$$\Delta_r S_T = 1447.6 - 10R \cdot \ln(\frac{p}{p^\ominus}) \qquad \text{(J/mol K)}$$

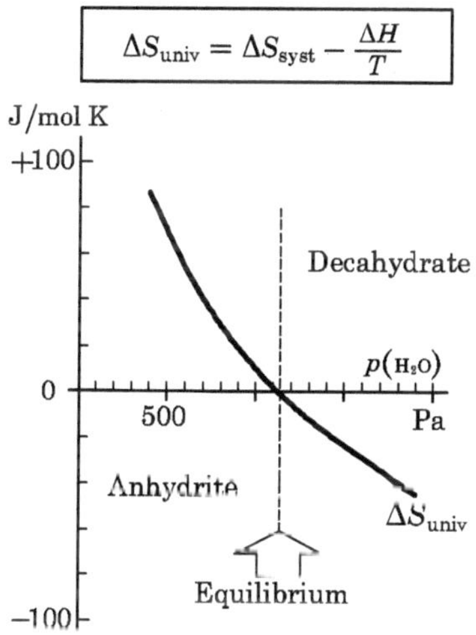

Figure 4.42. Determination of RH for salt pairs. For the transformation (a), ΔS_{univ} at 15 °C is calculated as a function of the partial pressure p of water vapour. Equilibrium corresponds to ΔS_{univ} being zero; this condition is fulfilled for $p \simeq 1129$ Pa.

According to (4.33), reversible transformation – and thus equilibrium – is reached when $\Delta S_{univ} = 0$

$$\Delta S_{univ} = \Delta S_{system} + \Delta S_{surr} = \Delta_r S_T - \frac{\Delta_r H_T}{T} = 0 \qquad \text{(c)}$$

$$1447.6 - 10R \cdot \ln(\frac{p}{p^{\ominus}}) - \frac{524240}{288.15} = 0 \qquad \text{(d)}$$

By introduction of the gas constant $R = 8.314$ J/mol K and the standard pressure $p^{\ominus} = 101325$ Pa, equation (d) determines the targeted partial pressure p of water vapour for equilibrium

Partial pressure of water vapour $p =$ **1159 Pa**

In a table of vapour pressure, the partial pressure of saturated water vapour at 15 °C is specified as: $p_s = 1705.1$ Pa. Thus, the anticipated, relative humidity RH at equilibrium in the salt pair is

Relative humidity $RH = \frac{p}{p_s} = \frac{1159}{1705.1} \simeq$ **68** %

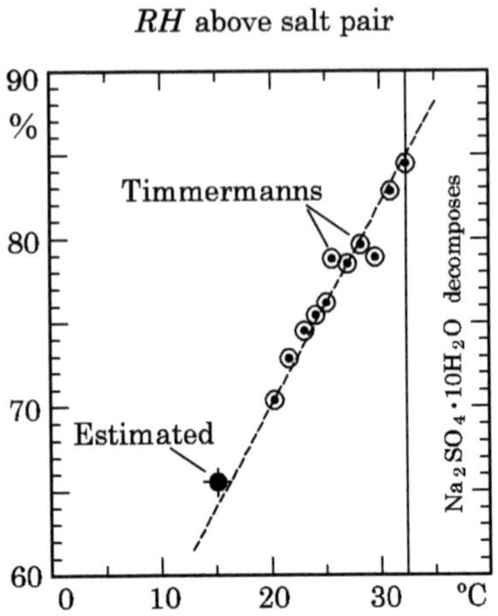

Figure 4.43. Calculated relative humidity for the salt pair sodium sulphate decahydrate and anhydrite, respectively, at 15 °C (•); for comparison, the test data (⊙) published by Timmermanns are shown.

Discussion. In Jean Timmermans: *The Physico-chemical Constants of Binary Systems in Concentrated Solution*, a number of experimentally determined values for the equilibrium (a) in the temperature range 20 °C to 32.4 °C are specified. By linear extrapolation from these data, the humidity RH at 15 °C can be found as 64-65 %.

Exercises

The following exercises can be used when learning the subjects addressed in Chapter 4. The times given are suggested guidelines for use in assessing the effectiveness of the teaching.

Exercise 4.1

☐ Thermal efficiency, the Carnot cycle *(4 min.)*

A heat engine is supplied with pressurized vapour at $\theta_1 = 330$ °C and releases heat to a condenser at the temperature $\theta_2 = 20$ °C. Calculate the theoretical *maximum* thermal efficiency η that can be obtained on these conditions, and specify the minimum amount of waste heat Q to be released from the engine if it shall be able to perform useful work of $W = 1000$ kJ!

Exercise 4.2

☐ Isothermal expansion of ideal gas *(8 min.)*

A cylinder contains 5 mol of ideal gas at a temperature of $\theta = 20$ °C in its initial state (1). Now, a reversible, isothermal change of state is performed whereby the gas is supplied with $Q = 8000$ J. Determine the work $W_{1.2}$ (J) done on the system, and calculate the increase in the molar entropy of the gas ΔS (J/mol K)!

Change of temperature

$$\Delta S_{1.2} = \int_{T_1}^{T_2} \frac{nc_p(T)}{T} dT$$

Exercise 4.3

☐ Change of entropy by heating of water *(3 min.)*

A closed system contains 5.00 kg of water at 20.0 °C. The specific heat capacity of the water is $c_p = 75.29$ J/mol K, and its molar mass is $M = 18.02$ g/mol. By heating, the water temperature is raised to 80.0 °C. Calculate the entropy change ΔS (J/K) occurring in the system!

Exercise 4.4

☐ Change of entropy by heating of metal *(10 min.)*

According to Barin & Knache: *Thermochemical Properties of Inorganic Substances*, the metal zinc Zn(s) has the following temperature-dependent heat capacity

$$c(T) = 22.40 + 10.05 \cdot 10^{-3} \cdot T \qquad \text{(J/mol K)} \qquad \text{(a)}$$

where T is the temperature in (K). Taking the temperature dependence of the heat capacity into account, calculate the increase in molar entropy ΔS (J/mol K) for Zn(s) by heating from 25 °C to the melting temperature 419 °C!

Exercise 4.5

☐ Irreversible expansion of ideal gas *(5 min.)*

A receptacle of the volume $V_1 = 5.00\,\ell$ contains ideal gas in equilibrium; the gas pressure is p = 1bar = 100000 Pa, and the gas temperature is 20.0 °C. Through a valve, the receptacle is connected with a *vacuum* receptacle of the volume $V_2 = 12\,\ell$. In this state, the valve is opened so that the gas flows into the empty receptacle. Calculate the change of entropy ΔS (J/K) in the total system when equilibrium has been obtained!

Exercise 4.6

☐ Heating with phase transformation *(8 min.)*

A sample of water is heated from 20.0 °C to 100 °C; by boiling, the water is transformed into water vapour at 100 °C. The boiling occurs at a pressure of 1 atm. Data for water: The heat of evaporation at 100 °C is 2257 J/g; specific heat capacity c_p = 75.29 J/mol K; the molar mass is M = 18.02 g/mol. Calculate the entropy increase in the water ΔS (J/mol K) for the entire process: heating + evaporation!

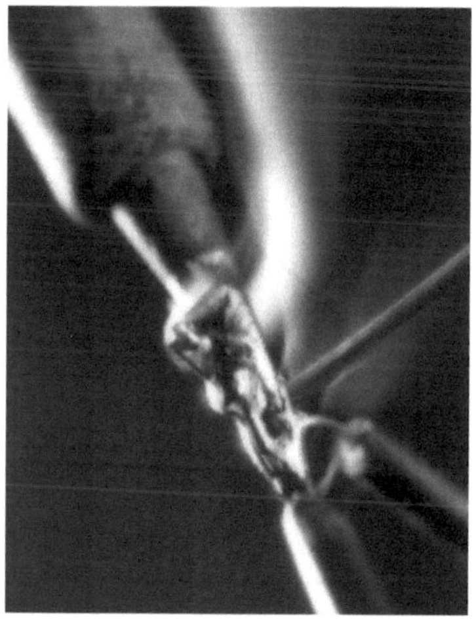

Figure 4.44. Soft soldering of metals is normally done with an alloy of tin Sn and lead Pb. Tin solder that comes into contact with food must, however, be unleaded, since Pb may form combinations that can be injurious to health.

Exercise 4.7

☐ Change of state, ideal gas *(10 min.)*

A receptacle contains 3 mol of ideal gas at 1 atm pressure; the gas temperature is 50 °C. By a reversible, isothermal expansion, the gas volume is changed to $V_2 = 500\,\ell$. Determine $Q_{1.2}$ (J), $W_{1.2}$ (J) and ΔS (J/K) for the given change of state!

Exercise 4.8

☐ Standard entropy of solid, liquid and gaseous substances *(12 min.)*

Calculate the molar standard entropy $S(T,p)$ (J/mol K) of the following substances under the given conditions of state: a) calcite $CaCO_3$(s) at θ = 150 °C; b) hydrogen $H_2(g)$ at θ = −10 °C, p = 1000 Pa; c) water $H_2O(\ell)$ at θ = 110 °C, p = 5 atm; d) water vapour H_2O(g) at θ = 80 °C, RH = 50 %; – in the calculations, c_p is assumed to be constant, and gases are assumed to be ideal.

Exercise 4.9

☐ Melting of tin Sn(s) *(10 min.)*

Tin Sn(s) and alloys of tin and lead Pb(s) are widely used for soft soldering of metals. For pure, metallic tin Sn(s), the following physical and thermodynamic values are tabulated: $S^{\ominus}_{298}$ = 51.5 J/mol K; c_p = 27.0 J/mol K; melting point θ_m = 232 °C; heat of fusion Q_m = 7.20 kJ/mol; molar mass M = 118.69 g/mol. Based on this information, calculate the standard entropy S_T (J/mol K) of molten tin Sn(ℓ) at 232 °C!

Figure 4.45. Iron Fe reacts spontaneously with the O_2 of the air and is covered with a surface layer of iron oxide. At high temperatures, for example hot-rolling, this oxide layer may become thick, and part of it may spall off. The layer is called mill scale.

Exercise 4.10

☐ Entropy of water, saturated water vapour, and ice at 0 °C *(10 min.)*

The following thermodynamic standard values are given for water $H_2O(\ell)$ and for water vapour H_2O(g)

$$H_2O(\ell): \quad S^{\ominus}_{298} = 69.91\,\text{J/mol K}; \quad c_p = 75.29\,\text{J/mol K} \qquad \text{(a)}$$

$$H_2O(g): \quad S^{\ominus}_{298} = 188.83\,\text{J/mol K}, \quad c_p = 33.58\,\text{J/mol K} \qquad \text{(a)}$$

At 0 °C and atmospheric pressure, the *heat of fusion* of ice is 6012 J/mol. The partial pressure of saturated water vapour at this temperature is 610.5 Pa. Calculate, based on this information, the entropy S_{273} for *water* $H_2O(\ell)$ at 0 °C, for *saturated water vapour* H_2O(g) at 0 °C and for *ice* H_2O(s) at 0 °C! – the water vapour is assumed to be in accordance with the ideal gas law.

Exercise 4.11

☐ Oxidation of iron Fe *(8 min.)*

Iron Fe(s) that comes into contact with atmospheric air forms a surface layer of iron oxide Fe_2O_3(s) by the following reaction

$$2\text{Fe(s)} + \tfrac{3}{2}\text{O}_2\text{(g)} \quad \rightarrow \quad \text{Fe}_2\text{O}_3\text{(s)} \qquad \text{(a)}$$

Calculate the reaction entropy $\Delta_r S_T$ (J/mol K) by the reaction (a) at 200 °C in normal atmospheric air!

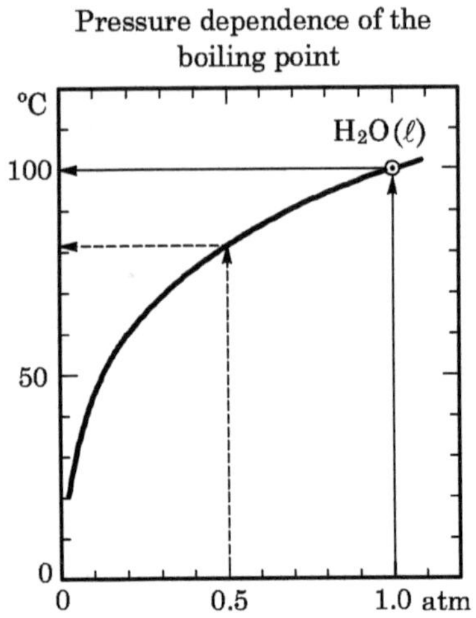

Figure 4.46. A liquid boils when its vapour pressure exceeds the pressure of the ambient atmosphere. Therefore, the boiling point of, for example, water is lowered with reduced atmospheric pressure as shown in the figure. For example, the boiling point has dropped to approximately 82 °C, when the atmospheric pressure is lowered to 0.5 atm.

Exercise 4.12

☐ Partial pressure of saturated water vapour at 25 °C *(8 min.)*

At 25 °C, the heat of evaporation of water is 2442 J/g. The thermodynamic standard entropy $S^{\ominus}_{298}$ for *water* $H_2O(\ell)$ and for *water vapour* $H_2O(g)$ is 69.91 J/mol K and 188.83 J/mol K, respectively. Based on these data, calculate the partial pressure p_s Pa of saturated water vapour at 25 °C! – the water vapour is assumed to be in accordance with the ideal gas law.

Exercise 4.13

☐ Entropy increase by irreversible heat flow *(8 min.)*

An *adiabatic* system contains two, thermally separated metal blocks A and B. Both of the blocks have the mass 1000 g and the specific heat capacity of the metal is $c_p = 0.38$ J/g K. In its initial state, block A is in thermal equilibrium at 0 °C and block B is in thermal equilibrium at 100 °C. The blocks are brought into thermal contact and after some time they have obtained equilibrium at the mutual temperature 50 °C. Calculate the total entropy increase ΔS_{univ} in the thermodynamic universe by this process and evaluate the result based on the *Clausius inequality*!

Exercise 4.14

☐ Heat of fusion for lead Pb(s) *(8 min.)*

Assume a system consisting of 5.00 kg of molten lead $Pb(\ell)$ at the melting temperature 328 °C. By reversible, isothermal solidification, the melt is transformed into solid lead Pb(s). During solidification, the increase in the system entropy is $\Delta S = -191.5$ J/K. The molar mass of lead is $M = 207.2$ g/mol. Based on this information, calculate the *heat of fusion* of lead Pb(s)!

Exercise 4.15

☐ Pressure dependence of the boiling point *(10 min.)*

Boiling of a liquid takes place at the temperature where the saturated vapour pressure of the corresponding gas phase increases the pressure in the ambient atmosphere. Calculate, based on thermodynamic standard values, the boiling point of water when the pressure in the ambient atmosphere is $p = 2000$ Pa!

Figure 4.47. Cross-section of aerated concrete with density 500 kg/m^3; the porous structure is produced by foaming the fresh concrete by a gasifying substance, e.g. aluminium powder. The heat insulating properties of the concrete are good due to the high content of air voids.

Exercise 4.16

☐ Autoclaving of aerated concrete *(15 min.)*

Aerated concrete is made from a mixture of water, Portland cement and a ground mix of lime, quartz sand and fly ash. The mixture is foamed by addition of aluminium powder by which a highly porous concrete with good heat insulating properties is produced (see example 1.3). After setting, the fresh aerated concrete is cut into building units or elements. The final hardening of the aerated concrete takes place in vapour saturated steel chambers at approximately 180 °C; this process is called *autoclaving*. During the autoclaving, the lime CaO reacts with quartz SiO_2 and forms strength-giving calcium silicate hydrates.

The autoclaving takes place under pressure; in the autoclaving, free water is in equilibrium with saturated water vapour at 180 °C. Calculate the approximate vapour pressure p in the atmosphere that arises during the autoclaving! – the following thermodynamic standard values are used in the calculation:

Component	$H^{\ominus}_{298}$ (kJ/mol)	$S^{\ominus}_{298}$ (J/mol K)	c_p (J/mol K)
$H_2O(\ell)$	−285.83	69.91	75.29
$H_2O(g)$	−241.82	188.83	33.58

Exercise 4.17

☐ Temperature resistance of concrete *(20 min.)*

In normal use, structural concrete is not exposed to temperatures above approximately 50 °C. In special structures, however, the exposure temperature may be considerably higher; this is for example the case in certain industrial plants such as concrete floors below furnaces, structures in retort houses and foundries as well as in certain chimneys and flues. In these cases, the concrete is exposed to high temperatures for a long time, which *may* lead to decomposition and loss of strength. Desiccation combined with high temperatures may, for example, lead to dehydration of *calcium hydroxide* $Ca(OH)_2$ in the concrete and thus to considerable reductions in strength. Dehydration of $Ca(OH)_2$ occurs by the reaction

$$Ca(OH)_2(s) \quad \rightarrow \quad CaO(s) + H_2O(g) \qquad \text{(a)}$$

The following thermodynamic standard values are given for components in reaction (a)

Component	$H^{\ominus}_{298}$ (kJ/mol)	$S^{\ominus}_{298}$ (J/mol K)	c_p (J/mol K)
$Ca(OH)_2(s)$	−986.6	76.1	84.5
$CaO(s)$	−635.5	39.7	42.8
$H_2O(g)$	−241.8	188.7	33.6

Calculate the approximate *transformation temperature* for calcium hydroxide in concrete in the reaction (a) if the concrete is exposed to long-term high temperatures at an ambient partial pressure of water vapour $p = 2338$ Pa!

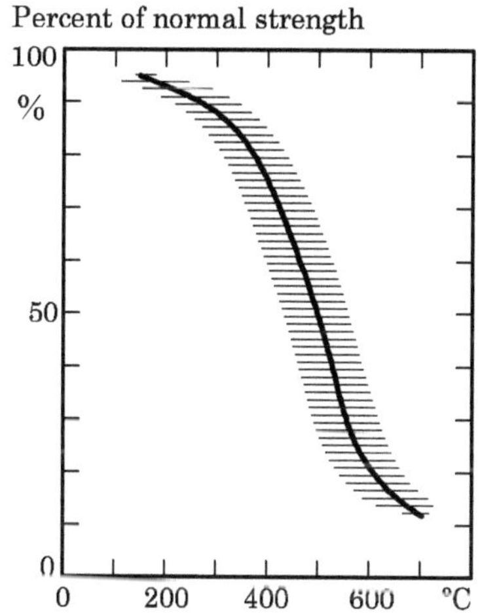

Figure 4.48. If concrete is exposed to high temperatures such as fire, the concrete strength can be drastically reduced; this loss of strength is due to decomposition of hydrates in the binder phase of the concrete.

Exercise 4.18

□ **Desiccation of hardening concrete – plastic shrinkage** *(20 min.)*

Heavy desiccation of newly cast concrete can result in plastic shrinkage cracks in the concrete surface. The desiccation rate of a *wet* concrete surface is generally determined by the wind velocity at the surface and of the difference between the partial pressure of water vapour at the wet concrete surface p_c and the partial pressure of water vapour in the ambient air p_a. For a concrete surface, p_c corresponds to the partial pressure of saturated water vapour at the concrete temperature θ_c. The partial pressure of water vapour in the ambient air is determined by the air temperature θ_a and the relative humidity of the air RH. Desiccation of the concrete takes place if $p_c > p_a$. The following thermodynamic standard values have been taken from the table

Component	$H^{\ominus}_{298}$ (kJ/mol)	$S^{\ominus}_{298}$ (J/mol K)	c_p (J/mol K)
$H_2O(\ell)$	−285.83	69.91	75.29
$H_2O(g)$	−241.82	188.83	33.58

A concrete with the temperature 20 °C is cast as a road surface which is fully exposed to wind action. The air temperature is $\theta_a = 24$ °C and the relative humidity of the air is $RH = 60$ %. Based on this information, calculate p_c (Pa) and p_a (Pa), and assess whether the concrete is subjected to desiccation at the time of casting!

Literature

The following literature and supplementary readings are recommended for further reference on the subjects covered in Chapter 4.

References

The following literature is a complete list of references made in Chapter 4.

- Holleman-Wiberg: *Lehrbuch der Anorganischen Chemie*, Walter de Gruyter, Berlin 1976.
- Karsten, R.: *Bauchemie für Schule und Baupraxis*, Strassenbau, Chemie und Technik Verlagsgesellshaft, Heidelberg 1960.
- Atkins, P.W.: *Physical Chemistry*, Oxford University Press, 1986.
- Timmermans, J.: *The Physico-chemical Constants of Binary Systems in Concentrated Solutions*, Vol.3, Interscience Publishers, Inc., New York 1960.
- Barin, I. & Knacke, O.: *Thermochemical Properties of inorganic Substances*, Springer-Verlag, Berlin 1973.
- Barin, I.: *Thermochemical Data of Pure Substances*, VCH, Basel 1989.

Measurement of vapour pressure, around 1850; the test setup is used for measuring the vapour pressure of water at temperatures above $100^{\circ}C$ where the vapour pressure exceeds the atmospheric pressure.

CHAPTER 5

Calculations of equilibrium

Dalton's atomic hypothesis (1808) explained the composition of substances, but not the chemical **affinity** of the driving forces that determine mutual transformation of the substances. The Dane **Julius Thomsen** was among the first to develop a thermochemical theory of affinity (1852); in this theory, the heat development during the chemical reaction is a measure of the affinity.

According to Thomsen's theory any spontaneous reaction had to be **exothermic** and to be followed by heat release. However, endothermic, chemical reactions that occurred spontaneously were soon recognized; in other words, Thomsen's simple concept of affinity was *not* universal.

At the end of the 1800s, works by **J.W. Gibbs** and **J.H. van't Hoff**, among others, showed that chemical affinity could be identified with losses of free energy (van't Hoff 1883). From these works, the thermodynamic theory of affinity in the form known today was developed.

Introduction of the concept of **free energy** was the origin of the development of a number of significant calculation methods for equilibrium systems. In this chapter, the concept of the **Gibbs free energy** is introduced and the practical application of this quantity is illustrated through problems.

J.H. van't Hoff (1852-1911)

Dutch physicist and chemist; professor of physical chemistry in Amsterdam, and later Berlin.

Contents

5.1 The Gibbs free energy G

When introduced through the **second law** (4.12), entropy S *is* the fundamental parameter for describing the **directionality** of spontaneous processes and for setting up **equilibrium criteria** for systems of matter. However, in order to develop efficient calculation methods it is appropriate to introduce a composed state function, the **Gibbs free energy** G. In the following it will be shown how the G function in a natural way summarizes the equilibrium criteria developed in Chapter 4.

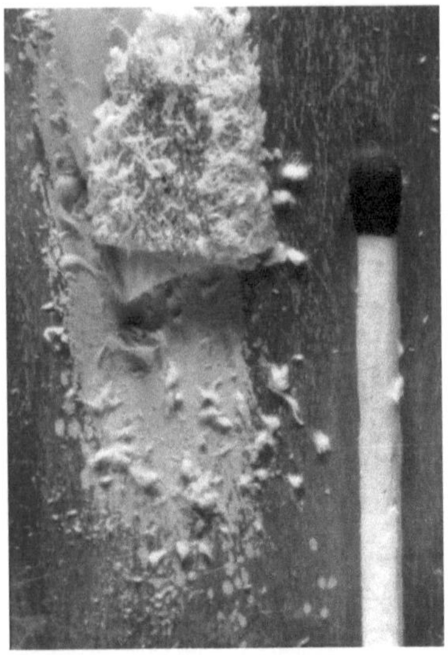

Figure 5.1. Aluminium reacts spontaneously with oxygen and forms Al_2O_3, which is deposited as a dense, passivating layer on the surface. Mercury cancels this passivation and produces pitting with expulsion of powdered Al_2O_3. The photo shows an aluminium sheet coated with mercury. The process is irreversible. It reduces the free energy G of the system.

Definition of free energy G

The Gibbs free energy G is a composite thermodynamic function, which is defined as follows

> **The Gibbs free energy G** (5.1)
>
> $$G \stackrel{\text{def}}{=} H - TS$$
>
> The **free energy** G of a system is defined as the system **enthalpy** H minus the product of the **temperature** T and the **entropy** S.

The equation of definition of the Gibbs free energy G *only* contains state functions (H, S) and state variables (T); therefore, the free energy G is a **state function** itself.

The definition of G includes the **enthalpy** H, which is a particularly suitable energy function for description of changes of state at constant pressure p (see section 3.4). As seen from the following, the G function is also adapted to the description of equilibrium in systems of matter at constant pressure p.

A similar energy function, the *Helmholtz free energy* A, can sometimes be found in the literature. The Helmholtz free energy is defined based on internal energy U in the same way as the G function is defined based on enthalpy H

$$A \stackrel{\text{def}}{=} U - TS \qquad \text{the Helmholtz free energy} \tag{5.2}$$

The A function, which is used within statistical thermodynamics in particular, will *not* be used in the following.

Figure 5.2. Glass with steel wool in water. When iron is in contact with pure oxygen free water, it does not rust (left). If oxygen of the air has access, rust forms spontaneously (right). The process is irreversible. It reduces the free energy G of the system.

The G function differential

To know the properties of the G function, it is useful to investigate the differential dG of the free energy, step by step. It is important to note the *assumptions* made with regard to process conditions. From the equation of definition (5.1) we get

$$dG = dH - d(TS) \qquad \text{since}: H = U + pV \text{ is}$$

$$dG = dU + d(pV) - d(TS) \qquad \text{and since}: dU = \delta Q + \delta W \text{ is}$$

$$dG = \delta Q + \delta W + pdV + Vdp - TdS - SdT \tag{5.3}$$

Now we focus on processes developing at **constant temperature** T and at **constant pressure** p; this entails that $Vdp = 0$ and $SdT = 0$, i.e.

$$dG = \delta Q + \delta W + pdV - TdS \tag{5.4}$$

It is then assumed that *only* **volume work** $\delta W = -pdV$ is performed on the system; introducing this into eqn. (5.4), we get

$$dG = \delta Q - TdS \qquad (T, p \text{ constant}) \tag{5.5}$$

In the above expression, δQ denotes the heat supplied to the system, δQ_{system}, and dS denotes the system increase in entropy, dS_{system}; in other words, the increase in the free energy dG of a system is equal to the heat δQ supplied *minus* the temperature T multiplied by the increase in the entropy dS of the system. However, according to the **Clausius inequality** (4.12), it generally applies that

$$TdS_{system} : \begin{cases} > \delta Q_{system} & \text{irreversible process} \\ = \delta Q_{system} & \text{reversible process} \\ < \delta Q_{system} & \text{impossible process} \end{cases} \qquad (5.6)$$

Introducing the condition in eqn. (5.6) of the second law into the differential eqn. (5.5) of the Gibbs free energy G, we get the following fundamental equilibrium condition

Reversible process

$\Delta G = \Delta H - T\Delta S = 0$

> **Thermodynamic equilibrium condition II** (5.7)
>
> In a process at constant **pressure** p and **temperature** T in which *only* **volume work** is performed on the system, the increase in **free energy** is
>
> $$dG = dH - TdS : \begin{cases} < 0 & \text{irreversible, spontaneous process} \\ = 0 & \text{reversible equilibrium processs} \\ > 0 & \text{termodynamically impossible process} \end{cases}$$
>
> where dH is the system increase in **enthalpy** and dS is the system increase in **entropy**.

Under the given process conditions, i.e. constant (T, p) and volume work as the sole work contribution, this can be expressed verbally as follows. Any spontaneous process will *reduce* the free energy G in a thermodynamic system; in reversible processes the free energy of the system is *unchanged*. The free energy G *cannot* increase spontaneously.

It is important to note that the equilibrium condition in eqn. (5.7) *assumes* that only **volume work** is performed on the system. If another kind of work, for example, **electrical work** δW_a, is performed on the system this contribution of work will be shown on the right-hand side in the three relations in eqn. (5.7). A reversible charge of an accumulator, for example, will *increase* the free energy of this system by $dG = \delta W_a$, where δW_a signifies the electrical work performed on the system. These conditions are further described in chapter 6: *Electrochemistry.*

Spontaneous processes

The equilibrium condition in eqn. (5.7) contains rather fundamental information of the reaction concepts of systems of matter. Any spontaneous reaction *reduces* the free energy of a system G – the system moves spontaneously towards a **state of equilibrium** with a **minimum of free energy** G_{min}. For a chemical reaction or phase transformation linking two states of equilibrium, ΔG is determined by

$$\Delta G = \Delta H - T\Delta S \qquad (T, p \text{ constant}) \qquad (5.8)$$

Consequently, two driving forces determine the behaviour of a system: A tendency to *reduce* the system **enthalpy** – and a tendency to *increase* the system **entropy**; both of these changes can result in a negative value of ΔG.

The weighting between these opposite tendencies is determined by the temperature T. At low temperatures, a negative ΔH contribution will normally be dominating since in this case the quantity $|T\Delta S|$ is small. Therefore, at low temperatures, systems of matter can typically be found in a state with **minimum enthalpy**, i.e. condensed in a **liquid** or a **solid state**.

At high temperatures, a positive ΔS contribution will normally be dominating because in this case $|T\Delta S|$ assumes a high value. Therefore, at high temperatures, systems of matter can typically be found in a state with **maximum entropy**, i.e. in a disorganized **gas state**.

Forthwith, this means that **exothermic** reactions, in which $\Delta H < 0$, promote a spontaneous process. On the other hand, **endothermic** reactions, in which $\Delta H > 0$, will counteract a spontaneous process. In 1852, The Dane **Julius Thomsen** put forward an affinity theory based on the notion that the enthalpy change *in itself*

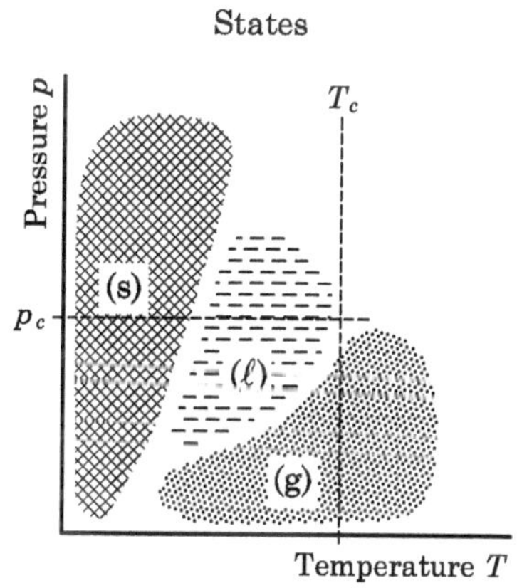

Figure 5.3. States, shown schematically. High temperature and low pressure give rise to formation of a gas phase (g) with high entropy. Low temperature and high pressure give rise to formation of a solid phase (s) with low entropy. The liquid state (ℓ) occurs as a transition phase between these two states. Note that the gas phase cannot liquefy above the critical temperature T_c.

was a measure of the affinity of substances. As can be seen from the following table 5.1, however, this measure of affinity is not universal.

Process type	Enthalpy	Entropy	Free energy	Spontaneous?
Exothermic	$\Delta H < 0$	$\Delta S > 0$	$\Delta G < 0$	always!
Exothermic	$\Delta H < 0$	$\Delta S < 0$	?	if: $\lvert T\Delta S\rvert < \lvert\Delta H\rvert$
Endothermic	$\Delta H > 0$	$\Delta S > 0$	?	if: $T\Delta S > \Delta H$
Endothermic	$\Delta H > 0$	$\Delta S < 0$	$\Delta G > 0$	never!

Table 5.1. Conditions for spontaneous chemical reactions of phase transformations, cf. the thermodynamic equilibrium condition in eqn. (5.7).

Free reaction energy

Irreversible process

$\Delta G = \Delta H - T\Delta S < 0$

Assume a chemical reaction or phase transformation occurring at constant *pressure* p and temperature T; during the process, only **volume work** is performed on the system. Then according to the equation of definition (5.1), the free reaction energy $\Delta_r G_T$ for the process is

$$\Delta_r G_T = \Delta_r H_T - T \cdot \Delta_r S_T \qquad (T, p \text{ constant}) \tag{5.9}$$

Especially, the **standard free reaction energy** $\Delta_r G_T^\ominus$ will be determined by

$$\Delta_r G_T^\ominus = \Delta_r H_T^\ominus - T \cdot \Delta_r S_T^\ominus \tag{5.10}$$

Therefore, for calculation of **free reaction energy** $\Delta_r G_T$ for a chemical reaction or phase transformation, we can immediately use the calculation rules that were previously set up for **reaction enthalpy** $\Delta_r H_T$ in (3.55) and for **reaction entropy** $\Delta_r S_T$ in (4.30). Thus we have the following procedure for calculation of free reaction energy

Free reaction energy (5.11)

For a reaction or phase transformation: $a\text{A} + b\text{B} \rightarrow c\text{C} + d\text{D}$ the free reaction energy $\Delta_r G_T$ at the temperature T is calculated by the following procedure

(a) Calculate the **reaction enthalpy** $\Delta_r H_T$ from (3.55).

(b) Calculate the **reaction entropy** $\Delta_r S_T$ from (4.30).

(c) Then the **free reaction energy** is: $\Delta_r G_T = \Delta_r H_T - T \cdot \Delta_r S_T$

A number of tables contain data for the standard free energy of substances $G_{298}^\ominus$ that *can* be used for calculation of $\Delta_r G_T$ values. In principle, however, all calculations can be performed using the procedure in eqn. (5.11).

When using table data for thermodynamic standard values, the following should be noted: **standard enthalpy** $H_{298}^\ominus$ and **standard free energy** $G_{298}^\ominus$ are absolute energy quantities defined from an **arbitrary zero point** (see e.g. (3.44)). Therefore, the standard values as such do *not* fulfil the equation of definition (5.1), i.e.

$$G_{298}^\ominus \neq H_{298}^\ominus - T \cdot S_{298}^\ominus \tag{5.12}$$

This is due to the previously mentioned fact that by experiments it is *only* possible to determine *changes* of the enthalpy H and the free energy G of a system; it is not possible to determine absolute values experimentally.

■ When lime is slaked, **quicklime** CaO is supplied with water H_2O and is thus transformed into **calcium hydroxide** $Ca(OH)_2$, which is an important raw material for the manufacture of lime mortar. Dry lime slaking without surplus of water follows the reaction equation

$$\text{CaO(s)} + \text{H}_2\text{O}(\ell) \quad \rightarrow \quad \text{Ca(OH)}_2\text{(s)} \tag{a}$$

The reaction components in (a) have the following thermodynamic standard values

CaO(s)	$H^{\ominus}_{298}$	$= -635.1$ kJ/mol	$S^{\ominus}_{298}$	$= 38.1$ J/mol K
$Ca(OH)_2(s)$		$= -986.1$ kJ/mol		$= 83.4$ J/mol K
$H_2O(\ell)$		$= -285.8$ kJ/mol		$= 69.9$ J/mol K

Calculate the free reaction energy $\Delta_r G_{298}$ (J/mol) for dry slaking of CaO(s) at atmospheric pressure and explain the result!

Solution. Determine $\Delta_r G_{298}$ using the procedure in eqn. (5.11); by insertion, the following is obtained

$$\Delta_r H_{298} = -986100 - (-635100 - 285800))\ \text{J/mol} = -65200\ \text{J/mol}$$

$$\Delta_r S_{298} = 83.4 - (38.1 + 69.9)\ \text{J/mol K} = -24.6\ \text{J/mol K}$$

$$\Delta_r G_{298} = -65200 - 298,15 \cdot (-24.6)\ \text{J/mol} = \mathbf{-57.9\ kJ/mol}$$

The reaction is strongly exothermic (heat-releasing), because the enthalpy content of the system is reduced by 65.2 kJ/mol. At the same time, the high negative value of $\Delta_r G_{298}$ shows that the reaction (a) is strongly shifted towards formation of calcium hydroxide $Ca(OH)_2$.

☐ **1.** For a given process, $\Delta_r H_T = -3000$ J/mol and $\Delta_r S_T = -12$ J/mol K; investigate whether this process can run spontaneously at $T = 323$ K?

☐ **2.** For a given process, $\Delta_r H_T = 4.7$ kJ/mol and $\Delta_r S_T = 17$ J/mol K; investigate whether this process can run spontaneously at $T = 278$ K?

☐ **3.** Given the reaction S(monoclinic) $\rightarrow$ S(rhombic) where monoclinic sulphur S is transformed into rhombic sulphur S, investigate whether the process is spontaneous at 25 °C!

☐ **4.** Calculate the partial pressure of saturated water vapour at 25 °C using the relations in equations (5.7) and (5.11)!

☐ **5.** Given the reaction $CaSO_4(s) + 2H_2O(\ell) \rightarrow CaSO_4 \cdot 2H_2O(s)$, calculate $\Delta_r G_{323}$ for the reaction and assess whether it can run spontaneously at 323 K!

Figure 5.4. At atmospheric pressure, ice and water are in thermodynamic equilibrium at 0 °C, i.e. that ΔG is zero when ice is transformed into water.

5.2 The Clapeyron equation

Assume a **homogeneous phase** in equilibrium at the temperature T and the pressure p. Let us now investigate how the Gibbs free energy G for this phase depends on pressure and temperature for a **reversible change of state**. Again, we focus on systems where the **volume work** $\delta W = -pdV$ is the *sole* work contribution. The differential of the Gibbs free energy G is first determined by eqn. (5.1)

$$dG = dH - d(TS) \qquad \text{however, since}: H = U + pV, \text{ is}$$

$$dG = dU + d(pV) - d(TS) \qquad \text{and since}: U = Q + W, \text{ is}$$

$$dG = \delta Q + \delta W + pdV + Vdp - TdS - SdT \tag{5.13}$$

Since we focused on systems where *only* volume work occurs, $\delta W = -pdV$. For a reversible change of state, $\delta Q = TdS$ according to eqn. (5.6); introducing this into eqn. (5.13) results in the important **Gibbs-Duhem equation.**

The Gibbs-Duhem equation (5.14)

For a reversible change of state (dT, dp) with a **homogeneous phase**, the increase in the molar free energy dG of the phase is determined by

$$dG = Vdp - SdT \qquad \text{(J/mol)}$$

where V is the **molar volume** (m^3/mol) and S is the **molar entropy** (J/mol K) of the phase.

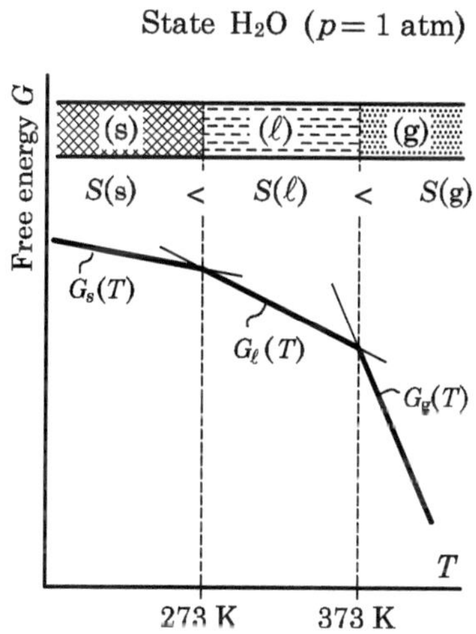

Figure 5.5. Homogeneous phase equilibrium in the system ice water - water vapour, shown schematically. At a given temperature T, the stable phase is the phase with the lowest free energy G. According to the Gibbs-Duhem equation, $(\partial G/\partial T)_p = -S$ at constant pressure. As $S(s) < S(\ell) < S(g)$, the increasing temperature will successively favour the states (s), (ℓ) and (g).

The expression (5.14) is, for example, used for describing the influence of the temperature and the pressure on phase equilibria. This is used in the following by derivation of the Clapeyron equation.

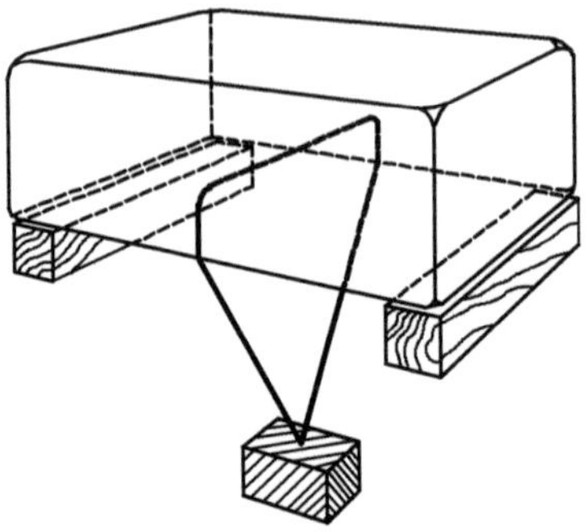

Figure 5.6. At 0°C, a metal wire with an attached weight can cut through a block of ice; the block is left intact. The phenomenon is explained by the Gibbs-Duhem equation: at the underside of the wire the melting point of the ice is lowered due to the pressure; here the ice melts and consumes heat in the process. At the upper side of the wire the liberated water freezes into ice anew and releases a corresponding amount of heat.

Phase equilibrium in single-component system

Consider a system of two **phases** a and b of the *same* substance; such a system is called a **single-component system**. In the initial situation the two phases are assumed to be in mutual **equilibrium** at the temperature T and the pressure p. As an example, let us consider a system containing **water** $H_2O(\ell)$ and **ice** $H_2O(s)$ in equilibrium at $\theta = 0\,°C$ and $p = 1\,atm$. Then, it follows from the equilibrium condition in eqn. (5.7) that

$$\Delta G = G_b - G_a = 0 \tag{5.15}$$

The state of the system is now changed infinitesimally by (dT, dp); thus, the free energy of the phases is changed by dG_a and dG_b, respectively. Then the condition to maintain equilibrium during this change of state is

$$\Delta G = (G_b + dG_b) - (G_a + dG_a) = dG_b - dG_a = 0 \tag{5.16}$$

When this condition is expressed using the Gibbs-Duhem equation (5.14), the following relation between change of pressure and change of temperature in the system is obtained

$$(V_b dp - S_b dT) - (V_a dp - S_a dT) = 0 \tag{5.17}$$

In this expression, V denotes **molar volume** (m^3/mol) and S denotes **mole-specific entropy** (J/mol K). When dp and dT are isolated, the following condition for phase equilibrium is obtained

$$\frac{dp}{dT} = \frac{S_b - S_a}{V_b - V_a} = \frac{\Delta S}{\Delta V} \tag{5.18}$$

According to eqn. (5.7), $\Delta G = \Delta H - T\Delta S = 0$ at phase equilibrium and thus $\Delta S = \Delta H/T$. Introducing this condition into eqn. (5.18), the important **Clapeyron equation** is obtained for equilibrium between phases in a single-component system.

> **The Clapeyron equation** (5.19)
>
> When two *arbitrary phases* a and b are in equilibrium in a **single-component system** the following condition is fulfilled
>
> $$\frac{dp}{dT} = \frac{\Delta H}{T\Delta V} \qquad \text{(Pa/K)}$$
>
> $\Delta H = H_b - H_a$ (J/mol) is the increase in **enthalpy** and $\Delta V = V_b - V_a$ (m^3/mol) is the increase in **molar volume** by the phase transformation $a \rightarrow b$.

The Clapeyron equation

$$\frac{dp}{dT} = \frac{\Delta H}{T\Delta V}$$

The Clapeyron equation (5.19) is valid for *all* forms of phase equilibrium in single-component systems, i.e. evaporation, condensation, melting, solidification, sublimation, and transformation (see table 4.1).

Integration of the Clapeyron equation

The Clapeyron equation of the form (5.19) denotes the slope of an equilibrium curve depicted in a pT diagram. For the phase equilibrium between, for example, **water** $H_2O(\ell)$ and **ice** $H_2O(s)$, eqn. (5.19) denotes the slope of the equilibrium curve at any point given by the temperature T.

To determine the shape of the equilibrium curve over a finite temperature interval ($T_0 \rightarrow T$), the Clapeyron equation is integrated. For **condensed phases**, ΔH and ΔV can to a good approximation be assumed constant in a limited temperature interval; integration of the Clapeyron equation gives

$$\int_{p_0}^{p} dp = \int_{T_0}^{T} \frac{\Delta H}{\Delta V}\frac{dT}{T} \simeq \frac{\Delta H}{\Delta V}\int_{T_0}^{T} d\ln(T) \tag{5.20}$$

$$p - p_0 \simeq \frac{\Delta H}{\Delta V} \cdot \ln(\frac{T}{T_0}) \qquad \text{(condensed phases)} \tag{5.21}$$

It should be noted when using the expression (5.21) that ΔH and ΔV are assumed to be *constant* in the temperature interval ($T_0 \rightarrow T$) considered.

■ At 0 °C and 1 atmosphere pressure, the thermodynamic data for **water** $H_2O(\ell)$ and **ice** $H_2O(s)$ are as follows

$H_2O(s)$:	$H^{\ominus}_{273}$	$= -293720\,\text{J/mol}$	V	$= 19.651 \cdot 10^{-6}\,\text{m}^3/\text{mol}$
$H_2O(\ell)$:		$= -287712\,\text{J/mol}$		$= 18.022 \cdot 10^{-6}\,\text{m}^3/\text{mol}$

Calculate (dp/dT) for the equilibrium between water and ice at 0°C and determine the change of pressure, Δp (Pa), which must be applied to water to lower the freezing point by 1.5 °C!

Solution. By freezing of water, $H_2O(\ell) \rightarrow H_2O(s)$, the given data result in

$$\Delta H = H_s - H_\ell = -293720 - (-287712) = -6008\,\text{J/mol}$$

$$\Delta V = V_s - V_\ell = 19.651 \cdot 10^{-6} - 18.022 \cdot 10^{-6} = 1.629 \cdot 10^{-6}\,\text{m}^3/\text{mol}$$

The slope of the equilibrium curve at 0 °C is determined by the Clapeyron equation (5.19)

$$\frac{dp}{dT} = \frac{\Delta H}{T\Delta V} = \frac{-6008}{273.15 \cdot 1.629 \cdot 10^{-6}} = \mathbf{-1.35 \cdot 10^7\,Pa/K}$$

It is now assumed that ΔH and ΔV are constant in a limited temperature range around 0 °C; using the integrated form of the Clapeyron equation (5.21), we get

$$\Delta p = p - p_0 \simeq \frac{\Delta H}{\Delta V} \cdot \ln(\frac{T}{T_0}) = \frac{-6008}{1.629 \cdot 10^{-6}} \cdot \ln(\frac{271.65}{273.15}) \simeq \mathbf{2.03 \cdot 10^7\,Pa}$$

To lower the freezing point of water by 1.5 K, the pressure must be increased by approximately $2.03 \cdot 10^7$ Pa, corresponding to approximately **200 atmospheres**!

Figure 5.7. A solid, coherent snowball can best be formed when the outside temperature is around 0 °C. At this temperature, a modest external pressure makes ice crystals melt at local contact points; when the pressure is relaxed the water freezes anew and leaves a coherent ice mass. In hard frost, there is no such melting, so the snowball remains loose and incoherent.

☐ **1.** Using the data from the above example, determine the freezing point of water when the system water + ice is subjected to a pressure of $p = 75$ atm!

☐ **2.** For substance A, $\Delta H > 0$ by melting; the solid phase sinks to the bottom in the melt. Will an increased pressure p raise or lower the melting point of the substance?

☐ **3.** Using the data from the above example, calculate the change of pressure Δp which is necessary for lowering the freezing point of water to −2.6 °C!

☐ **4.** Using the data from the above example, calculate the triple point temperature of water θ_t to 2 decimals and explain the result!

☐ **5.** At the melting point, −38.85 °C, ΔH for melting of mercury is 2.292 kJ/mol; at the melting, $\Delta V = 0.517\,\text{cm}^3/\text{mol}$. Calculate the freezing point for $Hg(\ell)$ at $p = 75$ atm!

5.3 The Clausius-Clapeyron equation

The Clapeyron equation (5.19) applies to **equilibrium** between two *arbitrary* phases in a **single-component system**. If one of the phases is an **ideal gas**, it can be useful to rewrite the Clapeyron equation. For the rewriting it shall be used that the molar volume of a condensed phase is normally negligible compared to the molar volume of a gas phase. If we consider, for example, the equilibrium between **water** and **water vapour** at 25 °C, we have the following numerical values. Molar volume of water $\simeq 18 \cdot 10^{-6}\,\text{m}^3/\text{mol}$; molar volume of saturated water vapour $\simeq 0.78\,\text{m}^3/\text{mol}$. In this system the molar volume of the gas phase is approximately 40000 times the volume of the condensed phase. Therefore, the difference ΔV included in the Clapeyron equation can to a good approximation be expressed as

States of water at 25 °C

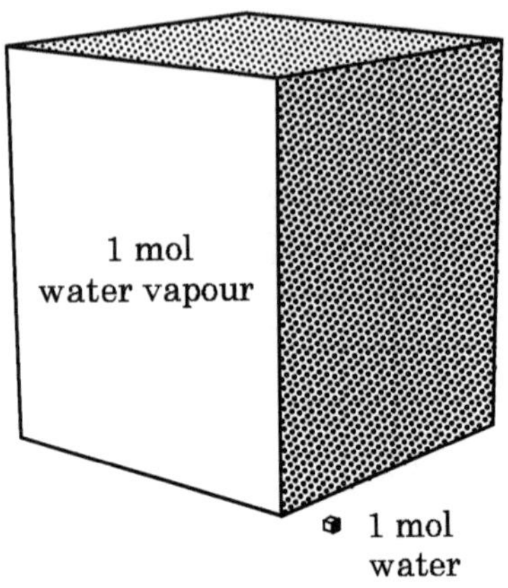

Figure 5.8. In the Clausius-Clapeyron equation it is assumed that the molar volume of the condensed phase is negligible. For example, the molar volume of saturated water vapour at 25°C is approximately 40000 times the molar volume of water!

$$\Delta V = V(\text{gas phase}) - V(\text{condensed phase}) \simeq V(\text{gas phase}) = \frac{RT}{p} \qquad (5.22)$$

Introducing this approximation into the Clapeyron equation (5.19), we get

$$\frac{dp}{dT} = \frac{\Delta H}{T\Delta V} = \frac{\Delta H\, p}{RT^2} \quad \Rightarrow \quad \frac{dp}{p} = \frac{\Delta H}{R} \cdot \frac{dT}{T^2} \qquad (5.23)$$

Using $dp/p = d\ln(p)$ and $dT/T^2 = -d(1/T)$, the connection between the equilibrium pressure of the gas phase p and the equilibrium temperature T can be expressed by the so-called **Clausius-Clapeyron equation**

The Clausius-Clapeyron equation (5.24)

At **equilibrium** between a **condensed phase** and an **ideal gas phase** in a **single-component system**, the following is fulfilled to a good approximation

$$\frac{d\ln(p)}{d(1/T)} = -\frac{\Delta H}{R}$$

where $\Delta H = H(\text{gas}) - H(\text{condensed phase})$, and R is the **gas constant**.

The Clausius-Clapeyron equation (5.24) applies to phase equilibrium in single-component systems where one phase is an **ideal gas**; thus, the expression can describe phase equilibrium at **evaporation** and **sublimation**. Phase equilibrium between condensed phases is described by the Clapeyron equations (5.19) and (5.21), respectively.

Integration of the Clausius-Clapeyron equation

The Clausius-Clapeyron equation on the form (5.24) denotes the slope of an equilibrium curve depicted in a pT diagram. If, for example, we consider the phase equilibrium between **water** $H_2O(\ell)$ and **water vapour** $H_2O(g)$, then eqn. (5.24) denotes the tangential slope of the vapour pressure curve for water at any point given by the temperature T.

The Clausius-Clapeyron equation shows the following characteristic property at the vapour pressure of the substances. Depicting the logarithm of the equilibrium vapour pressure (p) as ordinate and ($1/T$) as abscissa results in curves with the slope $-\Delta H/R$. The curves will be almost rectilinear since ΔH is only slightly dependent on temperature. Such form of depicting is, for example, useful for experimental determination of ΔH for phase transformations in systems of matter.

To determine the development of the equilibrium curve over a finite temperature interval ($T_0 \rightarrow T$), the equation (5.24) must be integrated. We consider a limited temperature range where ΔH can be assumed to be constant. If the equilibrium pressure p_0 of the gas phase at the temperature T_0 is known, the equilibrium pressure p at the temperature T is determined by the integration

$$\int_{p_0}^{p} d\ln(p) = -\int_{T_0}^{T} \frac{\Delta H}{R} d\left(\frac{1}{T}\right) \simeq -\frac{\Delta H}{R}\int_{T_0}^{T} d\left(\frac{1}{T}\right) \qquad (5.25)$$

$$\ln\left(\frac{p}{p_0}\right) \simeq -\frac{\Delta H}{R} \cdot \left(\frac{1}{T} - \frac{1}{T_0}\right) \qquad (5.26)$$

$$p \simeq p_0 \cdot \exp\left(-\frac{\Delta H}{R} \cdot \left(\frac{1}{T} - \frac{1}{T_0}\right)\right) \qquad (5.27)$$

Plot of vapour pressure curve

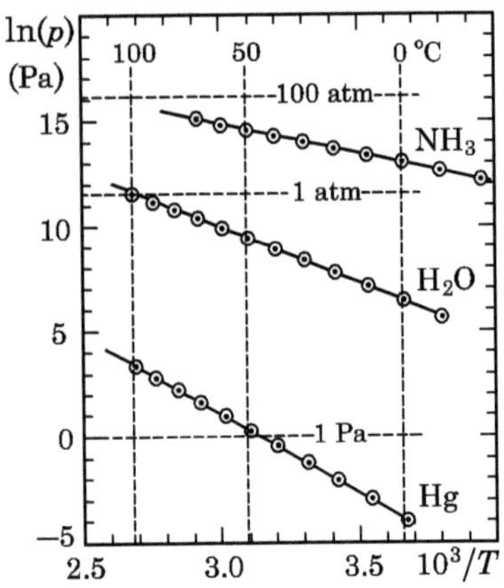

Figure 5.9. Depicting the logarithm of the vapour pressure of a substance ln(p) as a function of the reciprocal thermodynamic temperature 1/T, curves with the slope −ΔH/R result. The curves will be almost rectilinear since ΔH is only slightly dependent on temperature.

Using equations (5.26) and (5.27), it should be borne in mind that ΔH is assumed to be constant in the temperature interval ($T_0 \rightarrow T$) considered and that the equations describe equilibrium between a condensed phase and an ideal gas phase of a single-component system.

■ In section 3.8, the heat of sublimation for ice at 0 °C was calculated as 51076 J/mol. At 0 °C, the partial pressure of saturated water vapour in equilibrium with water and ice is 611.3 Pa. From this information, calculate the partial pressure of saturated water vapour in equilibrium with ice at −6 °C! The water vapour is assumed to be an ideal gas.

Solution. We use the integrated form of the Clausius-Clapeyron equation (5.27) introducing $p_0 = 611.3\,\text{Pa}$, $T_0 = 273.15\,\text{K}$, $T = 267.15\,\text{K}$ and $\Delta H = 51076\,\text{J/mol}$

$$p = 611.3 \cdot \exp\left(-\frac{51076}{8.314} \cdot \left(\frac{1}{267.15} - \frac{1}{273.15}\right)\right) = \mathbf{368.9\,Pa}$$

Table value: The partial pressure of water vapour in equilibrium with ice at −6 °C is 368.9 Pa.

□ **1.** The partial pressure of saturated water vapour is 1228 Pa at 10 °C and 1403 Pa at 12 °C; determine the evaporation heat of water (J/g) at 10 °C!

□ **2.** At the boiling point 100 °C, the evaporation heat of water is 2257 J/g; calculate the partial pressure of saturated water vapour at 108 °C!

□ **3.** At 0 °C, the partial pressure of saturated water vapour is 611.3 Pa and the evaporation heat of water is 2500 J/g; determine the partial pressure of saturated water vapour over supercooled water at −4 °C!

□ **4.** At 25 °C, the partial pressure of saturated water vapour is 3169.1 Pa and the evaporation heat is ca 44000 J/mol; determine the partial pressure of saturated water vapour at 30 °C !

□ **5.** At 80 °C, the evaporation heat of water is 2308 J/g and the partial pressure of saturated water vapour is 47373 Pa; calculate the increase in vapour pressure (Pa) per (°C) at 80 °C!

Vapour pressure curve for ice

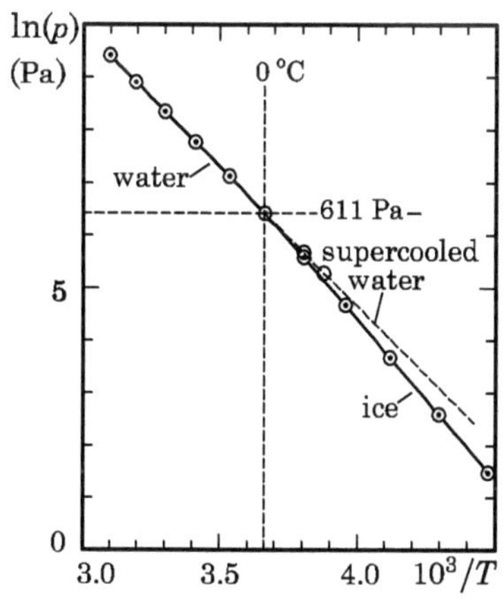

Figure 5.10. The logarithm of the saturated vapour pressure over water and ice, respectively, as a function of 1/T. The difference in the slope corresponds to the difference between $\Delta H/R$ for evaporation of water and sublimation of ice, respectively (see section 3.8).

5.4 Activity

In the above it was shown how to use data for the thermodynamic standard values of substances to solve construction materials problems. The tables of thermodynamic data used in the text conventionally refer to the following **standard state** $(p^\ominus, c^\ominus)$ at 25 °C:

$$p^\ominus = 1\,\text{atm} = 101325\,\text{Pa}; \quad c^\ominus = 1\,\text{mol}/\ell; \quad T = 298,15\,\text{K}\ (25\,°\text{C}) \qquad (5.28)$$

Reference state
$p = p^\ominus = 101325\,\text{Pa}$
$c = c^\ominus = 1\,\text{mol}/\ell$
$T = 298.15\,\text{K}\ (25\,°\text{C})$

With introduction of the concept of **molar free energy** G (J/mol) we have a measure of the tendency of the substances to react or to change state. Any system of matter will *spontaneously* seek towards an **equilibrium state** with a **minimum of free energy** G. Therefore, a substance can only be in equilibrium in a system if there is the *same* molar free energy G all over the system.

If we only consider systems of pure substances in their standard state, eqn. (5.28), it is immediately possible to use table data for thermodynamic standard values. However, we have already learned that, for example, the entropy S of an ideal gas depends on the pressure (see for example 4.26); therefore, the molar free energy of the gas must *also* depend on the pressure.

In the following we shall expand a set of practical calculation rules for the G function adjusted to the description of **ideal systems of matter**. These calculation rules are based on the introduction of a generalized concentration parameter, the **activity** a, defined by

Activity a (5.29)

The activity a_i of a component (i) in an *arbitrary* **solid**, **liquid** or **gaseous** mixture is *defined* by the equation

$$G_i = G_i^\ominus + RT \cdot \ln(a_i)$$

where G_i is the *actual*, molar free energy (J/mol) of the component and $G_i^\ominus$ is the molar free energy of the component in the **standard state**.

It is important to note that the activity a_i is **defined** from the quantities G_i and $G_i^\ominus$ and *not* the other way round. The reason for this somewhat peculiar equation of definition is given by the following reflection.

In a **reversible** change of state in a system where only volume work occurs, the differential of the free energy dG is determined by the Gibbs-Duhem equation (5.14); under **isothermal** conditions where $dT = 0$, then $dG = V dp$. For an **ideal gas** the molar free energy G can therefore be determined from $G^\ominus$ in the standard state

$$\int_{G^\ominus}^{G} dG = \int_{p^\ominus}^{p} V\,dp = RT \int_{p^\ominus}^{p} \frac{dp}{p} = RT \cdot \ln\left(\frac{p}{p^\ominus}\right) \qquad (5.30)$$

Rearranging the term $G - G^\ominus$ in the left-hand side, we obtain the following relation for an ideal gas

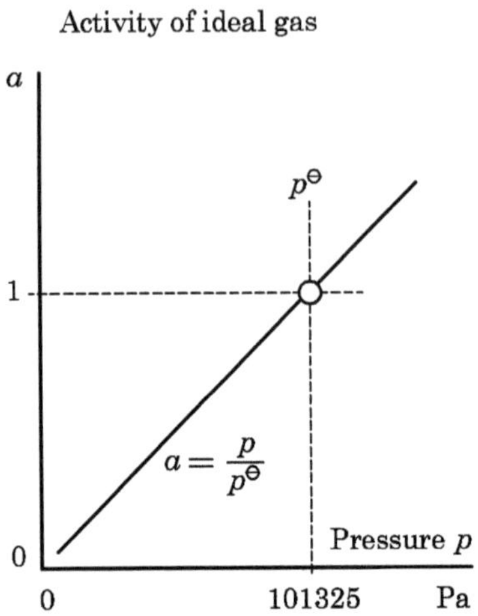

Figure 5.11. The activity a of an ideal gas can be perceived as a normalized pressure assuming the value $a = 1$ in the standard state $p = p^\ominus$.

Activity of ideal gas (5.31)

$$G = G^\ominus + RT \cdot \ln(a) = G^\ominus + RT \cdot \ln\left(\frac{p}{p^\ominus}\right) \qquad (T \text{ constant})$$

where $p^\ominus = 101325$ Pa is the **standard pressure.**

It is seen that the activity a for an **ideal gas** specifies exactly the previously used relation $p/p^\ominus$. In particular, it is seen from eqn. (5.31) that the activity of an ideal gas is $a = 1$ in the standard state $p = p^\ominus = 101325$ Pa.

For **ideal solutions** it is found that the molar free energy G of the substances can be expressed through a similar expression by the concentration

Activity of an ideal solution (5.32)

$$G = G^\ominus + RT \cdot \ln(a) = G^\ominus + RT \cdot \ln\left(\frac{c}{c^\ominus}\right) \qquad (T \text{ constant})$$

where $c^\ominus = 1\,\text{mol}/\ell$ is the **standard concentration.**

Therefore, the activity a of an **ideal solution** has the following simple meaning: $a = c/c^\ominus$, where $c^\ominus$ is the standard concentration, i.e. $1\,\text{mol}/\ell$. Thus, for dissolved substances, the activity is $a = 1$ in the standard state $c = c^\ominus = 1\,\text{mol}/\ell$.

For a solvent, i.e. the dominating component in a homogeneous mixture, the standard state refers to the pure solvent substance; in this case we obtain the following expression for the activity a of pure substances and solvents

Figure 5.12. All pure, solid and liquid substances have the activity $a = 1$.

Activity of pure substances and solvents (5.33)

$$G = G^\ominus + RT \cdot \ln(a) = G^\ominus + RT \cdot \ln\left(\frac{x}{x^\ominus}\right) \qquad (T \text{ constant})$$

where $x^\ominus = 1$ is the mole fraction of the pure substance.

The activity has a simple meaning, $a = x/x^\ominus$, in this system too, where $x^\ominus = 1$ is the mole fraction for the pure substance or solvent; in particular it is seen that **pure substances have the activity 1**!

From these examples it will be seen that the activity a is a non-dimensional quantity; the activity can be perceived as a generalized thermodynamic concentration parameter which for **ideal** systems of matter can be described by the relations (5.31)-(5.33).

The expressions set up for the activities apply to **ideal** gases and solutions. For gases, this assumption will normally be fulfilled. In diluted solutions of neutral substances the assumption of ideality is also reasonable. However, in ionic solutions, electrostatic interaction and complex formation may lead to considerable deviations from ideality.

For non-ideal mixtures, a correction factor, the **activity coefficient** γ is introduced into a further description. This factor formally states the deviation from ideality as follows (see Mathematical appendix, subchapter A9, *Debye-Hückel's law*)

$$a = \gamma \cdot \left(\tfrac{c}{c^{\ominus}}\right); \qquad a = \gamma \cdot \left(\tfrac{p}{p^{\ominus}}\right); \quad \text{etc.} \tag{5.34}$$

The value of the activity coefficient can be determined experimentally or predicted by calculations. In the following, however, we shall use the approximation of disregarding coefficients of activity and instead assume ideal systems of matter.

Activity, ideal systems of matter

Gas	$a = p/p^{\ominus}$
Dissolved substance	$a = c/c^{\ominus}$
Solvent	$a = x/x^{\ominus}$
Pure substance (s), (ℓ)	$a = 1$

■ During the hardening of concrete, a surplus of **calcium hydroxide** $Ca(OH)_2$ is precipitating in the pore water of the concrete, i.e. the solution is saturated with respect to $Ca(OH)_2$. At 25 °C, a saturated solution of $Ca(OH)_2$ contains approximately 1.8 g of dissolved $Ca(OH)_2$ per litre corresponding to 0.024 mol of $Ca(OH)_2$ per litre. The partial pressure of saturated water vapour p in equilibrium with this solution at 25 °C is 3160 Pa. The following table values are available

$Ca(OH)_2(s)$	$G^{\ominus} = -898.5\,\text{kJ/mol}$	$OH^-(aq)$	$G^{\ominus} = -157.3\,\text{kJ/mol}$
$H_2O(\ell)$	$G^{\ominus} = -237.2\,\text{kJ/mol}$	$H_2O(g)$	$G^{\ominus} = -228.6\,\text{kJ/mol}$

Assuming that solution, dissolved substances and gas phase are **ideal**, we want to calculate the **activity** a and **molar free energy** G for: $Ca(OH)_2(s)$, $OH^-(aq)$, $H_2O(\ell)$, and $H_2O(g)$!

Solution. 1) The precipitated solid **calcium hydroxide** $Ca(OH)_2(s)$ constitutes a **pure substance** with the mole fraction $x = x^{\ominus} = 1$; thus, from eqn. (5.33) we get

$$a = \frac{x}{x^{\ominus}} = \frac{1.0}{1.0} = \mathbf{1.0}$$

$$G = G^{\ominus} + RT \cdot \ln(a) = -898.5 + R \cdot 10^{-3} \cdot 298.15 \cdot \ln(1.0) = \mathbf{-898.5\,kJ/mol}$$

2) The molar concentration $[OH^-]$ of the dissolved hydroxide ions is twice the molar concentration $[Ca(OH)_2]$ of dissolved calcium hydroxide, because $Ca(OH)_2$ dissociates into 1 Ca^{++} and 2 OH^-; thus, $[OH^-] = 0.048\,\text{mol}/\ell$. Then, from eqn. (5.32) we get

$$a = \frac{c}{c^{\ominus}} = \frac{0.048}{1.0} = \mathbf{0.048}$$

$$G = G^{\ominus} + RT \cdot \ln(a) = -157.3 + R \cdot 10^{-3} \cdot 298.15 \cdot \ln(0.048) = \mathbf{-164.8\,kJ/mol}$$

3) One litre of solvent water $H_2O(\ell)$ weighs approximately 1000 g corresponding to the amount of substance $n = 1000/18.02 = 55.49\,\text{mol}$. As the content of ions (Ca^{++} and OH^- ions) in the saturated solution of $Ca(OH)_2$ is only $0.024 + 0.048 = 0.072\,\text{mol}/\ell$, we can, to a good approximation, take the mole fraction of the water as $x \simeq 1.00$. Then, from eqn. (5.33) we obtain the following for the solvent water

$$a = \frac{x}{x^{\ominus}} = \frac{1.00}{1.00} = \mathbf{1.00}$$

$$G = G^{\ominus} + RT \cdot \ln(a) = -237.2 + R \cdot 10^{-3} \cdot 298.15 \cdot \ln(1.00) = \mathbf{-237.2\,kJ/mol}$$

4) The activity and free energy of **water vapour** $H_2O(g)$ in equilibrium with the saturated solution of $Ca(OH)_2$ are determined from eqn. (5.31)

$$a = \frac{p}{p^{\ominus}} = \frac{3160}{101325} = \mathbf{0.0312}$$

$$G = G^{\ominus} + RT \cdot \ln(a) = -228.6 + R \cdot 10^{-3} \cdot 298.15 \cdot \ln(0.0312) = \mathbf{-237.2\,kJ/mol}$$

Note in particular that within the calculation accuracy we find the *same* free energy for **water** and **water vapour** in the system, corresponding to phase equilibrium. It should be noted that the pore solution in concretes contains other dissolved species, particularly alkali hydroxides; in consequence an appropriate thermodynamic analysis of actual pore solutions in concrete is more complicated than this.

☐ **1.** At 25 °C, an ideal aqueous solution contains 0.01 g of dissolved and dissociated NaCl per litre; calculate the activity a and the free energy G of Na^+ in the solution!

☐ **2.** At 25 °C, a saturated, ideal aqueous solution of $CaCO_3$ contains a deposit of solid calcite $CaCO_3(s)$; specify the activity a and free energy G for $CaCO_3(s)$!

☐ **3.** In a curing chamber, $\theta = 25\,^{\circ}C$ and $RH = 80$ % are given; calculate the activity a and the free energy G of water vapour $H_2O(g)$ in the air! – (assume ideal gas mixture).

☐ **4.** Calculate and specify the activity a and the free energy G for $O_2(g)$, $N_2(g)$, and $CO_2(g)$ in atmospheric air at 25 °C! – (assume an ideal gas mixture).

☐ **5.** An ideal aqueous solution of the temperature 25 °C contains 2.30 g of ethanol C_2H_5OH per litre; calculate the activity a and the free energy G of $H_2O(\ell)$ and $C_2H_5OH(aq)$ in the mixture considering the information that $G^{\ominus}_{298} = -182.0$ kJ/mol for $C_2H_5OH(aq)$!

5.5 Thermodynamic equilibrium constant

With the introduction of the concept of activity, the equilibrium conditions for chemical reactions and phase transformations can be treated systematically. Assume, for example, an **isothermal** reaction or transformation of the following kind

$$\begin{array}{ccc} a\text{A} + b\text{B} & \rightarrow & c\text{C} + d\text{D} \\ \text{(reactants)} & \rightarrow & \text{(products)} \end{array} \qquad (T, p \text{ constant}) \qquad (5.35)$$

The change $\Delta_r G_T$ in the free energy of the system per mol of reaction can now be expressed by the molar free energy of the individual reaction components; for simplicity, the subscript T is omitted in the following formula

$$\Delta_r G_T = \sum G(\text{products}) - \sum G(\text{reactants}) \qquad (5.36)$$

$$\Delta_r G_T = (c \cdot G_\text{C} + d \cdot G_\text{D}) - (a \cdot G_\text{A} + b \cdot G_\text{B}) \qquad (5.37)$$

Figure 5.13. Pure substances, precipitated from solutions, have the activity $a = 1$.

Now, expressing the molar free energy of the components as $G_T = G^{\ominus}_T + RT \cdot \ln(a)$ in accordance with the definition eqn. (5.29), we can rewrite the free reaction energy as

$$\begin{aligned}\Delta_r G_T = {} & (c \cdot G^{\ominus}_\text{C} + d \cdot G^{\ominus}_\text{D}) - (a \cdot G^{\ominus}_\text{A} + b \cdot G^{\ominus}_\text{B}) \\ & + RT \cdot ((c \cdot \ln(a_\text{C}) + d \cdot \ln(a_\text{D})) - (a \cdot \ln(a_\text{A}) + b \cdot \ln(a_\text{B})))\end{aligned} \qquad (5.38)$$

Analysing this expression, we see that the first line of the right-hand side of the equation denotes the standard free reaction energy $\Delta_r G^{\ominus}_T$ at the temperature T. Rearranging the logarithmic terms of the last bracket, we get a simple and systematic fraction for the activities a; this fraction defines the **thermodynamic equilibrium constant** K_a for the reaction

Thermodynamic equilibrium constant (5.39)

For a chemical reaction or a phase transformation $a\text{A} + b\text{B} \rightarrow c\text{C} + d\text{D}$, the **thermodynamic equilibrium constant** K_a is defined by

$$K_a \stackrel{\text{def}}{=} \frac{a^c_\text{C} \cdot a^d_\text{D}}{a^a_\text{A} \cdot a^b_\text{B}}$$

where $a_\text{A}, a_\text{B}, a_\text{C}$ and a_D denote the **activitetes** of the components.

Equilibrium constant

$$K_a \stackrel{\text{def}}{=} \frac{a^c_\text{C} \cdot a^d_\text{D}}{a^a_\text{A} \cdot a^b_\text{B}}$$

With the introduction of the thermodynamic equilibrium constant K_a, the free reaction energy for a chemical reaction or phase transformation of the form shown in eqn. (5.35) can now be expressed by

$$\Delta_r G_T = \Delta_r G^{\ominus}_T + RT \cdot \ln(K_a) \qquad (5.40)$$

The equilibrium constant K_a is the thermodynamic parallel to Guldberg and Waage's **law of mass action** put forward in 1864 as an empirical law.

Equilibrium condition

For the reaction considered in eqn. (5.35), equilibrium corresponds to $\Delta_r G_T = 0$; introducing this condition into eqn. (5.40), we get the following fundamental equilibrium condition

Reaction equilibrium

$$\Delta_r G_T^\ominus = -RT \cdot \ln(K_a)$$

Thermodynamic equilibrium condition III (5.41)

For a chemical reaction or phase transformation: $a\mathrm{A} + b\mathrm{B} \rightarrow c\mathrm{C} + d\mathrm{D}$, the following condition is fulfilled at **reaction equilibrium**

$$\Delta_r G_T^\ominus = -RT \cdot \ln(K_a) \qquad (T, p \text{ constant})$$

where $\Delta_r G_T^\ominus$ is the standard free reaction energy at the temperature T and K_a is the thermodynamic equilibrium constant.

In many ways, the equilibrium condition developed here is remarkable. The expression (5.41) is *universal* and *no requirements* for ideality are made. The equilibrium condition *only* includes thermodynamic **standard values** $G_T^\ominus$ that can be determined in a simple manner from table data. The thermodynamic **equilibrium constant** K_a can always be determined if either $\Delta_r G_T^\ominus$ is known or if the activities of the components are known.

Equilibrium: $K_a = \exp(-\frac{\Delta G^\ominus}{RT})$

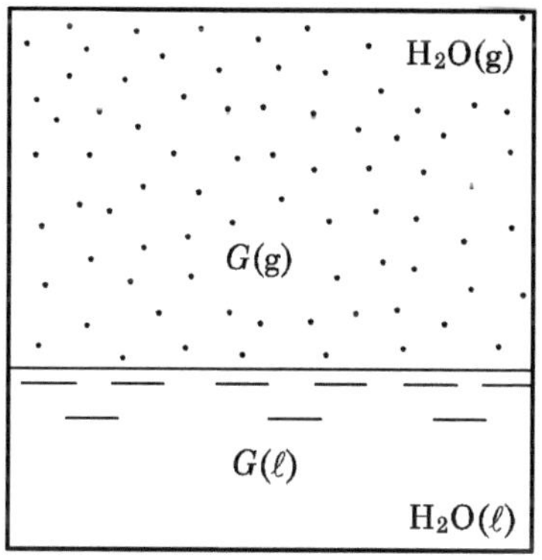

Figure 5.14. Equilibrium between water and water vapour means that $G(\ell) = G(g)$ so that $\Delta G = 0$ at phase transformation. Thus, according to eqn. (5.41), $\Delta G^\ominus = -RT \cdot \ln(K_a)$.

Determination of activity

In the practical application of the equilibrium condition eqn. (5.41), the activities for ideal (or almost ideal) systems of matter can be expressed by the **partial pressures** or **concentrations** of the components. The following expression of activities for **ideal** systems of matter is taken from section 5.4.

Component	Ideal system	Standard state
Gas	$a = p/p^\ominus$	$p^\ominus = 1\,\mathrm{atm} = 101325\,\mathrm{Pa}$
Dissolved substance	$a = c/c^\ominus$	$c^\ominus = 1\mathrm{mol}/\ell;\ p = p^\ominus$
Solvent	$a = x/x^\ominus$	$x^\ominus = 1;\quad p = p^\ominus$
Pure substance	$a = 1$	$p = p^\ominus$

Table 5.2. Overview of expressions for the activity a for ideal mixtures and solutions; the expressions given refer to section 5.4.

In simple technical calculations where the activities are determined by the **partial pressure** p of gases or the **concentration** c of dissolved substances, the following limitations should be noted

- For **gas phases**, the activity can, to a good approximation, be calculated as $p/p^\ominus$ for all commonly occurring pressures; only for high pressures or low temperatures, respectively, where the gas phase deviates notably from the ideal state, it is necessary to take corrective measures by introducing an activity coefficient.
- For **dissolved, neutral substances**, the activity can, to a good approximation, be calculated as $a = c/c^\ominus$ for concentrations less than 0.1 to 1.0 mol/ℓ.
- For **dissolved, dissociated salts**, the activity can, to an acceptable accuracy, be calculated as $a = c/c^\ominus$ for concentrations less than 0.01 to 0.02 mol/ℓ; for higher concentrations, this approximation can only be used to evaluate orders

of magnitude. For more exact calculations for higher concentrations, corrections with an experimentally determined or calculated activity coefficient are required.

■ The thermodynamic standard values for water $H_2O(\ell)$ and for water vapour $H_2O(g)$ are

$H_2O(\ell) : G^{\ominus}_{298} = -237.2\,\text{kJ/mol}$; $\quad H_2O(g) : G^{\ominus}_{298} = -228.6\,\text{kJ/mol}$;

Calculate from these data the partial pressure of saturated water vapour at 25 °C!

Solution. We consider an isothermal phase transition corresponding to evaporation of water

$$H_2O(\ell) \quad \rightarrow \quad H_2O(g) \qquad (T, p \text{ constant}) \qquad \text{(a)}$$

For this phase transformation, the standard free reaction energy is determined at 25 °C

$$\Delta_r G^{\ominus}_T = G^{\ominus}_T(g) - G^{\ominus}_T(\ell) = -228.6 - (-237.2) = 8.6\,\text{kJ/mol}$$

According to eqn. (5.41), the thermodynamic reaction equilibrium is

$$K_a = \exp(-\frac{\Delta_r G^{\ominus}_T}{RT}) = \exp(-\frac{8600}{8.314 \cdot 298.15}) = 0.03114$$

but according to eqn. (5.39) the thermodynamic equilibrium constant K_a is defined as

$$K_a = \frac{a(H_2O(g))}{a(H_2O(\ell))} = \frac{p/p^{\ominus}}{1} = \frac{p}{p^{\ominus}}$$

whereby

$$p = p^{\ominus} \cdot K_a = 101325 \cdot 0.03114\,\text{Pa} = \mathbf{3155\,Pa}$$

This value deviates by approximately 0.4 % from the table data for the partial pressure at 25 °C.

☐ **1.** Given the reaction $CaCO_3(s) \rightarrow CaO(s) + CO_2(g)$ at 25 °C, calculate K_a for the following partial pressure p of $CO_2(g)$: 30 Pa; 3000 Pa and 1 atm! – (an ideal gas shall be assumed).

☐ **2.** Given the reaction $2Fe(s) + \frac{3}{2}O_2(g) \rightarrow Fe_2O_3(s)$ at 25 °C, calculate what partial pressure p of $O_2(g)$ corresponds to the reaction equilibrium! – (an ideal gas shall be assumed).

☐ **3.** Given the reaction $CaSO_4 \cdot \frac{1}{2}H_2O(s) + 1\frac{1}{2}H_2O(g) \rightarrow CaSO_4 \cdot 2H_2O(s)$, calculate $p(H_2O(g))$ and RH for reaction equilibrium at 25 °C! – (an ideal gas shall be assumed).

Definition of pH

$\text{pH} \stackrel{\text{def}}{=} -\log_{10}(a(H^+))$

☐ **4.** Given the solution equilibrium $BaSO_4(s) \rightarrow Ba^{++}(aq) + SO_4^{--}(aq)$, calculate the solubility of $BaSO_4$ at 25 °C in (mol/ℓ) and (g/ℓ)! – (an ideal solution shall be assumed).

☐ **5.** Given the reaction equilibrium $H_2O(\ell) \rightarrow H^+(aq) + OH^-(aq)$, calculate $[H^+]$ and $\text{pH} = -\log_{10}(a(H^+))$ for water at 25 °C! – (an ideal state shall be assumed).

5.6 Temperature dependence of equilibrium

In the above section we have derived the thermodynamic condition for reaction equilibrium for chemical reactions or transformations

$$\Delta_r G^{\ominus}_T = -RT \cdot \ln(K_a) \qquad (T, p \text{ constant}) \qquad (5.42)$$

where K_a denotes the equilibrium constant for the actual reaction or transformation. As the molar free energy $G^{\ominus}_T$ of the components is defined by the **standard pressure** $p^{\ominus}$ and the **standard concentration** $c^{\ominus}$, $\Delta_r G^{\ominus}_T$ *only* depends on the temperature T. Consequently, the equilibrium constant K_a must be uniquely determined by the temperature T. For a given reaction or transformation, the standard free reaction energy $\Delta_r G^{\ominus}_T$ is determined by eqn. (5.10)

$$\Delta_r G^{\ominus}_T = \Delta_r H^{\ominus}_T - T \cdot \Delta_r S^{\ominus}_T \qquad (5.43)$$

Comparing this expression with eqn. (5.41), it is seen that the equilibrium constant K_a can be expressed by

$$\ln(K_a) = -\frac{\Delta_r H_T^\ominus}{RT} + \frac{\Delta_r S_T^\ominus}{R} \tag{5.44}$$

The temperature dependence of the equilibrium constant at a given temperature T can be expressed on the form

$$d\ln(K_a) = -\frac{\Delta_r H_T^\ominus}{R} \cdot d(\frac{1}{T}) \tag{5.45}$$

Depicting $\ln(K_a)$ for a given reaction in relation to $(1/T)$, the slope of the curve will at any point be $-\Delta_r H_T^\ominus/R$. A graph of $\ln(K_a)$ in relation to $(1/T)$ shows almost **rectilinear** curves since the reaction enthalpy for many reactions only varies slightly with temperature. Thus, this form of illustration is useful for a graphical determination of reaction enthalpy based on test data.

Temperature dependence of the equilibrium constant (5.46)

For a chemical reaction or transformation in **reaction equilibrium**, the equilibrium constant fulfils the following condition

$$\ln(K_a) = -\frac{\Delta_r H_T^\ominus}{R} \cdot (\frac{1}{T}) + \frac{\Delta_r S_T^\ominus}{R}$$

where $\Delta_r H_T^\ominus$ denotes the **standard reaction enthalpy** and $\Delta_r S_T^\ominus$ denotes the **standard reaction entropy** for the reaction concerned.

Thermodynamic equilibrium

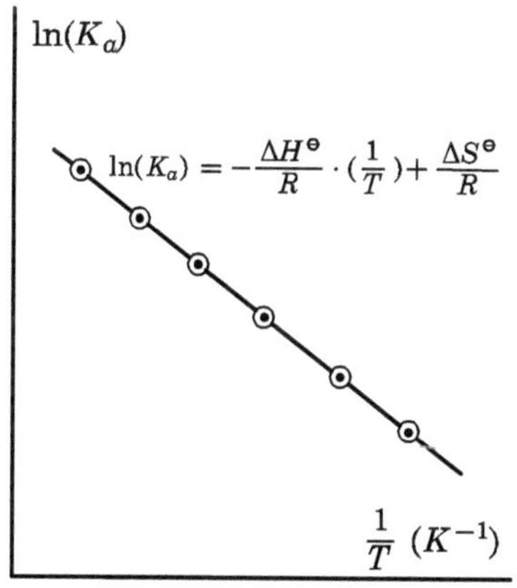

Figure 5.15. In a graph showing $\ln(K_a)$ in relation to the reciprocal thermodynamic temperature $1/T$, curves with the slope $-\Delta H/R$ are obtained, see equation (5.46). These curves are almost rectilinear, because the reaction enthalpy ΔH only varies slightly with temperature.

Important: Using the equilibrium relations in equations (5.41) and (5.46), it is important to note the following:

- Since we *are* dealing with **equilibrium relations**, we *have* thus *assumed* that $\Delta G = 0$ for the reactions considered!
- In the expressions, reaction enthalpy $\Delta H^\ominus$ and reaction entropy $\Delta S^\ominus$ are incorporated in their **standard state**; thus, in the calculation of these quantities, we have *assumed* that $p = p^\ominus$ and $c = c^\ominus$!

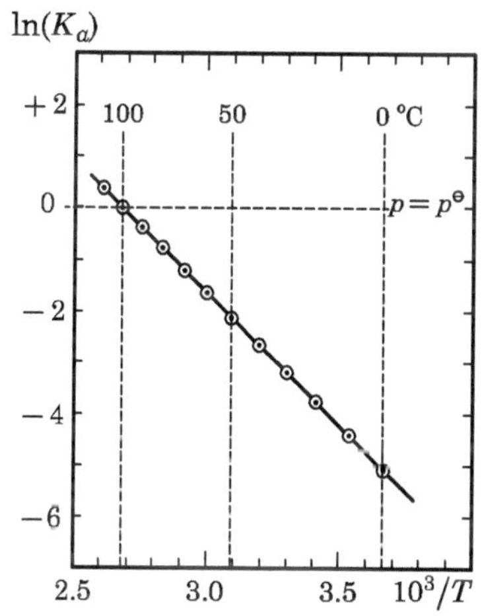

Figure 5.16. The graph for $\ln(K_a)$ at equilibrium between water and water vapour as a function of $1/T$. The curve represents a vapour pressure curve for water (see also figure 5.9).

■ The partial pressure of saturated water vapour is 2338.8 Pa at 20 °C and 4245.5 Pa at 30 °C. Calculate from these data, $\Delta_r H_T^\ominus$ and $\Delta_r S_T^\ominus$ for evaporation of water at 25 °C!

Solution. K_a for evaporation of water $H_2O(\ell) \rightarrow H_2O(g)$ is, according to eqn. (5.39),

$$K_a = \frac{a(H_2O(g))}{a(H_2O(\ell))} = \frac{p/p^\ominus}{1} = \frac{p}{p^\ominus} \tag{a}$$

assuming an ideal gas state and calculating the activity according to table 5.2. Then, for reaction equilibrium we have the following linear system of equations according to eqn. (5.46)

$$\begin{pmatrix} T_1^{-1} & 1 \\ T_2^{-1} & 1 \end{pmatrix} \begin{pmatrix} -\Delta H^\ominus/R \\ \Delta S^\ominus/R \end{pmatrix} = \begin{pmatrix} \ln(K_{a1}) \\ \ln(K_{a2}) \end{pmatrix} \tag{b}$$

$$\begin{pmatrix} 0.003411 & 1 \\ 0.003299 & 1 \end{pmatrix} \begin{pmatrix} -\Delta H^\ominus/R \\ \Delta S^\ominus/R \end{pmatrix} = \begin{pmatrix} -3.7689 \\ -3.1726 \end{pmatrix} \tag{c}$$

Solving these equations, we obtain $\Delta H^\ominus = 44.00$ kJ/mol and $\Delta S^\ominus = 118.8$ J/mol K which correspond exactly to the table values for $\Delta H^\ominus$ and $\Delta S^\ominus$ at the average temperature 25 °C.

□ **1.** The partial pressure of saturated water vapour is 8205.3 Pa at 42 °C and 11171 Pa at 48 °C; based on this, calculate $\Delta_r H_{318}^\ominus$ and $\Delta_r S_{318}^\ominus$ for evaporation of water at 45 °C!

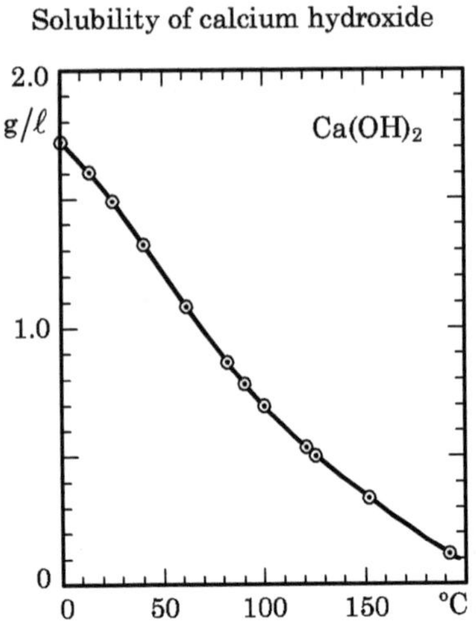

Figure 5.17. Solubility of calcium hydroxide as a function of temperature; the solubility decreases significantly with increasing temperatures.

☐ **2.** The solubility of $Ca(OH)_2$ into water decreases with increasing temperatures. Will the reaction $Ca(OH)_2(s) \rightarrow Ca^{++}(aq) + 2OH^-(aq)$ be a) exothermic, or b) endothermic?

☐ **3.** For pure water, $pH = 7.083$ at 20 °C and $pH = 6.917$ at 30 °C. Determine $\Delta_r H^{\ominus}_{298}$ for: $H_2O(\ell) \rightarrow H^+(aq) + OH^-(aq)$, when $pH = -\log_{10}(a(\mathrm{H}^+))$!

☐ **4.** What information does the expression (5.46) contain when applied to transformation of pure, condensed phases such as $H_2O(\ell) \rightarrow H_2O(s)$?

☐ **5.** For adsorbed water in cement paste it is found that at a relative humidity of 75.3 % RH, it is in equilibrium at 20.2 °C, and at a relative humidity of 78.2 % RH, it is in equilibrium at 25.2 °C. Calculate $\Delta_r H^{\ominus}$ and $\Delta_r S^{\ominus}$ for evaporation of adsorbed water in this temperature range!

List of key ideas

The following overview lists the key definitions, concepts and terms introduced in Chapter 5.

Gibbs free energy .... $G \overset{\text{def}}{=} H - TS$ (J/mol)

Helmholtz free energy ..... $A \overset{\text{def}}{=} U - TS$ (J/mol)

Equilibrium condition II ... During a process at constant **pressure** p and **temperature** T, during which *solely* **volume work** on a system is done, the following applies

$$dG = dH - TdS : \begin{cases} < 0 & \textbf{irreversible} \text{ process} \\ = 0 & \textbf{reversible} \text{ process} \\ > 0 & \textbf{impossible} \text{ process} \end{cases}$$

Spontaneous process ...... Any **spontaneous** process is **irreversible** and thus *reduces* the **free energy** G of a system; at **equilibrium**, the system has **minimum free energy** $G_{\min}$

Free reaction energy ...... For a process developing at constant **pressure** p and **temperature** T,

$$\Delta_r G_T = \Delta_r H_T - T\Delta_r S_T \qquad \text{(J/mol)}$$

The Gibbs-Duhem eqn. .... For **reversible** change of state in a homogeneous phase where solely **volume work** is done,

$$dG = V dp - S dT \qquad \text{(J/mol)}$$

Clapeyron eqn. .... For **equilibrium** between two arbitrary phases a and b in a **single-component system**,

$$\frac{dp}{dT} = \frac{\Delta H}{T\Delta V} \qquad \text{(Pa/K)}$$

Over a limited temperature range around an **equilibrium condition** (p_0, T_0), where $\Delta H \simeq$ constant, an **integrated** form of the Clapeyron equation applies

$$p - p_0 \simeq \frac{\Delta H}{\Delta V} \cdot \ln\left(\frac{T}{T_0}\right) \qquad \text{(Pa)}$$

Clausius-Clapeyron eqn. ... At **equilibrium** between a **condensed** phase and an **ideal gas phase** in a **single-component system** the following applies

$$\frac{d\ln(p)}{d(1/T)} = -\frac{\Delta H}{R} \qquad \text{(K)}$$

Over a limited temperature range around an **equilibrium condition** (p_0, T_0), where $\Delta H \simeq$ constant, an **integrated** form of the Clausius-Clapeyron equation applies

$$p \simeq p_0 \cdot \exp\Big(-\frac{\Delta H}{R} \cdot \big(\frac{1}{T} - \frac{1}{T_0}\big)\Big) \qquad \text{(Pa)}$$

Activity a .... The **activity** a_i of component (i) is **defined** by the equation

$$G_i = G_i^\ominus + RT \cdot \ln(a_i) \qquad \text{(J/mol)}$$

Activity, gas .... $a \simeq p/p^\ominus$, where $p^\ominus = 101325\,\text{Pa}$; applies, to a good approximation, to most gases at room temperature and atmospheric pressure

Activity, dissolved subst. .. $a \simeq c/c^\ominus$, where $c^\ominus = 1\,\text{mol}/\ell$; applies, to a good approximation, to dissolved, neutral substances up to 0.1 to 1.0 mol/ℓ; for charged ions, the approximation is only a guide for concentrations above approximately 0.01 mol/ℓ

Activity, solvent .... $a \simeq x/x^\ominus$, where $x^\ominus = 1$, corresponding to the mole fraction of a pure substance.

Activity, pure substance ... Any **pure substance** has the mole fraction $x = x^\ominus = 1$, and thus the activity $a = 1$

Activity coefficient γ ...... The activity coefficient γ adjusts for the deviation from the ideal state, for example,

gas : $a = \gamma \cdot (p/p^\ominus)$; dissolved substance : $a = \gamma \cdot (c/c^\ominus)$

Equilibrium constant K_a .. For a **reaction**: $a\text{A} + b\text{B} \rightarrow c\text{C} + d\text{D}$, the **thermodynamic equilibrium constant** K_a is defined by

$$K_a \overset{\text{def}}{=} (a_\text{C}^c \cdot a_\text{D}^d)/(a_\text{A}^a \cdot a_\text{B}^b)$$

Equilibrium condition III .. For a **reaction**: $a\text{A} + b\text{B} \rightarrow c\text{C} + d\text{D}$, the following condition is fulfilled for **reaction equilibrium** in a thermodynamic system

$$\Delta_r G_T^\ominus = -RT \cdot \ln(K_a) \qquad (T, p \text{ constant})$$

Temperature dependence .. In case of **reaction equilibrium** for a chemical reaction or transformation, the equilibrium constant fulfils the following condition

$$\ln(K_a) = -\frac{\Delta_r H_T^\ominus}{R} \cdot \big(\frac{1}{T}\big) + \frac{\Delta_r S_T^\ominus}{R}$$

Examples

The following examples illustrate how subject matter explained in Chapter 5 can be applied in practical calculations.

Figure 5.18. Chamber for steam curing of concrete elements. At the elevated temperature, the hardening of the concrete is faster, and elements can be stripped from the formwork and stored as early as 6-8 hours after casting.

Example 5.1

■ Loss of strength by high-temperature curing of concrete

In industrial concrete production, **high-temperature curing** of the cast concrete is often used. A high-temperature curing process may consist of the following process. The concrete is heated during mixing, for example by **adding steam**, and is cast when warm. After casting, the concrete members are conveyed through steam heated **curing chambers** where the concrete temperature is increased to 50 – 70°C; hardening at this temperature is 4-6 times faster than at room temperature. Therefore, the concrete can be **stripped from the formwork** and put into storage as early as 6-8 hours after casting.

In practice, it has been demonstrated that the quality of the finished concrete can be impaired by this type of high-temperature curing; for example, a high curing temperature reached at an early age may result in **lower final strengths** in the finished product. A contributory factor probably is that the chemical reactions during **setting** of the concrete and early **hardening** change somewhat at the elevated temperature.

The Portland cement used mainly consists of **calcium silicates, calcium aluminate**, and **calcium aluminoferrite**, the so-called **clinker minerals**. Further, the cement contains a small amount of gypsum added during grinding. The composition of a typical Portland cement is as follows:

Name	Chemical composition	Cement-chemical name	Typical content
Tricalcium silicate	$3CaO \cdot SiO_2$	C_3S	55 %
Dicalcium silicate	$2CaO \cdot SiO_2$	C_2S	20 %
Tricalcium aluminate	$3CaO \cdot Al_2O_3$	C_3A	7 %
Tetracalcium aluminoferrite	$4CaO \cdot Al_2O_3 \cdot Fe_2O_3$	C_4AF	9 %
Calcium sulphate (gypsum)	$CaO \cdot SO_3$	$C\overline{S}$	3 %

In the cement-chemical description the following notation or "abbreviations" are used for the oxides that form the clinker minerals

$$\text{'C'} = CaO; \quad \text{'S'} = SiO_2; \quad \text{'A'} = Al_2O_3; \quad \text{'F'} = Fe_2O_3; \quad \text{'}\overline{S}\text{'} = SO_3; \quad \text{'H'} = H_2O$$

During grinding of the cement, 2-5 wt-% of **gypsum** $CaSO_4 \cdot 2H_2O$ is added. The purpose of adding gypsum is, among others, to **retard** the hardening of tricalcium aluminate C_3A. The clinker mineral C_3A reacts very quickly in contact with water; a concrete made from cement without addition of gypsum will therefore set and stiffen within a few minutes - so-called "flash-setting" of C_3A. The added gypsum prevents flash-setting in that it reacts with C_3A and precipitates a retarding layer of "ettringite" on exposed C_3A surfaces. This reaction (a) of C_3A with gypsum may be written as

$$\begin{array}{ccccccc} C_3A & + & 3C\overline{S}H_2 & + & 25H & \rightarrow & C_3A \cdot 3C\overline{S} \cdot 31H \\ \textbf{Tricalcium aluminate} & + & \textbf{3 gypsum} & + & \textbf{25 water} & \rightarrow & \textbf{ettringite} \end{array}$$

The ettringite formed $C_3A \cdot 3C\overline{S} \cdot 31H$ is precipitated on the surface of the reactive C_3A and may thus retard the otherwise quick reaction. However, it is seen that ettringite is *not* stable at higher temperatures; when heated in hydrated cement systems, ettringite is transformed into a compound with a lower sulphate content, the so-called **monosulfate**. This reaction (b) may be written as

$$\begin{array}{ccccccc} C_3A \cdot 3C\overline{S} \cdot 31H & \rightarrow & C_3A \cdot C\overline{S} \cdot 12H & + & 2(C\overline{S} \cdot 2H) & + & 15H \\ \textbf{ettringite} & \rightarrow & \textbf{monosulphate} & + & \textbf{2 gypsum} & + & \textbf{15 water} \end{array}$$

A number of investigations indicate that this transformation is a contributory cause of the loss of strength at high temperatures during the early hardening of the concrete.

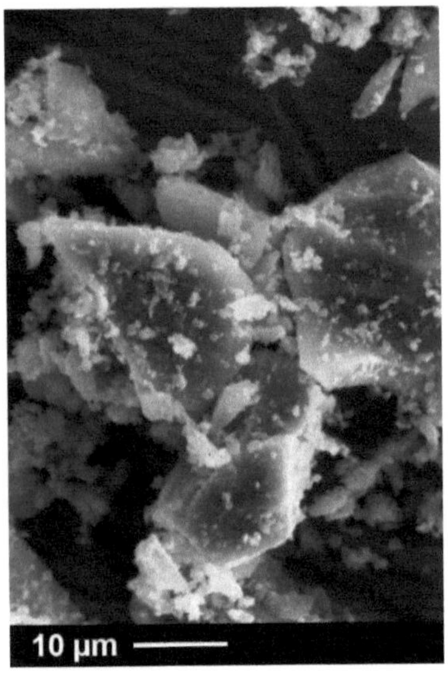

Figure 5.19. Grains of ground C_3A surrounded by small particles of gypsum $CaSO_4 \cdot 2H_2O$. Addition of gypsum delays the reaction of C_3A, so that freshly mixed concrete remains workable for several hours after mixing.

Problem. Make a rough determination of the temperature θ_{tr} (°C) at which it can be expected that the **ettringite** formed by the reaction (a) becomes **unstable** and is transformed into **monosulfate** by the reaction (b)!

Conditions. In: *Thermodynamik der Silikate*, Mchedlov-Petrossyan states the following thermodynamic standard values for the components in reaction equation (b).

Component	Formula	$H^{\ominus}_{298}$ (kJ/mol)	$S^{\ominus}_{298}$ (J/mol K)
Ettringite	$C_3A \cdot 3C\overline{S} \cdot 31H$	−17208	1688.6
Monosulfate	$C_3A \cdot C\overline{S} \cdot 12H$	−8719	719.2
Gypsum	$C\overline{S} \cdot 2H$	−2022	194.1
Water	H	−285.9	69.9

The above table does *not* contain data for the heat capacity c_p of ettringite and monosulfate.

Solution. We look into the transformation of ettringite by reaction (b). The reaction enthalpy is calculated according to eqn. (3.55) and the reaction entropy is calculated according to eqn. (4.30); both of the quantities are calculated for $T = 298.15\,\text{K}$ (25 °C). Introduction of the given data gives

$$\Delta_r H^{\ominus}_{298} = (-8719 - 2 \cdot 2022 - 15 \cdot 285.9) - (-17208) = 156.5\,\text{kJ/mol}$$

$$\Delta_r S^{\ominus}_{298} = (719.2 + 2 \cdot 194.1 + 15 \cdot 69.9) - 1688.6 = 467.3\,\text{J/mol K}$$

For $T = 298.15\,\text{K}$, the increase in free energy ΔG can now be determined from eqn. (5.10)

$$\Delta_r G^{\ominus}_{298} = \Delta_r H^{\ominus}_{298} - T\Delta_r S^{\ominus}_{298} = 156500 - 298.15 \cdot 467.3 = 17\,200\,\text{J/mol}$$

Since $\Delta G > 0$, the process is thermodynamically impossible at 25 °C; the process can only be spontaneous in the opposite direction, i.e. during formation of ettringite from monosulfate + gypsum + water. Therefore, ettringite can be expected to be the stable phase at room temperature as long as there is a surplus of gypsum.

Information of the specific heat c_p for components in reaction equation (b) is not given in the data; therefore, in a **rough calculation** we will assume that ΔH and ΔS are approximately constant in the temperature range considered. According to equation (5.7), equilibrium for the reaction (b) corresponds to $\Delta G = 0$; thus, we get the following condition for determination of the transformation temperature T_{tr}

$$T_{\text{tr}} \simeq \frac{\Delta_r H^{\ominus}_{298}}{\Delta_r S^{\ominus}_{298}} = \frac{156500\,\text{J/mol}}{467.3\,\text{J/mol K}} = 335\,\text{K}$$

Therefore, from the given data it can be expected that ettringite will decompose and be transformed into monosulfate if the temperature exceeds approximately 335 K corresponding to $\theta_{\text{tr}} \simeq 62\,°\text{C}$. Consequences of this transformation may therefore be expected if the concrete during the early setting and hardening is heated above this temperature.

Discussion. The assumption that ΔH and ΔS are $\simeq$ constant in the temperature range considered is normally reasonable in rough calculations. For example, for the transformation: $H_2O(\ell) \rightarrow H_2O(g)$, i.e. evaporation of water, $\Delta_r H^{\ominus}_{298} = 44000\,\text{J/mol}$ and $\Delta_r S^{\ominus}_{298} = 118.8\,\text{J/mol K}$. The boiling point of water corresponds to the partial pressure of saturated water vapour being exactly 1 atmosphere. Calculating roughly the temperature at the boiling point and assuming constant ΔH and ΔS, we get

$$T_{\text{bp}} \simeq \frac{\Delta_r H^{\ominus}_{298}}{\Delta_r S^{\ominus}_{298}} = \frac{44000}{118.8} = 370\,\text{K}$$

This value, approximately 97 °C, is a fully acceptable estimate of the boiling point of water in a simple rough calculation.

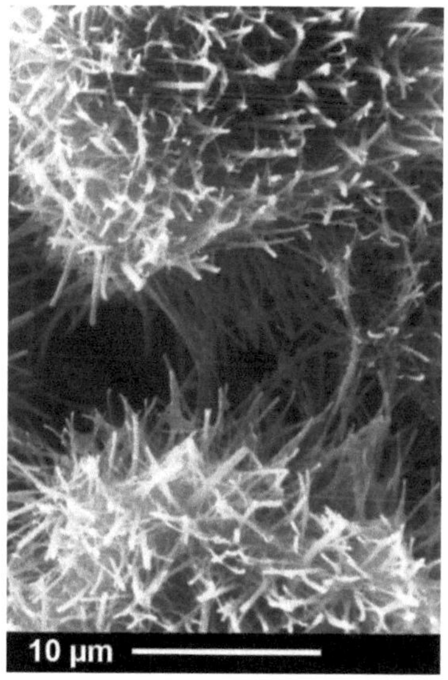

Figure 5.20. Needle-like ettringite formed on the surface of C_3A grains. The precipitation of ettringite shown retards the otherwise quick reaction of C_3A with water.

Figure 5.21. Hexagonal crystals of monosulfate may be formed by reaction of ettringite with C_3A; the investigation suggests that this transformation will be spontaneous at high temperatures during the early hardening of the concrete.

Example 5.2

■ Steel manufacture – reduction of iron ore in blast furnace

Steel is worked up from **iron ore**, mainly ferric oxides; examples of iron ores are **magnetite** ("lodestone") Fe_3O_4 and **hematite** Fe_2O_3. First, **pig iron** is made from these raw materials by reduction of the ores in a blast furnace. The pig iron contains a number of impurities, e.g. **sulphur**, **phosphorus** and **silicon**, and it has an unacceptably high carbon content. The contents of these undesired components are reduced afterwards by **refining** and then the raw steel is ready for processing. A modern **blast**

furnace works in the following way. The blast furnace is a shaft furnace that typically holds 2000 tons of raw material during processing. In the furnace, iron ore, carbon and a slag-forming material, generally limestone, are placed in layers. During the combustion process, preheated air is blown in at the bottom of the warmest zone in the shaft furnace, where the melted raw iron has been heated to a temperature of more than 1500 °C. During the process in the blast furnace, ferric oxide is reduced to iron Fe by two different reactions. At moderate temperatures, **carbon monoxide** CO(g) (formed from partial oxidation of the carbon) functions as a reducing agent; for example, the net reaction by reduction of hematite is

$$Fe_2O_3(s) + 3CO(g) \quad \rightarrow \quad 2Fe(s) + 3CO_2(g) \tag{a}$$

At higher temperatures, the reduction can be performed by direct action of the carbon on the treated iron ore.

The reduction of ferric oxide to iron Fe by action of the carbon monoxide CO(g) can actually take place at temperatures far below the melting point of iron. Thus, the iron can be transformed into a so-called **sponge iron** of Fe, without having been melted at any time. By subsequent forging, this sponge iron can be transformed into solid iron units. In earlier times, where metal workers were not able to heat the iron to its melting point, iron was manufactured based on this principle. During the process in the blast furnace, the reduction (a) occurs in parallel with a process that **regenerates** carbon monoxide CO(g). The carbon dioxide CO_2(g) formed by the reduction reacts with carbon and is hereby again transformed into carbon monoxide CO(g).

$$CO_2(g) + C(s) \quad \rightarrow \quad 2CO(g) \tag{b}$$

whereupon further transformation of hematite Fe_2O_3 into iron Fe may take place according to (a). During the process in the blast furnace, this transformation of carbon dioxide into carbon monoxide is vital. The concentration of the reduced CO gas is directly determined by the **temperature** in the blast furnace. At room temperature, CO_2(g) is stable together with carbon C(s); with increasing temperatures, the amount of CO(g) will increase in relation to CO_2(g). The reduction of ferric oxide to iron, therefore, is connected to a certain temperature range in the blast furnace.

Figure 5.22. An old engraving showing a primitive process where iron ore is worked up into iron; when the carbon is reduced, the ore can be transformed into a forgeable sponge iron of Fe without being melted at any time.

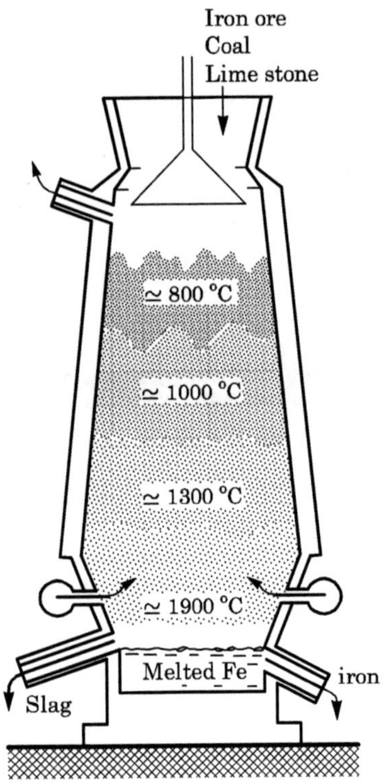

Figure 5.23. Blast furnace for manufacture of raw iron from iron ore. Iron ore is reduced through action of carbon monoxide CO on the unmelted ore and by direct reduction with carbon at higher temperatures.

Problem. In a closed tank, carbon C(s), carbon dioxide CO_2(g), and carbon monoxide CO(g) are in mutual equilibrium at the following reaction

$$CO_2(g) + C(s) \quad \rightleftarrows \quad 2CO(g) \tag{c}$$

The gas pressure in the tank is kept constant at $p = p^\ominus = 101325$ Pa. Make a rough calculation of the carbon monoxide partial pressure, $p(\text{CO})$, in the tank as a function of the temperature in the range: 25 °C – 1000 °C.

Conditions. The following table data for the thermodynamic standard values of the components are used in the calculations

Component		$H^\ominus_{298}$ (kJ/mol)	$S^\ominus_{298}$ (J/mol K)	c_p (J/mol K)
Carbon monoxide	CO(g)	−110.5	197.6	29.1
Carbon dioxide	CO_2(g)	−393.5	213.6	37.1
Carbon	C(s)	0	5.7	8.5

In calculations, CO_2(g) and CO(g) are assumed to be ideal gases; the specific heat c_p is assumed to be constant in the temperature range considered.

Solution. We determine the standard free reaction energy $\Delta_r G^\ominus_{298}$ by the procedure in eqn. (5.11)

$$\Delta_r H^\ominus_{298} = 2 \cdot (-110.5) - (0 + (-393.5)) = 172.5 \text{ kJ/mol}$$

$$\Delta_r S^\ominus_{298} = 2 \cdot 197.6 - (5.7 + 213.6) = 175.9 \text{ J/mol K}$$

$$\Delta_r c_p = 2 \cdot 29.1 - (8.5 + 37.1) = 12.6 \text{ J/mol K}$$

Then, we can express $\Delta_r H^\ominus_T$ and $\Delta_r S^\ominus_T$ as functions of the temperature; *no* correction for pressure is included, the gases are assumed to be maintained at the standard pressure $p^\ominus$

$$\Delta_r H^\ominus_T = 172500 + 12.6 \cdot (T - 298.15) = 168743 + 12.6 \cdot T \quad \text{(J/mol)} \tag{d}$$

$$\Delta_r S_T^{\ominus} = 175.9 + 12.6 \cdot \ln(T/298.15) = 104.1 + 12.6 \cdot \ln(T) \quad \text{(J/mol K)} \qquad \text{(e)}$$

$$\Delta_r G_T^{\ominus} = 168743 + 12.6 \cdot T - T(104.1 + 12.6 \cdot \ln(T)) \quad \text{(J/mol)}$$

$$\Delta_r G_T^{\ominus} = 168743 - 91.5 \cdot T - 12.6 \cdot T \cdot \ln(T) \quad \text{(J/mol)} \qquad \text{(f)}$$

Reaction equilibrium

$$K_a = \exp(-\frac{\Delta_r G_T^{\ominus}}{RT})$$

Using the equilibrium condition in equation (5.41), we can now express the thermodynamic equilibrium constant K_a as a function of the temperature T

$$K_a = \exp(-\frac{\Delta_r G_T^{\ominus}}{RT}) = \exp(-\frac{168743 - 91.5 \cdot T - 12.6 \cdot T \cdot \ln(T)}{RT}) \qquad \text{(g)}$$

The expression (g) determines the equilibrium constant K_a as a function of the temperature T. Then, K_a is related to the activity of CO(g) in the system. When there is equilibrium for the reaction (c), the activities fulfil the following relations in accordance with table 5.2

$$a(\mathrm{C}) = 1; \quad a(\mathrm{CO_2}) = \frac{p(\mathrm{CO_2})}{p^{\ominus}}; \quad a(\mathrm{CO}) = \frac{p(\mathrm{CO})}{p^{\ominus}} \qquad \text{(h)}$$

Since the total pressure in the tank is assumed to be constant at 1 atmosphere, we have the following relation between pressures and activities in the system considered

$$p(\mathrm{CO_2}) + p(\mathrm{CO}) = p^{\ominus} \Rightarrow a(\mathrm{CO_2}) + a(\mathrm{CO}) = 1 \Rightarrow a(\mathrm{CO_2}) = 1 - a(\mathrm{CO}) \qquad \text{(i)}$$

The thermodynamic equilibrium constant K_a can then be expressed by eqn. (5.39)

$$K_a = \frac{a(\mathrm{CO})^2}{1 \cdot a(\mathrm{CO_2})} = \frac{a(\mathrm{CO})^2}{1 - a(\mathrm{CO})} \qquad \text{(j)}$$

from which the activity $a(\mathrm{CO})$ of carbon monoxide can be calculated for all known values of K_a in the following way

$$a(\mathrm{CO})^2 + K_a \cdot a(\mathrm{CO}) - K_a = 0$$

$$a(\mathrm{CO}) = -\tfrac{1}{2}K_a + \sqrt{(\tfrac{1}{2}K_a)^2 + K_a} \qquad \text{(k)}$$

We can now perform the calculation of $a(\mathrm{CO})$; for a *given* temperature T, K_a is determined from the expression (g), and the activity $a(\mathrm{CO})$ is determined from (k). The following table can be set up

θ	25	400	500	600	700	800	900	1000	°C
T	298	673	773	873	973	1073	1173	1273	K
K_a	$9 \cdot 10^{-22}$	$9 \cdot 10^{-5}$	0.006	0.14	1.77	14.4	82.5	364	cf. (g)
$a(\mathrm{CO})$	$3 \cdot 10^{-11}$	0.01	0.07	0.31	0.71	0.94	0.99	1.00	cf. (k)

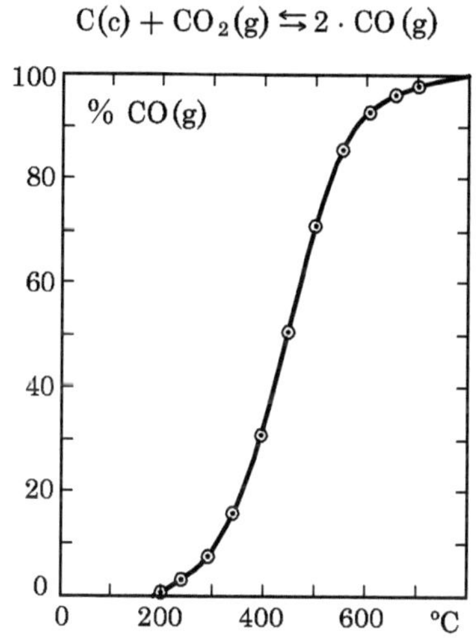

Figure 5.24. Equilibrium between carbon C, carbon dioxide CO_2, and carbon monoxide CO as a function of temperature. With surplus of carbon, the content of carbon monoxide of the gas phase is increased to almost 100 % at temperatures above approximately 700 °C. This transformation is used by reduction of iron ore in the blast furnace process where CO serves as a reducing agent.

Discussion. An approximate calculation shows that the content of reduced carbon monoxide CO(g) of the gas is strongly increased when the temperature exceeds 5-600 °C; at 8-900 °C, an almost complete transformation of carbon dioxide CO_2(g) into carbon monoxide CO(g) takes place when in contact with carbon.

Example 5.3

■ Capillary condensation in porous construction materials – the Kelvin equation

In normal use, porous construction materials contain large or small amounts of **moisture** in the form of evaporable water. After direct water exposure, the materials can contain capillary water in pores open to the atmosphere; this water has the same properties as free water. At relative humidities **below** 100 % *RH*, the materials can absorb the water vapour of the air and fix it in the internal structure by

- *Capillary condensation*, i.e. binding by surface tension in internal pores.
- *Surface adsorption*, i.e. binding by molecular adsorption on internal surfaces.

Surface adsorption will especially apply to moderate and low relative humidities; **capillary condensation** takes place at relatively high relative humidities. The mechanism in capillary condensation of water vapour in porous materials is as follows.

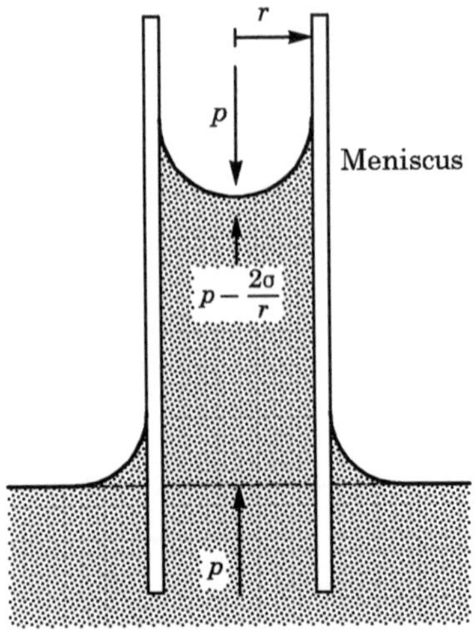

Figure 5.25. Capillary suction in cylindrical pore. At equilibrium the surface tension results in reduced pressure under the curved meniscus. The reduced pressure, $\Delta p = -2\sigma/r$, is related to the ability of porous materials to absorb moisture from the air by capillary condensation.

Assume a long, cylindrical pore with radius r; the pore is partially filled with water. If the water is able to wet the pore wall completely, a **concave** surface will be formed in the pore – a so-called **meniscus.** The **surface tension** σ (N/m) along the pore wall will influence this meniscus with a circumferential tension parallel with the longitudinal direction of the pore. The resulting force F is the circumference of the pore $2\pi r$ times the surface tension σ. At equilibrium, the liquid pressure p under the curved meniscus is changed by

$$\Delta p = \frac{\text{Force}}{\text{Area}} = -\frac{2\pi r\sigma}{\pi r^2} = -\frac{2\sigma}{r} \qquad \text{(Pa)} \tag{a}$$

in relation to the pressure of the ambient atmosphere. The minus sign denotes a *negative* increase of pressure; in other words, a **hydrostatic underpressure** is created below the curved meniscus. According to the **Gibbs-Duhem equation** eqn. (5.14), this results in a reduction of the molar free energy G of the water at constant temperature T with

$$dG = V \cdot dp \qquad (T \text{ constant}) \tag{b}$$

where V denotes the molar volume of the liquid phase and dp denotes the pressure increase. Again, this means that the equilibrium between **water** and **water vapour** is changed.

Problem. Use the **Gibbs-Duhem equation** to derive an expression that determines the **relative humidity** RH in atmospheric air which is in equilibrium with the capillary-bound water in a cylindrical pore with radius r! – based on this, calculate and tabulate the relation $RH = f(r)$ at 25 °C for $r > 1$ nm!

Conditions. In the calculation, the water is assumed to wet the pore wall completely. In case of partial wetting, the surface tension σ is simply replaced in all expressions by the component $\sigma \cdot \cos(\theta)$ in the longitudinal direction of the pore, where θ denotes the contact angle. At 25 °C, the following data for water are used in the numerical calculation of the vapour pressure reduction over capillary pores: density $\varrho = 997.0\,\text{kg/m}^3$; surface tension $\sigma = 0.072\,\text{N/m}$; molar mass $M = 18.02\,\text{g/mol}$. Water vapour is assumed to be an ideal gas.

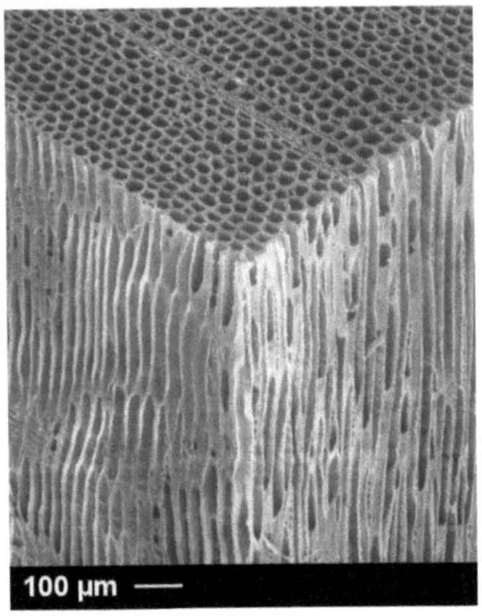

Figure 5.26. Wood is composed on long xylem cells that mainly consist of chain molecules of cellulose. The xylem cells are hygroscopic, since the cellulose can absorb and bind the moisture of the air by adsorption and capillary condensation.

Solution. We consider a reversible, isothermal phase transformation where water is transformed into water vapour by the process

$$H_2O(\ell) \rightarrow H_2O(g) \qquad (T \text{ constant}) \tag{c}$$

Since the process is reversible (equilibrium), the increase in free energy is $\Delta G = 0$, i.e. H_2O has the *same* molar free energy in the two states

$$\Delta G = G(g) - G(\ell) = 0 \quad \Rightarrow \quad G(\ell) = G(g) \tag{d}$$

Changing the pressure in the water by dp, the molar free energy of the water is changed according to the Gibbs-Duhem equation (5.14) by $dG(\ell) = V(\ell) \cdot dp$. To **maintain** the state of equilibrium, the molar free energy of the water must be changed by a similar value; thus, we have the following condition for maintaining equilibrium

$$dG(g) = V(g) \cdot dp(g) = dG(\ell) = V(\ell) \cdot dp(\ell) \tag{e}$$

where $V(g), V(\ell)$ denotes the molar volume of the respective phases. The water vapour is assumed to be an ideal gas: $V(g) = RT/p(g)$. Introducing this term into (e) and rearranging the terms, we get

$$\frac{dp(g)}{p(g)} = \frac{V(\ell)}{RT} \cdot dp(\ell) \tag{f}$$

It is now assumed that the water is incompressible, i.e. $V(\ell)$ is constant. By integration from the initial state $p(g) = p_s$ and $p(\ell) = 0$ to the final state $p(g) = p$ and $p(\ell) = -2\sigma/r$, and using the rewriting $dp/p = d\ln(p)$, we obtain

$$\int_{p_s}^{p} d\ln(p(g)) = \frac{V(\ell)}{RT} \int_0^{-2\sigma/r} dp(\ell) \tag{g}$$

$$\ln\left(\frac{p}{p_s}\right) = -\frac{2\sigma V(\ell)}{rRT} \tag{h}$$

Using that $RH = p/p_s$ and $V(\ell) = M/\varrho$, where ϱ denotes the density of the water, we obtain the **Kelvin equation** for vapour pressure reduction over capillary pores

$$\ln(RH) = -\frac{2M\sigma}{\varrho rRT} \quad \text{(The Kelvin equation)} \tag{i}$$

where ϱ is the density of the water (kg/m^3), M is the molar mass H_2O (kg/mol), σ is the surface tension of the water at T, (N/m), r is the capillary radius (m), R is the gas constant (J/mol K), and T is the thermodynamic temperature (K). Inserting the given data for water at $T = 298.15\,K$ ($25\,^\circ C$), the Kelvin equation can be reduced to the form

$$\ln(RH) = -\frac{2 \cdot 18.02 \cdot 10^{-3} \cdot 0.072}{997.0 \cdot r \cdot 8.314 \cdot 298.15} = -\frac{1.05 \cdot 10^{-9}}{r} \tag{j}$$

By table calculation we find the desired relation between the relative humidity RH in equilibrium over capillary pores and the radius r in the pore considered.

r	1	2	4	8	10	20	40	100	1000	nm
RH	35.0	59.2	76.9	87.7	90.0	94.9	97.4	99.0	100.0	%

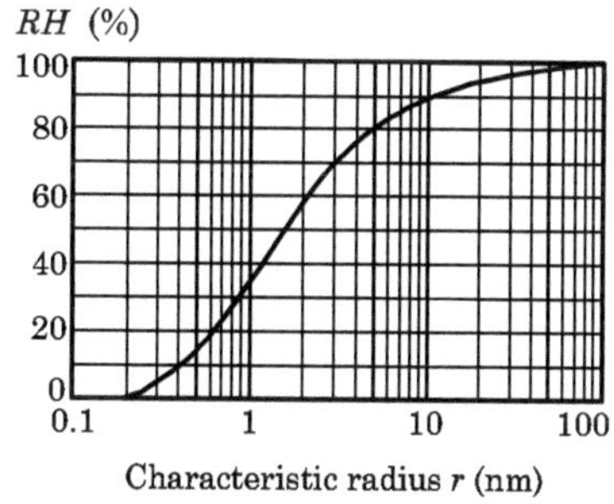

Figure 5.27. The figure shows the relative humidity RH at 25 °C over capillary condensed water as a function of the characteristic capillary radius. The RH reduction is due to the surface tension producing a hydrostatic underpressure in the absorbed water.

Discussion. Capillary condensed water in pores is in a state of tension. Due to this, the free energy of the water G is reduced in relation to the free energy of water at the standard pressure of 1 atmosphere. This change of pressure results in a reduction of the equilibrium vapour pressure over capillary-bound water. In practice this phenomenon is seen by the fact that the water vapour of the air is able to condense and be bound in narrow pores although the relative humidity $RH < 100$ %. As seen from the calculations, the vapour pressure reduction is only significant at pore radii r of the order of magnitude 100 mm and below.

Example 5.4

■ Computerized calculation of the partial pressure of saturated water vapour

Calculations of moisture is an important discipline within materials technology. Water and moisture, for example, are considered to be one common factor within such important and very different areas as

- electrolytic *corrosion* of metals
- *freezing damage* of concrete structures
- *fungal attack* in timber structures
- cracking due to *plastic shrinkage* in hardening concrete The partial pressure of saturated water vapour is often used as a calculation parameter when dealing with moisture problems. Rough calculations can simply be based on **tables** of the pressure of water vapour as a function of temperature. In connection with computer calculations of moisture, however, it is desirable to have access to exact **analytical expressions** for the vapour pressure as a function of the temperature: $p_s = f(T)$. The following example illustrates how simple computer functions suitable for moisture calculation can be developed and adjusted based on the **Clausius-Clapeyron equation**.

Figure 5.28. Parquet floor decomposed by dry rot. Cause: an external wall has absorbed moisture whereby the relative humidity under the flooring has increased to a level where there is a risk of dry rot attack. After a few years, the underside of the floor has been decomposed by the mycelia of the fungus.

Problem. We want to develop a computer function *pvap(T)* for moisture calculations; by input of the function with the argument T (K), the function should return the partial pressure of saturated water vapour p_s in the unit (Pa). The function shall cover a temperature range from $0\,^\circ C$ to $80\,^\circ C$. Make an evaluation of the exactness of the calculation.

Conditions. In the calculations, water vapour is assumed to be an ideal gas. As a reference table of the vapour pressure curve of water, Haar, Gallagher & Kell: *NBS/ NRC Wasserdampftafeln*, (1988) is used.

Solution. We look into the temperature dependence of the phase equilibrium between water $H_2O(\ell)$ and water vapour $H_2O(g)$ in the temperature range $0 - 80\,^\circ C$

$$H_2O(\ell) \leftrightarrows H_2O(g) \qquad (a)$$

For this single-component system with one **condensed phase** and one **ideal gas phase**, and according to the **Clausius-Clapeyron** equation (5.24), we obtain

$$d\ln(p) = -\frac{\Delta H}{R} \cdot d\left(\frac{1}{T}\right) \qquad (b)$$

The Clausius-Clapeyron equation

$$\frac{d\ln(p)}{d(1/T)} = -\frac{\Delta H}{R}$$

In this expression, the gas phase, i.e. the water vapour, is assumed to be an ideal gas. Introducing ΔH for the phase transformation (a) as a function of the temperature T into this, we obtain from equation (3.55)

$$\Delta_r H_T = \Delta_r H^\ominus_{298} + \Delta_r c_p \cdot (T - 298.15) \qquad (\mathrm{J/mol}) \qquad (c)$$

We introduce this relation into the Clausius-Clapeyron equation (b) and obtain the following expression for $d\ln(p)$ after rearranging the terms

$$d\ln(p) = -\frac{\Delta_r H^\ominus_{298} - \Delta_r c_p \cdot 298.15}{R} \cdot d\left(\frac{1}{T}\right) + \frac{\Delta_r c_p}{R} \cdot \frac{dT}{T} \qquad (d)$$

In the reference state $T_0 = 298.15$ K, the partial pressure of saturated water vapour is $p_0 = 3169.1$ Pa. By integration of (d) based on this reference state (T_0, p_0) and after rearranging the terms, we obtain

$$\ln\left(\frac{p}{p_0}\right) = \mathrm{A} \cdot \left(1 - \frac{T_0}{T}\right) + \mathrm{B} \cdot \ln\left(\frac{T_0}{T}\right) \qquad (e)$$

where the constants are $\mathrm{A} = (\Delta_r H^\ominus_{298} - \Delta_r c_p \cdot T_0)/RT_0$ and $\mathrm{B} = -\Delta_r c_p/R$. In principle, the constants A and B can be determined from table values of the thermodynamic standard values of the phases. As shown in example 4.3, this will typically give deviations of approximately 1 % in the calculated vapour pressure. To achieve more accurate calculations, the constants A and B, and thus $\Delta_r H^\ominus_{298}$ and $\Delta_r c_p$, are determined by **linear regression** (see *Mathematical appendix, subchapter A5*).

The Clausius-Clapeyron equation

Describes equilibrium between two phases in a single-component system; it is assumed that one of the phases is an ideal gas.

In the expression (e), $Y = \ln(p/p_0)$ can be taken as a variable depending linearly on $X_1 = (1 - T_0/T)$ and $X_2 = \ln(T_0/T)$. Then the regression equation becomes

$$Y = \mathrm{A} \cdot X_1 + \mathrm{B} \cdot X_2 + \varepsilon \qquad (f)$$

The following table values for the vapour pressure curve of water can be used to determine the constants A and B in this regression equation:

No.	1	2	3	4	5	6	7	8	9
θ (°C)	0	10	20	30	40	50	60	70	80
p (Pa)	611.3	1228.1	2338.8	4245.5	7381.4	12344	19932	31176	47373

With this set of data we now determine the coefficients in accordance with the procedure described in *Mathematical appendix, subchapter A5*. When we develop from the point: $(T_0, p_0) = (298.15, 3169.1)$, the following matrices are formed in accordance with item (1) in the procedure

$$\mathbf{Y} = \begin{pmatrix} \ln(p_1/p_0) \\ \ln(p_2/p_0) \\ \vdots \\ \ln(p_9/p_0) \end{pmatrix} ; \quad \mathbf{X} = \begin{pmatrix} (1 - T_0/T_1) & \ln(T_0/T_1) \\ (1 - T_0/T_2) & \ln(T_0/T_2) \\ \vdots & \vdots \\ (1 - T_0/T_9) & \ln(T_0/T_9) \end{pmatrix} ; \quad \mathbf{k} = \begin{pmatrix} A \\ B \end{pmatrix} \qquad (g)$$

Following (2) in the procedure, we then determine the **normal equation** for the equations determined above

$$\mathbf{X^T X\, k} = \mathbf{X^T Y} \qquad (h)$$

Inserting the table data, the following numerical expression for the normal equation is obtained

$$\begin{pmatrix} 0.0725153097 & -0.0763547933 \\ -0.0763547933 & 0.0805311004 \end{pmatrix} \begin{pmatrix} A \\ B \end{pmatrix} = \begin{pmatrix} 1.2707454050 \\ -1.3373985277 \end{pmatrix} \qquad (i)$$

The solution to the normal equation (i) determines A and B so that the sum of the square of the deviations $\sum \varepsilon_i^2$ is minimum. Solving the system of equations, we obtain

$$\mathrm{A} = 22.48604854; \qquad \mathrm{B} = 4.71270177 \qquad (j)$$

To check this result, the corresponding thermodynamic standard values are determined; from the substitution we have

$$\Delta_r c_p = -\mathrm{B} \cdot R = -4.7127 \cdot 8.314 = -39.2\,\mathrm{J/mol\,K}$$

$$\Delta_r H^{\ominus}_{298} = (\mathrm{A} - \mathrm{B}) \cdot RT_0 = (22.4860 - 4.7127) \cdot 8.314 \cdot 298.15 = 44.06\,\mathrm{kJ/mol}$$

The calculated standard values agree reasonably well with the table values for the same quantities: $\Delta_r H^{\ominus}_{298} = 44.0\,\mathrm{kJ/mol}$ and $\Delta_r c_p = -41.7\,\mathrm{J/mol\,K}$, respectively. It shall be remembered that the value determined for $\Delta_r c_p$ is a characteristic mean value in the temperature range 0 – 80 °C considered.

The Clapeyron equation

Describes equilibrium between two phases in a single-component system; applies generally to all states.

Now the computer function *pvap(T)* can be made in a pseudo code which in a simple way can be rewritten to, for example, *Matlab* or *C*.

```
FUNCTION pvap(T:real):real;                       (* vapour pressure function *)
const A  = 22.48604854;  B = 4.71270177;                       (* constants *)
      T0 = 298.15;       p0 = 3169.1;                          (* ref. point *)
begin
  if (T < 273.15) or (T > 353.15)                              (* range test *)
    then writeln('Out of range')                               (* error note *)
   else
    pvap:= p0*exp(A*(1-T0/T) + B*ln(T0/T))                (* calculate value *)
end;                                                        (* return value *)
```

We can then test the computer function developed. Calculating the data used above in the temperature range 0 – 80 °C, we obtain the following result

No.	1	2	3	4	5	6	7	8	9
θ (°C)	0	10	20	30	40	50	60	70	80
p (Pa)	611.3	1228.1	2338.8	4245.5	7381.4	12344	19932	31176	47373
pvap	611.5	1228.1	2338.8	4245.9	7383.2	12349	19939	31179	47352
$\lvert\Delta\rvert$%	0.03	0.00	0.00	< 0.01	0.02	0.04	0.04	< 0.01	0.04

The largest deviation between table value and functional value in temperature range concerned can thus be expected to be approximately 0.04 %.

Discussion. The kind of vapour pressure functions as outlined here can be developed for temperatures below 0 °C; in this temperature range, only thermodynamic standard values are used for the reaction

$$H_2O(s) \quad \rightarrow \quad H_2O(g) \qquad \text{(k)}$$

corresponding to sublimation of ice. The procedure for determining constants will correspond to calculations in the example shown above.

Example 5.5

■ Adsorption of water in hardened cement paste - shrinkage and swelling

Porous construction materials such as wood, concrete and bricks are **hygroscopic**, i.e. these materials can take up moisture from the ambient air. If a hygroscopic material is surrounded by constant relative humidity RH, it will take up or liberate water vapour until moisture equilibrium with the surroundings is achieved. At moisture equilibrium, the adsorbed water in the material and the water vapour in the ambient air have the *same* molar free energy G so that $\Delta G = 0$ for the transformation

$$H_2O(\text{adsorbed}) \quad \rightarrow \quad H_2O(\text{water vapour}) \qquad (T, p \text{ constant}) \qquad \text{(a)}$$

For a given material at a given temperature T, the equilibrium moisture content u (gram moisture per gram of dry weight) is determined by the relative humidity RH of the air in the surroundings. This relation: $u = f(RH)$ at constant temperature is called a **sorption isotherm**.

Under service conditions where the moisture condition of the surroundings is fluctuating, the moisture content in hygroscopic materials will fluctuate continuously. In

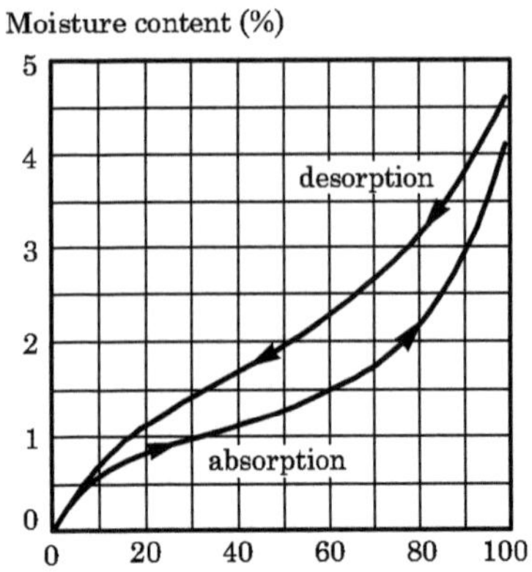

Figure 5.29. Typical moisture sorption isotherm for concrete with a cement content of 240 kg/m^3. The ordinate signifies the moisture content of the concrete which is the ratio between the content of evaporable water and the dry mass of the concrete. Note the hysteresis effect which is characteristic for sorption of moisture in porous materials.

practice, this results in a number of problems because a hygroscopic change of moisture will always be connected to a change of dimension.

A hygroscopic material expands - **swells** – when adsorbing moisture from the air, and it contracts – **shrinks** – when it gives off moisture. These moisture movements can provoke **stresses** that may result in **shrinkage cracks** in materials that are embedded in a structure. This phenomenon is well-known from concrete and other cement-bound materials where the binder phase consists of **cement paste**.

Hardened cement paste mainly consists of a porous, **semi-colloid** phase of calcium silicate hydrates physically incorporating crystalline calcium aluminate sulphate hydrates, the so-called **cement gel**. The cement gel has a very large active internal surface area that can adsorb water molecules; it has been found through measurements that the **specific surface** S of hardened cement paste is of the magnitude of 200 000 m^2/kg of the material. The specific surface is a measure of the internal, free surface in m^2 per kg of the material.

The cement gel can take up moisture from the surrounding environment by a combination of several different physical adsorption processes, including **hydrogen bonds** (see section 1.10). During adsorption of moisture, the gel expands – it swells. If swelling is prevented during adsorption, the cement gel can exert a considerable **swelling pressure** on the environments; this phenomenon is analogous to the pressure that can arise when water is frozen in a closed tank. As seen from the Gibbs-Duhem equation (5.14), the issue in both cases is the influence of the pressure on the free energy G of a condensed phase. Therefore, analyses of shrinkage and swelling phenomena in hardened cement paste are closely connected with determination of the thermodynamic state of the adsorbed water in the cement gel.

Problem. In a preliminary investigation of a **fibre reinforced** cement-bound material, the following measurement was performed in a laboratory. A fibre reinforced sample of cement mortar is cast with the w/c ratio 0.25. The sample is hardened for 28 days at 20 °C. After hardening the sample is crushed and placed in a closed glass container equipped with a sensor to measure the relative humidity RH and temperature T. The relation between the relative humidity RH in the closed container and the temperature T is determined by measurement. The following results of this measurement are available

Moisture content u

$$u \stackrel{\text{def}}{=} \frac{\text{evaporable water}}{\text{dry mass}}$$

Measurement no.		1	2	3	4	
Temperature	θ	20.2	25.2	30.2	35.1	°C
Temperature	T	293.35	298.35	303.35	308.25	K
Relative humidity	RH	75.3	78.2	81.2	83.8	%
Partial vapour pressure	p	1781.7	2506.5	3485.8	4740.2	Pa

Based on these measurements, make a rough determination of the following molar thermodynamic data for the adsorbed water phase $H_2O(a)$ in the hardened cement paste

$\Delta_r H^\ominus_{298}$ (kJ/mol) for the equilibrium transformation: $H_2O(a) \rightarrow H_2O(g)$ (b)

$\Delta_r S^\ominus_{298}$ (J/mol K) for the equilibrium transformation: $H_2O(a) \rightarrow H_2O(g)$ (c)

$H^\ominus_{298}$ (kJ/mol) for the adsorbed water phase $H_2O(a)$ in the cement gel (d)

$S^\ominus_{298}$ (J/mol K) for the adsorbed water phase $H_2O(a)$ in the cement gel (e)

Compare the calculated thermodynamic properties of adsorbed water $H_2O(a)$ with the corresponding data for free water $H_2O(\ell)$ in the standard state, and discuss the reason for the difference between them!

Reaction equilibrium

$$\Delta_r G^\ominus_T = -RT \cdot \ln(K_a)$$

Conditions. It is assumed that water vapour $H_2O(g)$ is an **ideal gas**. The given data are assumed to represent **equilibrium conditions** and all **hysteresis effects** are disregarded. The following thermodynamic standard values are assigned to water and water vapour

$H_2O(\ell)$	$H^\ominus_{298}$	$= -285.83$ kJ/mol	$S^\ominus_{298}$	$= 69.91$ J/mol K
$H_2O(g)$		$= -241.82$ kJ/mol		$= 188.83$ J/mol K

The mean value of $H^{\ominus}$ and $S^{\ominus}$ for adsorbed water in the temperature range 20.2 – 35.1 °C is determined by the calculation; these values are assumed to represent the state at 25 °C, which is: $H^{\ominus}_{298}$, $S^{\ominus}_{298}$.

Solution. We consider the following phase equilibrium in a **single-component system** consisting of adsorbed water $H_2O(a)$ and water vapour $H_2O(g)$

$$H_2O(a) \leftrightarrows H_2O(g) \qquad (T, p \text{ constant}) \tag{f}$$

At equilibrium, $\Delta G = 0$ for the transformation. In the limited temperature range around the standard temperature, this condition, according to equations (5.41) and (5.10), can be expressed by

$$\Delta_r G^{\ominus}_{298} = -RT \cdot \ln(K_a) = \Delta_r H^{\ominus}_{298} - T\Delta_r S^{\ominus}_{298} \qquad (T, p \text{ constant}) \tag{g}$$

The thermodynamic equilibrium constant K_a can be expressed using eqn. (5.39). If the water vapour is assumed to be an ideal gas, the **activity** of water vapour is $a(g) = p(g)/p^{\ominus}$, where $p^{\ominus}$ is the standard pressure 101325 Pa. By definition, the condensed phase has the activity $a(a) = 1$. We can thus rewrite (g) to the form

$$\Delta_r H^{\ominus}_{298} - T \cdot \Delta_r S^{\ominus}_{298} = -RT \cdot \ln\left(\frac{a(g)}{a(a)}\right) = -RT \cdot \ln\left(\frac{p(g)}{p^{\ominus}}\right) \tag{h}$$

In this expression, the target quantities $\Delta_r H^{\ominus}_{298}$ and $\Delta_r S^{\ominus}_{298}$ are the only unknowns in relation to the given test data. Therefore, these two unknowns can be determined by linear regression (see *Mathematical appendix, subchapter A5*). The calculations are performed with, for example, 6 significant digits according to the procedure given in the mathematical appendix. Thus, the unknowns are determined in such a way that the sum of the squared deviations *SSD* is at a minimum. First, the following matrices are formed

$$\mathbf{Y} = \begin{pmatrix} -RT_1 \cdot \ln(p_1/p^{\ominus}) \\ -RT_2 \cdot \ln(p_2/p^{\ominus}) \\ -RT_3 \cdot \ln(p_3/p^{\ominus}) \\ -RT_4 \cdot \ln(p_4/p^{\ominus}) \end{pmatrix}; \quad \mathbf{X} = \begin{pmatrix} 1 & -T_1 \\ 1 & -T_2 \\ 1 & -T_3 \\ 1 & -T_4 \end{pmatrix}; \quad \mathbf{k} = \begin{pmatrix} \Delta_r H^{\ominus}_{298} \\ \Delta_r S^{\ominus}_{298} \end{pmatrix} \tag{i}$$

With this substitution, the given test data can be represented by the following matrices

$$\mathbf{Y} = \begin{pmatrix} 9855.07 \\ 9176.41 \\ 8498.40 \\ 7847.92 \end{pmatrix}; \quad \mathbf{X} = \begin{pmatrix} 1 & -293.35 \\ 1 & -298.35 \\ 1 & -303.35 \\ 1 & -308.25 \end{pmatrix}; \quad \mathbf{k} = \begin{pmatrix} \Delta_r H^{\ominus}_{298} \\ \Delta_r S^{\ominus}_{298} \end{pmatrix} \tag{j}$$

and the regression equation is obtained as the following linear equation system where the quantity $\boldsymbol{\varepsilon}$ signifies the **residual**

$$\mathbf{X}\,\mathbf{k} = \mathbf{Y} + \boldsymbol{\varepsilon} \tag{k}$$

The normal equation for this equation system is obtained when the left-hand side and the right-hand side are multiplied by the transposed matrix $\mathbf{X}^{\mathrm{T}}$

$$\mathbf{X}^{\mathrm{T}}\,\mathbf{X}\,\mathbf{k} = \mathbf{X}^{\mathrm{T}}\,\mathbf{Y} \tag{l}$$

Introducing the matrices from (j), we obtain the following normal equation for the given data

$$\begin{pmatrix} 4.00000 & -1203.30 \\ -1203.30 & -362106 \end{pmatrix} \begin{pmatrix} \Delta_r H^{\ominus}_{298} \\ \Delta_r S^{\ominus}_{298} \end{pmatrix} = \begin{pmatrix} 35377.8 \\ -10025000 \end{pmatrix} \tag{m}$$

When the solution to this equation system is determined by, for example, **Cramer's method** (see *Mathematical appendix, subchapter A4*), we obtain the determinants

$$D = 493.110; \quad D_1 = 24368200; \quad D_2 = 66506.7 \tag{n}$$

The target values for **standard reaction enthalpy** $\Delta_r H^{\ominus}_{298}$ and **standard reaction entropy** $\Delta_r S^{\ominus}_{298}$ can now be determined by

$$\Delta_r H^{\ominus}_{298} = D_1/D = 24368200/493.110 = \mathbf{49417\,J/mol}$$

$$\Delta_r S^{\ominus}_{298} = D_2/D = 66506.7/493.110 = \mathbf{134.87\,J/mol\,K}$$

Figure 5.30. Shrinkage movements in structural members can give rise to a destructive decomposition of joints and supports. The photo shows a concrete beam where movements due to moisture and temperature have damaged the column support. If the structure is exposed to moisture, the damage can result in insufficient load-carrying capacity of the support.

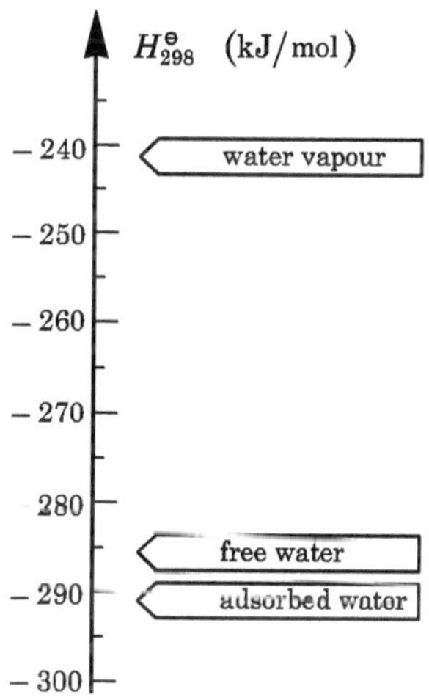

Figure 5.31. Standard enthalpy $H^{\ominus}_{298}$ for water vapour, free water and adsorbed water in hardened fibre reinforced concrete. When free water is adsorbed on the cement gel, its enthalpy content is reduced; physically, this means that the water molecules are transferred to a state with lower potential energy. For the adsorption process considered, $\Delta H \simeq -5$ kJ/mol. This is of the same magnitude as ΔH for freezing of free water.

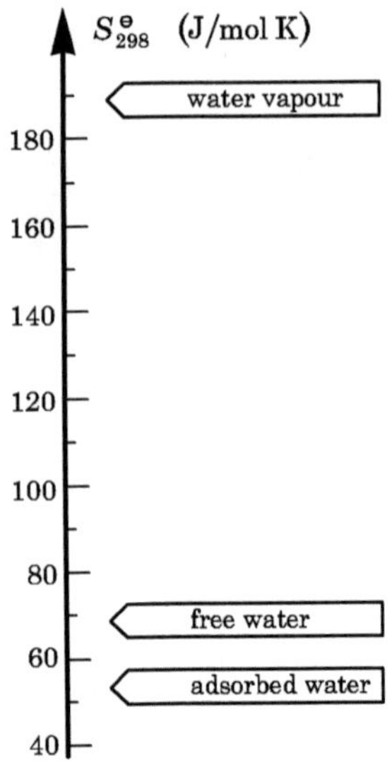

Figure 5.32. Standard entropy $S^{\ominus}_{298}$ for water vapour, free water and adsorbed water in fibre reinforced concrete. When free water is adsorbed on the cement gel, its entropy is reduced; physically this means that the water is transferred to a more organized, molecular state. For the adsorption process concerned, $\Delta S \simeq -16$ J/mol K. This is of the same magnitude as ΔS for freezing of water.

Based on these reaction values, the **standard enthalpy** $H^{\ominus}_{298}$ and the **standard entropy** $S^{\ominus}_{298}$ can now be calculated for the adsorbed phase

$$H^{\ominus}_{298}(\mathrm{a}) = H^{\ominus}_{298}(\mathrm{g}) - \Delta_r H^{\ominus}_{298} = -241.82 - 49.42 = \mathbf{-291.24\,kJ/mol}$$
$$S^{\ominus}_{298}(\mathrm{a}) = S^{\ominus}_{298}(\mathrm{g}) - \Delta_r S^{\ominus}_{298} = 188.83 - 134.87 = \mathbf{53.96\,J/mol\,K}$$

Discussion. Comparing these results with the thermodynamic standard values for water and water vapour, we obtain the following situation

$$H^{\ominus}_{298}(\mathrm{a}) < H^{\ominus}_{298}(\ell) \ll H^{\ominus}_{298}(\mathrm{g}) \qquad \text{(o)}$$
$$S^{\ominus}_{298}(\mathrm{a}) < S^{\ominus}_{298}(\ell) \ll S^{\ominus}_{298}(\mathrm{g}) \qquad \text{(p)}$$

The adsorbed water has a **lower enthalpy** and a **lower entropy** than free water. The standard entropy of the adsorbed phase is very close to the value found for **ice** H_2O(s), viz. $S^{\ominus} \simeq 69.9 - 22.0 \simeq 50$ J/mol K (see the example in section 4.6). This suggests that the adsorbed water has a more organized molecular structure and a lower potential energy than found for free water.

Example 5.6

■ Precipitation of salt in porous materials – salt damages

Porous construction materials such as **brick, natural stone** and **concrete** that are exposed to saline groundwater can by capillary suction pull a suction front up above the ground water table and into a structure. When the water evaporates from the moist structural members, the contained salts remain in the material; thus, the pore system can gradually be filled with solid salt deposits.

In the structure the phenomenon is seen by **salt precipitation** which taints surfaces in evaporation zones. In certain cases, the deposited salts in the pores of the material can result in serious salt damage in the form of **spalling** and **decomposition** of the material exposed to moisture.

Salt precipitation from capillary suction of groundwater mainly consists of **sodium sulphate** and **sodium carbonate**. Development of salt damage can, for example, be connected with phase equilibrium in the salt system: *sodium sulphate – water*; in this system the following phases occur

$$Na_2SO_4(s) \quad - \quad Na_2SO_4 \cdot 10H_2O(s) \quad - \quad H_2O \qquad \text{(a)}$$

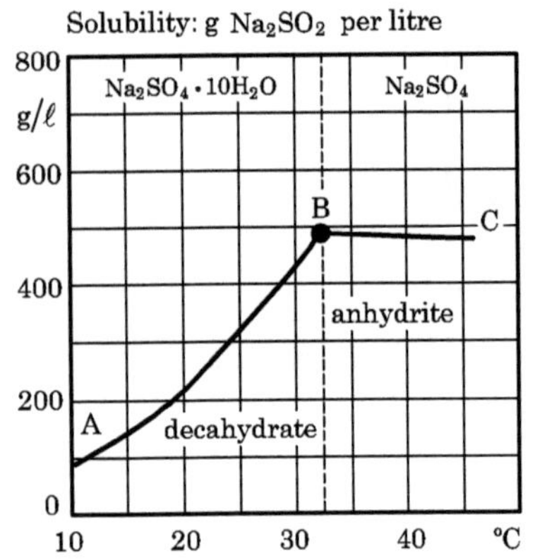

Figure 5.33. Solubility of sodium sulphate into water. At temperatures below 32.4 °C, decahydrate is a solid deposit. In this range, the solubility is strongly increased with rising temperatures (A–B). At temperatures above 32.4 °C, decahydrate is transformed into anhydrite. Above this temperature, the solubility is slightly reduced with rising temperatures (B–C).

In the equilibrium system, solid sodium sulphate can thus occur as an **anhydrite** Na_2SO_4(s) and as a **decahydrate** $Na_2SO_4 \cdot 10H_2O$(s). The decahydrate $Na_2SO_4 \cdot 10H_2O$(s), in particular, *cannot* exist at temperatures above 32.4 °C, because the hydrate is then transformed into the anhydrite Na_2SO_4(s).

In the range $\theta \leq 32.4$ °C the solubility of sodium sulphate increases with rising temperatures θ. In this temperature range, the solubility is determined by the equilibrium

$$Na_2SO_4 \cdot 10H_2O(s) \leftrightarrows 2Na^+(aq) + SO_4^{--}(aq) + 10H_2O(\ell) \qquad (\theta \leq 32.4\,°C)\,\text{(b)}$$

where the decahydrate is the stable, solid phase for $\theta \leq 32.4$ °C. At temperatures above 32.4 °C, the solubility of the solid sodium sulphate decreases with rising temperatures. In this temperature range, the solubility at equilibrium is determined by

$$Na_2SO_4(s) \leftrightarrows 2Na^+(aq) + SO_4^{--}(aq) \qquad (\theta \geq 32.4\,°C) \qquad \text{(c)}$$

since anhydrite is the stable, solid phase for $\theta \geq 32.4$ °C. Therefore, at 32.4 °C, solid anhydrite and decahydrate can exist in mutual equilibrium in a saturated aqueous solution.

Transformation of anhydrite into decahydrate, however, can also occur as a solid-phase reaction: solid anhydrite Na_2SO_4(s) can absorb water vapour from the air and be transformed into solid decahydrate $Na_2SO_4 \cdot 10H_2O$(s). This solid-phase transformation of anhydrite into decahydrate follows the reaction equation

$$Na_2SO_4(s) + 10H_2O(g) \quad \rightarrow \quad Na_2SO_4 \cdot 10H_2O(s) \qquad (\theta \leq 32.4\,°C) \qquad \text{(d)}$$

The transformation (d) is limited to temperatures ≤ 32.4 °C, because the decahydrate will melt at higher temperatures. In contrast to the solution equilibrium between anhydrite and decahydrate at 32.4 °C, the transformation (d) does *not* require an aqueous phase – it can occur in a "dry" system.

- Salt damage is caused by precipitation of $Na_2SO_4 \cdot 10H_2O$ crystals from a supersaturated solution in liquid-filled pores (reaction equations (b) and (c) above). During their growth, the crystals are assumed to exert a **mechanical pressure** on the surrounding pore walls; by a sufficient supersaturation in the liquid phase this pressure can result in **cracks** and **spalling** of the material concerned. Such transformations can, for example, occur if the temperature of the wet salt-saturated material fluctuates around 32.4 °C.

Problem. In the literature, the mechanism behind salt damage and spalling is discussed based on two different hypotheses:

- Build-up of pressure and salt spalling arise due to solid-phase transformation of anhydrite into decahydrate according to reaction equation (d) because this transformation is connected with a considerable increase of the solid-matter volume. When a porous salt-saturated material is exposed to fluctuating temperatures θ and relative humidity RH, such solid-phase transformations in the pore system of the otherwise "dry" material will take place under certain conditions. For further explanation of these different hypotheses, we want to estimate by calculation what changes in temperature θ and relative humidity RH can be expected to generate the solid-phase transformation (d) in a porous salt saturated material. Also we want to make an estimate of the increase in solid-matter volume by the transformation (d) and an estimate of the influence of the pressure on the phase equilibrium (d).

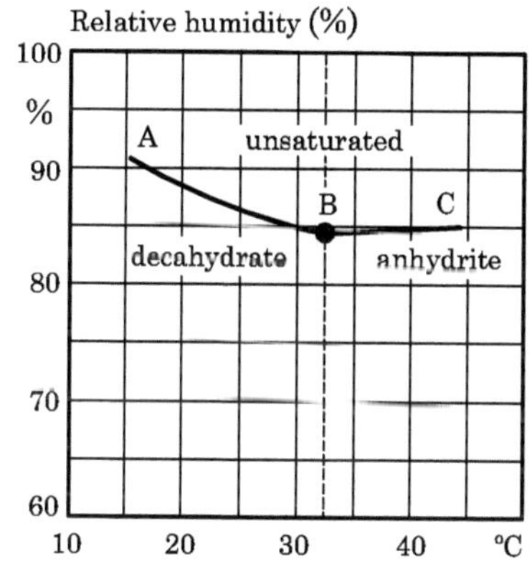

Figure 5.34. Relative humidity RH over a saturated solution of sodium sulphate. At temperatures below 32.4°C, the solid deposit is decahydrate. In this range, RH decreases with rising temperature, because the solubility of decahydrate increases (A–B). Above 32.4°C the decahydrate is transformed into anhydrite, and RH over the saturated solution increases slightly with the temperature, because the solubility of anhydrite decreases with temperature (B–C).

Conditions. For calculations, water vapour $H_2O(g)$ is assumed to be an ideal gas. Calculations are performed with the following data and thermodynamic standard values that are determined from tables.

Substance	M (g/mol)	ϱ (g/cm^2)	$H^\ominus_{298}$ (kJ/mol)	$S^\ominus_{298}$ (J/mol K)	c_p (J/mol K)
$Na_2SO_4 \cdot 10H_2O(s)$	322.21	1.46	−4326.1	593.2	587.7
$Na_2SO_4(s)$	142.05	2.70	−1385.1	149.6	127.7
$H_2O(g)$	18.02	–	− 241.8	188.7	33.6

Formation of heptahydrate $Na_2SO_4 \cdot 7H_2O(s)$ is not considered. This can sometimes occur as a metastable intermediary product by the transformation (d).

Solution. We first determine the increase in the solid-matter volume by transformation of anhydrite into decahydrate according to reaction equation (d). By the reaction, 1 mol of anhydrite is transformed into 1 mol of decahydrate – we can therefore determine the increase in molar volume: $V_m = M/\varrho$ for the transformation concerned.

$$\text{Final volume : } V_m(\text{decahydrate}) = 322.21/1.46 = 220.7\,\text{cm}^3/\text{mol}$$

$$\text{Initial volume : } V_m(\text{anhydrite}) = 142.05/2.70 = 52.6\,\text{cm}^3/\text{mol}$$

Thus, the target relative volume increase in % is

$$\frac{\Delta V}{V_0} = \frac{V(\text{decahydrate}) - V(\text{anhydrite})}{V(\text{anhydrite})} = \frac{220.7 - 52.6}{52.6} \cdot 100\% = \mathbf{320}\,\%$$

For a reversible solid-phase transformation (equilibrium), $\Delta G = 0$ for the reaction (d); then, according to equilibrium condition III in eqn. (5.41)

$$K_a = \exp\left(-\frac{\Delta_r G^\ominus_T}{RT}\right) \qquad (T, p \text{ constant}) \qquad \text{(e)}$$

We first determine $\Delta_r G^\ominus_T$ according to the procedure in eqn. (5.11); thus, the temperature T is introduced as a variable

$$\Delta_r H^\ominus_{298} = -4326.1 - (-1385.1 + 10 \cdot (-241.8)) = -523.0\,\text{kJ/mol}$$

$$\Delta_r S^\ominus_{298} = 593.2 - (149.6 + 10 \cdot 188.7) = -1443.4\,\text{J/mol K}$$

$$\Delta_r c_p = 587.7 - (127.7 + 10 \cdot 33.6) = 124.0\,\text{J/mol K}$$

$$\Delta_r G^\ominus_T = -523.0 \cdot 10^3 + 124.0 \cdot (T - 298.15) - T \cdot (-1443.4 + 124.0 \cdot \ln(\frac{T}{298.15}))$$

$$\Delta_r G^\ominus_T = -559971 + 2273.9 \cdot T - 124.0 \cdot T \ln(T) \quad \text{(J/mol K)} \qquad \text{(f)}$$

Then, the thermodynamic equilibrium constant K_a is determined by the definition in eqn. (5.39); hereby the partial pressure of water vapour p is introduced as a variable

$$K_a = \frac{a(\mathrm{Na_2SO_4 \cdot 10H_2O(s)})}{a(\mathrm{Na_2SO_4(s)}) \cdot a(\mathrm{H_2O(g)})^{10}} = \frac{1}{1 \cdot (p/p^{\ominus})^{10}} = \left(\frac{p}{p^{\ominus}}\right)^{-10} \qquad \text{(g)}$$

In the equilibrium state the partial pressure of water vapour p (Pa) is now expressed as a function of the temperature T; introducing (f) and (g) into the equilibrium expression (e), we obtain the following connection

$$p = p^{\ominus} \cdot \left(\exp\left(-\frac{-55971 + 2273.9 \cdot T - 124.0 \cdot T \ln(T)}{RT}\right)\right)^{-0.10} \quad \text{(Pa)} \qquad \text{(h)}$$

where $p^{\ominus}$ denotes the standard pressure 101325 Pa. By calculations shown in the table below, the partial pressure p (Pa) of water vapour for the reaction (d) for selected temperatures $\leq 32.4\,^{\circ}\mathrm{C}$ is found. Dividing these values by the partial pressure of saturated water vapour p_s at the same temperature, we obtain the target relative humidity RH at equilibrium

θ	10	15	20	25	30	32.4	°C
p	787.3	1159	1683	2412	3416	4019	Pa
p_s	1228.1	1705.6	2338.8	3169.1	4245.5	4867.7	Pa
RH	64.1	68.0	72.0	76.1	80.5	82.6	%

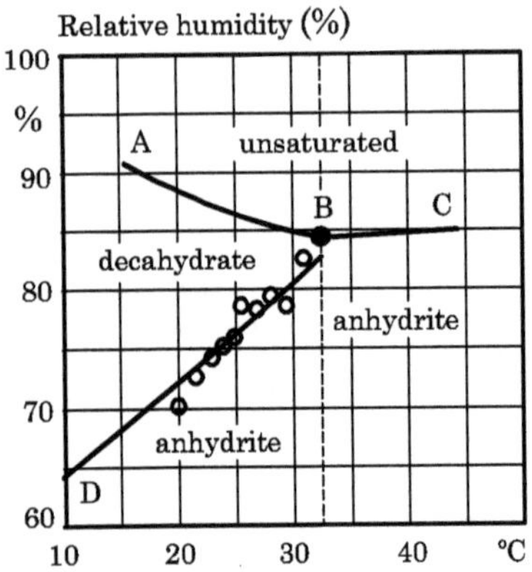

Figure 5.35. If RH over a saturated solution of sodium sulphate decahydrate (A–B) is reduced, the solution is dehydrated and solid decahydrate is formed. When below a certain RH value, decahydrate emits the chemically bound hydrate water and is transformed into anhydrite (B–D). The boundary curve denotes RH over a decahydrate-anhydrite salt pair in thermodynamic equilibrium. The curve B–D corresponds to this expression (h); measurement values denote the experimental data, cf. Timmermanns.

Any passage of the phase boundary $RH(\theta)$ determined will change the **equilibrium composition** of the system considered

$$\mathrm{Na_2SO_4(s)} \quad - \quad \mathrm{Na_2SO_4 \cdot 10H_2O(s)} \quad - \quad \mathrm{H_2O(g)} \qquad \text{(i)}$$

When in a state of equilibrium, a porous material containing pure sodium sulphate will react in the following way to changes in temperature $\theta < 32.4\,^{\circ}\mathrm{C}$ and relative humidity RH.

- If the phase boundary is exceeded at **constant temperature** θ during **increasing** RH, anhydrite is transformed into decahydrate; the transformation is connected with an increase of solid-matter volume of approximately 320 %.
- If the phase boundary is exceeded at **constant** RH during **decreasing temperature** θ, anhydrite is transformed into decahydrate; the transformation is connected with an increase of solid-matter volume of approximately 320 %.

From these considerations it is not possible to conclude **that** the transformation from anhydrite into decahydrate causes damages and spalling in a material. In conclusion, however, it can be illustrative to evaluate the magnitude of the pressures that can be built up by the transformation discussed here.

Assume an isothermal state where anhydrite, decahydrate and water vapour are in mutual equilibrium, i.e.

$$\mathrm{Na_2SO_4(s) + 10H_2O(g)} \leftrightarrows \mathrm{Na_2SO_4 \cdot 10H_2O(s)} \qquad \text{(j)}$$

If the partial pressure on the solid-matter system is increased by $dp(\mathrm{s})$, the partial pressure of water vapour changes by a certain quantity $dp(\mathrm{g})$ to maintain the equilibrium in the system. According to the Gibbs-Duhem equation (5.14), equilibrium can be maintained under isothermal conditions ($dT = 0$), provided that

$$dG_a + 10 \cdot dG_g = dG_d \quad \Rightarrow \quad V_a dp(\mathrm{s}) + 10 \cdot V_g dp(\mathrm{g}) = V_d dp(\mathrm{s}) \qquad \text{(k)}$$

The Gibbs-Duhem equation

$$dG = V dp - SdT$$

where G_a, G_g and G_d denote the molar free energy, and V_a, V_g and V_d denote the molar volume of anhydrite, water vapour or decahydrate, respectively. Thus, at equilibrium, the pressure increase $dp(\mathrm{s})$ in the solid phases is connected with the increase in the partial pressure of water vapour $dp(\mathrm{g})$ as follows

$$dp(\mathrm{s}) = \frac{10 \cdot V_g}{V_d - V_a} \cdot dp(\mathrm{g}) = \frac{10 \cdot RT}{V_d - V_a} \cdot d\ln(p(\mathrm{g})) \qquad \text{(l)}$$

where in the latter rewriting we have utilized that the water vapour complies with the ideal gas law: $V_g = RT/p$. When, for example, we look into the state of equilibrium at $25\,^{\circ}\mathrm{C}$, we obtain the following by insertion

$$\frac{10 \cdot RT}{V_d - V_a} = \frac{10 \cdot 8.314 \cdot 298.15}{220.7 \cdot 10^{-6} - 52.6 \cdot 10^{-6}}\ \mathrm{Pa} \qquad \simeq 1450\,\mathrm{atm}$$

Introducing this value into eqn. (l) and integrating based on the state of equilibrium at 25 °C, we see that the increase in solid-matter pressure $\Delta p(\mathrm{s})$ is connected with the increase in RH as follows

$$\Delta p(\mathrm{s}) \simeq 1450 \cdot \Delta \ln(RH) \quad \mathrm{atm} \tag{m}$$

The above expression shows that theoretically the solid-matter system can build up a pressure of 18-20 atmospheres for each % of RH by which the relative humidity exceeds the equilibrium value at atmospheric pressure. A solid-matter pressure of this magnitude will, for example, be critical for bricks with a typical ultimate tensile strength of 15-30 kg/cm^2.

Discussion. In conclusion, the calculated curve of state $RH = RH(\theta)$ at equilibrium in the salt system, eqn. (i), shall be compared with experimentally determined values. In: *The Physico-Chemical Constants of Binary Systems in Concentrated Solutions*, J.Timmermanns has published the following, measured equilibrium humidities over the salt system (i) concerned

θ	20.2	21.6	23.0	24.1	25.0	25.6	27.0	28.1	29.5	31.0	32.4	°C
p	12.5	14.1	15.7	17.0	18.1	19.4	21.0	22.7	25.0	27.9	30.8	Torr
RH	70.4	72.9	74.5	75.5	76.2	78.8	78.5	79.6	78.9	82.8	84.4	%

These experimentally determined values are seen to be in satisfactory correspondence with the calculated curve of $RH = RH(\theta)$ for the equilibrium system (j).

Exercises

The following exercises can be used when learning the subjects addressed in Chapter 5. The times given are suggested guidelines for use in assessing the effectiveness of the teaching.

Exercise 5.1

☐ **Reversible evaporation of water** *(4 min)*

Assume a system where water evaporates reversibly at constant pressure p and temperature T. For the evaporation process considered, $\Delta_r H_T = 2401$ J/g and $\Delta_r S_T = 137.3$ J/mol K. Specify $\Delta_r G_T$ for this evaporation process and determine the temperature θ (°C) at which the process occurs!

Reversible process

$$\Delta G = \Delta H - T\Delta S = 0$$

Exercise 5.2

☐ **Assessment of process conditions** *(4 min.)*

Assume a process in which: $\Delta_r H_T = 1280$ kJ/mol and $\Delta_r S_T = -112.1$ J/mol K. a) specify whether the process is *endothermic* or *exothermic*! b) Is the process *irreversible*, *reversible* or *impossible*?

Exercise 5.3

☐ **Stable form of aluminium oxide Al_2O_3(s)** *(5 min.)*

Aluminium oxide Al_2O_3(s) can be found in four different modifications named α-Al_2O_3, γ-Al_2O_3, δ-Al_2O_3 and κ-Al_2O_3, respectively. Determine from the equilibrium condition II in eqn. (5.7), which of these modifications is the equilibrium form at 25 °C!

Exercise 5.4

☐ **Equilibrium temperature for phase transformation** *(5 min.)*

Assume a substance A, which can exist in two different crystalline modifications A(α) and A(β). For the transformation: $\alpha \rightarrow \beta$ it applies that $\Delta_r H = -3.41$ kJ/mol and $\Delta_r S = -12.21$ J/mol K; it is assumed that $\Delta_r H$ and $\Delta_r S$ are independent of the temperature T. Determine the temperature θ (°C) at which A(α) and A(β) are in thermodynamic equilibrium.

Exercise 5.5

☐ Pressure dependence of the melting point *(8 min.)*

Assume a single-component system where there is thermodynamic equilibrium between the melt and the solid phase of a substance A at the temperature T, i.e. A(s) $\leftrightarrows$ A(ℓ). In this system, the solid phase floats on the surface of the melt. Explain whether an increased pressure p will increase or decrease the melting point of substance A!

Exercise 5.6

☐ Heat of evaporation of water *(10 min.)*

The partial pressure of saturated water vapour at 14 °C is $p_s = 1598.8\,\text{Pa}$, and the partial pressure of saturated water vapour at 16 °C is $p_s = 1818.5\,\text{Pa}$. Based on this information, determine the evaporation enthalpy ΔH for water at 15 °C; the result shall be given in (kJ/kg)!

Exercise 5.7

☐ Free energy and activity of ideal gas *(8 min.)*

An ideal system contains a total of 4.60 mol of N_2(g) at 25 °C. The total free energy of the system is $G = -18352\,\text{J}$. Calculate for this system: a) the molar free energy G (kJ/mol) of N_2(g); b) the activity a of N_2(g), and c) the pressure p (Pa) of N_2(g)!

Activity, ideal systems of matter

Gas	$a = p/p^\ominus$
Dissolved substance	$a = c/c^\ominus$
Solvent	$a = x/x^\ominus$
Pure substance (s), (ℓ)	$a = 1$

Exercise 5.8

☐ Free energy and activity of dissolved substance *(10 min.)*

At 25 °C, an ideal aqueous solution contains 0.320 g of dissolved and dissociated $Ca(OH)_2$ per litre. Calculate: a) the activity a of Ca^{++}(aq) and OH^-(aq) in the solution! b) the molar free energy G (kJ/mol) of Ca^{++}(aq) and OH^-(aq) in the solution!

Exercise 5.9

☐ The activity of solvent *(8 min.)*

An ideal aqueous solution of methanol $CH_3OH(\ell)$ contains 6.80 g of methanol per kg of water $H_2O(\ell)$. Calculate the activity a of the solvent $H_2O(\ell)$!

Exercise 5.10

☐ Thermodynamic equilibrium constant *(8 min.)*

The partial pressure of saturated water vapour at 40 °C is $p_s = 7381.4\,\text{Pa}$. From this information, calculate: a) the thermodynamic equilibrium constant for the transformation $H_2O(g) \rightarrow H_2O(\ell)$ at 40 °C and b) $\Delta_r G^\ominus$ for the same transformation at 40 °C! (ideality is assumed).

Exercise 5.11

☐ Modifications of lead oxide PbO *(4 min.)*

Lead oxide PbO (*plumbous oxide*) can exist in two different modifications. Red lead oxide PbO precipitates in **tetragonal** crystals. Yellow lead oxide precipitates in **rhombic** crystals. Yellow lead oxide is also called Litharge (="Stone Silver"); this is because this oxide precipitates as slags when plumbiferous silver is refined. According to Atkins: *Physical Chemistry*, these modifications of lead oxide have the following thermodynamic standard values:

$$\text{Red PbO}: G^\ominus_{298} = -188.93\,\text{kJ/mol}; \quad \text{Yellow PbO}: G^\ominus_{298} = -187.89\,\text{kJ/mol}$$

Calculate $\Delta_r G^\ominus_{298}$ for transformation of red PbO(s) into yellow PbO(s) at 25 °C, and state which modification is stable at room temperature!

Exercise 5.12

☐ Stable form of calcium carbonate at room temperature *(8 min.)*

Natural occurrence of calcium carbonate $CaCO_3$ is used as raw material for producing **burnt lime** CaO(s) and **calcium hydroxide** $Ca(OH)_2$(s), which are used as binders in **lime mortar** for masonry. Calcium carbonate can exist in two modifications: a hexagonal form, **calcite** ("calcspar") and a rhombic form **aragonite**. In his book *The*

Second Law, H.A. Bent specifies the following thermodynamic standard values for these modifications of $CaCO_3(s)$

Calcite	$H^{\ominus}_{298}$	$= -288.45$ kcal/mol	$S^{\ominus}_{298}$	$= 22.2$ cal/mol K
Aragonite		$= -288.49$ kcal/mol		$= 21.2$ cal/mol K

Calculate $\Delta_r G^{\ominus}_{298}$ in the unit (J/mol) for transformation of calcite into aragonite at 25 °C

$$CaCO_3(s, calcite) \rightarrow CaCO_3(s, aragonite)$$

and specify what modification of $CaCO_3(s)$ is the stable form at 25 °C! (1 cal = 4.186 J).

Exercise 5.13

☐ **Freezing damage of porous construction materials – freezing pressure of ice** *(10 min.)* When water is transformed into ice, its **volume expands**. This expansion is the reason why water saturated porous construction materials can burst and disintegrate when exposed to freezing; for example, **freezing damage** occurs in moisture saturated **concrete** and **brick** that are exposed to repeated freezing. The maximum freezing pressure Δp that can arise during the formation of ice depends on the supercooling ΔT in relation to the freezing point of water at atmospheric pressure. The following data are given for water $H_2O(\ell)$ and for ice $H_2O(s)$ at 0 °C

$H_2O(\ell)$	$H^{\ominus}_{273}$	$= -287.71$ kJ/mol	V_{molar}	$= 18.022 \cdot 10^{-6}$ m^3/mol
$H_2O(s)$		$= -293.72$ kJ/mol		$= 19.651 \cdot 10^{-6}$ m^3/mol

Calculate using the **Clapeyron equation**, the maximum freezing pressure Δp (atm) that can arise at a temperature of −2 °C!

Figure 5.36. Freezing of porous humid construction materials can produce scaling and bursting. Decomposition is caused by the volume expansion of water. The photo shows a frost damaged brick wall. Due to faulty covering of the wall, the bricks have attained a critical degree of water saturation.

Exercise 5.14

☐ **Temperature influence on the partial pressure of saturated water vapour** *(10 min.)*

The partial pressure of saturated water vapour is significantly increased with rising temperature. This fact is, for example, important for concrete casting at high temperatures. The desiccation rate of a fresh concrete depends on the vapour pressure at the concrete surface; therefore, damage due to desiccation most often arises in connection with casting of warm concrete.

The partial pressure of saturated water vapour at 20 °C is 2338 Pa. In the temperature range from 20 °C to 40 °C, the **heat of evaporation** of water is 2430 J/g on average. Calculate from the **Clausius-Clapeyron equation** the partial pressure of saturated water vapour at 30 °C and at 40 °C!

Exercise 5.15

☐ **Free energy and activity of water vapour over saline solution** *(10 min.)*

When a bowl with a **saturated saline solution** is placed in a closed room at constant temperature, the relative humidity *RH* in the room will adjust to a constant value. The relative humidity at equilibrium will depend on the salt used. Such setup is often used to control the relative humidity during laboratory tests.

The relative humidity at 25 °C over a saturated solution of **sodium chloride** NaCl is $RH = 75.5$ %. At this temperature, the partial pressure of saturated water vapour is 3169 Pa. Water vapour $H_2O(g)$ at 25 °C has a standard free energy $G^{\ominus} = -228.6$ kJ/mol. Calculate the **activity** a and the free energy G (kJ/mol) of water vapour in equilibrium with a saturated NaCl solution at 25 °C!

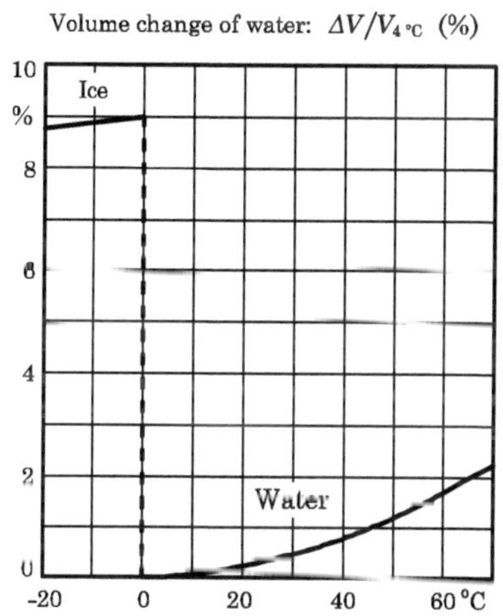

Figure 5.37. When water freezes, its volume expands by approximately 10 %. If this expansion is prevented, a severe freezing pressure can arise. The effect of this freezing pressure is known from, for example, water pipes and from decomposition of humid, porous construction materials.

Exercise 5.16

☐ **Formation of aluminium oxide on the surface of Al** *(10 min.)*

Pure aluminium Al is a base metal; when it is exposed to the oxygen $O_2(g)$ of the air, an **oxide film** of $Al_2O_3(s)$ ("corundum") is formed on the surface of the metal. The thickness of the film is about 0,01 μm; the film is very **hard** and adheres tightly - a passivation of the metal surface is said to occur during the oxide formation. In normal circumstances, this passivation is desirable because the film of Al_2O_3 acts as **corrosion protection** and gives the surface high wear resistance. However, the oxide film complicates any form of **soldering** of the metal aluminium. The oxide film is formed by the reaction

$$4Al(s) + 3O_2(g) \rightarrow 2Al_2O_3(s) \qquad (a)$$

Figure 5.38. The photo shows natural aluminium oxide Al_2O_3 (corundum). Aluminium oxide is only surpassed in hardness by diamond. Artificially made aluminium oxide is widely used as an abrasive.

In atmospheric air the partial pressure of oxygen $O_2(g)$ is approximately 22 000 Pa. We want to investigate whether formation of oxide according to (a) can be prevented by storing the metal in an atmosphere that is very poor in oxygen. Calculate $\Delta_r G_{298}$ for the formation of oxide: 1) when $p(O_2) = 22000$ Pa, and 2) when $p(O_2) = 1$ Pa, and assess whether the formation of oxide can be prevented by the reduction of oxygen pressure mentioned above!

Exercise 5.17

☐ Dissociation constant for water – pH *(12 min.)*

Pure water has a natural content of ions formed by the following **dissociation**:

$$2H_2O(\ell) \leftrightarrows H_3O^+(aq) + OH^-(aq) \qquad \text{(a)}$$

The ions $H_3O^+(aq)$ and $OH^-(aq)$ are assumed to form an ideal solution, and the water activity is assumed to be $a = 1$. Calculate the thermodynamic equilibrium constant K_a (to 3 decimals) for the dissociation equilibrium in eqn. (a) at 25 °C, and determine $pH = -log_{10}(a(H_3O^+))$ of pure water at 25 °C (to 2 decimals)!

Exercise 5.18

☐ Assessment of desiccation conditions for fresh concrete *(6 min.)*

A newly cast concrete is surrounded by humid atmospheric air. The temperature of the concrete and the air is 25 °C. The partial pressure of water vapour in the air is 2870 Pa. The following thermodynamic standard values are given

$$H_2O(\ell) : G^{\ominus}_{298} = -237.13\,\text{kJ/mol}; \qquad H_2O(g) : G^{\ominus}_{298} = -228.57\,\text{kJ/mol}$$

The fresh concrete surface is assumed to be wet, with a free water film. Under these circumstances, calculate $\Delta_r G_{298}$ for evaporation of water: $H_2O(\ell) \rightarrow H_2O(g)$ and specify whether or not desiccation of the concrete will occur!

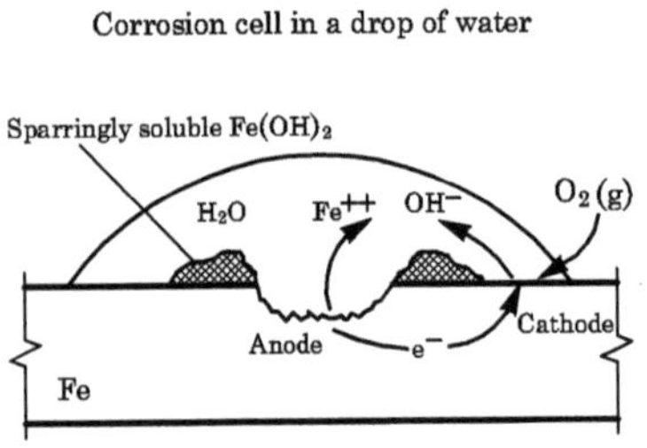

Figure 5.39. Example of corrosion cell in a drop of water. During corrosion, dissolved oxygen acts as an electron acceptor. The cathode area is formed along the edge of the drop with the easiest access of O_2. The anode area is formed in the middle of the drop. During the anode process, Fe is oxidized into Fe^{++} and is dissolved. The liberated electrons are included in the cathode process where dissolved oxygen O_2 is reduced and forms hydroxide ions OH^- with the water. The product of the reaction is the well-known "rust" which is hydrates of $Fe(OH)_2$.

Exercise 5.19

☐ Prevention of corrosion by cathodic inhibition *(15 min.)*

Steel exposed to water and moisture can decay in a few years due to electrolytic corrosion that corrodes the iron through the following **anode process**

$$Fe(s) \rightarrow Fe^{++}(aq) + 2e^- \qquad \text{(anode process)} \qquad \text{(a)}$$

By the process, the dissolved $Fe^{++}(aq)$ ions are liberated leaving the free electrons in the metal that thus gets a negative potential. Deep corrosion requires that the free electrons are continuously removed by an **electron acceptor** during a concurrent **cathode process**. The most frequent electron acceptor is dissolved oxygen; in this case, the cathode process is

$$2e^- + O_2(aq) + H_2O(\ell) \rightarrow 4OH^-(aq) \qquad \text{(cathode process)} \qquad \text{(b)}$$

The ions OH^- formed react with Fe^{++} to form $Fe(OH)_2$, which then oxidizes further to produce **rust**.

If the water contains dissolved salts that can react with OH^- and form a **dense, sparingly soluble** coating on the cathode, the corrosion will stop completely or it will be reduced. Salts with this property are used as **cathodic inhibitors** for prevention of corrosion. As an example, mention can be made of highly soluble **zinc salts**, which can precipitate as **zinc hydroxide**

$$Zn^{++}(aq) + 2OH^-(aq) \rightarrow Zn(OH)_2(s) \qquad \text{(c)}$$

Given: $G^{\ominus}_{298}(Zn(OH)_2(s)) = -555.1$ kJ/mol. Calculate the thermodynamic equilibrium constant K_a for solution of $Zn(OH)_2(s)$ into water and determine the solubility of $Zn(OH)_2(s)$ into water with the unit g/ℓ! – assess whether zinc hydroxide is a highly soluble or sparingly soluble salt!

Exercise 5.20

☐ Temperature dependence of sorption isotherms *(15 min.)*

The moisture content of a given wood sample is $u = 0.10$ g/g dry weight. The sample is placed in a moisture saturated measuring chamber at 30 °C. After reaching the moisture equilibrium between the sample and the air in the chamber, a humidity of 57.5% RH in the chamber is measured. The heat of evaporation for the adsorbed water is 47.0 kJ/mol at the given moisture content u. Calculate the relative humidity RH, which will be in equilibrium with the sample at 50 °C! – the water vapour is assumed to be an ideal gas.

Literature

The following literature and supplementary readings are recommended for further reference on the subjects covered in Chapter 5.

References

In Chapter 5, a number of references to the literature have been made; the following is a complete list of citations of the references used.

- Mchedlov-Petrossyan: *Thermodynamik der Silikate*, VEB Verlag für Bauwesen, Berlin 1966.
- Weast, R.C. (ed.): *CRC Handbook of Chemistry and Physics*, CRC Press, Inc., Florida 1983.
- Timmermans, J.: *The Physico-chemical Constants of Binary Systems in Concentrated Solutions*, Vol.3, Interscience Publishers, Inc., New York 1960.
- Bent, H.A.: *The Second Law*, Oxford University Press, New York 1965.

Laboratory table from around 1900 with drying racks for glass ware; after cleaning, the equipment is flushed with distilled water before drying on the racks.

CHAPTER 6

Electrochemistry

The theory on positive and negative electricity and the laws of electrostatic attraction were in the main developed through the 18th century. The first systematic investigations of the chemical effects of the electric current were performed by **Luigi Galvani** (1737-1798) and **Alessandro Volta** (1745-1828). Through their observations, the basis of the science of *Electrochemistry* was formed.

The development of electrochemistry in the early 19th century is especially connected with a number of experimental and theoretical works by the Swedish chemist **Jacob Berzelius** (1779-1848) and the English chemist **Humphry Davy** (1778-1829). In continuation, Davy's apprentice and successor **Michael Faraday** presented his famous thesis on *chemical electrolysis* in 1835.

Especially, Faraday's merit was that through introduction of precise concepts and definitions, he formed the basis of the quantitative description of the electrochemical phenomena known today. Thus, Faraday introduced such concepts as *ion, anion, cation, anode, cathode* and *electrolyte.*

Today electrochemistry has a number of significant applications within materials technology, for example, manufacture and refining of metals, corrosion and corrosion protection, and chemical treatment and shaping of metals.

Michael Faraday (1791-1867)

English scientist, professor of chemistry at The Royal Institution.

Contents

6.1 Electric current and charge

As the name suggests, electrochemistry is a *multidisciplinary* science. This borderland between chemistry and electrotechnology has played an important role for the development of the materials technology known today. For example, electrochemistry is used within such areas as

- *Manufacture* of important metals such as aluminium and magnesium by electrolysis of melted metal oxides.
- Electrolytic *refining* of metals to high purity; this is, for example, used for manufacture of copper with a high electric conductivity.
- *Corrosion protection* by galvanic electroplating with metal, for example, nickel-plating, chromium-plating, and zinc-galvanizing.
- *Electro-forming* of complicated specimens of not readily workable metals by electrolytic metal precipitation.
- Electrochemical *working* and *shaping* of complicated metal specimens.

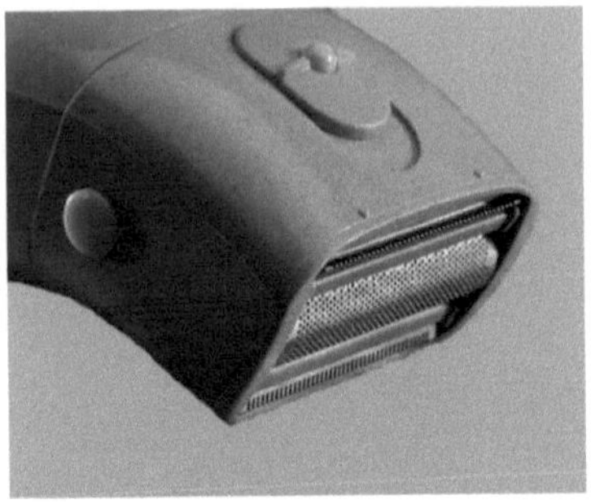

Figure 6.1. Example of an electro-formed specimen of nickel: the perforated metal foil for hair shawers. Electro-forming allows manufacture of complicated specimens of not readily workable metals. In electro-forming the metal is precipitated electrolytically on a form with electric conductivity. After the process, the form is removed.

Also, within theoretical and experimental materials technology, electrochemistry has contributed to the development in many fields. Thus, today's knowledge of corrosion of metals is to a wide extent based on the use of electrochemical measuring methods and calculation principles.

In this chapter, the most important **definitions** and **concepts** within the field that can be called **equilibrium electrochemistry** will be described. It is the intention, among other things, to present the computational foundation for the later tuition on the **science of corrosion**.

In civil engineering structures exposed to moisture, **electrochemical corrosion** can lead to extensive damage on metal members; this damage is due to galvanic connections – so-called **corrosion cells** – that produce local electric currents in the metal. From equilibrium considerations, the risk of formation of electrochemical corrosion cells in a given civil engineering structure can be predicted and in many cases the risk can be limited. On the other hand, one must realize that equilibrium considerations *cannot* give information of corrosion rate and thus the extent of a foreseeable corrosion attack. The latter subject is addressed separately under "Corrosion kinetics".

The reason that electrochemistry is normally associated with a certain degree of complexity is partly due to the special definitions and concepts used in this part of chemistry. In the following explanation, an effort has been made to carefully explain the fundamental definitions and concepts and illustrate their use in solving practical problems.

Figure 6.2. Electrochemical stress corrosion in a bellows of 18-8 austenitic steel. Especially, stress corrosion arises in metals exposed to tension in a corrosive environment.

At the same time, the examples have been chosen to show the *wide application* of electrochemistry as a discipline. Thus, the principles of the important **measurement methods** used for material investigations are illustrated by practical calculation examples. As far as possible, the calculations have been based on **thermodynamic standard values** rather than tables of standard potentials. Therefore, this chapter 6 on "Electrochemistry" presents a natural extension of chapter 5 on "Calculation of equilibrium".

Electric current

Electric current I in the unit **ampere** (A) forms one of the 7 **base units** of the SI-system, as stated in Mathematical appendix subchapter A2. The unit for amperage has been determined by the following accurate, but not very operational definition.

Amperage unit ampere (A) (6.1)

Definition: the unit **ampere** is the constant **electric current** which, if maintained in two straight parallel conductors of infinite length, of negligible circular cross-section, and placed 1 metre apart in vacuum, would produce between these conductors a force equal to $2 \cdot 10^{-7}$ newton per metre of length. The symbol or amperage is I.

Electric charge

Electric current is by definition the flow of **electric charge** Q. Therefore, the unit of electric charge **coulomb** (C) is simply determined from the unit ampere by the following definition.

> **Electric charge unit coulomb (C)** (6.2)
> *Definition:* 1 coulomb is the amount of electric charge carried by a current of $I = 1$ ampere flowing for $t = 1$ second. The symbol for electric charge is Q.

Electric elementary charge
The electric elementary charge e specifies the charge of the proton

$$e = +1.602 \cdot 10^{-19}\, C$$

The transmission of the charge Q through a conductor with an amperage $I(t)$ that varies over time is determined as the integral of the electric current over time τ

$$dQ = I(t) \cdot dt \qquad (C) \qquad (6.3)$$

$$Q = \int_0^{\tau} I(t) \cdot dt \qquad (C) \qquad (6.4)$$

The transmission of an electric charge involves a physical transmission of charged particles such as **electrons**, **protons**, or **ions**. The numerical charge of these particles will always be a multiple of the **electric elementary charge** $e = 1.602 \cdot 10^{-19}$ coulomb (see Chapter 1.1). For molecular considerations, therefore, transmission of electric charge must be described in terms of units of this minimum charge.

The Faraday constant

Generally within chemistry, an **amount of substance** is given in the SI unit (mol). Therefore, for describing electrochemical phenomena it is expedient to introduce the electric charge quantity the **Faraday constant** $\mathcal{F}$ that is associated with the mol unit for amount of substance. This association is described in the following definition.

> **The Faraday constant $\mathcal{F}$** (6.5)
> *Definition:* The **Faraday constant** $\mathcal{F}$ is defined as the electric charge of 1 mol of elementary charge
>
> $\mathcal{F} = \mathcal{N} \cdot e =$(the Avogadro constant)·(the elementary charge) (C/mol)
>
> Numerical value: $\mathcal{F} = 96485$ C/mol $\simeq 96500$ C/mol.

Note: The elementary charge $e = 1.602 \cdot 10^{-19}$ C denotes the electric charge of a **proton**; therefore, the electric charge of an **electron** is $-1.602 \cdot 10^{-19}$ C.

■ Steel specimens can be protected against corrosion by application of a metallic cover of zinc. By electro-galvanizing of steel articles, the zinc is precipitated on them by electrolysis through the following **cathode reaction**

$$Zn^{++} + 2e^{-} \rightarrow Zn \qquad (a)$$

During a given process, electro-galvanizing at constant amperage I is performed; 2000 g of Zn per hour is precipitated. Determine the amperage I (A) and the total charge transmission Q (C) for the specified process period of 1 hour! The molar mass is $M(Zn) = 65.38\,g/mol$.

Solution. The precipitated amount of substance Zn is determined:

$$n(Zn^{++}) = \frac{m}{M} = \frac{2000\,g}{65.36\,g/mol} = 30.6\,mol$$

Figure 6.3. Electro-galvanized samples of steel. The metallic cover of zinc acts as a corrosion inhibitor because it forms the sacrificial anode to the steel. When exposed to corrosion, the steel is protected until full or partial corrosion of the zinc cover has occurred.

The total charge transmission Q is determined from the **Faraday constant** $\mathcal{F}$ cf. eqn. (6.5). It shall be noted that 2 electrons are required to reduce each Zn atom; therefore, the charge supplied becomes

$$Q = 2 \cdot n(\mathrm{Zn}^{++}) \cdot \mathcal{F} = 2 \cdot 30.6\,\mathrm{mol} \cdot 96500\,\mathrm{C/mol} = \mathbf{5.91 \cdot 10^6\ C}$$

Finally, the constant amperage I corresponding to transmission of the calculated charge Q during 1 h = 3600 s is determined; from equations (6.3) and (6.4) we obtain

$$I = \frac{Q}{t} = \frac{5.91 \cdot 10^6\,\mathrm{C}}{3600\,\mathrm{s}} = \mathbf{1.64 \cdot 10^3\ A}$$

The Faraday constant

The Faraday constant $\mathcal{F}$ denotes the electric charge of 1 mol of elementary charge e

$$\mathcal{F} = \mathcal{N} \cdot e \simeq 96500\ C/mol$$

☐ **1.** Specify the electric charge Q (C) that can be transmitted during 4 minutes by an electric current with an amperage of $I = 600$ A!

☐ **2.** An electric current in a conductor transmits the charge $Q = 800$ C per minute; determine the amperage I (A) in the conductor!

☐ **3.** The current $I = 5$ A runs through a metallic conductor; how many electrons pass through a cross-section in the conductor per second?

☐ **4.** What amperage I (A) is required to precipitate 150 g of zinc per hour from a solution of zinc sulphate ?

☐ **5.** One litre of aqueous solution contains 200 g of dissolved and dissociated NaCl; calculate the total positive electric charge Q of Na^+ ions in the solution!

6.2 Electric potential

An **electrostatic field** in space is understood as a field in which an electric charge Q at rest is influenced by an electric force.

Field strength

An electrostatic field is a **vectorial field** where the field vector $\mathbf{E}(x, y, z)$ at a given point denotes the **magnitude** and **direction** of the electric **field strength** in space. The electric field strength **E** is defined as follows.

Electric field strength E (6.6)

Definition: The electric field strength $\mathbf{E}(x, y, z)$ at a point is defined as the ratio between the force $\mathbf{F}(x, y, z)$ acting on a point charge at the point and the magnitude Q of this charge, calculated with signs.

Electric field strength: $\mathbf{E} = \dfrac{\mathbf{F}}{Q}$ (N/C) = (V/m)

The electric field strength $\mathbf{E}(x, y, z)$ constitutes a **vectorial field** in space.

Field strength: $\mathbf{E} = \frac{\mathbf{F}}{Q}$

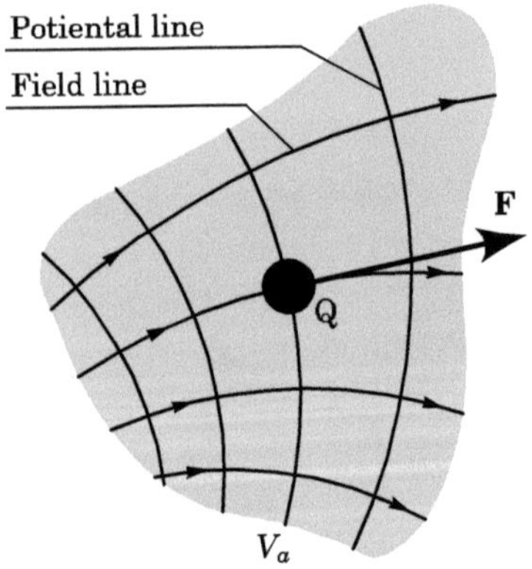

*Figure 6.4. The electric field strength **E** at a point (a) is defined as the ratio between the force **F** acting on a point-shaped charge at (a) and the magnitude of this charge Q, calculated with signs.*

Electric potential

An electrostatic field is a **conservative field**; the work W done by the field on an electric charge Q, which is moved from point a to another point b in the field, is therefore *independent* of the path. Due to this important property of the field, any point of the electrostatic field can unambiguously be given an **electric potential** V by the following definition.

Electric potential V (6.7)

Definition: The electric potential $V(x, y, z)$ at a point a in an electric field specifies the relation between the work $\delta W_{\infty.a}$ to be done on a charge and the magnitude dQ of this charge, calculated with signs, when the charge is reversibly moved from infinity to the point a

Electric potential: $V_a = \dfrac{\delta W_{\infty.a}}{dQ}$ (J/C) = (V)

The electric potential $V(x, y, z)$ constitutes a **scalar field** in space.

Note that the electric potential V_∞ at an infinite distance from the point considered has arbitrarily been given the value of 0 volt.

It follows from the definition of electric potential that the **potential difference** ΔV between a point a and a point b in a field is determined by

Electric potential difference ΔV (6.8)

The electric potential difference $\Delta V = V_b - V_a$ specifies the relation between the work $\delta W_{a.b}$ to be done on a charge and the magnitude dQ of this charge, calculated with signs, when the charge is reversibly moved from point a to point b by an arbitrary path.

Potential difference: $\Delta V_{a.b} = V_b - V_a = \dfrac{\delta W_{a.b}}{dQ}$ (J/C) = (V)

The E-field is conservative

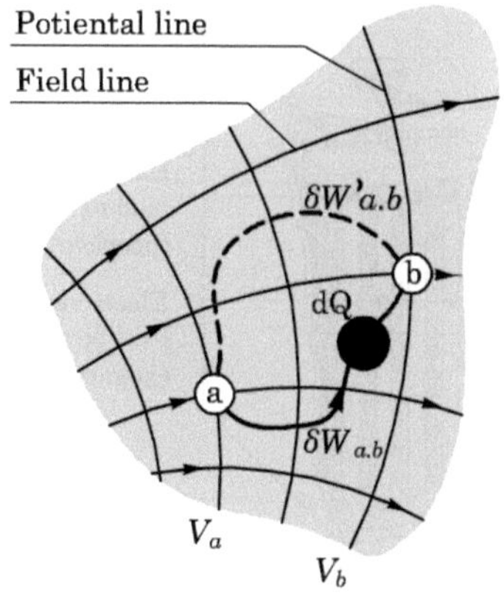

Figure 6.5. The electric field is conservative; when an electric charge dQ is moved from the potential V_a to the potential V_b the work $\delta W_{a.b}$ done is therefore independent of the path, i.e. $\delta W_{a.b} = \delta W'_{a:b}$

As will be seen, these fundamental definitions link the thermodynamics concepts to electrochemical phenomena in practical systems.

Before moving away from this general description of electric fields and potentials, the following should be noted. As seen from eqn. (6.7), the electric field $V(x, y, z)$ constitutes a **scalar field** derived from the **vectorial field** that is formed by the field strength $\mathbf{E}(x, y, z)$. The mathematical connection between electric field strength and electric potential is

$$\mathbf{E}(x, y, z) = -\mathbf{grad}\, V(x, y, z) \qquad (6.9)$$

In other words, the field strength $\mathbf{E}(x, y, z)$ is the **gradient field** of the electric potential $V(x, y, z)$ with a negative sign (see *Mathematical Appendix, subchapter A7: Gradient field*).

Electrical work

Transferring an electrically charged particle dQ reversibly from a point a to a point b within the **E** field, the field will according to eqn. (6.6) induce a force on the charged particle or body of magnitude

$$\mathbf{F} = \mathbf{E} \cdot dQ \qquad (6.10)$$

Thus, the total work δW_{field} done by the field on the charge dQ during a displacement is determined by

$$\delta W_{\text{field}} = \int_a^b \mathbf{F} \cdot ds = dQ \int_a^b \mathbf{E} \cdot ds = -dQ \int_a^b \mathbf{grad}\, V \cdot ds = -(V_b - V_a)dQ \qquad (6.11)$$

Since the external work on the charge during the displacement $\delta W_{a.b}$ is $-\delta W_{\text{field}}$, we obtain the following expression for the external work in accordance with eqn. (6.8)

Electrical work (6.12)

for a **reversible** displacement of an electric charge dQ from an electric potential V_a to an electric potential V_b, the **external work** $\delta W_{a.b}$ on the charge is determined by

$\delta W_{a.b} = (V_b - V_a)dQ$ (J)

where the electric charge dQ is calculated with signs.

Work: $\delta W_{a.b} = (V_b - V_a) \cdot dQ$

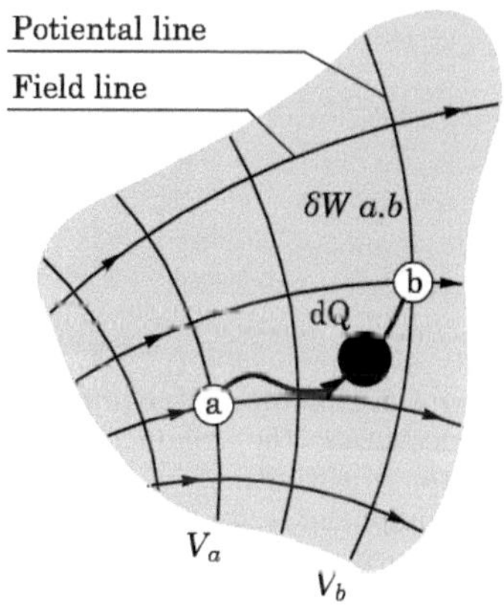

Figure 6.6. When an electric charge dQ is transmitted from an electric potential V_a to a potential V_b, the external work $\delta W_{a.b}$ on the charge is $\delta W_{a.b} = (V_b - V_a) \cdot dQ$ (J).

By use of eqn. (6.12) it should be remembered that the electrostatic field is a **conservative field** so that the work $W_{a.b}$ done is *independent* of the path.

■ Assume a thermodynamic system consisting of an accumulator; one terminal on the accumulator has the potential $V_a = 0\,\mathrm{V}$ and the other terminal has the potential $V_b = 6\,\mathrm{V}$. At constant terminal voltage, an **electric charge** $dQ = -2\,\mathrm{C}$ is transmitted from terminal a to terminal b. Calculate the **electrical work** δW done by the surroundings on the system!

Solution. The work δW can be directly calculated from eqn. (6.12) inserting the actual quantity with signs

$$\delta W_{a.b} = (V_b - V_a)dQ = (6\,\mathrm{V} - 0\,\mathrm{V}) \cdot (-2\,\mathrm{C}) = \mathbf{-12\,J}$$

In this rewriting it shall be remembered that the unit [V] corresponds to [J/C], cf. eqn. (6.8). The negative sign for δW means that the accumulator has done positive work on its surroundings during the process, because **chemical energy** has been transformed into **electrical work**.

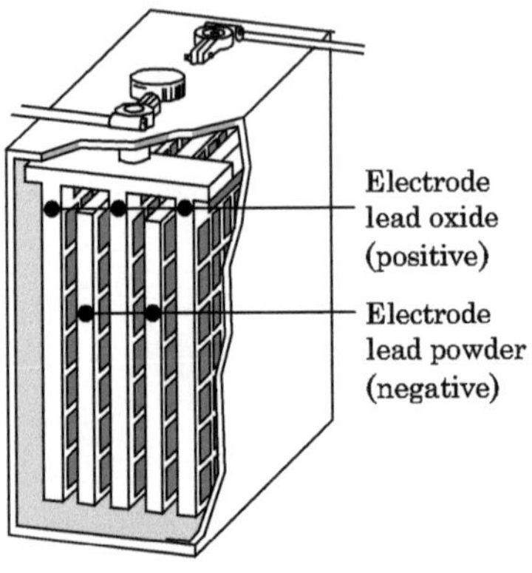

Figure 6.7. A lead accumulator can be perceived as a thermodynamic system. By interaction with the terminals of the accumulator, the surroundings can perform work W on this system. When the accumulator is charged, W > 0, and when it is discharged, W < 0.

☐ **1.** Determine the electrical work $W_{a.\infty}$, if the charge $dQ = +5\,\mathrm{C}$ is reversibly transmitted from a point a with the potential $V_a = 12\,\mathrm{V}$ to infinity!

☐ **2.** Electrical work $W_{a.b} = 110\,\mathrm{J}$ is done when the charge $dQ = -2\,\mathrm{C}$ is transmitted from a to b in an electric field; determine the potential difference $\Delta V_{a.b}$!

☐ **3.** 1 eV is the kinetic energy accumulated in an electron during a potential increase of 1 V; determine the conversion factor from (eV) to (J)!

☐ **4.** Calculate the magnitude of the force $|\mathbf{F}|$ (N) acting on an electron in an electric field with the field strength 1000 V/m!

☐ **5.** An electric current $I = 2\,\mathrm{A}$ is transmitted over a potential difference of $\Delta V = 25\,\mathrm{V}$; calculate the electrical work W (J) done per hour during this operation!

6.3 Electric conductivity

Electric current represents a *systematic* movement of electrically charged particles. If we confine ourselves to substances in either a solid or a liquid state at normal temperatures, the electric flow of power can occur in the form of

- *Metallic charge*, which is movement of negatively charged **electrons** in an electric field.
- *Electrolytic charge*, which is movement of positively and negatively charged **ions** in an electric field.

Electrons constitute a universal component of all substances; therefore, a steady current carried by electrons does *not* change the chemical state of the substances. Thus, metallic conduction is not connected with transport of substances in the general sense.

On the other hand, a steady current carried by ions will represent an actual **transport of substances**; therefore, electrolytic conduction is always connected with changes of the chemical state of substances. The latter circumstance is an essential feature of the electrochemical phenomena addressed in the following sections.

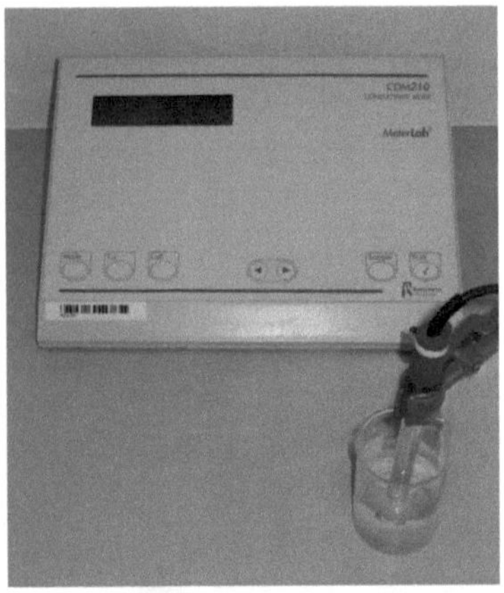

Figure 6.8. Laboratory equipment for measuring the conductivity σ and the resistivity ρ of solvents. Such equipment is, for example, used for measuring the salt concentration in solutions by titration and for determining the purity of demineralized water.

Conductivity

The ability of a substance to conduct an electric current depends on the *number* of charge-carrying particles and their *mobility* in the substance. As a measure of this ability, the **conductivity** σ is used.

Assume a homogeneous cubical sample of an electrically conducting substance. The two opposite end surfaces of the sample are provided with electrodes; the area of the end surface is A (m^2), and their mutual distance is L (m). The electrodes are now provided with an electric potential difference of ΔV volt, and the current I through the sample is determined. In such experiments it is found that the current I through the sample is proportional with the field strength $|\mathbf{E}| = \Delta V/L$ and with the charged area A

$$I = \sigma \cdot \frac{\Delta V}{L} \cdot A = \sigma \cdot \frac{A}{L} \cdot \Delta V \qquad \text{(A)} \tag{6.13}$$

The proportionality constant σ denotes the **conductivity** ("specific conductance") of the substance with the unit $(\Omega^{-1}\text{m}^{-1}) = (\text{A/V m})$. It shall be noted that the conductivity σ is a **substance property** that is *independent* of the geometry of the conductor concerned.

Conductance

The ability of a given **conductor** to transmit electric current depends on the geometrical form of the conductor *as well as* the conductivity of the materials used. As a measure of this complex property, the quantity **conductance** G with the unit $(\Omega^{-1}) = (\text{A/V})$ is used. The unit (Ω^{-1}) is also denoted *siemens*. For a given electric conductor with the conductance G, the following relation applies

$$I = G \cdot \Delta V \qquad \text{(A)} \qquad (6.14)$$

Often, in practical calculations, the reciprocal values **resistivity** $\rho = 1/\sigma$ and **resistance** $R = 1/G$ are used. In the SI unit system the following symbols, units and designations are used for the quantities mentioned.

Quantity	Symbol	Unit	Relation	Former designation
Conductivity	σ	$\Omega^{-1}\text{m}^{-1}$	$\sigma = 1/\rho$	"specific conductance"
Resistivity	ρ	Ωm	$\rho = 1/\sigma$	"specific resistance"
Conductance	G	Ω^{-1}	$G = 1/R$	
Resistance	R	Ω	$R = 1/G$	

Table 6.1. Overview of physical quantities used for calculation of the electric flow of power in conducting substances and electric conductors.

In conclusion, we thus obtain the following expression for calculating the properties of the flow of power in a **homogeneous substance**.

Flow of power in homogeneous substances (6.15)

$$I = \sigma \cdot \frac{A}{L} \cdot \Delta V \qquad \text{(A)}$$

$$\Delta V = \rho \cdot \frac{L}{A} \cdot I \qquad \text{(V)}$$

σ : **conductivity** $(\Omega^{-1}\text{m}^{-1})$

ρ : **resistivity** (Ωm)

By analogy, for calculation of flow of power in an **electric conductor** the following relations are used

Flow of power in electric conductor (6.16)

$$I = G \cdot \Delta V \qquad \text{(A)}$$

$$\Delta V = R \cdot I \qquad \text{(V)}$$

G : **conductance** (Ω^{-1})

R : **resistance** (Ω)

From the expressions in equations (6.15) and (6.16), it is seen that **conductivity** σ and **resistivity** ρ are substance constants, whereas **conductance** G and **resistance** R are system-dependent quantities that depend on the geometry of the conductor *as well as* the properties of the conductive substance.

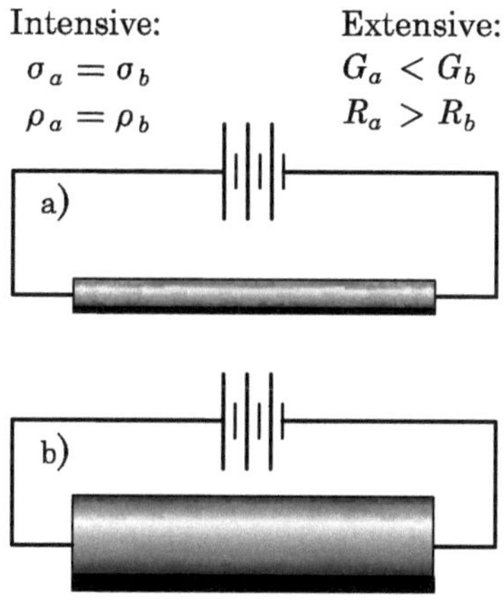

Figure 6.9. Conductivity σ and resistivity ρ are intensive quantities that are independent of the geometry of a system. Conductance G and resistance R are extensive quantities that depend on the geometry of the system.

Figure 6.10. At normal temperatures, concrete remains workable and castable for 3-5 hours after the time of mixing. After this period of time the concrete sets and stiffens. Irregularities during the setting progress can cause reduced castability and may lead to rejection of the concrete delivered (see example).

■ In the laboratory we want to investigate whether the time for setting of a concrete can be determined by measurement of changes in the electric conductivity of the concrete. A cylindrical measuring cell is made from a plastic pipe; the inner diameter of the cell D is 100 mm and the length L is 200 mm. The end surfaces of the measuring cell consists of metallic electrodes. During measurement, the electrodes are provided with a low-frequency alternating voltage. From the initial test we have a related measurement of ΔV and I

$$\Delta V = 6.0\,\text{volt}; \qquad I = 50\,\text{milliampere}$$

Determine the **conductance** G and **resistance** R of the measuring cell and calculate the **conductivity** σ and **resistivity** ρ of the concrete! Any self-induction and electrolytic polarization in the system shall be disregarded in the calculations.

Solution. The **conductance** G of the measuring cell is determined from eqn. (6.16) as

$$G = \frac{I}{\Delta V} = \frac{50 \cdot 10^{-3}\,\text{A}}{6.0\,\text{V}} = \mathbf{8.33 \cdot 10^{-3}\,\Omega^{-1}}$$

In accordance with table 6.1, the **resistance** R of the measuring cell is determined as

$$R = \frac{1}{G} = \frac{1}{8.33 \cdot 10^{-3}\,\Omega^{-1}} = \mathbf{120\,\Omega}$$

From eqn. (6.15), the **conductivity** σ of the concrete is determined as

$$\sigma = \frac{I \cdot L}{A \cdot \Delta V} = \frac{50 \cdot 10^{-3}\text{A} \cdot 0.20\text{m}}{\pi \cdot 0.05^2\text{m}^2 \cdot 6.0\text{V}} = \mathbf{0.21\,\Omega^{-1}m^{-1}}$$

Finally, in accordance with table 6.1, the **resistivity** ρ of the concrete is determined as

$$\rho = \frac{1}{0.21\Omega^{-1}\text{m}^{-1}} = \mathbf{4.7\,\Omega m}$$

☐ **1.** In a metallic conductor the amperage of the electric current is $I = 5.0\,\text{A}$; how many electrons will pass a cross-section in the conductor per second ?

☐ **2.** Specify the magnitude of the electric charge Q (C) transmitted by electrolytic conduction when 5.3 mol of Al^{+++} pass a cross-section !

☐ **3.** In a table, the conductivity σ of aluminium is given as 36 $\text{m}/(\Omega \cdot \text{mm}^2)$; convert this conductivity σ to the unit $\Omega^{-1}\text{m}^{-1}$!

☐ **4.** Assume a cylindrical conductor of copper with a length of 100 mm and a diameter of 0.10 mm with the conductivity $\sigma = 58 \cdot 10^6\,\Omega^{-1}\text{m}^{-1}$. Calculate its resistance R and conductance G !

☐ **5.** A cube with an edge length of 10.0 cm of a homogeneous substance is provided with electrodes on two opposite sides. At the voltage $\Delta V = 5.0$ between the electrodes, a current of $I = 0.10$ ampere is transmitted. Calculate the conductivity σ and resistivity ρ of the substance !

6.4 Electrochemical reaction

In the following sections, the special thermodynamic system that is called an **electrochemical cell** is addressed. First we shall specify this concept and describe the special electrochemical transformations that are characteristic of this system.

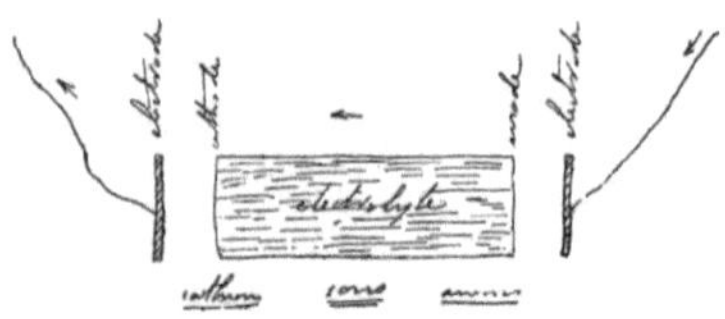

Figure 6.11. In his thesis on chemical electrolysis (1835), Michael Faraday introduced a number of concepts and designations that are now universally used. In this sketch made by Faraday, concepts as electrolyte, cathode, anode, ion, anion, and cation are recognized.

Electrochemical cell

An electrochemical cell consists of two **electrodes** that are electrically interconnected by an **electrolyte**. Each electrode consists of an electric conductor in which the electric current is transmitted by electrons. As examples of electrode material, metallic copper, Cu, and zinc, Zn, can be mentioned. The electrolyte connecting the electrodes contains mobile electrically charged ions that serve as carriers of electric current. The positively charged ions are named **cations** and the negatively charged ions are named **anions**. As examples of electrolytes, water, aqueous salt solutions, and melts of ionically bound substances can be mentioned.

It is characteristic of the electrochemical cell that passage of **electric current** in the cell is coupled with the occurrence of a **chemical reaction**. During such reaction, an electrochemical cell can have a passive function as an **electrolytic cell** or an active function as a **galvanic cell**; the significance of this distinction is as follows:

Electrolytic cell: In the electrolytic cell the surroundings perform electrical work on the cell and thereby generate a chemical reaction. In a thermodynamic sense, *positive* work is done on the cell. As examples of this, mention can be made of splitting up water $H_2O(\ell)$ into hydrogen $H_2(g)$ and oxygen $O_2(g)$ by electrolysis.

Galvanic cell: In the galvanic cell, a chemical reaction generates an electric current that can do electrical work on its surroundings. In a thermodynamic sense, the surroundings thereby perform *negative* work on the cell. Examples of galvanic cells are the Daniell cell and the lead battery.

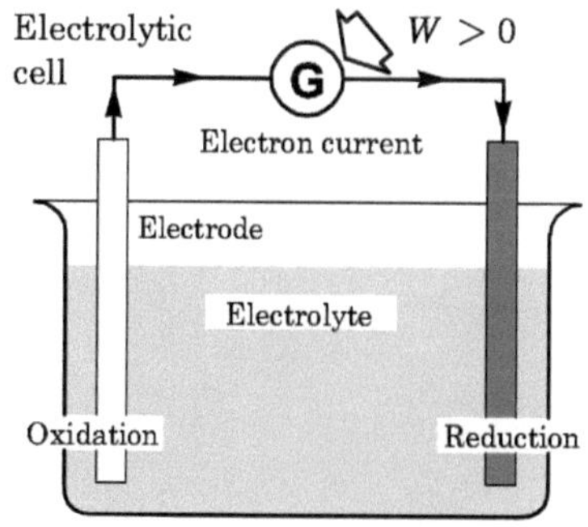

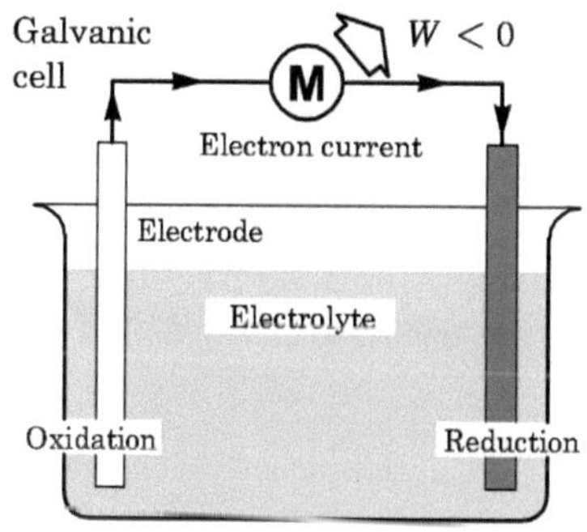

Figure 6.12. In the electrolytic cell the surroundings perform electrical work $W > 0$ on the cell and thus bring about a chemical reaction in the cell. In the galvanic cell, a chemical reaction produces a current that can perform electrical work on the surroundings; for the galvanic cell, $W < 0$.

Redox reaction

In an electrochemical cell the coupling between electric current and a chemical reaction occurs in a **boundary layer** between electrolyte and electrode. Assume an electrochemical cell through which a constant electric current passes. In the electrolyte the electric current is carried by ions, the movement of which involves transport of matter at the same time. In the electrodes the electric current is carried by electrons. Therefore, when the current passes through the boundary layer between electrolyte and electrode, a change occurs in the nature of the charge-carrying particles; as shall be seen, this change involves **oxidation** or **reduction** of compounds.

The charge transferring reaction in the boundary layer between electrode and electrolyte involves a change in a compound, which either *gains* or *loses* electrons. In a chemical sense, this corresponds to changing the oxidation state by reduction or oxidation.

Reduction and oxidation (6.17)

Reduction: $A + e^- \rightarrow A^-$ example.: $Cl_2 + 2e^- \rightarrow 2Cl^-$

Oxidation: $B \rightarrow B^+ + e^-$ example.: $Zn \rightarrow Zn^{++} + 2e^-$

By **reduction** of a compound, electrons are *gained*; by **oxidation** of a compound, electrons are *lost*.

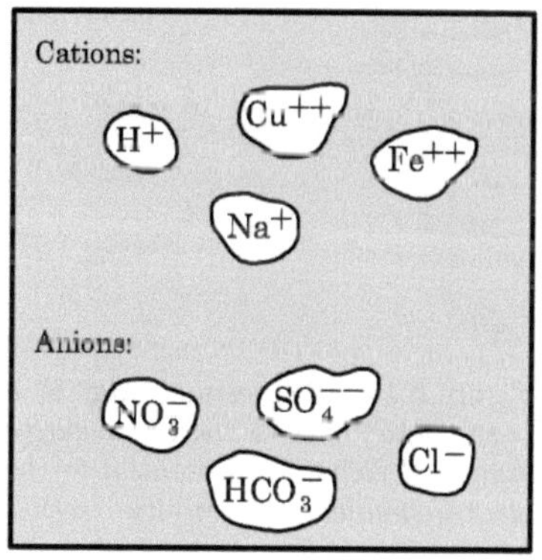

Figure 6.13. Positively charged ions are called "cations" and negatively charged ions are called "anions".

In the overall electrochemical cell, a neutral state of electrons is always maintained; therefore, reduction and oxidation always take place as a connected reaction, the so-called **redox reaction**.

Electrochemical process

The electrode that *liberates* electrons to the electrolyte is named the **cathode**. Therefore, in an electrochemical cell, the cathode process is always a reduction process. The electrode that *gains* electrons from the electrolyte is named the **anode**; therefore, the corresponding anode process is always an oxidation process. Thus, in conclusion:

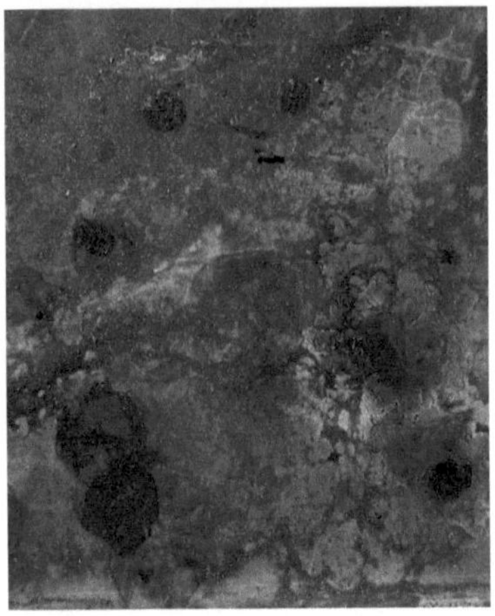

Figure 6.14. Galvanic corrosion of steel is an example of an anode process during which Fe(s) is oxidized and forms ferrous ions Fe^{++} or ferric ions Fe^{+++}.

Electrochemical reaction (6.18)

An electrochemical reaction includes a *simultaneous* **cathode reaction** and **anode reaction** characterized by

Cathode reaction: The electrode that *loses* electrons to the electrolyte is the **cathode**; at the cathode, a **reduction** always takes place in accordance with the following formal reaction equation

$$A + e^- \rightarrow A^-$$

Anode reaction: The electrode that *gains* electrons from the electrolyte, is the **anode**; at the anode, an **oxidation** always takes place in accordance with the following formal reaction equation

$$B \rightarrow B^+ + e^-$$

As a whole, the electrochemical reaction constitutes a so-called **redox reaction**.

In a chemical sense, reduction and oxidation mean change of **oxidation state**; an electrochemical redox reaction, therefore, has a further significance beyond that of formation or neutralization of simple ions. As examples, the following can be mentioned:

$Fe^{+++} + e^- \rightarrow Fe^{++}$ (reduction of *ferric* ion to *ferrous* ion)
$Sn^{++} \rightarrow Sn^{++++} + 2e^-$ (oxidation of *stannous* ion to *stannic* ion)

Further, it shall be noted that even though all ions included in an electrolyte participate in the transport of the charge, they are not all necessarily included in the resulting redox reaction.

■ Magnesium, Mg, is made industrially from the double salt **carnalite** with the composition $KMgCl_3 \cdot 6H_2O$. It is made by electrolysis of a melt of anhydrous carnalite during which process the net reaction is

$$MgCl_2(s) \rightarrow Mg(s) + Cl_2(g)$$

Identify and describe the **cathode reaction** and the **anode reaction** in the electrolytic manufacture of magnesium!

Solution. Magnesium chloride $MgCl_2$ is an ioncally-bound salt formed by Mg^{++} and Cl^-; in a melt of the salt, these ions carry the electric current in the electrolyte. Formation of metallic magnesium requires the reaction

$$Mg^{++} + 2e^- \rightarrow Mg \quad \text{(absorption of electrons)}$$

corresponding to **reduction** of Mg^{++} during absorption of electrons. According to eqn. (6.18), free Mg is therefore formed by a cathode reaction during the electrolysis.

Formation of chlorine gas $Cl_2(g)$ requires the reaction

$$2Cl^- \rightarrow Cl_2(g) + 2e^- \quad \text{(liberation of electrons)}$$

which according to eqn. (6.18) corresponds to **oxidation** of Cl^- during liberation of electrons; therefore, free chlorine is developed by an anode reaction during the electrolysis.

Figure 6.15. Magnesium, Mg, is a base metal with a highly negative standard reduction potential in the electrochemical reactivity series. Magnesium is therefore widely used as a sacrificial anode in the corrosion protection of steel. The photo shows preparation of sacrificial anodes - in this case of zinc - to be mounted on offshore steel structures.

☐ **1.** What is the difference between a galvanic cell and an electrolytic cell when they are considered as thermodynamic systems?

☐ **2.** Find the reaction equation for a) reduction of Fe^{++} to Fe b) oxidation of Mn to Mn^{++++} c) reduction of Cu^{++} to Cu^+ d) reduction of $2H^+$ to $H_2(g)$!

☐ **3.** Specify the reaction type (reduction or oxidation) for the following electrochemical reactions: a) $Mg^{++} + 2e^- \rightarrow Mg$ b) $Cr^{++} \rightarrow Cr^{+++} + e^-$ c) $H^+ + e^- \rightarrow \frac{1}{2}H_2(g)$!

☐ **4.** Specify the ion type (cation or anion) for the following ions: a) OH^- b) Zn^{++} c) NH_4^+ d) SO_4^{--} e) H^+ f) NO_3^- g) O^{--} h) Li^+!

☐ **5.** During an electrochemical reaction, 1.00 kg of Al is generated by reduction of Al^{+++}; calculate the necessary transfer of charge Q (C) by the process!

6.5 Electrochemical potential

When a metallic **zinc electrode** is immersed in an acid aqueous solution of **copper sulphate**, $CuSO_4$, the electrode is instantly covered with metallic copper. This copper coating is generated by the following reaction in the boundary layer between electrode and electrolyte

$$Zn(s) + Cu^{++}(aq) \quad \rightarrow \quad Zn^{++}(aq) + Cu(s) \tag{6.19}$$

In the boundary layer between electrode and electrolyte, there is a simultaneous **reduction** of Cu^{++} and **oxidation** of Zn. According to eqn. (6.17), the individual steps in this **redox reaction** can be written as follows

$$\text{Reduction of } Cu^{++}: \quad Cu^{++}(aq) + 2e^- \quad \rightarrow \quad Cu(s) \tag{6.20}$$

$$\text{Oxidation of Zn} \quad : \quad Zn(s) \quad \rightarrow \quad Zn^{++}(aq) + 2e^- \tag{6.21}$$

Figure 6.16. When a zinc plate is immersed in an acid solution of copper sulphate, it is spontaneously coated with metallic copper Cu(s). In the process, Zn(s) is oxidized to Zn^{++} and dissolves, and the dissolved cupric ions Cu^{++} are reduced to Cu(s).

Metallic copper precipitates *spontaneously*; therefore, the reaction according to the equilibrium condition in eqn. (5.7) necessarily involves a decrease of free energy ($\Delta G < 0$). The reaction is qualitatively explained in the following way: Zinc atoms at the surface of the electrode show a certain tendency to dissolve into ions $Zn^{++}(aq)$ and leave 2 free electrons in the electrode metal. The corresponding tendency to dissolve is less for copper atoms. Copper ions Cu^{++} in the solution, therefore, take up the liberated electrons and are reduced to Cu(s) that is precipitated on the electrode. In the following, we shall see how this phenomenon is decisive to the build-up of an **electrochemical potential** in galvanic cells.

The electrochemical potential

Assume a **zinc** electrode surrounded by an aqueous solution of zinc sulphate, $ZnSO_4$. Because of the tendency of the zinc atoms to dissolve into ions, the following **equilibrium** is present in the boundary layer between electrode and electrolyte

$$Zn(s) \quad \leftrightarrows \quad Zn^{++}(aq) + 2e^- \tag{6.22}$$

The liberated electrons remain in the electrode and thus produce a negative, electric potential in relation to the surrounding electrolyte. At equilibrium, an electric double layer $[e^- \| Zn^{++}]$ is created over the boundary layer between electrode/electrolyte. The resulting potential difference over the boundary layer is an indication of the tendency of the zinc metal to dissolve.

Correspondingly, a **copper** electrode surrounded by an aqueous solution of copper sulphate $CuSO_4$ will build up an electric potential difference over the boundary layer between electrode and electrolyte; in this case, the potential difference is an indication of the tendency of the copper to dissolve, determined by the equilibrium

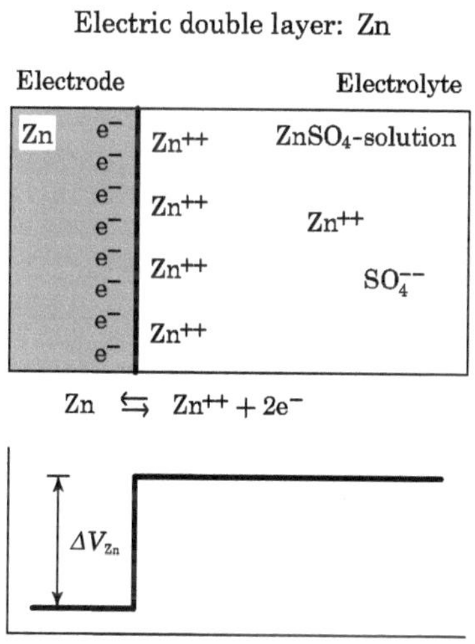

Figure 6.17. At the interface between a zinc electrode and a surrounding electrolyte, equilibrium $Zn \leftrightarrows Zn^{++} + 2e^-$ is obtained as the zinc atoms have a certain tendency to dissolve into ions; the liberated electrons remain in the zinc electrode. Due to electrostatic attraction between ions and electrons, an electric double layer over the interface is built up. The magnitude of the resulting potential difference is a measure of the oxidation tendency of the zinc.

$$Cu(s) \quad \leftrightarrows \quad Cu^{++}(aq) + 2e^- \tag{6.23}$$

In principle, the electric potential difference over the boundary layer *cannot* be measured for an individual electrode; Such measurement would mean that electrode and electrolyte have simultaneous access to metallic wires from a voltmeter, whereby a new, unknown electrode is introduced into the measurement circuit. However, for a **pair of electrodes**, the measurement can be made as described. Assume that the electrolytes described above are in mutual electric contact, for example, achieved by inserting a porous wall between the solution of zinc sulphate and the solution of copper sulphate. In this way, electric contact between the two electrolytes is achieved without mixing them. Designating the electric

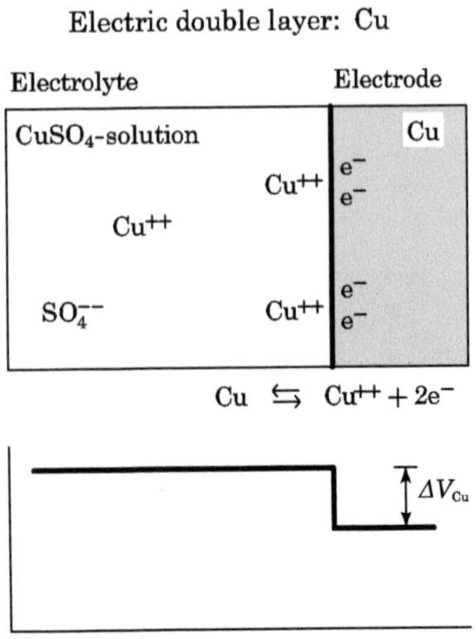

Figure 6.18. At the interface between a copper electrode and the surrounding electrolyte, equilibrium: $Cu \leftrightarrows Cu^{++} + 2e^-$ is obtained. Since copper is less prone to oxidation than zinc, the potential leap over the interface becomes less than that arising at a zinc electrode.

potential of the zinc electrode $V(\text{Zn})$ and of the copper electrode $V(\text{Cu})$, the measurable potential becomes

$$\Delta V = V(\text{Cu}) - V(\text{Zn}) \tag{6.24}$$

For the **electrode pair** Zn||Cu considered, a characteristic potential difference (electro motive force, emf) of 1.10 volt is found at standard conditions.

Daniell cell

If the two electrodes are connected through an external metallic conductor, a steady current of electrons will flow from the Zn electrode to the Cu electrode. At the Zn **anode**, a steady **oxidation** of Zn(s) takes place, cf. eqn. (6.21), and at the Cu **cathode**, a simultaneous **reduction** of Cu^{++} takes place, cf. eqn. (6.20). In the electrolyte the electron flow is balanced by transfer of ions through the porous partition. The net reaction, therefore, corresponds to the redox reaction, eqn. (6.19): the Zn electrode is dissolved by an anode process and metallic copper is precipitated on the Cu electrode through a cathode process.

The galvanic cell with Zn || Cu electrodes described here is called a **Daniell cell**, named after the English physicist *J.F. Daniell* (1836). The Daniell cell illustrates the underlying principle for build-up of the electrochemical potentials in galvanic cells. Similar potentials can arise locally in moisture exposed metal members in building structures and cause **electrolytic corrosion**. In the following sections, therefore, we shall look further into this phenomenon.

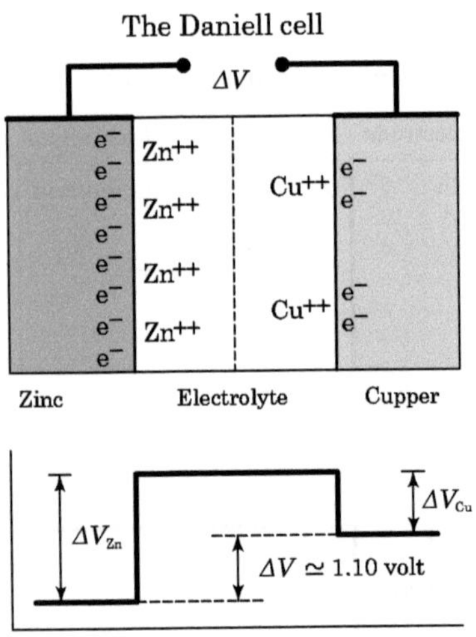

Figure 6.19. In the Daniell cell the electrode pair Zn–Cu is in conductive electrolytic contact through a porous wall; thus, the same electric potential in both of the electrolytes is ensured. Since the potential difference over the interface between electrode and electrolyte is different for the Zn electrode and the Cu electrode, an electrochemical potential difference of $V \simeq 1.10$ volt arises between the electrodes in the unloaded cell.

■ The tendency to give off metal ions to a solution varies from one metal to another. Therefore, each pair of electrodes will build up a characteristic electrochemical potential ΔV. When unloaded and at standard conditions, the Daniell cell with the electrode pair Zn||Cu in a zinc sulphate solution and copper sulphate solution, respectively, has a potential of $V = 1.10$ volt. If the Cu electrode in the Daniell cell is replaced by an electrode of iron, Fe, immersed in a 1 molar iron sulphate solution, the potential $\Delta V = 0.32$ volt for the Zn||Fe electrode pair can be measured when unloaded.

Determine the electrochemical potential ΔV for an unloaded cell with the electrode pair Fe||Cu in a 1 molar iron sulphate solution and 1 molar copper sulphate solution, respectively!

Solution. For the electrode pair Zn || Cu in the Daniell cell we have

$$V(\text{Cu}) - V(\text{Zn}) = +1.10 \text{ volt}$$

For the electrode pair Zn || Fe, the electrochemical potential is

$$V(\text{Fe}) - V(\text{Zn}) = +0.32 \text{ volt}$$

By subtraction of these contributions the targeted quantity can now be isolated

$$V(\text{Cu}) - V(\text{Fe}) = (V(\text{Cu}) - V(\text{Zn})) - (V(\text{Fe}) - V(\text{Zn})) = 1.10 - 0.32 \text{ volt}$$

$$\Delta V = V(\text{Cu}) - V(\text{Fe}) = \mathbf{0.78\ volt}$$

☐ **1.** The electrochemical potential is $\Delta V = -0.46$ volt for the electrode pair Ag||Cu and $\Delta V = +1.10$ volt for Zn||Cu. Determine ΔV for the electrode pair Zn||Ag!

☐ **2.** Specify and explain which of the metals copper Cu or zinc Zn can be expected to have the highest resistance against corrosion!

☐ **3.** Set up the redox equation for the reaction in a galvanic cell formed by the electrode pair Zn||Ag!

☐ **4.** Set up the redox equation for the reaction in a galvanic cell formed by the electrode pair Zn||Fe!

☐ **5.** Specify the direction in which the ions Cu^{++}(aq), Zn^{++}(aq) and SO_4^{--}(aq) are transferred through the porous wall in a Daniell cell!

6.6 The Nernst equation

In the following it will be shown how thermodynamic standard values can be used to determine the electrochemical **equilibrium potential** for a given galvanic cell.

In chapter 5 we have formulated the condition for thermodynamic reaction equilibrium in **isothermal** systems at **constant pressure**, if *only* volume work $\delta W_{vol} = -p \cdot dV$ occurs during the reaction. Furthermore, **electrical work** δW_{el} occurs in galvanic systems during the reaction; we shall now see how this contribution of work is introduced into the thermodynamic equilibrium conditions.

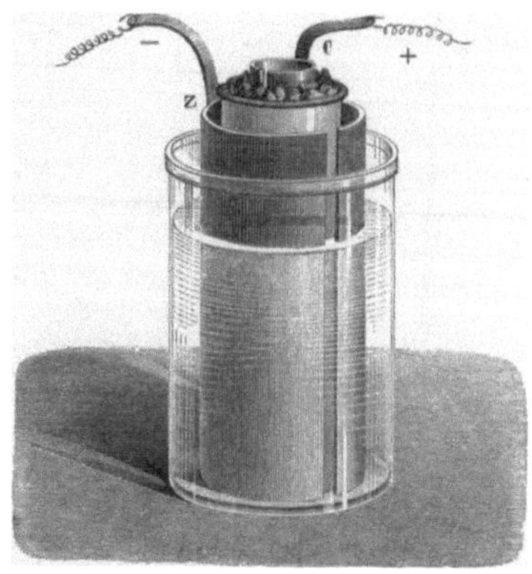

Figure 6.20. The Daniell cell from 1836; the element is named after its inventor, J.F. Daniell, who was professor of chemistry at King's College in London.

Electrical work contribution

Assume a thermodynamic system comprising a galvanic cell; using the definitions for **free energy** G, eqn. (5.1), for **enthalpy** H, eqn. (3.11), and for **internal energy** U, eqn. (3.2), we find the following differentials of the G function for the system

$$dG = dH - d(TS) = dU + d(pV) - d(TS) \qquad (6.25)$$

$$dG = \delta Q + \delta W + pdV + Vdp - TdS - SdT \qquad (6.26)$$

We consider this system on the following assumptions

constant **pressure** and **temperature** : $Vdp = 0; \quad SdT = 0$ (6.27)

reversible change of state : $\delta Q = \delta Q_{rev} = TdS$ (6.28)

electrical work contribution is included : $\delta W = -pdV + \delta W_{el}$ (6.29)

and obtain by introduction of these terms into eqn. (6.25) that $dG = \delta W_{el}$. Comparing this with the expression for electrical work in eqn. (6.12), we obtain

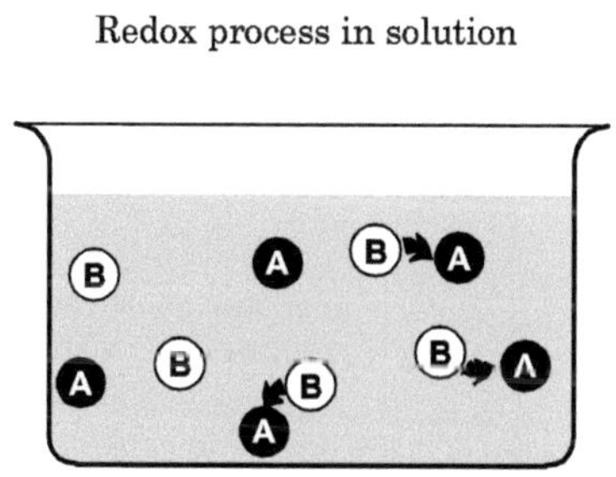

Figure 6.21. If a redox reaction develops in a solution of A and B, the components exchange electrons. During the process, the electric charge between A and B is transferred in random directions and, therefore, it does not represent an electric current that is able to do work on the environments.

> **Galvanic cell** (6.30)
>
> For a **reversible**, **isothermal** change of state in a galvanic cell working at **constant pressure**, the change of free energy is
>
> $$dG = \delta W_{el} = (V_b - V_a) \cdot dQ \qquad (T, p \text{ constant})$$
>
> where δW_{el} is the external electrical work on the cell when the charge dQ, calculated with signs, is **reversibly** moved from the potential V_a to V_b.

For a given galvanic cell, the cathode reaction and anode reaction can formally be written in the following way

$$\text{Cathode reaction}: \quad \mathrm{a}A(ox) + ze^- \rightarrow \alpha A(red) \qquad (6.31)$$

$$\text{Anode reaction}: \quad \mathrm{b}B(red) \rightarrow \beta B(ox) + ze^- \qquad (6.32)$$

where (red) denotes components in a **reduced** state and (ox) denotes components in an **oxidized** state. With this notation, the reaction equation for the compound **redox reaction** in the cell becomes

$$\mathrm{a}A(ox) + \mathrm{b}B(red) \rightarrow \alpha A(red) + \beta B(ox) \qquad (6.33)$$

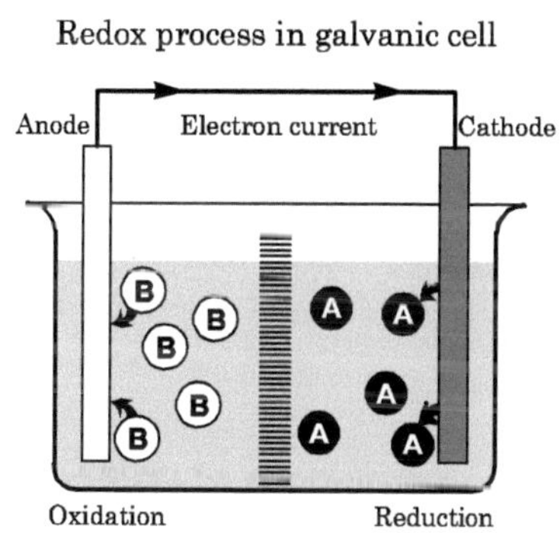

Figure 6.22. In the electrochemical redox reaction in a galvanic cell, the components A and B are physically separated. Transfer of electrons from A to B takes place by a systematic charge transport through an external circuit; the electron transfer, therefore, represents an electric current that can be used to do work on the environments.

For a mole of the specified reaction equation, $z \cdot \mathcal{N}$ electrons are transferred from anode to cathode; this corresponds to an electric charge transfer of Q_{el}

$$Q_{el} = z \cdot \mathcal{N} \cdot e^- = -z \cdot \mathcal{F} \qquad \text{(coulomb)} \qquad (6.34)$$

where $\mathcal{F}$ according to eqn. (6.5) denotes the **Faraday constant** $\mathcal{F} \sim 96500\,\mathrm{C/mol}$. It shall be noted that the transfer of charge from anode to cathode has a negative sign since we are dealing with transfer of negatively charged electrons.

The Nernst equation

At **constant pressure** we now perform the following **isothermal** change of state in the galvanic cell considered. By a **reversible** work process $z \cdot \mathcal{N}$, electrons are now transferred from the anode to the cathode; according to eqn. (6.30) we have thus changed the free energy of the system by

$$\Delta G = W_{el} = (V_b - V_a) \cdot Q_{el} = -(V_b - V_a) \cdot z\mathcal{F} \qquad (6.35)$$

where V_a and V_b denote the constant, electric potential of the anode and the cathode, respectively, during the process. The change of state results in a **chemical transformation** in the cell corresponding to "1 mol of reaction equation", see eqn. (6.33). Since the G function is a **state function**, ΔG can for this change of state be expressed by eqns. (5.36) and (5.40)

$$\Delta G = \sum G(\text{prod}) - \sum G(\text{react}) = \Delta_r G^{\ominus}_T + RT \cdot \ln(K_a) \qquad (6.36)$$

where K_a designates the **thermodynamic equilibrium constant** for the redox reaction, eqn. (6.33). From the identity between eqns. (6.35) and (6.36) it follows that

$$\Delta V = (V_b - V_a) = -\Big(\frac{\Delta_r G^{\ominus}_T}{z\mathcal{F}} + \frac{RT}{z\mathcal{F}} \cdot \ln(K_a)\Big) \qquad (T, p \text{ constant}) \qquad (6.37)$$

If all reactants and products are in their **standard state**, the equilibrium constant is $K_a = 1$; in this case, the electrochemical potential becomes equal to the **standard potential** $\Delta V^{\ominus}$, determined by

$$\Delta V^{\ominus} = -\frac{\Delta_r G^{\ominus}_T}{z\mathcal{F}} \qquad (T, p \text{ constant}) \qquad (6.38)$$

Summarizing this result, we obtain the important **Nernst equation** of electrochemical potentials (W.H. Nernst, 1889)

> The **Nernst equation** (6.39)
> In the galvanic **redox reaction**
>
> $$\text{a}A(ox) + \text{b}B(red) \quad \rightarrow \quad \alpha A(red) + \beta B(ox) \quad (T, p \text{ constant})$$
>
> the **electrochemical equilibrium potential** is $\Delta V = V(\text{cathode}) - V(\text{anode})$:
>
> $$\Delta V = \Delta V^{\ominus} - \frac{RT}{z\mathcal{F}} \cdot \ln(K_a) = -\Big(\frac{\Delta_r G^{\ominus}_T}{z\mathcal{F}} + \frac{RT}{z\mathcal{F}} \cdot \ln(K_a)\Big) \qquad (\text{volt})$$

The Nernst equation

$$\Delta V = \Delta V^{\ominus} - \frac{RT}{z\mathcal{F}} \cdot \ln(K_a)$$

The **Nernst equation** (5.39) is fundamental for understanding a number of electrochemical phenomena. The universal nature of the equation is due to the fact that it describes electrochemical equilibria from general thermodynamic principles of energy. In the following, therefore, we shall often revert to this important equation.

Written with a formal notation, the Nernst equation (6.39) can seem unapproachable; however, the practical use of the equation is in many cases rather simple, as seen from the following example.

■ A galvanic cell consists of a Zn electrode in a $ZnSO_4$ solution which through a porous wall is in electrolytic contact with a Cu electrode in a solution of $CuSO_4$ ("Daniell cell"). The following thermodynamic standard values are given in the unit kJ/mol

$$G^{\ominus}(\text{Zn}) = G^{\ominus}(\text{Cu}) = 0.0; \quad G^{\ominus}(\text{Zn}^{++}) = -147.0; \quad G^{\ominus}(\text{Cu}^{++}) = 65.5$$

Calculate the electrochemical standard potential $\Delta V^{\ominus}$ for the electrode pair Zn||Cu at 25 °C!

Solution. The galvanic redox reaction in the cell is written according to eqn. (6.33)

$$\text{Cathode}: \ \text{Cu}^{++}(\text{aq}) + 2\text{e}^- \ \rightarrow \text{Cu(s)} \qquad (\text{reduction of Cu}^{++}(\text{aq}))$$

Anode	: Zn(s)	$\rightarrow Zn^{++}(aq) + 2e^-$	(oxidation of Zn(s))
Net	: $Zn(s) + Cu^{++}(aq) \rightarrow Zn^{++}(aq) + Cu(s)$		(redox reaction in cell)

The standard free reaction energy $\Delta_r G_T^\ominus$ is determined for the net reaction at 25 °C

$$\Delta_r G_{298}^\ominus = G^\ominus(\mathrm{Zn^{++}(aq)}) + G^\ominus(\mathrm{Cu(s)}) - (G^\ominus(\mathrm{Zn(s)}) + G^\ominus(\mathrm{Cu^{++}(aq)}))$$
$$\Delta_r G_{298}^\ominus = (-147.0 + 0.0) - (0.0 + 65.5)\ \text{kJ/mol} = -212\,500\ \text{J/mol}$$

Since $K_a = 1$ and thus $\ln(K_a) = 0$, the standard potential $\Delta V^\ominus$ is calculated from eqn. (6.38)

$$\Delta V^\ominus = -\frac{\Delta G^\ominus}{z\mathcal{F}} = -\frac{-212\,500}{2 \cdot 96500} = \mathbf{+1.10\,volt}$$

☐ **1.** Given the redox reaction $Zn + Fe^{++} \rightarrow Fe + Zn^{++}$ for the electrode pair Zn||Fe, set up the cathode and anode reactions and specify (*red*) and (*ox*) states for components!

☐ **2.** In the Nernst equation (6.39), constant T and p are assumed. Discuss briefly whether a change of pressure will have significant effect on the emf of an electrode pair!

☐ **3.** Calculate from thermodynamic standard values the electrochemical standard potential $\Delta V^\ominus$ for the electrode pair Ag||Zn!

☐ **4.** Calculate from thermodynamic standard values the electrochemical standard potential $\Delta V^\ominus$ for the electrode pair Zn||Pb and Pb||Zn!

☐ **5.** Calculate the total charge transfer Q (C) by reduction of 100 g of iron Fe at the reaction $Fe^{++}(aq) \rightarrow Fe(s)$; molar mass $M(\mathrm{Fe}) = 55.85$ g/mol!

Figure 6.23. The Faraday constant, $\mathcal{F} \simeq 96500$ C/mol, signifies the electric charge of 1 mol of elementary charges. To transfer 96500 C through the filament, a 40 W electric bulb needs to be lit for approximately 36 hours at 110 volt direct current.

6.7 Temperature dependence of the potential

If we calculate the electrochemical **equilibrium potential** for a given cell reaction, it applies according to eqn. (6.35) that

$$\Delta V = V_b - V_a = -\frac{\Delta_r G_T}{z\mathcal{F}} \tag{6.40}$$

where ΔV denotes equilibrium potentials in volt. From eqn. (5.9) we use that the increase in free energy can be expressed by $\Delta G = \Delta H - T \cdot \Delta S$ and thus we obtain

$$\Delta V = -\frac{\Delta_r H_T}{z\mathcal{F}} + T \cdot \frac{\Delta_r S_T}{z\mathcal{F}} \tag{6.41}$$

From the above, the temperature influence on the equilibrium potential at **constant pressure** is obtained. If eqn. (6.41) is partially differentiated with regard to the temperature T, we find

Temperature dependence on the equilibrium potential (6.42)

$$\left(\frac{\partial(\Delta V)}{\partial T}\right)_p = \frac{\Delta_r S_T}{z\mathcal{F}} \qquad \text{(volt/K)}$$

For components in **standard state**, the temperature coefficient is

$$\left(\frac{\partial(\Delta V^\ominus)}{\partial T}\right)_p = \frac{\Delta_r S_T^\ominus}{z\mathcal{F}} \qquad \text{(volt/K)}$$

The electrochemical equilibrium potential ΔV has a **temperature coefficient** in (volt/K) that is equal to $\Delta_r S_T$ for the cell reaction divided by $z\mathcal{F}$.

Temperature coefficient

$$\left(\frac{\partial(\Delta V)}{\partial T}\right)_p = \frac{\Delta_r S_T}{z\mathcal{F}}$$

Since $\Delta S/z\mathcal{F}$ is normally a small numerical quantity, the equilibrium potential only depends *weakly* on temperature. For galvanic cells, temperature coefficients around 10^{-4} volt/K are found in tests. In rough calculations, the temperature influence on the equilibrium potential ΔV can normally be disregarded. However, the following example shows how to estimate the effect.

■ Assume a **galvanic cell** in which a zinc electrode in 1 molar $ZnSO_4$ solution through a porous wall is in conductive contact with a copper electrode in one molar $CuSO_4$ solution ("Daniell cell"). The following thermodynamic standard values at 298.15 K are given as

Component:	Zn(s)	Zn^{++}(aq)	Cu(s)	Cu^{++}(aq)
$H^\ominus$ (kJ/mol)	0.0	−153.9	0.0	64.8
$S^\ominus$ (J/mol K)	41.6	−112.1	33.2	−99.6

Determine the computational value of the **equilibrium potential** $\Delta V^\ominus$ for the cell at 25 °C and at 0 °C. *Ideal* solutions are assumed in the calculations.

Thermogalvanic corrosion

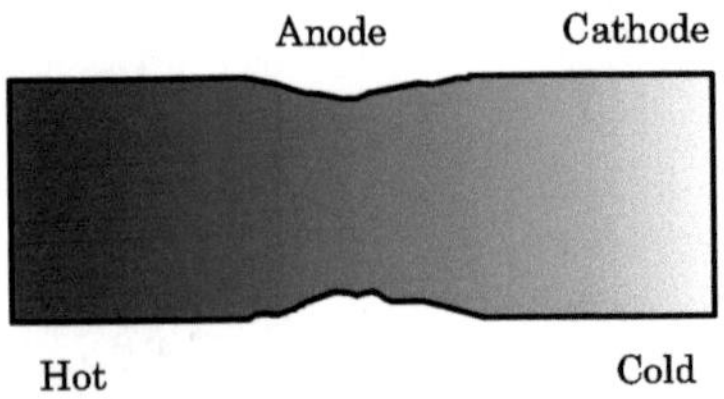

Figure 6.24. If a metal is exposed to large temperature gradients in a corrosive environment, thermogalvanic corrosion can arise. Normally, hot areas ("hot spots") form including anodes and corrode. The reason for thermogalvanic corrosion is the temperature influence on the equilibrium potential of the metal and the temperature-induced changes of the corrosive environment.

Solution. The galvanic cell reaction is written; it is a redox reaction where Zn(s) is oxidized and Cu^{++} is reduced according to the following equation

$$Zn(s) + Cu^{++}(aq) \rightarrow Zn^{++}(aq) + Cu(s)$$

From eqns. (3.50) and (4.28), reaction enthalpy and reaction entropy at 25 °C are calculated

$$\Delta_r H^\ominus_{298} = (-153.9 + 0.0) - (0.0 + 64.8)\ \text{kJ/mol} = -218\,700\ \text{J/mol}$$
$$\Delta_r S^\ominus_{298} = (-112.1 + 33.2) - (41.6 - 99.6)\ \text{J/mol K} = -20.9\ \text{J/mol K}$$

The equilibrium potential $\Delta V^\ominus$ at 25 °C is determined from (6.38)

$$\Delta V^\ominus_{298} = -\left(\frac{\Delta_r H^\ominus_{298}}{z\mathcal{F}} - T \cdot \frac{\Delta_r S^\ominus_{298}}{z\mathcal{F}}\right) = \mathbf{1.1008\ volt}$$

The temperature coefficient of the equilibrium potential is determined by (6.42)

$$\frac{\partial(\Delta V^\ominus)}{\partial T} = \frac{\Delta_r S^\ominus_{298}}{z\mathcal{F}} = \frac{-\,20.9}{2 \cdot 96\,500} = -1.1 \cdot 10^{-4}\ \text{volt/K}$$

If the temperature coefficient is assumed to be *constant* in the temperature range considered, the equilibrium potential at 0 °C is determined from

$$\Delta V^\ominus_{273} = \Delta V^\ominus_{298} + \frac{\partial(\Delta V^\ominus)}{\partial T}\Delta T = 1.1008 - 1.1 \cdot 10^{-4} \cdot (298 - 273) = \mathbf{1.1053\ volt}$$

□ **1.** Specify the numerical value and unit for the quantity ($z\mathcal{F}$) for the following values of z: a) $z = 1$ b) $z = 2$, and c) $z = 3$!

□ **2.** For an ideal cell reaction, $\Delta_r G^\ominus_{298} = -150.0$ kJ/mol; determine the electrochemical equilibrium potential $\Delta V^\ominus$ (V) at 25 °C, when $z = 2$ for the reaction!

□ **3.** The electrochemical equilibrium potential is $\Delta V = +0.25$ V at 300 K; determine $\Delta_r G_{300}$ (kJ/mol) for the reaction when $z = 3$!

□ **4.** For an ideal cell reaction with $z = 2$, $\Delta_r S^\ominus_{298} = +12.5$ J/mol K; determine $(\partial V^\ominus/\partial T)_p$ for the cell reaction at 25 °C!

□ **5.** The equilibrium potential ΔV for a given cell reaction with $z = 2$ is 0.2135 V at 25 °C and 0.2152 V at 15 °C; determine $\Delta_r S_{298}$ for the reaction!

6.8 Notation rules

When describing electrochemical cell reactions it is expedient to use a standardized notation form; in the following, the generally applied formalism for setting up an **electrochemical cell diagram** is addressed.

Cell diagram

An electrochemical reaction *always* occurs in an interface between two phases. In a working galvanic cell, for example, **oxidation** occurs in the interface between anode and electrolyte, and **reduction** occurs in the interface between cathode and electrolyte. These interfaces – reaction surfaces – are symbolized by a vertical line " | " in a cell diagram. If, for example, we consider the anode reaction, eqn. (6.21), in the *Daniell* cell, it is given by the following cell diagram

$$\mathrm{Zn(s)\,|\,Zn^{++}(aq)} \qquad \text{(oxidation of Zn to Zn}^{++}\text{)} \tag{6.43}$$

Similarly, the concurrent cathode reaction, eqn. (6.20), is written as

$$\mathrm{Cu^{++}(aq)\,|\,Cu(s)} \qquad \text{(reduction of Cu}^{++}\text{ to Cu)} \tag{6.44}$$

A conductive connection between two electrolytes of different composition is designated by two vertical lines " || ". The electrode pair is *always* given in such a way that the electrons flow *from* left *to* right in the equation. According to these rules, therefore, the composed redox reaction in the *Daniell* cell is given by the following cell diagram

$$\underbrace{\mathrm{Zn(s)\,|\,Zn^{++}(aq)}}_{\text{anode reaction}} \;||\; \underbrace{\mathrm{Cu^{++}(aq)\,|\,Cu(s)}}_{\text{cathode reaction}} \tag{6.45}$$

In the electrochemical cell **electrons** are transferred from anode to cathode through a metallic conductor. Therefore, the anode reaction is always given to the left and the cathode reaction to the right in a reaction equation.

Electrochemical redox reaction

At reduction, electrons are gained

example : $Cu^{++} + 2e^- \rightarrow Cu$

At oxidation, electrons are lost

example : $Fe \rightarrow Fe^{++} + 2e^-$

Cell reaction

On the other hand, when we come across an electrochemical cell diagram set up in this form, it is "read" in the following way illustrated by the *Daniell* cell mentioned above.

Cell reaction: active/active electrode (6.46)

For a galvanic cell diagram with **active/active** electrode, the cell reaction is to be read in the following way

$$\overbrace{\underbrace{\mathrm{Zn(s)\,|\,Zn^{++}(aq)}}_{\mathrm{Zn\rightarrow Zn^{++}+2e^-}}}^{\text{anode reaction}} \;||\; \overbrace{\underbrace{\mathrm{Cu^{++}(aq)\,|\,Cu(s)}}_{\mathrm{Cu^{++}+2e^-\rightarrow Cu}}}^{\text{cathode reaction}}$$

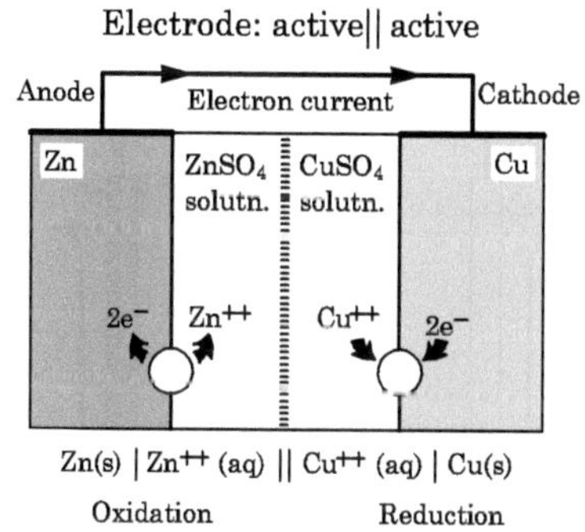

Figure 6.25. Physical significance of a galvanic cell for a cell with an active Zn-anode and an active Cu cathode. It shall be noted that electrochemical cell reactions are always connected with the interface between electrode and electrolyte.

In the example mentioned, the electrode metal takes an *active* part in the electrochemical reaction: Zn(s) is oxidized at the anode and Cu(s) is precipitated on the cathode. Later, we shall deal with *passive* electrodes that only function as catalysts for the electrochemical redox reactions. In an electrolyte that contains dissolved hydrogen, $H_2(aq)$ is, for example, readily oxidized on the surface of the platinum electrode by the following reaction

$$\mathrm{H_2(aq)} \;\rightarrow\; \mathrm{2H^+(aq) + 2e^-} \tag{6.47}$$

The electrons thus liberated are lost to the metallic platinum electrode. A **passive electrode** that is not included in the electrochemical reaction is shown in parenthesis, for example, (Pt), (Fe) etc. Oxidation of dissolved hydrogen $H_2(aq)$ on the surface of a passive platinum electrode is thus given by the following cell diagram

$$\mathrm{(Pt), H_2(aq)\,|\,2H^+(aq)} \qquad \text{("Hydrogen electrode")} \tag{6.48}$$

If, for example, we replace the zinc electrode in the *Daniell* cell by a hydrogen electrode, the cell diagram is read in the following way

Cell reaction: passive/active electrode (6.49)

For a galvanic cell diagram with **passive/active** electrode, the cell reaction is read in the following way

$$\mathrm{(Pt),}\overbrace{\underbrace{\mathrm{H_2(aq)\,|\,2H^+(aq)}}_{\mathrm{H_2\rightarrow 2H^+ +2e^-}}}^{\text{anode reaction}} \;||\; \overbrace{\underbrace{\mathrm{Cu^{++}(aq)\,|\,Cu(s)}}_{\mathrm{Cu^{++}+2e^-\rightarrow Cu}}}^{\text{cathode reaction}}$$

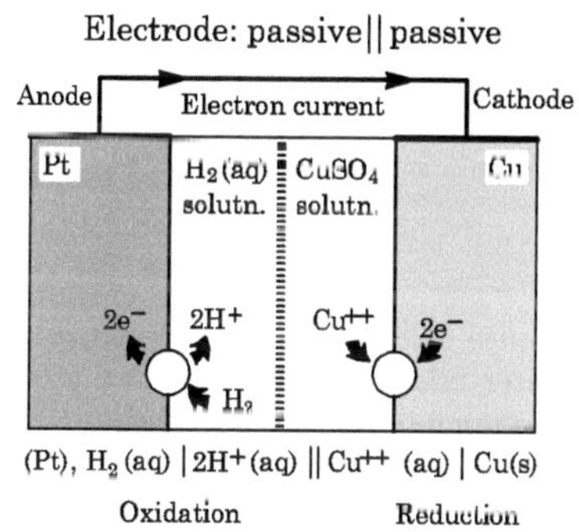

Figure 6.26. Physical significance of galvanic cell diagram for a cell with a passive Pt anode and an active Cu cathode. At the anode, dissolved hydrogen is oxidized into hydrogen ions and the lost electrons are gained by the passive Pt electrode.

In an *adjusted* cell diagram the number of electrons formed by the anode process will be equal to the number gained in the cathode process.

An electrochemical cell diagram is conventionally arranged in such a way that electrons flow *from* left *to* right in the equation. Comparing this with eqn. (6.30), we obtain the following sign convention for a **galvanic cell** and an **electrolytic cell**, respectively.

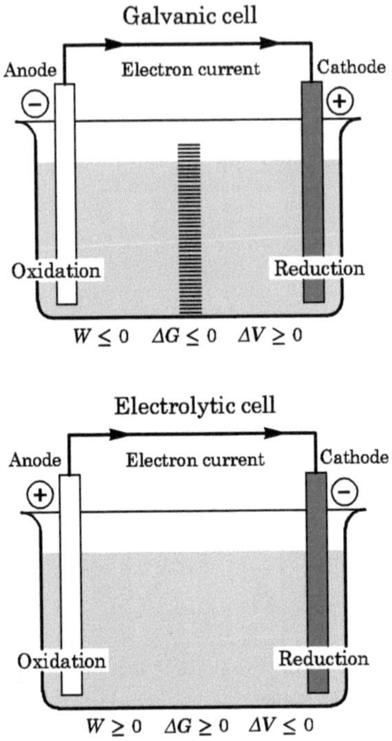

Figure 6.27. Sign convention for electrolytic and galvanic cell reaction. Note in particular the signs for W, ΔG, and ΔV.

Sign convention for cell reactions (6.50)

Galvanic cell:	$\delta W_{el} \leq 0;\ \Delta G \leq 0;\ \Delta V \geq 0$	
Electrodes:	*anode* ‖ *cathode*	
Processes:	*oxidation*	*reduction*
Potentials:	$V(anode) \leq V(cathode)$	
Electrolytic cell:	$\delta W_{el} \geq 0;\ \Delta G \geq 0;\ \Delta V \leq 0$	
Electrodes:	*anode* ‖ *cathode*	
Processes:	*oxidation*	*reduction*
Potentials:	$V(anode) \geq V(cathode)$	

During a cell reaction the electron flow in the external conductor is *always* counteracted by a concurrent transfer of ions through the cell electrolyte.

■ In weather-exposed steel structures subjected to permanent moisture and salt actions, so-called **crevice corrosion** is a general phenomenon. Crevice corrosion especially occurs in water-filled cracks at joints; the corrosion is caused by differences in dissolved, atmospheric oxygen $O_2(aq)$ in the water. The following two electrochemical reactions are known to be included in the corrosion process

$$Fe(s) \rightarrow Fe^{++}(aq) + 2e^- \quad (a)$$

$$\tfrac{1}{2}O_2(aq) + H_2O + 2e^- \rightarrow 2OH^-(aq) \quad (b)$$

The liberated ions Fe^{++} and OH^- precipitate as **ferrous hydroxide** $Fe(OH)_2$, which may oxidize further to **ferric hydroxide** $Fe(OH)_3$. Precipitation of these hydroxides is seen as the well-known rust formation.

Write the galvanic cell diagram for crevice corrosion and indicate the area that is decomposed during the process!

Solution. In process (a), Fe is oxidized to Fe^{++} by losing 2 electrons; this is an **anode process** occurring on an active electrode of iron Fe. In process (b), dissolved oxygen $\frac{1}{2}O_2$ is reduced to O^{--} when 2 electrons are gained; this is a **cathode process**, occurring on a passive electrode of iron (Fe). With the notation introduced, we thus obtain the following galvanic cell diagram

$$Fe(s)\,|\,Fe^{++}(aq) \parallel \tfrac{1}{2}O_2(aq), H_2O\,|\,2OH^-(aq), (Fe)$$

The cathode reaction (b) is promoted by a high oxygen content in the water; the cathode area is therefore formed at the opening of the crack where there is unrestricted access of O_2 from the air. The anode reaction (a) is independent of the oxygen content of the water; therefore, the anode area is formed in the inner crack. The iron is decomposed at the anode, i.e. inside the crack.

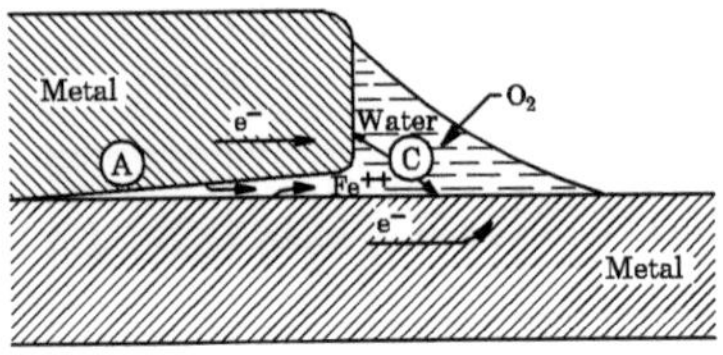

Figure 6.28. Crevice corrosion in a water-filled joint between two steel specimens (shown schematically). At the cathode, dissolved oxygen O_2 is reduced and at the anode, iron Fe is oxidized. The cathode (C) is created at the opening of the crack where there is unrestricted access of oxygen from the ambient atmosphere. The iron is decomposed in the anode area (A) that is created in the inner crack which is poor in oxygen.

☐ **1.** Determine the cell reaction for an anode process that includes: a) Na(s) and $Na^+(aq)$ b) Cl(aq) and $Cl^-(aq)$ c) $Cu^+(aq)$ and $Cu^{++}(aq)$!

☐ **2.** Determine the cell reaction for a cathode process that includes: a) Ag(s) and $Ag^+(aq)$ b) $Fe^{++}(aq)$ and $Fe^{+++}(aq)$ c) $Cu^+(aq)$ and $Cu^{++}(aq)$!

☐ **3.** Determine the cell diagram for an anode process that includes: a) Ca(s) and $Ca^{++}(aq)$ b) $H_2(aq)$, $H^+(aq)$ and (Pt) c) $Sn^{++}(aq)$ and Sn(s)!

☐ **4.** Determine the cell diagram for a cathode process that includes: a) Fe(s) and $Fe^{+++}(aq)$ b) $O_2(aq)$, H_2O, $OH^-(aq)$ and (Pt) c) $Sn^{++}(aq)$, $Sn^{++++}(aq)$ and (Pt)!

☐ **5.** Determine the cell diagram for galvanic decomposition of a sacrificial anode of Zn(s) which is in conductive contact with Fe(s) in a wet environment that includes $O_2(aq)$!

6.9 Standard potential

The electrochemical potential difference ΔV between two electrodes in a galvanic cell can be determined by experiments; however, it is *not* possible to *measure* the absolute potential V of a **single electrode** ("half-cell"). In many calculation problems, however, it is expedient to be able to assign an **absolute** electrochemical potential to a single electrode. Conventionally, therefore, an arbitrary zero in the electrochemical series has been established by the following **definition**

Standard hydrogen electrode (SHE) (6.51)

The electrochemical **equilibrium potential** for the hydrogen electrode at the **standard state**: $p^{\ominus} = 101325\,\mathrm{Pa}$; $c^{\ominus} = 1\,\mathrm{mol}/\ell$, and $T = 298.15\,\mathrm{K}$

$$H_2(g, p^{\ominus}) \leftrightarrows 2H^+(aq, c^{\ominus}) + 2e^-$$

is given the value of $V^{\ominus} = 0.0000$ volt by **definition.**

Compared to the standard hydrogen electrode (SHE), the **absolute** electrochemical potential of other **single electrodes** can now be determined by measurement.

Standard hydrogen electrode

The standard hydrogen electrode (SHE) consists of a *passive* platinum electrode (Pt) embedded in a 1 molar HCl solution; pure hydrogen H_2(g) bubbles around the Pt electrode so that the equilibrium, eqn. (6.51), adjusts itself at the interface between electrode and electrolyte. Thus, it is ensured that the Pt electrode is in equilibrium at a partial pressure of H_2(g) of $p^{\circ} = 101325\,\mathrm{Pa}$. By measurement with (SHE) against another electrode the two electrolytes are connected by a conductive "salt bridge" consisting of a U-pipe filled with a gelled aqueous salt solution. The salt bridge allows passage of ions without mixing of the electrolyte solutions used.

For practical calculations, the **standard potential** $V^{\ominus}$ has been measured and tabulated for a considerable number of electrode reactions (see the tables "Electrochemical standard potential" in Appendix B). Standard potentials specify the electrochemical equilibrium potential $V^{\ominus}$ measured against the standard hydrogen electrode (SHE), when all components are in their standard state.

According to eqn. (6.50), electrodes with a *positive* standard potential $V^{\ominus} > 0$ will form a **cathode** in relation to the standard hydrogen electrode (SHE); on the other hand, electrodes with a *negative* standard potential $V^{\ominus} < 0$ will form an **anode** in relation to (SHE).

Measurement of standard potential

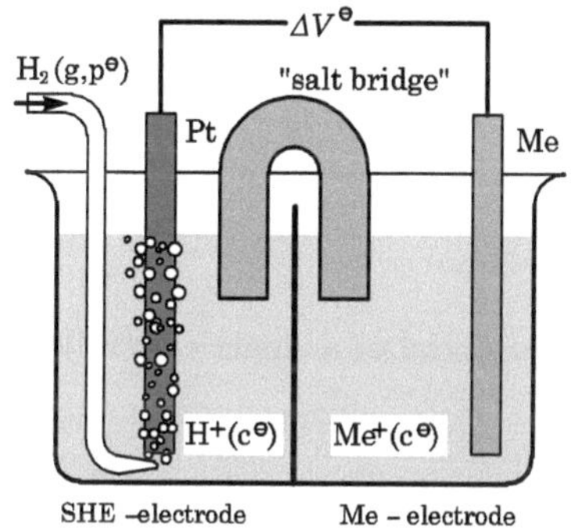

Figure 6.29. Set-up for measurement of standard potential $V^{\ominus}$ (shown schematically). The reference electrode is a standard hydrogen electrode (SHE) which by definition has been given the potential of 0 volt. A "salt bridge" formed by gelled aqueous salt solution ensures conductive contact between the two electrolytes.

Potential of single electrode

Through eqn. (6.51), the standard hydrogen electrode (SHE) defines an arbitrary zero in the electrochemical series. In relation to (SHE) any other single electrode can be given an **absolute** electrochemical equilibrium potential. In the following it is shown how the equilibrium potential in a simple way is connected with ΔG for the electrode reaction.

Initially, it is noted that by definition, the standard free energy $G^{\ominus}_{298}$ for $H^+(aq, c^{\ominus})$ and for $H_2(g, p^{\ominus})$ is given the value 0 kJ/mol; for the standard hydrogen electrode, therefore, the following condition applies

$$V^{\ominus} = 0 \text{ volt}; \qquad \Delta_r G^{\ominus}_{298} = 0 \text{ kJ/mol} \tag{6.52}$$

According to the Nernst equation (6.39), this entails that the **equilibrium potential** V for a given electrode is *solely* determined by ΔG for the reaction of this electrode. Assume, for example, an electrode pair in which the standard hydrogen electrode is included as **anode**; from eqns. (6.39) and (6.52) we then obtain the following

$$\Delta V = V - 0 = -\frac{\Delta_r G_{cathode} - 0}{z\mathcal{F}} = -\frac{\Sigma G_{red} - \Sigma G_{ox}}{z\mathcal{F}} \tag{6.53}$$

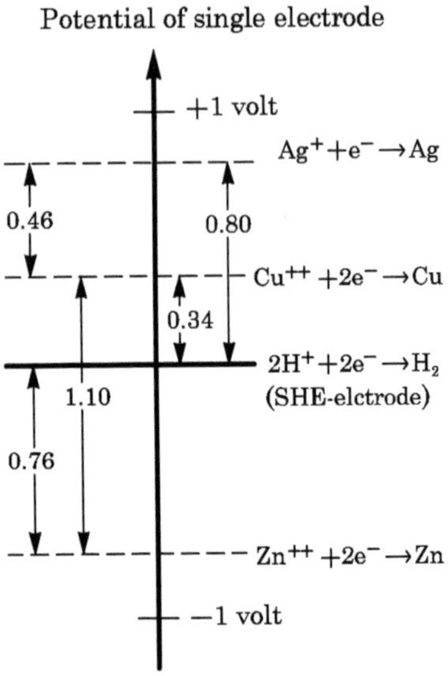

Figure 6.30. An arbitrary zero in the electrochemical series is defined by the standard hydrogen electrode (SHE). Other single electrodes can thus be given an absolute, electrochemical potential V.

where (*red*) and (*ox*) denote the components of the cathode reaction in the *reduced* and the *oxidized* state. If, on the other hand, the standard hydrogen electrode is included as **cathode** in an electrode pair, we obtain from eqns. (6.39) and (6.52)

$$\Delta V = 0 - V = -\frac{0 - \Delta_r G_{anode}}{z\mathcal{F}} = -\frac{\Sigma G_{red} - \Sigma G_{ox}}{z\mathcal{F}} \tag{6.54}$$

where (*red*) and (*ox*) now denote the components of the anode reaction in a *reduced* and an *oxidized* state, respectively. Therefore, from thermodynamic standard values, we can in a simple way calculate the electrochemical **equilibrium potential** V for a given single electrode. Introducing the definition (5.29): $G_i = G_i^\ominus + RT \cdot \ln(a_i)$ into the expression for V, we obtain the following expression for determination of the potential of the single electrode

Potential of single electrode (6.55)

For the electrode reaction $aA(ox) + ze^- \rightarrow \alpha A(red)$, the standard potential $V^\ominus$ and the electrochemical equilibrium potential V are determined by

$$\text{Standard potential: } V^\ominus = -\frac{\Sigma G^\ominus_{red} - \Sigma G^\ominus_{ox}}{z\mathcal{F}} = -\frac{\Delta_r G^\ominus}{z\mathcal{F}}$$

$$\text{Equilibrium potential: } V = V^\ominus - \frac{RT}{z\mathcal{F}} \cdot \ln(K_a)$$

where (*red*), (*ox*) denote reaction components in a **reduced** and an **oxidized** state, respectively, and K_a is the **thermodynamic equilibrium constant** for the reaction written as a **reduction process** (cathode reaction).

It is noted that the equilibrium potential for a single electrode refers to a matched electrochemical reaction during which the standard hydrogen electrode (SHE) constitutes the other electrode.

Potential of a single electrode

The electrochemical reduction potential of a single electrode against a standard hydrogen electrode

$$V = V^\ominus - \frac{RT}{z\mathcal{F}} \cdot \ln(K_a)$$

When calculating the potential of single electrodes according to eqn. (6.55) it is important that we obey the sign convention used; the following procedure will ensure this.

- state the actual cell reaction as a **reduction process** (cathode reaction), so that oxidized components are on the left-hand side and reduced components are on the right-hand side in the reaction equation
- calculate $\Delta_r G^\ominus$ and K_a in the usual way for the reaction; in this calculation, *all substance components* including those that do *not* change their oxidation state during the reaction must be included. Electrons are *not* included in the calculation of $\Delta_r G^\ominus$ and K_a
- finally, by introduction into eqn. (6.55) the targeted potentials will be found.

■ The following table values for thermodynamic standard values $G^\ominus_{298}$ (kJ/mol) are given:

$Zn(s)$	$Zn^{++}(aq)$	$O_2(g)$	$H_2O(\ell)$	$OH^-(aq)$
0.0	−147.0	0.0	−237.2	−157.3

Calculate from these data the **equilibrium potential** V for the electrode reaction:

$$Zn^{++}(aq) + 2e^- \rightarrow Zn(s) \tag{a}$$

for $[Zn^{++}] = 10^{-5}$ mol/ℓ at 25 °C, and calculate at the same temperature, the **standard potential** $V^\ominus$ for the electrode reaction

$$O_2(g) + 2H_2O(\ell) + 4e^- \rightarrow 4OH^-(aq) \tag{b}$$

Solution. It shall be noted that (a) and (b) are reduction processes, cf. eqn. (6.17). Therefore, the left-hand side contains **oxidized** components and the right-hand side contains **reduced** components. An ideal solution with the ion activity $a = c/c^{\ominus}$ is assumed; the **standard potential** $V^{\ominus}$ at 25 °C for the electrode reaction (a) is determined by eqn. (6.55)

$$V^{\ominus} = -\frac{G^{\ominus}(\mathrm{Zn}) - G^{\ominus}(\mathrm{Zn^{++}})}{z\mathcal{F}} = -\frac{0-(-147\,000)}{2\cdot 96\,500} = -0.76\,\mathrm{V}$$

Thus, the equilibrium potential at the actual concentration of zinc ions is

$$V = V^{\ominus} - \frac{RT}{z\mathcal{F}}\cdot \ln(K_a) = -0.76 - \frac{8.314\cdot 298.15}{2\cdot 96\,500}\cdot \ln\left(\frac{1}{10^{-5}}\right) = \mathbf{-0.91\,volt}$$

Correspondingly, the **standard potential** $V^{\ominus}$ for (b) is determined by eqn. (6.55)

$$V^{\ominus} = -\frac{4\cdot G^{\ominus}(\mathrm{OH^-}) - (G^{\ominus}(\mathrm{O_2(g)}) + 2\cdot G^{\ominus}(\mathrm{H_2O}(\ell)))}{z\mathcal{F}}$$

$$V^{\ominus} = -\frac{4\cdot(-157\,300) - (0 + 2\cdot(-237\,200))}{4\cdot 96\,500} = \mathbf{+0.40\,volt}$$

☐ **1.** For the electrode reaction $Cu(s) \rightarrow Cu^{++}(aq) + 2e^-$, $V^{\ominus} = +0.34$ volt. Will Cu(s) form anode or cathode against (SHE)?

☐ **2.** Calculate from thermodynamic data the standard potential $V^{\ominus}$ for the single electrode reaction $Cu(s) \rightarrow Cu^{+}(aq) + e^-$ at 25 °C!

☐ **3.** $V^{\ominus} = -2.34$ V for the reaction $Mg^{++}(aq) + 2e^- \rightarrow Mg(s)$, and $V^{\ominus} = +0.80$ V for $Ag^{+}(aq) + e^- \rightarrow Ag(s)$; calculate $V^{\ominus}$ for $Mg(s)\,|\,Mg^{++}\,\|\,2Ag^{+}(aq)\,|\,2Ag(s)$!

☐ **4.** Calculate the equilibrium potential for $Fe^{++}(aq) + 2e^- \rightarrow Fe(s)$ at 25 °C, if $[Fe^{++}] = 10^{-6}$ mol/ℓ and an ideal solution is assumed!

☐ **5.** Calculate from thermodynamic data the standard potential $\Delta V^{\ominus}$ at 25 °C for the cell reaction $Fe(s)\,|\,Fe^{++}(aq)\,\|\,2Ag^{+}(aq)\,|\,2Ag(s)$!

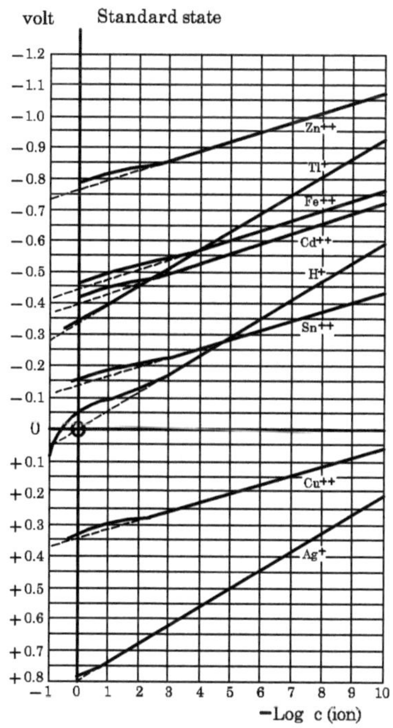

Figure 6.31. Equilibrium potential for single electrodes as a function of the ion concentration c. The standard state for an ideal solution is log(c) = 0 corresponding to 1-molar solution. The full lines signify measured potentials; the broken lines signify measured potentials of ideal solutions (cf. Ewing: "Instrumental Methods of Chemical Analysis", McGraw-Hill, New York 1969).

The electrochemical series

Mg^{++}/Mg	− 2.37 volt
Al^{+++}/Al	− 1.66 volt
Zn^{++}/Zn	− 0.76 volt
Fe^{++}/Fe	− 0.44 volt
Ni^{++}/Ni	− 0.23 volt
Sn^{++}/Sn	− 0.14 volt
Pb^{++}/Pb	− 0.13 volt
$2H^{+}/H_2$ (SHE)	0.00 volt
Cu^{++}/Cu	+ 0.34 volt
Ag^{+}/Ag	+ 0.80 volt
Au^{+++}/Au	+ 1.69 volt

Standard potentials $V^{\ominus}$, 25 °C

6.10 Passivation

Electrochemical redox reactions are characterized by changing of the oxidation state. As examples of electrochemical oxidation and reduction, we have previously looked into the following processes:

Anode process :	$Zn(s) \rightarrow Zn^{++}(aq) + 2e^-$	(oxidation)
Cathode process :	$Cu^{++}(aq) + 2e^- \rightarrow Cu(s)$	(reduction)

In both cases, the substances are in their oxidized state as dissolved ions in an electrolyte; in their reduced form, the substances at the same time constitute the parent metal in an electrode. Evidently, these kinds of processes can develop unhindered as long as electrodes and electrolyte remain unchanged.

Anode reaction with passivation

During other electrode reactions, however, sparingly soluble *non*-metallic oxides are precipitated. If the precipitation forms a *dense* coating it may result in **passivation** of the electrode. An important example of this phenomenon is passivation of iron by electrochemical precipitation of iron oxide Fe_2O_3 ("hematite") or Fe_3O_4 ("magnetite").

Assume an iron electrode, Fe(s), in equilibrium with an electrolyte at pH = 7; the electrolyte is assumed to have a constant concentration of ferrous ions $Fe^{++}(aq)$ of $1\cdot 10^{-6}$ mol/ℓ. In this state, according to eqn. (6.55) the electrochemical equilibrium potential of the electrode is $V \simeq -0.60$ volt at 25°C. We now impose a slowly increasing electric potential $V \geq -0.60$ volt, on the electrode and at the same time we measure the current I through the electrode. Due to the current, the following **anode reaction** with oxidation of Fe to Fe^{++} develops

$$Fe(s) \rightarrow Fe^{++}(aq) + 2e^- \qquad (6.56)$$

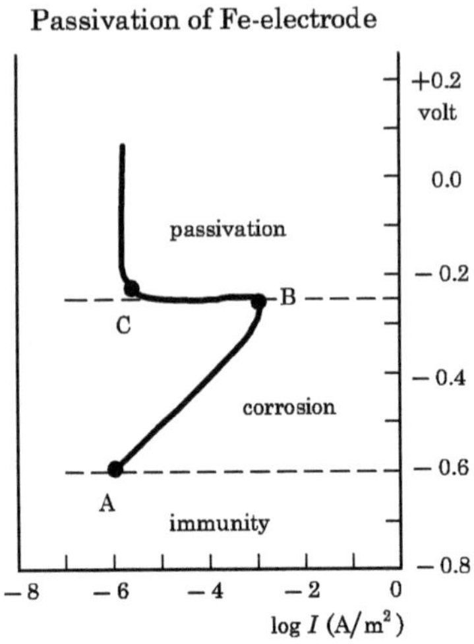

Figure 6.32. Fe electrode in water at pH = 7 and [Fe++] = 10−6 mol/ℓ. The equilibrium potential is −0.60 volt (A). When an increased potential A–B is imposed on the electrode, the current density is increased until the potential reaches approximately −0.25 volt. At this potential, passivation B–C is obtained and the current density decreases abruptly to almost 0. The passivation is due to precipitation of a dense oxide layer on the electrode (see example 6.2).

Figure 6.33. Iron is a base metal that decomposes by electrochemical corrosion in a humid environment. If iron (or steel) is embedded in concrete which is strongly alkaline, however, it forms a passivating surface layer of dense iron oxides. This passivation is decisive of the corrosion resistance of reinforcing bars.

From experiments it is seen that the amperage is increased with growing potential until $V \simeq -0.25$ volt; if this potential is exceeded, the electrode is passivated and the amperage decreases abruptly to almost zero. At the potential −0.25 volt, a dense, covering oxide layer of Fe_2O_3 is precipitated on the surface of the electrode. The oxide layer, which has a thickness of approximately 30 Å, prevents transfer of liberated electrons to the electrode. The oxide layer is produced by the following anode reaction

$$2Fe^{++}(aq) + 3H_2O(\ell) \rightarrow Fe_2O_3(s) + 6H^+(aq) + 2e^- \tag{6.57}$$

By the anode reaction, eqn. (6.57), ferrous ions Fe^{++} are oxidized to ferric ions Fe^{+++} that are included in the oxide Fe_2O_3 formed.

It is important to note the following: The passivated state is characterized by an extremely slow electrode reaction; however, the system is not in thermodynamic equilibrium with the original electrode metal Fe(s). The electrochemical equilibrium potential for the passivated electrode can be calculated from eqn. (6.55) in the normal way assuming the anode reaction to be the oxide producing reaction, see eqn. (6.57).

Reinforcement corrosion

In practice, electrochemical passivation often determines the corrosion resistance of base metals. Passivation of steel by the reaction in eqn. (6.57), for example, is decisive of the durability of reinforcement bars in concrete. As will be shown later in a calculation example, the strongly alkaline environment in concrete with pH ≃ 12.5 furthers the formation of a passivating oxide layer of Fe_2O_3 on the embedded reinforcement.

■ The following table values for thermodynamic standard values $G^\ominus_{298}$ (kJ/mol) are given:

Fe(s)	Fe_2O_3(s)	Fe^{++}(aq)	$H_2O(\ell)$	H^+(aq)
0.0	−742.3	−78.9	−237.2	0.0

By electrochemical corrosion of iron Fe(s), the iron is usually decomposed due to the anode reaction given in eqn. (6.56). For **passivated** iron with oxide coating of Fe_2O_3, the anode reaction is assumed to correspond to the reaction equation (6.57). Based on thermodynamic standard values for $G^\ominus_{298}$, determine the electrochemical potential for the electrode reactions, eqns. (6.56) and (6.57), at 25 °C. The following assumptions are made: pH = 7.0, $[Fe^{++}(aq)] = 1 \cdot 10^{-6}$ mol/ℓ; ideal solutions: $a = c/c^\ominus$.

Solution. It is noted that eqns. (6.56) and (6.57) are oxidation processes (anode processes); the left-hand side denotes components in a reduced state and the right-hand side denotes components in an oxidized state. Both of the reaction equations can be reversed and written as reduction processes (cathode processes) as follows:

$$Fe^{++}(aq) + 2e^- \rightarrow Fe(s) \tag{a}$$

$$Fe_2O_3(s) + 6H^+(aq) + 2e^- \rightarrow 2Fe^{++}(aq) + 3H_2O(\ell) \tag{b}$$

Next, we can proceed to determine the equilibrium potential V for the electrode reaction (a) utilizing eqn. (6.55)

$$V^\ominus = -\frac{G^\ominus(\text{Fe}) - G^\ominus(\text{Fe}^{++})}{z\mathcal{F}} = -\frac{0-(-78\,900)}{2 \cdot 96\,500} = -0.41 \text{ volt}$$

$$V = V^\ominus - \frac{RT}{z\mathcal{F}} \cdot \ln(K_a) = -0.41 - \frac{8.314 \cdot 298.15}{2 \cdot 96\,500} \cdot \ln\left(\frac{1}{1 \cdot 10^{-6}}\right) = \mathbf{-0.59\ volt}$$

Correspondingly, the equilibrium potential V for the electrode reaction (b) can be determined utilizing eqn. (6.55)

$$V^\ominus = -\frac{2 \cdot G^\ominus(\text{Fe}^{++}) + 3 \cdot G^\ominus(\text{H}_2\text{O}) - (G^\ominus(\text{Fe}_2\text{O}_3) + 6 \cdot G^\ominus(\text{H}^+))}{z\mathcal{F}}$$

$$V^{\ominus} = -\frac{2 \cdot (-78\,900) + 3 \cdot (-237\,200) - (-742\,300 + 6 \cdot 0)}{2 \cdot 96\,500} = +0.66 \text{ volt}$$

$$K_a = \frac{a(\text{Fe}^{++})^2 \cdot a(\text{H}_2\text{O})^3}{a(\text{Fe}_2\text{O}_3) \cdot a(\text{H}^+)^6} = \frac{(10^{-6})^2 \cdot 1}{1 \cdot (10^{-7})^6} = 1 \cdot 10^{30}$$

$$V = V^{\ominus} - \frac{RT}{z\mathcal{F}} \cdot \ln(K_a) = +0.66 - \frac{8.314 \cdot 298.15}{2 \cdot 96\,500} \cdot \ln(1 \cdot 10^{30}) = \mathbf{-0.23\ volt}$$

Therefore, imposing a potential of $V \geq -0.59$ volt on the Fe electrode, then

$-0.59 \text{ volt} \leq \text{V} \leq -0.23 \text{ volt}$	**active** Fe electrode cf. (6.56)
$-0.23 \text{ volt} \leq \text{V}$	**passive** Fe electrode cf. (6.57)

☐ **1.** Calculate the electrochemical standard potential $V^{\ominus}$ (V) at 25 °C for the following cell reaction: $Al^{+++}(aq) + 3e^- \rightarrow Al(s)$!

☐ **2.** Calculate the electrochemical standard potential $V^{\ominus}$ (V) at 25 °C for the following cell reaction: $Ag(aq)^+ + e^- \rightarrow Ag(s)$!

☐ **3.** Calculate the electrochemical potential V (V) at 25 °C for the following cell reaction: $Zn^{++}(aq) + 2e^- \rightarrow Zn(s)$ when $[Zn^{++}] = 10^{-2}$ mol/ℓ in an ideal solution !

☐ **4.** Calculate the electrochemical potential V for the reaction in eqn. (6.57) at 25 °C, if $\text{pH} = -\log_{10}(a(\text{H}^+)) = 12.5$ and $[Fe^{++}] = 10^{-6}$ mol/ℓ (ideal solution) !

☐ **5.** Why are the two electrolytes in a galvanic cell connected by a "salt bridge" of stiffened gel and not by a conductive metal rod ?

Definition of pH

$$pH \stackrel{\text{def}}{=} -\log_{10}(a(\text{H}^+))$$

List of key ideas

The following overview lists the key definitions, concepts and terms introduced in Chapter 6.

Electric current I .... Symbol: I; unit: **ampere** (A); def.: (6.1)

Electric charge Q .... Symbol: Q; unit: **coulomb** (C); def.: (6.2)
1 coulomb is the **electric charge** transferred by a **current** of 1 A in 1 s, i.e. (C) = (A s).
$dQ = I(t) \cdot dt; \quad Q = \int I(t)\, dt$

The Faraday constant $\mathcal{F}$ ... $\mathcal{F} = \mathcal{N} \cdot e$ (C/mol); denotes the electric charge of 1 mol of **elementary charges**. $\mathcal{F} \simeq 96\,500$ C/mol

Electric field strength E.... $\mathbf{E} = \mathbf{F}/Q$; unit: (N/C) = (V/m) The electric field strength denotes the relation between the **force** **F** acting on a **point charge** Q and the magnitude of this charge, calculated with signs. Force **F** and field strength **E** are vectors.

Electric potential V .... $V_a = \delta W_{\infty.a}/dQ$; unit: (J/C) = (V) The electric potential denotes the **external work** on a charge unit taken **reversibly** from infinity to a point a; Q is calculated with signs.

Potential difference ΔV ... $\Delta V_{a.b} = V_b - V_a = \delta W_{a.b}/dQ$; unit: (V) The electric potential difference $\Delta V_{a.b}$ specifies the **external work** on a charge unit that is moved **reversibly** from point a to point b; Q is calculated with signs.

Electrical work δW .... $\delta W_{a.b} = (V_b - V_a) \cdot dQ$ (J); $\Delta W_{a.b}$ is the **external work** on the charge dQ moved **reversibly** from point a to point b; Q is calculated with signs.

Conductivity σ .... Symbol: σ; unit: $(\Omega^{-1}\text{m}^{-1})$ cf.: eqn. (6.15)
Conductivity σ is a **substance property**; the conductivity is a measure of the electric conductivity of a homogeneous substance ("specific conductance").

$\sigma = \rho^{-1}$; $\Delta V = \sigma \cdot (A/L) \cdot \Delta V$

Resistivity ρ .... Symbol: ρ; unit: $(\Omega\,\text{m})$ cf.: eqn. (6.16)
Resistivity ρ is a **substance property**; the resistivity is a measure of the electric resistance in a homogeneous substance ("specific resistance").

$\rho = \sigma^{-1}$; $\Delta V = \rho \cdot (L/A) \cdot I$

Conductance G .... Symbol: G; unit: (Ω^{-1}) cf: eqn. (6.15)
The conductance G specifies the electric conductivity of a system, for example, a metal conductor of a given cross-section and length; G depends on system geometry *as well as* the properties of the conducting substance.

$G = R^{-1}$; $I = G \cdot \Delta V$

Resistance R .... Symbol: R; unit: (Ω) cf.: eqn. (6.16)
The resistance R specifies the electric resistance in a system, for example, a metal conductor of a given cross-section and length; R depends on system geometry *as well as* the properties of the conducting substance.

$R = G^{-1}$; $\Delta V = R \cdot I$

Cation .... **Positive** ion; example: $H^+(aq)$; $Ca^{++}(aq)$

Anion .... **Negative** ion; example: $OH^-(aq)$; $SO^{--}(aq)$

Reduction .... A process during which a substance **gains** electrons and thus is transformed to a lower oxidation state.

Example: $Zn^{++} + 2e^- \rightarrow Zn$; $Cl_2 + 2e^- \rightarrow 2Cl^-$

Oxidation .... A process during which a substance **loses** electrons and thus is transformed to a higher oxidation state.

Example: $Ag \rightarrow Ag^+ + e^-$; $2O^{--} \rightarrow O_2 + 4e^-$

Cathode reaction .... Electrode reaction during which a substance is **reduced** when gaining electrons from the electrode.

Anode reaction .... Electrode reaction during which a substance is **oxidized** when losing electrons to the electrode.

The Nernst equation ...... Expresses the **equilibrium potential** ΔV for an electrode pair in a **galvanic cell**.

$\Delta V = -(\Delta_r G^\ominus_T + RT \cdot \ln(K_a))/z\mathcal{F}$ (volt)

Hydrogen electrode .... By **definition**, the standard hydrogen electrode (SHE) is given the equilibrium potential $V^\ominus$ = 0.0000 volt by the reaction

$H_2(g, p^\ominus) \leftrightarrows 2H^+(aq, c^\ominus) + 2e^-$ (298.15 K)

Single electrode .... The electrochemical potential of a single electrode against the standard hydrogen electrode

$$V = -((\Sigma G^{\ominus}_{red} - \Sigma G^{\ominus}_{ox}) + RT\ \ln(K_a))/z\mathcal{F}\ \text{(volt)}$$

Examples

The following examples illustrate how subject matter explained in Chapter 6 can be applied in practical calculations.

Example 6.1

■ Oxidation of metals in water – pH dependence

The **corrosion rate** is an important factor in estimations of the danger of corrosion processes. A *first* estimate of the risk of corrosion attack can in many cases be made by investigating the electrochemical **potential differences** in a system. The following example shows the procedure for producing such estimates.

It is emphasized that the *extent* of corrosion *cannot* be estimated on this basis only. As shown in the following example, the corrosion products formed can **passivate** the attacked surface and thus suppress further corrosion (see section 6.10 *Passivation*).

Figure 6.34. Iron is a base metal with a negative standard reduction potential in the electrochemical series. Therefore, iron often forms the anode in corrosion cells and thus corrodes due to galvanic action. Iron structural elements that are exposed to moisture and air can decompose in a short time due to galvanic corrosion. The photo shows a major corrosion hole in a cantilevered beam from a balcony structure.

Problem. Investigate within what pH range pure metals such as **gold** Au, **copper** Cu, **iron** Fe, and **zinc** Zn can be oxidized at a temperature of 25 °C by the following **anode processes**

(a) $Au(s) \rightarrow Au^{+++}(aq) + 3e^-$ (b) $Cu(s) \rightarrow Cu^{++}(aq) + 2e^-$

(c) $Fe(s) \rightarrow Fe^{++}(aq) + 2e^-$ (d) $Zn(s) \rightarrow Zn^{++}(aq) + 2e^-$

against the following **cathode processes**

(1) $2H^+(aq) + 2e^- \rightarrow H_2(g, p^{\ominus})$ (reduction of H^+)

(2) $O_2(g, p^{\ominus}) + 4H^+(aq) + 4e^- \rightarrow 2H_2O(\ell)$ (acid reduction of O_2)

$O_2(g, p^{\ominus}) + 2H_2O(\ell) + 4e^- \rightarrow 4OH^-(aq)$ (alkaline reduction of O_2)

Conditions. By experience, actual corrosion can only develop at electrochemical potentials that build up a concentration of metal ions at the anode surface of at least $1 \cdot 10^{-6}$ mol/ℓ. In the subsequent calculations, therefore, it is assumed that metal ions in the electrolyte solution have a molar concentration of $c = 1 \cdot 10^{-6}$ mol/ℓ. Ideal solutions are assumed so that ion activities can be expressed as $a = c/c^{\ominus}$.

Solution. Data for the standard free energy of the substances $G^{\ominus}_{298}$ (kJ/mol) can be obtained from tables; the following table values are used for the calculations

Au(s)	0.0;	Au^{+++}(aq)	433.7;	Cu(s)	0.0;	Cu^{++}(aq)	65.5;
Fe(s)	0.0;	Fe^{++}(aq)	−78.9;	Zn(s)	0.0;	Zn^{++}(aq)	−147.0;
H_2(g)	0.0;	H^+(aq)	0.0;	O_2(g)	0.0;	OH^-(aq)	−157.3;
$H_2O(\ell)$	−237.2;						

OXIDATION OF METAL DURING HYDROGEN DEVELOPMENT

Reduction potentials for single electrodes (a)...(d) are calculated using eqn. (6.55). Since all of the reactions stated are oxidation processes, they will be inverted before calculating ΔG and K_a. The specified limit value $c = 1 \cdot 10^{-6}$ mol/ℓ is used as metal ion concentration, and ideality is assumed. By introduction into eqn. (6.55) we obtain the following equilibrium potentials for the metal electrodes given.

(a) $$V(\text{Au}) = -\left(\frac{0 - 433\,700}{3 \cdot 96\,500} + \frac{8.314 \cdot 298.15}{3 \cdot 96\,500} \cdot \ln\left(\frac{1}{1 \cdot 10^{-6}}\right)\right) = \mathbf{+1.39\ volt}$$

(b) $$V(\text{Cu}) = -\left(\frac{0 - 65\,500}{2 \cdot 96\,500} + \frac{8.314 \cdot 298.15}{2 \cdot 96\,500} \cdot \ln\left(\frac{1}{1 \cdot 10^{-6}}\right)\right) = \mathbf{+0.16\ volt}$$

$$\text{(c)}\quad V(\text{Fe}) = -\Big(\frac{0-(-78\,900)}{2\cdot 96\,500} + \frac{8.314\cdot 298.15}{2\cdot 96\,500}\cdot \ln\Big(\frac{1}{1\cdot 10^{-6}}\Big)\Big) = \mathbf{-0.59\,volt}$$

$$\text{(d)}\quad V(\text{Zn}) = -\Big(\frac{0-(-147\,000)}{2\cdot 96\,500} + \frac{8.314\cdot 298.15}{2\cdot 96\,500}\cdot \ln\Big(\frac{1}{1\cdot 10^{-6}}\Big)\Big) = \mathbf{-0.94\,volt}$$

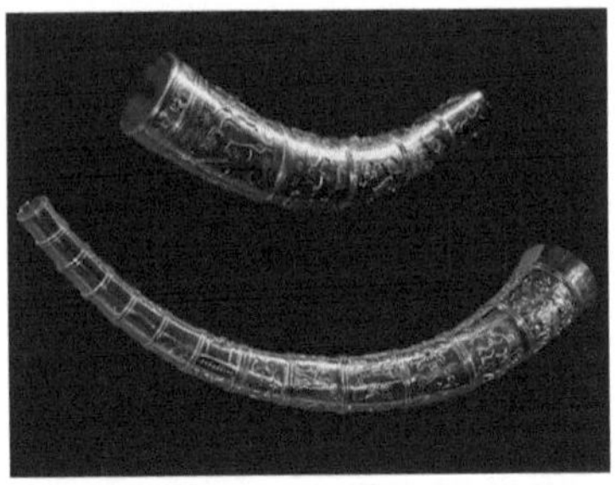

Figure 6.35. Gold is a noble metal with a highly positive standard reduction potential in the electrochemical series. Therefore, gold will normally form the cathode in corrosion cells. The photo shows the Danish so-called "golden horns" which were produced around 450 AD. Despite exposure to earth, moisture and air for more than 1000 years, the metal showed no signs of corrosion when the horns were found.

Next, we determine the equilibrium potential for the cathode reaction (1) describing the reduction of $H^+(aq)$ during the formation of hydrogen $H_2(g)$ at atmospheric pressure. The **standard potential** for (1) is given by the definition in eqn. (6.51): $V^\ominus = 0\,\text{volt}$.

The **equilibrium potential** for the reaction depends on the hydrogen ion concentration $[H^+]$ in the electrolyte; for an ideal solution there is the following relation between $[H^+]$ and the pH value of the solution

$$\text{pH} \stackrel{\text{def}}{=} -\log_{10}(a(\text{H}^+)) \quad\Rightarrow\quad a(\text{H}^+) = 10^{-\text{pH}}$$

Utilizing this relation, the equilibrium potential $V(1)$ can be expressed as a function of pH; since we are dealing with a reduction process, we obtain directly using eqn. (6.55)

$$V(1) = V^\ominus - \frac{RT}{z\mathcal{F}}\cdot \ln\Big(\frac{1}{a(\text{H}^+)^2}\Big) = V^\ominus - \frac{RT}{z\mathcal{F}}\cdot \ln(10^{2\cdot\text{pH}})$$

$$V(1) = 0 - \frac{8.314\cdot 298.15}{2\cdot 96\,500}\cdot \ln\big(10^{2\cdot\text{pH}}\big) = \mathbf{-0.059\cdot\text{pH}\ volt}$$

Now, the equilibrium potentials of the metals (a)...(d) are plotted together with $V(1)$ in a pH voltage chart. The pH ranges in which there is a potential risk of oxidation, can then be identified in the diagram. According to eqn. (6.50), pure metals can be oxidized in pH ranges where $V(1) > V(metal)$. Thus, we obtain the following from the chart,

Au(s): is not oxidized	**Cu(s)**: not oxidized
Fe(s): oxidized for pH < approximately 10	**Zn(s)**: oxidized at all pH values

OXIDATION OF METAL DURING OXYGEN ABSORPTION

First, we determine the equilibrium potential $V(2)$ for the cathode reaction (2) occurring during absorption of oxygen $O_2(g)$. Initially, it is noted that ΔG is the same for the reduction of oxygen in both acid and alkaline environments. This is realized by adding $4H^+(aq)$ on both sides of the last-mentioned reaction equation in (2); thus the reaction equation in (2) as shown above is obtained.

The electrochemical potential $V(2)$ is calculated using the expression for the potential of a single electrode, eqn. (6.55). Since we are dealing with a reduction equation, the standard potential $V^\ominus(2)$ can be directly determined by

$$V^\ominus(2) = -\frac{\Sigma G^\ominus_{red} - \Sigma G^\ominus_{ox}}{z\mathcal{F}} = -\frac{2\cdot(-237\,200) - 0 - 4\cdot 0}{4\cdot 96\,500} = +1.23\ \text{volt}$$

Utilizing the above relation: $a(H^+) = 10^{-\text{pH}}$, the equilibrium potential $V(2)$ can now be expressed as a function of the pH value of the solution

$$V(2) = V^\ominus - \frac{RT}{z\mathcal{F}}\cdot \ln\Big(\frac{1}{a(\text{H}^+)^4}\Big) = V^\ominus - \frac{RT}{z\mathcal{F}}\cdot \ln\big(10^{4\cdot\text{pH}}\big) \qquad \text{(volt)}$$

$$V(2) = 1.23 - \frac{8.314\cdot 298.15}{4\cdot 96\,500}\cdot \ln\big(10^{4\cdot\text{pH}}\big) = \mathbf{1.23 - 0.059\cdot pH\ volt}$$

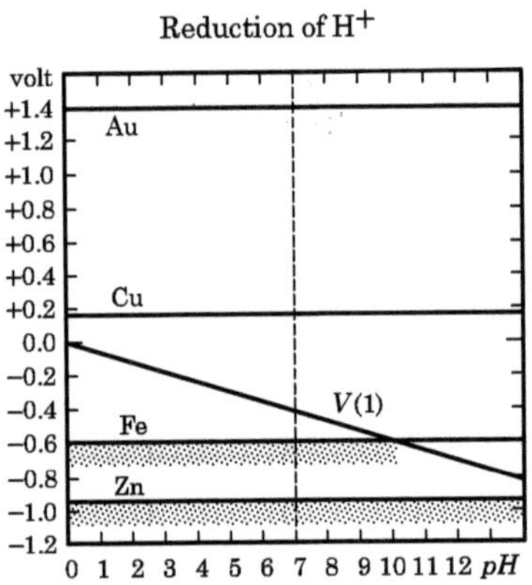

Figure 6.36. Electrode potentials for the metals Au, Cu, Fe, and Zn against the potential V(1) for reduction of H^+ during development of hydrogen. Oxidation of a metal can take place spontaneously, if V(1) exceeds V(metal). It is seen that Zn can be oxidized at all values of pH and that Fe can be oxidized if pH < approximately 10.

The equilibrium potential of the metal electrodes can be plotted in a pH volt chart together with $V(2)$. According to eqn. (6.50), the pure metals can be oxidized in pH areas where $V(2) > V(metal)$; thus, we obtain the following from the chart

Au(s): not oxidized	**Cu(s)**: oxidized at all pH values
Fe(s): oxidized at all pH values	**Zn(s)**: oxidized at all pH values

Discussion. The example shows how significant oxygen is for the tendency of metals to corrode in a humid environment. However, it shall be repeated once more that an oxidation process will not necessarily develop to deep pitting. In many cases, the oxides formed *passivate* the metal surface whereby harmful pitting is prevented.

Example 6.2

■ Calculating and producing a Pourbaix diagram for iron Fe

The hydrogen ion concentration - expressed by pH - is an important parameter for calculating electrochemical equilibrium potentials. The **Pourbaix diagram** maps in a clear way the influence of the pH value on complicated electrochemical equilibria. The diagram form, which is named after the Belgian corrosion researcher *Marcel Pourbaix*, was developed in the 1940s.

The Pourbaix diagram specifies electrochemical equilibrium curves for metals and metal oxides in a voltage vs. pH coordinate system. These curves delimit areas where the metal is **immune**, where the metal is **passivated**, and areas where the metal is **corrosion active** at equilibrium conditions. These equilibrium curves can usually be calculated from the thermodynamic data of the substances; areas with passivation or corrosion are determined by tests and from practical experience.

The Pourbaix diagram is an important tool for mapping the corrosion properties of metals. However, since the Pourbaix diagram is an **equilibrium diagram**, it is *not* possible to estimate the rate of a corrosion attack based on it. The following calculations outline the procedure for calculating and producing a Pourbaix diagram for iron, Fe.

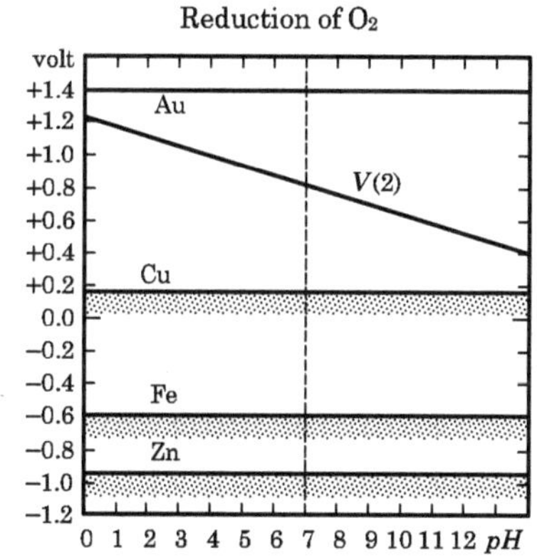

Figure 6.37. Electrode potentials for the metals Au, Cu, Fe, and Zn against the potential V(2) for reduction of dissolved O_2 at atmospheric pressure. Oxidation of a metal can occur spontaneously if V(2) exceeds V(metal). It is seen that Zn, Fe, and Cu can oxidize at all pH values; gold Au cannot oxidize.

Problem. During electrochemical corrosion of iron, precipitation of the **oxides** $Fe_2O_3(s)$ and $Fe_3O_4(s)$ will usually passivate the surface of the iron and thus prevent pitting. In contrast, precipitation of **ferrous hydroxide** $Fe(OH)_2(s)$ will not form a passivating layer on the surface of the iron. Calculate and plot into a pH volt diagram the equilibrium potentials for the following electrode reactions and give an outline of a Pourbaix diagram of iron!

$$Fe^{++}(aq) + 2e^- \rightarrow Fe(s) \qquad (a)$$
$$Fe^{+++}(aq) + e^- \rightarrow Fe^{++}(aq) \qquad (b)$$
$$Fe_3O_4(s) + 8H^+(aq) + 8e^- \rightarrow 3Fe(s) + 4H_2O(\ell) \qquad (c)$$
$$Fe_2O_3(s) + 6H^+(aq) + 6e^- \rightarrow 2Fe(s) + 3H_2O(\ell) \qquad (d)$$
$$Fe_3O_4(s) + 8H^+(aq) + 2e^- \rightarrow 3Fe^{++}(aq) + 4H_2O(\ell) \qquad (e)$$
$$Fe_2O_3(s) + 6H^+(aq) + 2e^- \rightarrow 2Fe^{++}(aq) + 3H_2O(\ell) \qquad (f)$$

Conditions. The diagram is set up for $[Fe^{++}] = [Fe^{+++}] = 1 \cdot 10^{-6}$ mol/ℓ. Solutions are assumed to be **ideal** and in accordance with eqn. (5.32), activities are expressed as: $a = c/c^{\ominus}$.

Solution. Data for the standard free energy of the substances $G^{\ominus}_{298}$ (kJ/mol) are looked up in a table; the following numerical values are used

$Fe(s)$ 0.0; $Fe^{++}(aq)$ −78.9; $Fe^{+++}(aq)$ −4.6; $H^+(aq)$ 0.0;
$Fe_2O_3(s)$ 742.2; $Fe_3O_4(s)$ −1015.5; $H_2O(\ell)$ −237.2;

It shall be noted that all of the electrode reactions (a)...(f) considered are reduction processes, cf. eqn. (6.17). Therefore, without rewriting the reaction equations, we can calculate the equilibrium potentials V for these reactions using eqn. (6.55).

EQUILIBRIUM POTENTIALS V(a) AND V(b)

The electrochemical potentials $V(a)$ and $V(b)$ are independent of pH because neither $H^+(aq)$ nor $OH^-(aq)$ are included as reactants. Using eqn. (6.55), the single potentials $V(a)$ and $V(b)$ are determined at equilibrium

$$V(a) = -\left(\frac{0-(-78\,900)}{2 \cdot 96\,500} + \frac{8.314 \cdot 298.15}{2 \cdot 96\,500} \cdot \ln\left(\frac{1}{1 \cdot 10^{-6}}\right)\right) = \mathbf{-0.59\ volt}$$

$$V(b) = -\left(\frac{-78\,900-(-4\,600)}{1 \cdot 96\,500} + \frac{8.314 \cdot 298.15}{1 \cdot 96\,500} \cdot \ln\left(\frac{1 \cdot 10^{-6}}{1 \cdot 10^{-6}}\right)\right) = \mathbf{+0.77\ volt}$$

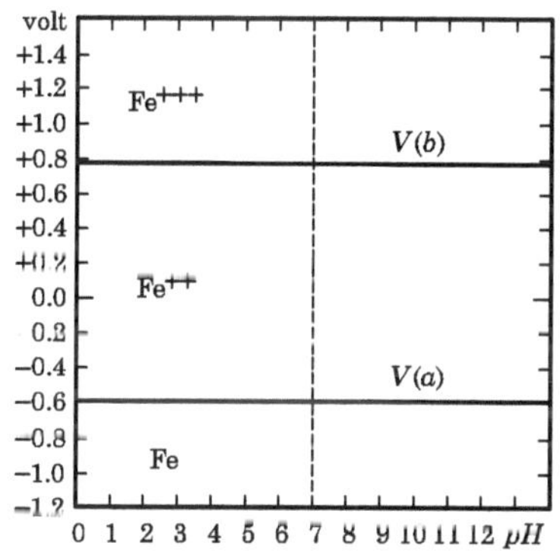

Figure 6.38. Pourbaix diagram: Areas where the oxidation state of iron corresponds to Fe, Fe^{++} and Fe^{+++}, are determined based on equilibrium potentials V(a) and V(b).

Since the equilibrium potentials $V(a)$ and $V(b)$ are independent of pH, they are represented as horizontal lines in the Pourbaix diagram. Thus, the areas are shown in which $Fe(s)$, $Fe^{++}(aq)$, and $Fe^{+++}(aq)$ are stable, if there are *no other* reaction components in the system.

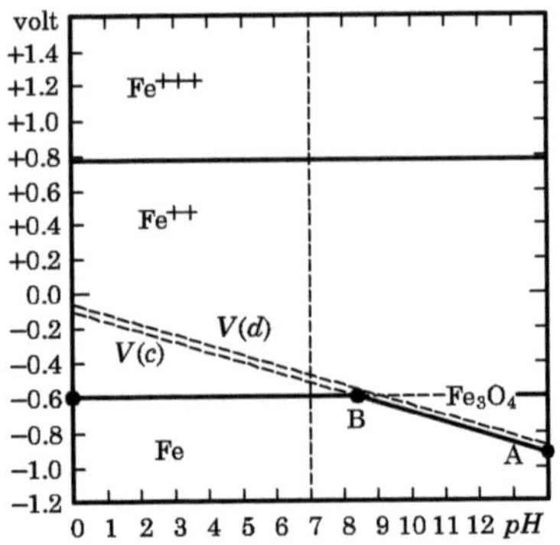

Figure 6.39. Pourbaix diagram: The boundary line A–B for oxidation of Fe to passivating Fe_3O_4 is determined by the equilibrium potential V(c).

EQUILIBRIUM POTENTIALS $V(c)$ AND $V(d)$

We then determine the equilibrium potentials $V(c)$ and $V(d)$ for the electrode reactions (c) and (d). These potentials represent equilibrium between Fe(s) and the oxides Fe_3O_4(s) and Fe_2O_3(s), respectively. Since H^+(aq) is included as a reactant in (c) and (d), the equilibrium potentials depend on the hydrogen ion concentration $[H^+]$, and thus on the pH value of the electrolyte. To introduce pH as a variable, the following relation is used

$$\mathrm{pH} \stackrel{\text{def}}{=} -\log_{10}(a(\mathrm{H}^+)) \quad \Rightarrow \quad a(\mathrm{H}^+) = 10^{-\mathrm{pH}}$$

Then, for the electrode reactions (c) and (d), we can introduce pH into the thermodynamic equilibrium constant K_a as follows

$$K_a(c) = \frac{a(\mathrm{Fe})^3 \cdot a(\mathrm{H_2O})^4}{a(\mathrm{Fe_3O_4}) \cdot a(\mathrm{H}^+)^8} = \frac{1^3 \cdot 1^4}{1 \cdot (10^{-\mathrm{pH}})^8} = \frac{1}{10^{-8 \cdot \mathrm{pH}}} = 10^{8 \cdot \mathrm{pH}}$$

$$K_a(d) = \frac{a(\mathrm{Fe})^2 \cdot a(\mathrm{H_2O})^3}{a(\mathrm{Fe_2O_3}) \cdot a(\mathrm{H}^+)^6} = \frac{1^2 \cdot 1^3}{1 \cdot (10^{-\mathrm{pH}})^6} = \frac{1}{10^{-6 \cdot \mathrm{pH}}} = 10^{6 \cdot \mathrm{pH}}$$

Next, the equilibrium potentials $V(c)$ and $V(d)$ are calculated using eqn. (6.55); thus, the potentials are expressed as functions of pH in the electrolyte

$$V(c) = -\Big(\frac{0 + 4 \cdot (-237\,200) - (-1015\,500 + 0)}{8 \cdot 96\,500} + \frac{8.314 \cdot 298.15}{8 \cdot 96\,500} \cdot \ln\big(10^{8 \cdot \mathrm{pH}}\big)\Big)$$

$$V(c) \qquad = \mathbf{-0.09 - 0.059 \cdot pH\ volt}$$

$$V(d) = -\Big(\frac{0 + 3 \cdot (-237\,200) - (-742\,200 + 0)}{6 \cdot 96\,500} + \frac{8.314 \cdot 298.15}{6 \cdot 96\,500} \cdot \ln\big(10^{6 \cdot \mathrm{pH}}\big)\Big)$$

$$V(d) \qquad = \mathbf{-0.05 - 0.059 \cdot pH\ volt}$$

The equilibrium potentials $V(c)$ and $V(d)$ are plotted in the Pourbaix diagram. In the pH area, where $V(\mathrm{Fe}) > V(c)$, Fe(s) can spontaneously oxidize to Fe_3O_4(s); this is seen to be the case for pH $>$ ca. 8.5. The equilibrium potential $V(c)$ is lower than $V(d)$; the lower boundary curve A–B for oxidizing of Fe(s), therefore, corresponds to formation of Fe_3O_4(s) during the process (c).

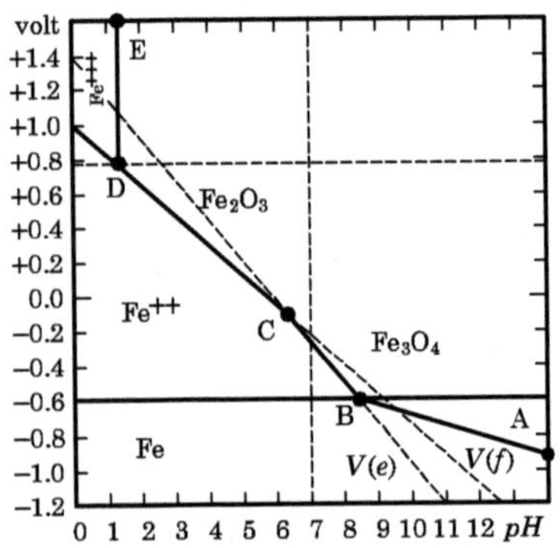

Figure 6.40. Pourbaix diagram: The boundary line B–C for oxidation of Fe^{++} to Fe_3O_4 is determined by the equilibrium potential V(e); the boundary line C–D for oxidation of Fe^{++} to Fe_2O_3 is determined by the equilibrium potential V(f).

EQUILIBRIUM POTENTIALS $V(e)$ AND $V(f)$

The equilibrium potentials $V(e)$ and $V(f)$ represent thermodynamic equilibrium between Fe^{++}(aq) and the oxides Fe_3O_4(s) and Fe_2O_3(s), respectively. Calculation of $V(e)$ and $V(f)$ is performed in analogy with the calculation of $V(c)$ and $V(d)$ shown above

$$V(e) = -\Big(\frac{3 \cdot (-78\,900) + 4 \cdot (-237\,200) - (-1015\,500 + 0)}{8 \cdot 96\,500} + \frac{RT}{z\mathcal{F}} \cdot \ln\Big(\frac{(10^{-6})^3}{10^{-8 \cdot \mathrm{pH}}}\Big)\Big)$$

$$V(e) \qquad = \mathbf{+1.41 - 0.237 \cdot pH\ volt}$$

$$V(f) = -\Big(\frac{2 \cdot (-78\,900) + 3 \cdot (-237\,200) - (-742\,200 + 0)}{6 \cdot 96\,500} + \frac{RT}{z\mathcal{F}} \cdot \ln\Big(\frac{(10^{-6})^2}{10^{-6 \cdot \mathrm{pH}}}\Big)\Big)$$

$$V(f) \qquad = \mathbf{+1.01 - 0.177 \cdot pH\ volt}$$

The equilibrium potentials $V(e)$ and $V(f)$ are plotted in the Pourbaix diagram. In areas where $V > V(e)$, Fe^{++}(aq) will be able to oxidize spontaneously and precipitate as Fe_3O_4(s); correspondingly, Fe^{++}(aq) can be oxidized and precipitated as Fe_2O_3(s) in areas where $V > V(f)$. Comparing potential curves for $V(e)$ and $V(f)$, it is seen that the lowest oxidation potential is the boundary curve B–C with formation of Fe_3O_4(s) and the boundary curve C–D with formation of Fe_2O_3(s). In areas where $V > V(b)$, Fe^{++}(aq) is oxidized to Fe^{+++}(aq), corresponding to the boundary line D–E in the Pourbaix diagram.

Practical experience shows that electrochemical precipitation of the oxides Fe_2O_3(s) and Fe_3O_4(s) passivates the surface of the iron so that harmful corrosion is suppressed. Therefore, in the Pourbaix diagram, the line A–B–C–D–E delimits the area in which passivation occurs.

When pH $\leq$ ca. 13.5, the passivating oxide film of Fe_3O_4(s) can be decomposed during formation of a non-passivating hydroxide. Therefore, a complete Pourbaix diagram will contain a small area of corrosion risk when pH exceeds approximately 13.5.

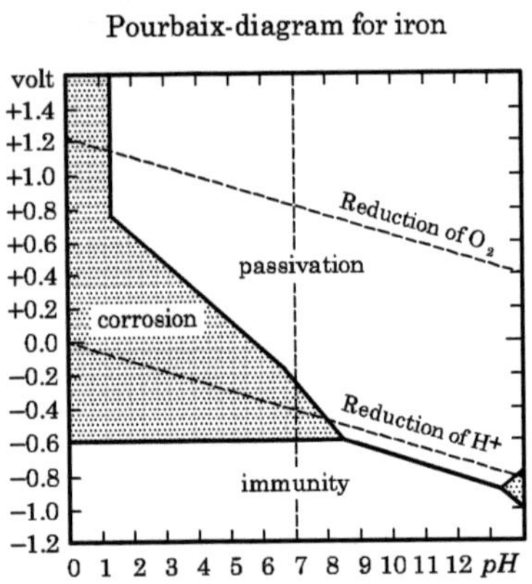

Figure 6.41. Pourbaix diagram for Fe where the areas of immunity, passivation, and corrosion are indicated. The diagram has been calculated for an ion concentration of 10^{-6} mol/ℓ.

Discussion. The pore solution in hardened, non-carbonated concrete normally contains dissolved alkali hydroxide as well as dissolved calcium hydroxide $Ca(OH)_2$; therefore, the pore water of the concrete is strongly alkaline with pH values in the range of 12-14. From the Pourbaix diagram for iron it is seen that embedded reinforcement will be passivated in this pH area. The natural alkaline environment in concrete thus provides conditions necessary for the corrosion resistance of the reinforcement.

In a concrete exposed to the carbon dioxide of the air reaction with CO_2 slowly neutralizes the dissolved alkali hydroxides and transforms the $Ca(OH)_2$ into calcium carbonate $CaCO_3$; the transformation is caused by the following reaction

$$Ca(OH)_2(s) + CO_2(g) \quad \rightarrow \quad CaCO_3(s) + H_2O(\ell)$$

The concrete is said to "carbonate" by this process. In permeable, poor-quality concrete, the concrete cover over the embedded reinforcement can sometimes carbonate in the course of a few years. The carbonated concrete is almost neutral when its pH range is reduced to ca. 8; thus, the pH-provided protection of the reinforcement is removed. Therefore, deep carbonation may result in reinforcement corrosion, and spalling and discolouration of the concrete over the corroded reinforcement may occur.

The Pourbaix diagram sketched here for iron gives a useful overview of the interaction between the environment and corrosion possibilities. However, one should remember that the validity of such equilibrium diagrams is limited to pure systems that form the basis of the diagram. In certain cases, modest quantities of alloying elements in the metal considered, or the presence of foreign ions in the electrolyte solution can result in radical changes in the equilibrium conditions. For example, the described passivation of iron in an alkaline environment can be terminated due to chloride ions $Cl^-(aq)$ in the electrolyte solution.

Figure 6.42. When de-icing salts with NaCl are applied to concrete, Cl^- can penetrate into the concrete by diffusion; the presence of Cl^- terminates the passivation that is normal for iron embedded in concrete. Thus, de-icing salts can be the cause of corrosion damages on the reinforcement, as shown here for a balcony structure.

Example 6.3

■ Measurement of pH with glass electrode – membrane potential

Measuring the pH value of liquids is a frequent task when materials are subjected to laboratory investigations. Today, pH is routinely measured with modern **glass electrodes** that are robust and reliable electrochemical cells.

In the **glass electrode** we utilize the property that certain alkaline glass types can act as **semipermeable** membranes against hydrogen ions $H^+(aq)$. Assume that a thin membrane of such a type of glass is placed between two electrolytes of different hydrogen ion concentration $[H^+]$. Then, a diffusion of H^+ from the electrolyte with high concentration to the electrolyte with low concentration of H^+ will take place. Since this transfer only includes positive hydrogen ions, an electric potential difference - the so-called **membrane potential** ΔV - is built up between the two electrolytes.

For electrochemical equilibrium, the membrane potential ΔV is determined by the *relation* between the hydrogen ion activity $a(\mathrm{H}^+)$ in the two electrolytes. If the pH, i.e. the value $-\log_{10}(a(\mathrm{H}^+))$ is known in one electrolyte (the reference), the pH can therefore be determined in the other electrolyte, by simply measuring the membrane potential ΔV against a semipermeable glass membrane between the two electrolytes.

The quantity of substance $H^+(aq)$ transferred by adjustment of equilibrium over the membrane is so small that it cannot be measured by the usual analytical methods. Therefore, the measuring cell does *not* affect the system being investigated.

Figure 6.43. pH meter to measure the activity of H^+ in liquids. Together with the surrounding electrolyte, the sensor forms a galvanic cell whose equilibrium potential is determined by the hydrogen ion activity.

Problem. Derive, using thermodynamic equilibrium calculations, an expression for the **membrane potential**: $\Delta V = f(\mathrm{pH}, \mathrm{pH_{ref}})$ between an electrolyte with unknown pH and a reference electrolyte with known $\mathrm{pH} = \mathrm{pH_{ref}}$; based on this, give a rough outline of the principle in a setup for measuring pH using a **glass electrode**.

Conditions. Ideal, semipermeable glass membrane allowing free passage of hydrogen ions $H^+(aq)$, but is fully impermeable to other ions.

Solution. An electrochemical cell is assumed to be divided into two compartments 1 and 2 by a glass membrane that is **semipermeable** to hydrogen ions $H^+(aq)$. Compartment 1 contains an electrolyte with hydrogen ion activity $a1$, and compartment 2 contains an electrolyte with hydrogen ion activity $a2$. By diffusion of $H^+(aq)$ through the glass membrane, the system adjusts itself to **electrochemical equilibrium** with the potentials $V1$ and $V2$ in electrolyte 1 and electrolyte 2, respectively. In this state, the chemical and the electric potentials *balance* exactly. Then the electrochemical equilibrium is

$$H^+(a1, V1) \leftrightarrows H^+(a2, V2) \qquad \text{(a)}$$

Now, assume a reversible transfer of $H^+(aq)$ from compartment 1 with the potential $V1$ to compartment 2 with the potential $V2$; according to eqn. (6.30), this results in the following change ΔG in the free energy of the system

$$\Delta G = W_{el} = (V2 - V1) \cdot Q_{el} = (V2 - V1) \cdot 1 \cdot \mathcal{F} \qquad \text{(b)}$$

because the electric elementary charge e is equal to the charge of the proton H^+. Since $\Delta G^\ominus$ is zero for the process (a), it is evident that the following also applies

$$\Delta G = G(H_2^+) - G(H_1^+) = \Delta G^\ominus + RT \cdot \ln\left(\frac{a2}{a1}\right) = RT \cdot \ln\left(\frac{a2}{a1}\right) \qquad \text{(c)}$$

where $a1$ and $a2$ denote the activity $a(H^+)$ of hydrogen ions in compartments 1 and 2, respectively. The identity between the expressions (b) and (c) for ΔG shows that

$$\Delta V = V2 - V1 = \frac{RT}{\mathcal{F}} \cdot \ln\left(\frac{a2}{a1}\right) \qquad \text{(d)}$$

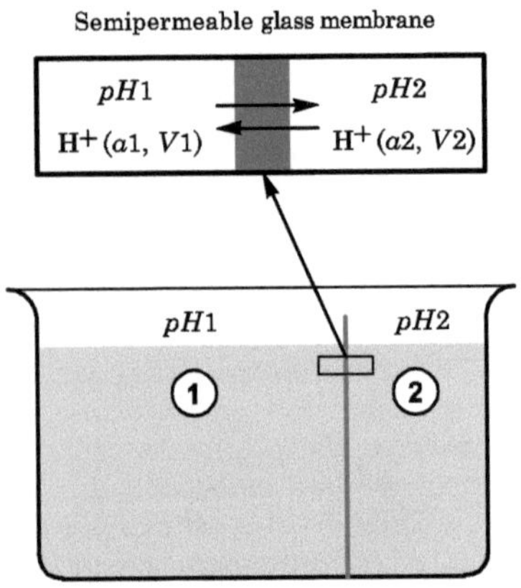

Figure 6.44. Membrane potential ΔV over a glass membrane that is semipermeable to hydrogen ions H^+. At equilibrium, the chemical and the electric potential balance.

Using that $\text{pH} \stackrel{\text{def}}{=} -\log_{10}(a(H^+))$ and thus $a(H^+) = 10^{-\text{pH}}$, (d) can be rewritten to the targeted expression for the membrane potential ΔV

$$\Delta V = V2 - V1 = -\frac{RT \cdot \ln(10)}{\mathcal{F}} \cdot (\text{pH2} - \text{pH1}) \qquad \text{(e)}$$

Assume, for example, that electrolyte 2 is a reference with known $\text{pH2} = \text{pH}_{\text{ref}} = 4.000$; the membrane potential ΔV at 25 °C is then determined by introduction into eqn. (e)

$$\Delta V = -\left(\frac{8.314 \cdot 298.15 \cdot \ln(10)}{96\,500}\right) \cdot (4.000 - \text{pH1}) \qquad = \mathbf{0.0592 \cdot pH1 - 0.237\ volt}$$

To measure the membrane potential ΔV it is necessary to make electric contact to the unknown solution in compartment 1 and to the known solution in compartment 2 (the glass electrode). This cannot be achieved by placing a metal electrode in the unknown solution which would merely introduce a new electrode with an unknown electrochemical potential in the measuring circuit. Therefore, it is necessary to establish a potential free electrolytic contact to the unknown liquid.

In practice, the membrane potential ΔV is determined as follows: compartment 1 with the unknown liquid sample is put in electrolytic contact with a reference electrode with a known, *constant* electrode potential $Ve1$. In compartment 2 with the known electrolyte in the glass electrode, another corresponding reference electrode also with a known, *constant* potential $Ve2$ is placed. The electrode pair and membrane thus form a galvanic cell with the following electrochemical potential

$$\Delta V_{\text{cell}} = \Delta V_{\text{membrane}} + Ve2 - Ve1 \qquad = 0.0592 \cdot \text{pH} + \text{k volt}$$

where the constant k can be determined by calibration. Normally, the membrane of the glass electrode is shaped as a thin spherical shell of the semipermeable glass. Often, a **silver electrode** or a **calomel electrode** with the cell diagram is used

silver electrode : $(Ag), AgCl \,|\, Ag, Cl^- \,||\, \cdots$ $\qquad V^\ominus \simeq 0.22$ volt

calomel electrode : $(Pt), Hg_2Cl_2 \,|\, 2Hg, 2Cl^- \,||\, \cdots$ $\qquad V^\ominus \simeq 0.28$ volt

Discussion. Since the electric resistance over the glass membrane is significant - up to 100 MΩ - the measurement requires that a special voltmeter with particularly high input impedance be used.

Example 6.4

■ Electrochemical measurement of thermodynamic standard values

Electrochemical measurements are extensively used for experimental determination of table data for thermodynamic standard values. Such measurements are highly suitable for mapping thermodynamic data for ions.

In *"Journal of Research", National Bureau of Standards No. 53 (1954), R.G. Bates* and *V.E. Bowel* specify measured values for the electrochemical **standard potential** $V^\ominus(\theta)$ for the following electrode pairs

$$(Pt), \tfrac{1}{2}H_2(g, p^\ominus) \,|\, H^+(aq) \,||\, AgCl(s) \,|\, Cl^-(aq), Ag(s), (Ag) \qquad \text{(a)}$$

As seen, it is a **standard hydrogen electrode** (SHE) against a **silver electrode**. For comprehensive measurements, *Bates & Bowel* have found the following expressions for the standard potential $V^{\ominus}(\theta)$ of the electrode pair as a function of the temperature θ (°C)

$$\Delta V^{\ominus} = 0.23659 - 4.8564 \cdot 10^{-4} \cdot \theta - 3.4205 \cdot 10^{-6} \cdot \theta^2 + 5.869 \cdot 10^{-9} \cdot \theta^3 \text{ volt} \quad \text{(b)}$$

Problem. The following thermodynamic standard values for solids and gases are known

	Ag(s)	AgCl(s)	H^+(aq)	H_2(g)
$H^{\ominus}_{298}$ (kJ/mol)	0.0	−127.1	0.0	0.0
$G^{\ominus}_{298}$ (kJ/mol)	0.0	−109.8	0.0	0.0
$S^{\ominus}_{298}$ (J/mol K)	42.6	96.2	0.0	130.6

Calculate the following thermodynamic standard values for the **chloride ion** Cl^-(aq): $G^{\ominus}_{298}$ (kJ/mol); $S^{\ominus}_{298}$ (J/mol K) and $H^{\ominus}_{298}$ (kJ/mol).

Conditions. The measured standard potential $V^{\ominus}(\theta)$ refers to thermodynamic equilibrium in an ideal electrochemical cell corresponding to (a).

Solution. The cell diagram (a) denotes **oxidation** of H_2(g) to H^+(aq) at the anode and **reduction** of Ag^+(aq) to Ag(s) at the cathode; the following electrochemical reactions occur

$$\text{Anode reaction}: \quad \tfrac{1}{2}H_2(g, p^{\ominus}) \rightarrow H^+(aq) + e^- \quad \text{(c)}$$
$$\text{Cathode reaction}: \quad AgCl(s) + e^- \rightarrow Ag(s) + Cl^-(aq) \quad \text{(d)}$$
$$\text{Net reaction}: \quad AgCl(s) + \tfrac{1}{2}H_2(g, p^{\ominus}) \rightarrow Ag(s) + H^+(aq) + Cl^-(aq) \quad \text{(e)}$$

We now consider the net reaction (e). Utilizing eqns. (6.38) and (6.42), $\Delta G^{\ominus}$ and $\Delta S^{\ominus}$ can for this reaction be expressed by $\Delta V^{\ominus}$ and $\partial \Delta V^{\ominus}/\partial T$, respectively. Finally, in accordance with eqn. (5.10), $\Delta H^{\ominus}$ can be expressed by $\Delta G^{\ominus}$ and $\Delta S^{\ominus}$. Thus, we have the following relations

$$(G^{\ominus}(\text{Ag}) + G^{\ominus}(\text{H}^+) + G^{\ominus}(\text{Cl}^-)) - (G^{\ominus}(\text{AgCl}) + \tfrac{1}{2}G^{\ominus}(\text{H}_2)) = -\Delta V^{\ominus} \cdot z\mathcal{F} \quad \text{(f)}$$
$$(S^{\ominus}(\text{Ag}) + S^{\ominus}(\text{H}^+) + S^{\ominus}(\text{Cl}^-)) - (S^{\ominus}(\text{AgCl}) + \tfrac{1}{2}S^{\ominus}(\text{H}_2)) = \frac{\partial \Delta V^{\ominus}}{\partial T} \cdot z\mathcal{F} \quad \text{(g)}$$
$$(H^{\ominus}(\text{Ag}) + H^{\ominus}(\text{H}^+) + H^{\ominus}(\text{Cl}^-)) - (H^{\ominus}(\text{AgCl}) + \tfrac{1}{2}H^{\ominus}(\text{H}_2)) = \Delta G^{\ominus} + T\Delta S^{\ominus} \quad \text{(h)}$$

The targeted thermodynamic standard values for Cl^-(aq) can thus be determined by

$$G^{\ominus} = G^{\ominus}(\text{AgCl}) + \tfrac{1}{2}G^{\ominus}(\text{H}_2) - G^{\ominus}(\text{Ag}) - G^{\ominus}(\text{H}^+) - \Delta V^{\ominus} \cdot z\mathcal{F} \quad \text{(i)}$$
$$S^{\ominus} = S^{\ominus}(\text{AgCl}) + \tfrac{1}{2}S^{\ominus}(\text{H}_2) - S^{\ominus}(\text{Ag}) - S^{\ominus}(\text{H}^+) + \frac{\partial \Delta V^{\ominus}}{\partial T} \cdot z\mathcal{F} \quad \text{(j)}$$
$$H^{\ominus} = H^{\ominus}(\text{AgCl}) + \tfrac{1}{2}H^{\ominus}(\text{H}_2) - H^{\ominus}(\text{Ag}) - H^{\ominus}(\text{H}^+) + \left(T \cdot \frac{\partial \Delta V^{\ominus}}{\partial T} - \Delta V^{\ominus}\right) \cdot z\mathcal{F} \quad \text{(k)}$$

The **temperature coefficient** of the equilibrium potential $\partial \Delta V^{\ominus}/\partial T$ is determined by differentiation of eqn. (b)

$$\frac{\partial \Delta V^{\ominus}}{\partial T} = -4.8564 \cdot 10^{-4} - 6.8410 \cdot 10^{-6} \cdot \theta + 17.607 \cdot 10^{-9} \cdot \theta^2 \quad \text{volt/K} \quad \text{(l)}$$

Introducing $\theta = 25\,°C$ into eqns. (b) and (l), we can now determine the standard potential and the temperature coefficient of the standard potential at 298.15 K

$$\Delta V^{\ominus}(\theta) = +0.2224 \text{ volt} \qquad \frac{\partial \Delta V^{\ominus}(\theta)}{\partial T} = -6.457 \cdot 10^{-4} \text{ volt/K}$$

whereupon the targeted thermodynamic standard values for Cl^-(aq) are determined by introduction of table data and calculated values into the relations (i),(j), and (k)

$$G^{\ominus} = -109\,800 + 0 - 0 - 0 - 0.224 \cdot 1 \cdot 96\,500 \quad \text{J/mol}$$
$$S^{\ominus} = 96.2 + 0.5 \cdot 130.6 - 42.6 - 0 - 6.457 \cdot 10^{-4} \cdot 1 \cdot 96\,500 \quad \text{J/mol K}$$
$$H^{\ominus} = -127100 + 0 - 0 - 0 + (298.15 \cdot (-6.457 \cdot 10^{-4}) - 0.2224) \cdot 1 \cdot 96\,500 \quad \text{J/mol}$$

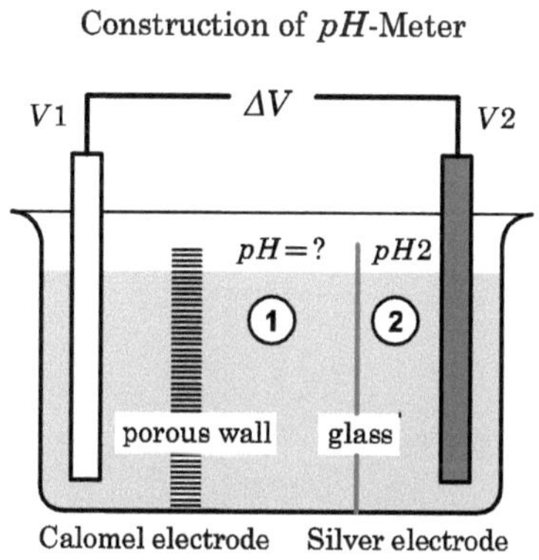

Figure 6.45. Schematic illustration of a pH-meter with calomel electrode (left) and silver electrode (right). The semipermeable glass membrane surrounds the silver electrode that is immersed in an electrolyte with known pH. Through a porous wall, the calomel electrode is in electric contact with the electrolyte on which the measurements are made.

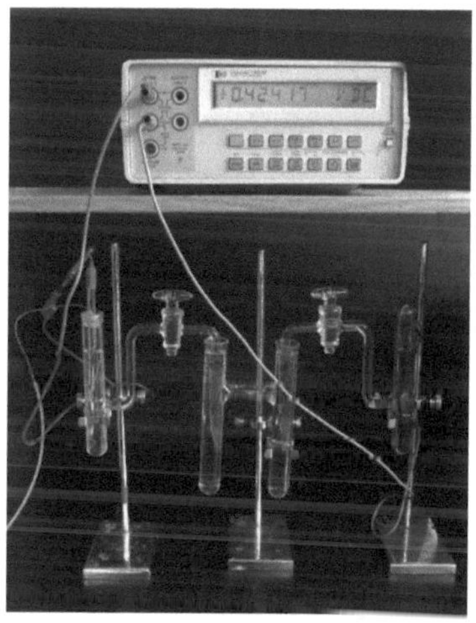

Figure 6.46. Laboratory setup for measuring electrochemical equilibrium potential against a standard hydrogen electrode. The compartment to the left is the hydrogen electrode that is in electrolytic contact with the measuring cell to the right through a "salt bridge".

CONCLUSION: the measuring data stated determine the following thermodynamic standard values for the chloride ion $Cl^-(aq)$

$$G^{\ominus}_{298} = \mathbf{-131.3\,kJ/mol}; \quad S^{\ominus}_{298} = \mathbf{+56.6\,J/mol\,K}; \quad H^{\ominus}_{298} = \mathbf{-167.1\,kJ/mol}$$

Discussion. Comparing the calculated thermodynamic standard values with table data for $Cl^-(aq)$, we find the following table

	$H^{\ominus}_{298}$ (kJ/mol)	$G^{\ominus}_{298}$ (kJ/mol)	$S^{\ominus}_{298}$ (J/mol K)
Table value	-167.2	-131.3	56.5
Calculated value	-167.1	-131.3	56.6

The example illustrates how thermodynamic data for ions in a solution can be determined by measuring the electrochemical potentials. Similarly, by measuring electrochemical potentials, we can determine the **activity** a of ions in solutions with known molar concentration c; thus, this is an accurate method for determining **activity coefficients**, cf. eqn. (5.34).

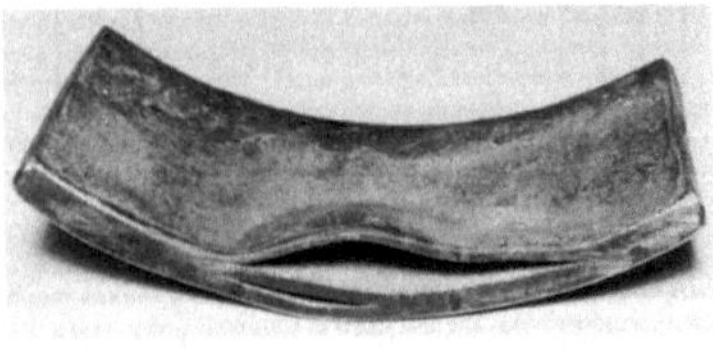

Figure 6.47. Cut-out segment from a steel tank exposed to corrosion showing blistering of the steel due to internal hydrogen gas pressure. Such hydrogen damages particularly arise due to galvanic corrosion in an acid environment.

Example 6.5

■ Hydrogen reduction with polarization - hydrogen brittleness

The hydrogen atom H(g) is willingly *adsorbed* on metal surfaces, and by diffusion it can penetrate into the solid metal. In contrast, metals are *not* penetrable to the hydrogen molecule $H_2(g)$.

When steel absorbs large quantities of atomic hydrogen, its properties are drastically changed - the steel becomes **brittle** and spontaneous **cracking** and perhaps **blistering** of the surface can occur. Hydrogen damages can have catastrophic consequences, since failure of load-carrying capacity normally occurs suddenly without any warning. Hydrogen brittleness and hydrogen damage are well known in connection with galvanic corrosion in an acid environment; the reason for this is briefly explained in the following example.

By galvanic corrosion in acid liquids with a high content of hydrogen ions, the cathode reaction will often be in the form of **reduction** of $H^+(aq)$ during hydrogen development

$$2H^+(aq) + 2e^- \;\rightarrow\; H_2(aq) \;\rightarrow\; H_2(g) \qquad \text{(a)}$$

At the anode, **oxidation** necessarily occurs at the same time; iron Fe(s), for example, can oxidize due to the anode reaction

$$Fe(s) \;\rightarrow\; Fe^{++}(aq) + 2e^- \qquad \text{(b)}$$

A well-defined electrochemical **equilibrium potential** V can be assigned to the single electrodes (a) and (b); this potential corresponds to the electrodes that are *not* carrying any current. During a corrosion process, however, the current will change the electrode potentials so that $\Delta V = V(\text{cathode}) - V(\text{anode})$ is reduced. A **polarization** of the electrodes is said to occur. The **overvoltage** η is used as a measure of the polarization, which specifies

$$\eta = V(\text{actual}) - V(\text{equilibrium}) \qquad (c)$$

For the anode, the overvoltage η is always **positive** and for the cathode the overvoltage η is always **negative**. Both of these changes contribute to reducing ΔV. Polarization caused by the internal resistance of the cell system against flow of current can be divided into **concentration polarization** and **activation polarization**. The significance of this is

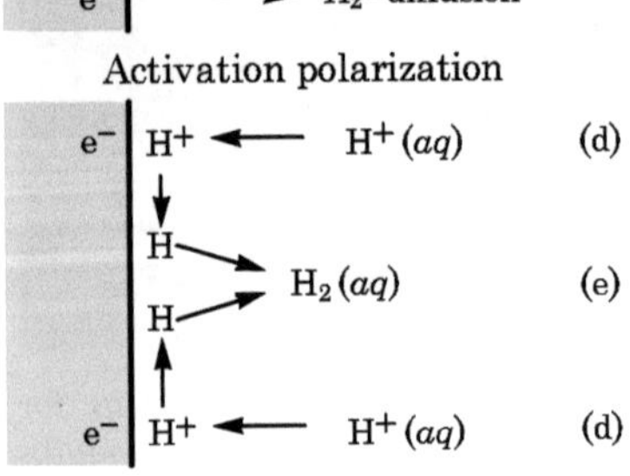

Figure 6.48. Concentration polarization is caused by the concentration gradients that arise in a charged electrolyte due to diffusion resistance. Activation polarization is caused by the delayed reaction connected with the electrochemical reactions on the surface of the electrode.

CONCENTRATION POLARIZATION is due to the **diffusion resistance** to be overcome by the reacting ions during the process. By the hydrogen reduction (a), $H^+(aq)$, for example, is transported to the cathode by diffusion. However, such diffusion can only take place if the ion concentration at the cathode is lower than in the surrounding electrolyte; according to eqn. (6.55), this means that the electrochemical potential of the cathode is less than that of its equilibrium potential.

ACTIVATION POLARIZATION represents the resistance to be overcome by ions and electrons during the **chemical reactions** at the interface between the electrode and the electrolyte.

Activation polarization plays a special role in the cathodic reduction of hydrogen. In fact, the reduction process of hydrogen (a) comprises two separate sub-reactions:

$$\begin{cases} 2H^+ + 2e^- \rightarrow 2H & \text{formation of atomic hydrogen} \quad (d) \\ 2H \rightarrow H_2 & \text{formation of molecular hydrogen} \quad (e) \end{cases}$$

In *"An Introduction to Electrochemical Science"*, *Wykeham, London (1974)*, this composite hydrogen reduction process has been analyzed by *Bockris, Bonciocat & Gutmann*. Based on electrochemical measurements, sub-process (e) is in most cases found to be the *slowest*, and thus, this process is the rate-determining process in the cathodic reaction of $H^+(aq)$. This applies in particular to cases where the cathode is one of the **transition metals**, for example, Pt, Ni, W or Cu of the periodic table. For these metals, *Bockris, Bonciocat & Gutmann* show that by activation polarization, the following **pseudo-equilibrium** occurs during the reduction process

$$\underbrace{2H^+(aq) + 2e^- \leftrightarrows 2H(ads)}_{\text{pseudo-equilibrium}} \rightarrow H_2(aq) \quad (f)$$

During the flow of current at a steady rate, a *constant* concentration of adsorbed, atomic hydrogen H(ads) is built up on the surface of cathode metals. Therefore, the pseudo-equilibrium (d) can be perceived as the electrochemical "equilibrium reaction" for the cathode.

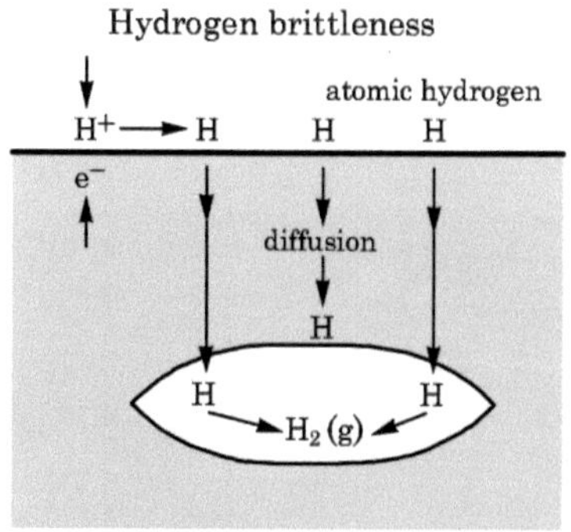

Figure 6.49. Schematic illustration of a cathode process with hydrogen adsorption in an electrode. On the cathode, a layer of adsorbed atomic hydrogen is built up; the hydrogen penetrates into the cathode metal by diffusion. In internal pores, the atomic hydrogen combines to hydrogen molecules. By thermodynamic equilibrium, a given concentration of atomic hydrogen on the surface will have a corresponding equilibrium gas pressure in the internal pores.

Problem. Assume a metal electrode where the following thermodynamic equilibrium has been established in the initial state

$$2H^+(aq, c^\ominus) + 2e^- \leftrightarrows 2H(ads) \leftrightarrows H_2(aq) \leftrightarrows H_2(g, p^\ominus) \quad (g)$$

with the equilibrium potential $V0$ volt. Now, a steady current is imposed on the electrode with the result that a cathodic overvoltage of η (volt) is produced as a result of activation polarization. The slowest process, and thus the one that determines the rate, is sub-process (e). Therefore, the following pseudo-equilibrium adjusts itself at the electrode

$$2H^+(aq, c^\ominus) + 2e^- \leftrightarrows 2H(ads) \rightarrow H_2(aq) \quad (h)$$

with the equilibrium potential $V1 = V0 + \eta$ (volt). During this process, surface adsorbed atomic hydrogen penetrates into the metal electrode by **diffusion** into internal pores in the electrode; in these internal pores the atomic hydrogen H is transformed into molecular hydrogen H_2. Calculate and draw the approximate theoretical **equilibrium expression** for $H_2(g)$ in the internal pore as a function of the **activation overvoltage** η

$$p(H_2(g)) = f(\eta) \quad (i)$$

Conditions. The electrode metal is assumed to be *impenetrable* to molecular hydrogen $H_2(g)$, and the gas is assumed to be ideal. Hydrogen ions in the electrolyte form an ideal solution with constant concentration: $[H^+] = c^\ominus = 1\,mol/\ell$.

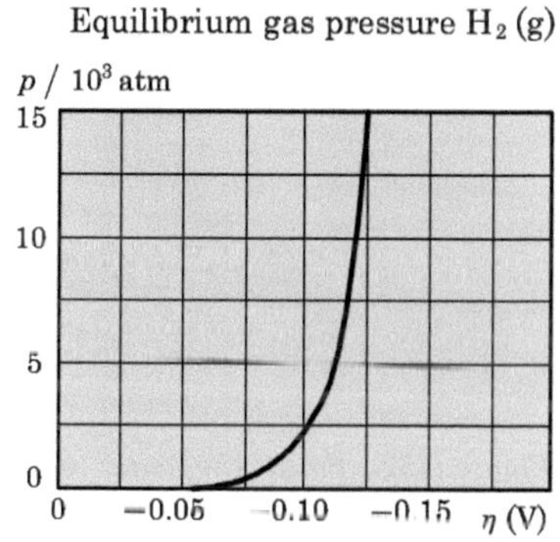

Figure 6.50. Theoretical equilibrium gas pressure of hydrogen H_2 in internal pores as a function of the activation overvoltage η by cathodic reduction of hydrogen.

Solution. The unknown activities of adsorbed atomic hydrogen H(ads) are designated $a0(H)$ in their initial state (g) and $a1(H)$ in their final state (h). The electrode potential $V0$ for reduction of $H^+(aq)$ is in its initial state (g) determined by eqn. (6.55)

$$V0 = -\Big(\frac{2 \cdot G^\ominus(H(ads)) - 2 \cdot G^\ominus(H^+(aq))}{z\mathcal{F}} + \frac{RT}{z\mathcal{F}} \cdot \ln\Big(\frac{a0(H)}{a(H^+)}\Big)^2\Big) \quad (j)$$

Correspondingly, the electrode potential $V1$ in its final state (h) is determined by eqn. (6.55)

$$V1 = -\Big(\frac{2 \cdot G^\ominus(H(ads)) - 2 \cdot G^\ominus(H^+(aq))}{z\mathcal{F}} + \frac{RT}{z\mathcal{F}} \cdot \ln\Big(\frac{a1(H)}{a(H^+)}\Big)^2\Big) \quad (k)$$

Since the hydrogen ion activity $a(H^+)$ is constant, the **overvoltage** η can be expressed as

$$\eta = V1 - V0 = -\frac{RT}{z\mathcal{F}} \cdot \ln\Big(\frac{a1(H)}{a0(H)}\Big)^2 \quad (l)$$

During the stationary pseudo-equilibrium (h), adsorbed atomic hydrogen penetrates into the metal by diffusion. In the internal pores, it is transformed into hydrogen gas $H_2(g)$. For diffusion equilibrium, the activity $a(\mathrm{H})$ of atomic hydrogen is the same all over the metal; therefore, in its final state (h), there is thermodynamic equilibrium between $H_2(g)$ in pores and adsorbed hydrogen H(ads)

$$2\mathrm{H(ads)} \leftrightarrows \mathrm{H_2(g)} \quad \Rightarrow \quad K_{a1} = \frac{a1(\mathrm{H_2})}{a1(\mathrm{H})^2} \qquad \text{(m)}$$

In the initial state (g), a corresponding thermodynamic equilibrium was established between adsorbed hydrogen and $H_2(g)$ at the standard pressure $p^\ominus = 1$ atm, i.e.

$$2\mathrm{H(ads)} \leftrightarrows \mathrm{H_2(g)} \quad \Rightarrow \quad K_{a0} = \frac{a0(\mathrm{H_2})}{a0(\mathrm{H})^2} = \frac{1}{a0(\mathrm{H})^2} \qquad \text{(n)}$$

Setting $K_{a1} = K_{a0} = \exp(-\Delta G^\ominus/RT)$, we can eliminate the unknown activities in the expression for the overvoltage η. Summation of (m) and (n) and introduction into eqn. (l) lead to

$$a1(\mathrm{H_2}) = \left(\frac{a1(\mathrm{H})}{a0(\mathrm{H})}\right)^2 \quad \Rightarrow \quad \eta = -\frac{RT}{z\mathcal{F}} \cdot \ln(a1(\mathrm{H_2})) \qquad \text{(o)}$$

If the activity of an ideal gas $a = p/p^\ominus$ is introduced into the above, we obtain the following expression for the equilibrium pressure $p = f(\eta)$ in the internal pores as a function of the overvoltage η

$$p = p^\ominus \cdot \exp\left(-\frac{z\mathcal{F}\eta}{RT}\right) = 1 \cdot \exp\left(-\frac{2 \cdot 96500 \cdot \eta}{8.314 \cdot 298.15}\right) \text{ atm} \qquad = \exp(-\mathbf{77.9} \cdot \eta) \textbf{ atm}$$

The equilibrium gas pressure $p = f(\eta)$ in internal pores is calculated for different values of the overvoltage η (volt); the result is given in the following table

η (volt)	−0.010	−0.025	−0.050	−0.075	−0.100	−0.125	−0.150
p (atm)	2	7	49	345	2416	16941	118777

Figure 6.51. In a laboratory experiment hydrogen was introduced into the metal in atomic form by electrolysis in sulfuric acid. When the atoms entered cavities, gas molecules were formed and became trapped, permitting the gas pressure to reach high levels.

Discussion. The calculations show that even modest cathodic activation overvoltages can theoretically build op internal hydrogen gas pressure that can result in **blistering** and even **rupture** of the cathode metal. In practice, such hydrogen damages are well known within the galvanic industry; cathodic reduction of hydrogen in galvanic baths that are maintained at low pH levels often results in blistering of the base metal. Hydrogen brittleness in combination with blistering and rupture of steel plates is also a well-known phenomenon associated with galvanic **acid corrosion** in container plants. In these cases, the cause will often be cathodic reduction of hydrogen.

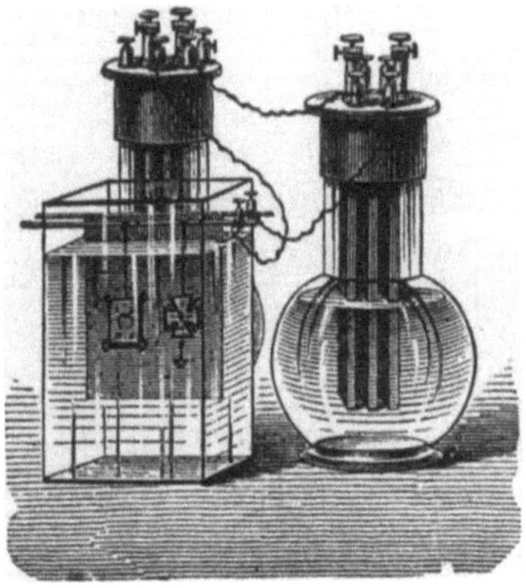

Figure 6.52. From the early days of electrochemistry, galvanoplastic-electrolytic metal coating has been used for solution of different technical and artistic tasks.

Exercises

The following exercises can be used when learning the subjects addressed in Chapter 6. The times given are suggested guidelines for use in assessing the effectiveness of the teaching.

Exercise 6.1

☐ Conversion factor between electron volt and Joule *(4 min.)*

The unit eV ("electron volt") designates the **work** done when an **elementary charge** e is conducted reversibly over an electric potential difference of $\Delta V = +1$ volt. Determine the numerical value for the conversion factor $k1$ (J/eV) and the conversion factor $k2$ (eV/J) that are are used for conversion of the unit (eV) into (J) and from the unit (J) into (eV), respectively!

Exercise 6.2

☐ Charge and current *(8 min.)*

During start-up of a certain industrial plant, the amperage $I(t)$ increases from $I = 0$ A to the final value $I = 1000$ A according to the following relation (a)

$$I(t) = 1000 \cdot (1 - \exp(-\frac{t}{60})) \quad \text{A} \qquad \text{(a)}$$

where the time t is given in seconds. Calculate the total charge increase Q (C) to the plant at the time $t = 40$ s after start-up !

Exercise 6.3

☐ Galvanic precipitation of copper, Cu *(8 min.)*

In a plant with galvanic bath for electrolytic copper-plating of steel specimens, 5.00 kg of copper, Cu(s), per hour are precipitated. The precipitation takes place due to reduction of Cu^{++}(aq) from a sulphuric solution. Calculate the amperage I (A) supplied during operation of the plant!

Exercise 6.4

☐ Electrical work *(8 min.)*

Assume a thermodynamic system consisting of an electric resistance of $R = 20.0\,\Omega$. A constant current of $I = 2.00$ A runs through the resistance. Calculate: a) the drop in voltage ΔV over the resistance, b) the total charge transfer Q (C) per hour, and c) the magnitude of the thermodynamic work W (J) done on the system in the course of 1 hour!

Exercise 6.5

☐ Battery as a thermodynamic system *(8 min.)*

Assume a thermodynamic system consisting of a battery with a **constant** terminal voltage of $V1 = 0$ volt and $V2 = 12$ volt. Through an external conductor an electric resistance with $R\,(\Omega)$ between terminal (1) and terminal (2) is plugged in. At the specified terminal voltage, a constant current of $I = 0.1$ A runs through the resistance for a total of 10 min.

Calculate a) the **resistance** $R\,(\Omega)$, b) the **charge** Q (C) transferred, and c) the **work** W (J), done by the surroundings on the system !

Figure 6.53. Aluminium is extensively used for aerial cables in high-voltage plants; this is due to a favourable combination of the density, strength, conductivity, and price of the metal.

Exercise 6.6

☐ Aluminium for aerial cables *(8 min.)*

Aluminium is extensively used as a live conductor in, for example, aerial cables for high-voltage plants. Aluminium is in many cases preferred to copper because of the favourable *combination* of **conductivity**, **density**, **strength**, and **price** of aluminium.

By measuring a 100.0 m long aluminium wire of circular cross-section of 2.00 mm the **resistance** of the wire has been determined as $R = 0.901\,\Omega$. Calculate a) the **conductance** $G\,(\Omega^{-1})$ of the wire, b) the **conductivity** $\sigma\,(\Omega^{-1}\mathrm{m}^{-1})$ of the wire metal, and c) the **resistivity** $\rho\,(\Omega\mathrm{m})$ of the wire metal!

Exercise 6.7

☐ Electrical work on elementary charge *(4 min.)*

A **positive** elementary charge e is transferred from infinity to a point (a) with electric potential of $V_a = -12.0$ volt. Calculate the **electrical work** $\delta W_{\infty.a}$ done on the charge; the result shall be given in electron volt (eV) and in (J)!

Exercise 6.8

☐ Electric transfer of charge *(6 min.)*

Consider a galvanic bath in operation. Due to the current in the electrolyte, 8.50 mol Zn^{++}(aq) of ions per hour are transferred through an imaginary sectional plane in the bath. Calculate a) the **electric charge** Q (C) transferred per hour by Zn^{++} ions, and b) the **amperage** I (A) related to this transfer of Zn^{++} ions!

Figure 6.54. Electrolysis hall for production of aluminium, Ørdal and Sunndal, Ltd., Trondheim, Norway. In each electrolysis cell, 24 graphite electrodes transfer a current of approximately 128 000 A, when the plant is operating.

Exercise 6.9

☐ Electrolytic conductivity *(12 min.)*

In a laboratory investigation, a measuring cell is built for determination of electrolytic conductivity. The cell consists of a glass basin with plane parallel sides; the measurements of the basin are: 100 mm × 100 mm × 30 mm. On the opposite sides with the dimension 100 mm × 100 mm, plane metal electrodes that fully cover the sides are applied to the inside of the basin. The electrode spacing is assumed to be 30 mm.

Then the measuring cell is filled up with the electrolyte solution to be investigated. The resistance R between the electrodes is determined by conducting an alternating current through the electrolyte; in this way, measuring errors due to polarization (see example 6.5) are eliminated. The measurement shows a resistance of $R = 32.5\,\Omega$

between the electrodes. Calculate the **conductivity** σ ($\Omega^{-1}\text{m}^{-1}$) and **resistivity** ρ (Ωm) of the electrolyte solution!

Figure 6.55. In coastal areas, the wind can carry sea salt in the form of microscopic particles. The salt stems from the surf zone where atomized water in the air evaporates and leaves microscopic salt particles. Metal exposed to this wind-induced salt are especially prone to galvanic corrosion.

Figure 6.56. Certain base metals form a dense passivating oxide layer on the surface when they are exposed to atmospheric air. This applies to, for example, aluminium that forms a wear-resistant layer of aluminium oxide Al_2O_3 when exposed to the air. Therefore, aluminium is far more corrosion resistant than indicated by its position in the electrochemical series.

Exercise 6.10

☐ **Production of aluminium by electrolysis** *(12 min.)*

Aluminium is made by electrolysis of an approximately 1000 °C melt of **cryolite** Na_3AlF_6 to which 2 - 8 % **bauxite** Al_2O_3 is added; the cryolite, which is more easily fusible, serves as **solvent** and **catalyst** during the process. Large-scale electrolysis of a melt of pure bauxite is not possible since the melting point of bauxite is above 2000 °C. Electrolysis of the cryolite-bauxite melt is performed in basin-like cells; a typical cell is operated by a maintained voltage of 4 - 5 volts and an amperage of approximately 100 000 A. Graphite is used as the cathode material; the melt formed by the reduced aluminium accumulates at the bottom of the basin. During the process, oxygen is liberated and reacts with the graphite cathode whereby CO_2(g) is formed. The net reaction is

$$2\text{Al}_2\text{O}_3(\text{l}) + 3\text{C(s)} \rightarrow 4\text{Al}(\ell) + 3\text{CO}_2(\text{g}) \qquad \text{(a)}$$

Calculate, how many kg of Al(ℓ) are produced in an electrolysis cell operated at an amperage of $I = 100\,000$ A !

Exercise 6.11

☐ **Equilibrium potential for concentration cell** *(12 min.)*

Electrochemical potential differences can arise between two electrodes of the *same* metal if the electrodes are in contact with electrolytes of *different* composition. This is, for example one of the reasons why deposited salt particles on a moist metal surface can induce **local** galvanic corrosion cells on the surface. Approximate from the **Nernst equation** the electrochemical **equilibrium potential** $\Delta V = V_2 - V_1$ (volt) at 25 °C for a Zn-Zn electrode pair with the following cell diagram

$$\text{Zn(s)} \,|\, \text{Zn}^{++}(\text{aq}, 0.005\,\text{mol}/\ell) \,\|\, \text{Zn}^{++}(\text{aq}, 1.5\,\text{mol}/\ell) \,|\, \text{Zn(s)} \qquad \text{(a)}$$

assuming the solutions to be ideal; then determine $\Delta_r G_{298}$ (kJ/mol) for the electrode reaction concerned!

Exercise 6.12

☐ **Standard potential and temperature coefficient** *(20 min.)*

Calculate from thermodynamic data for $G^\ominus_{298}$ (kJ/mol) the electrochemical **standard potential** $\Delta V^\ominus$ (volt) for the following electrochemical cells

$$\text{Zn(s)} \,|\, \text{Zn}^{++}(\text{aq}) \,\|\, 2\text{H}^{+}(\text{aq}) \,|\, \text{H}_2(\text{g}), (\text{Pt}) \qquad \text{(a)}$$

$$\text{Zn(s)} \,|\, \text{Zn}^{++}(\text{aq}) \,\|\, \text{Fe}^{++}(\text{aq}) \,|\, \text{Fe(s)} \qquad (b)$$

$$\text{Fe(s)} \,|\, \text{Fe}^{+++}(\text{aq}) \,\|\, 3\text{Ag}^{+}(\text{aq}) \,|\, \text{Ag(s)} \qquad \text{(c)}$$

and determine based on thermodynamic data for $S^\ominus_{298}$ (J/mol K), the **temperature coefficient** $\partial V^\ominus/\partial T$ (volt/K) for the three cell schemes specified!

Exercise 6.13

☐ **Electrochemical equilibrium potential** *(15 min.)*

Calculate from thermodynamic data for $G^\ominus_{298}$ (kJ/mol) the electrochemical **equilibrium potential** V (volt) for the following electrochemical cells

$$\text{Zn(s)} \,|\, \text{Zn}^{++}(\text{aq}, 10^{-2}\,\text{mol}/\ell) \,\|\, \text{Fe}^{++}(\text{aq}, 10^{-3}\,\text{mol}/\ell) \,|\, \text{Fe(s)} \qquad \text{(a)}$$

$$\text{Ni(s)} \,|\, \text{Ni}^{++}(\text{aq}, 0.01\,\text{mol}/\ell) \,\|\, 2\text{H}^{+}(\text{aq}, \text{c}^\ominus) \,|\, \text{H}_2(\text{g}), (\text{Pt}) \qquad \text{(b)}$$

assuming the metal ions to form ideal solutions with the activity: $a = c/c^\ominus$!

Exercise 6.14

☐ **Concentration dependence of the potential** *(10 min.)*

Set up, based on thermodynamic data for $G^\ominus_{298}$ (kJ/mol), an expression for the electrochemical equilibrium potential $V = V(c)$ (volt) for the electrode reaction

$$\text{Zn}^{++}(\text{aq}, \text{c}) \,|\, \text{Zn(s)} \qquad (\text{volt}) \qquad \text{(a)}$$

where $V(c)$ expresses the potential as a function of the molar concentration of zinc ions $c = [Zn^{++}]$; draw up a table of the potential $V(c)$ in the range $10^{-6}\,\text{mol}/\ell \le c \le c^{\ominus}$!

Exercise 6.15

☐ Electrochemical series *(12 min.)*

By galvanic contact corrosion, base metals will often form **anodes** and decompose by oxidation while more noble metals function as cathodes. The **electrochemical series** specifies the reduction potential $V^{\ominus}$ (volt) for single electrodes; in this series, base anodic metals have a *low* reduction potential and noble cathodic metals have a *high* reduction potential. The following thermodynamic standard values are given for $G^{\ominus}_{298}$ (kJ/mol)

Ca^{++}(aq)	Ag^{+}(aq)	Na^{+}(aq)	Cr^{++}(aq)	Al^{+++}(aq)	Ni^{++}(aq)
−553.6	+77.1	−261.9	−176.2	−485.3	−45.6

Calculate, based on these data *alone*, the **standard potential** $V^{\ominus}$ (volt) for reduction of the specified metal ions to metal and range them by increasing "nobleness" in the electrochemical series! (hint: $G^{\ominus} = 0$ for pure metals).

Exercise 6.16

☐ Calculation of electrochemical equilibrium potential *(15 min.)*

From calculated electrochemical equilibrium potentials it can often be decided whether galvanic corrosion of a metal is *possible* under given circumstances. On the other hand, the equilibrium calculations *cannot* predict anything about the rate of corrosion, and thus, the extent of corrosion attack, if any. Calculate the electrochemical **equilibrium potential** ΔV (volt) at 25 °C for the following cell diagram

$$Fe(s)\,|\,Fe^{++}(aq, 1\cdot 10^{-3}\,\text{mol}/\ell)\,||\,\tfrac{1}{2}O_2(g, 0.20\,\text{atm}), 2H^{+}(aq)\,|\,H_2O(\ell), (Fe) \qquad (a)$$

for pH = 3.0 and specify whether *spontaneous* oxidation of iron Fe(s) can take place under the given conditions! – ideal ion solutions shall be assumed.

Exercise 6.17

☐ pH dependence of the standard potential *(15 min.)*

The **standard potential** for $Sn^{++}(aq) + 2e^{-} \rightarrow Sn(s)$ is $\Delta V^{\ominus} = -0.14$ volt. The following cell diagram shows **oxidation** of Sn(s) under hydrogen development

$$Sn(s)\,|\,Sn^{++}(aq,c)\,||\,2H^{+}(aq)\,|\,H_2(g,p^{\ominus}), (Sn) \qquad (a)$$

Definition of pH

$$\text{pH} \overset{\text{def}}{=} -\log_{10}(a(H^{+}))$$

Draw up a calculation expression $c(Sn^{++}) = f(\text{pH})$ stating the molar concentration $c = [Sn^{++}]$ of stannic ions in the electrolyte at **reaction equilibrium** in the cell ($\Delta V = 0$) as a function of the *pH* of the electrolyte ! – solutions are assumed to be ideal.

Exercise 6.18

☐ Calculation of standard free energy $G^{\ominus}$ *(16 min.)*

The standard potential at 25 °C, $V^{\ominus}$ for **reduction** of the **cupric ions** Cu^{++}(aq) and the cuprous ion Cu^{+}(aq), respectively, is given according to the following reaction equations

$$Cu^{++}(aq) + 2e^{-} \rightarrow Cu(s) \qquad V^{\ominus} = +0.342\,\text{volt}$$

$$Cu^{+}(aq) + e^{-} \rightarrow Cu(s) \qquad V^{\ominus} = +0.521\,\text{volt}$$

Calculate from this information, the **standard free energy** $G^{\ominus}_{298}$ (kJ/mol) for Cu^{++}(aq) and Cu^{+}(aq), and calculate the molar concentration of **cuprous ions**: $c = [Cu^{+}]$ in an electrolyte where the following electrochemical equilibrium has been established at 25 °C

$$Cu^{++}(aq, 0.1\,\text{mol}/\ell) + Cu(s) \leftrightarrows 2Cu^{+}(aq,c) \qquad (a)$$

By calculation, ions are assumed to form an ideal solution so that $a = c/c^{\ominus}$.

Exercise 6.19

☐ Equilibrium concentration at contact corrosion *(12 min.)*

At pH 7, an electrolyte contains Fe(s), Zn(s), Fe^{++}(aq,c1) and Zn^{++}(aq,c2), where $c1 = [Fe^{++}]$ denotes the molar concentration of **ferrous ions**, and $c2 = [Zn^{++}]$ denotes the molar concentration of **zinc ions** in the solution. The temperature is $\theta = 25\,°C$. The following **standard potentials** are given

$$V^{\ominus}(Fe^{++} + 2e^- \to Fe) = -0.41\,\text{volt} \qquad V^{\ominus}(Zn^{++} + 2e^- \to Zn) = -0.76\,\text{volt}$$

Calculate the molar equilibrium concentration $c1$ of ferrous ions in the solution if the concentration of zinc ions is $c2 = [Zn^{++}] = 0.1\,\text{mol}/\ell$!

Exercise 6.20

☐ Pourbaix diagram for zinc *(20 min.)*

Zinc is a **base** metal with a negative standard reduction potential in the electrochemical series; the metal is *amphoteric* and is thus attacked by both **acids** and **bases**. The Pourbaix diagram for zinc reflects these special properties. In a narrow pH area about 10, zinc is passivated by precipitation of the very sparingly soluble **zinc hydroxide** $Zn(OH)_2$ on the surface. In an alkaline environment at pH > circa 11, this passivating layer is decomposed during formation of $HZnO_2^-$(aq) and ZnO_2^{--}(aq). Figure 6.57 shows a Pourbaix diagram for zinc; the diagram is shown for $[Zn^{++}] = 10^{-6}\,\text{mol}/\ell$. The lines delimiting areas with **immunity**, **passivation**, and **corrosion** have been determined by the following electrochemical equilibrium reactions

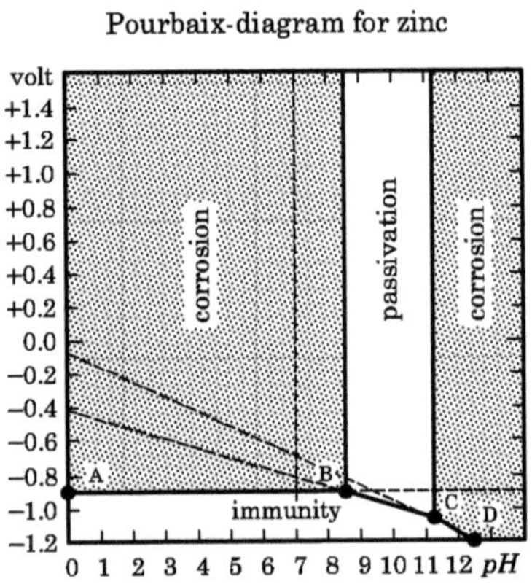

Figure 6.57. Pourbaix diagram, zinc, at 25 °C. The figure is an enclosure to exercise no. 6.20.

$$Zn^{++}(aq) + 2e^- \leftrightarrows Zn(s) \qquad \text{(A–B)}$$

$$Zn(OH)_2(s) + 2H^+(aq) + 2e^- \leftrightarrows Zn(s) + 2H_2O(\ell) \qquad \text{(B–C)}$$

$$HZnO_2^-(aq) + 3H^+(aq) + 2e^- \leftrightarrows Zn(s) + 2H_2O(\ell) \qquad \text{(C–D)}$$

Above A–B, Zn is oxidized to Zn^{++}, above B–C, sparingly soluble, passivating $Zn(OH)_2$ is precipitated, and above C–D, $HZnO_2^-$ and ZnO_2^{--}, respectively, are formed at very high pH values. The following table data have been given for the standard free energy $G^{\ominus}_{298}$ (kJ/mol) of the components

Zn(s)	Zn^{++}(aq)	$HZnO_2^-$(aq)	$Zn(OH)_2$(s)	$H_2O(\ell)$	H^+(aq)
0.0	−147.0	−457.4	−555.1	−237.2	0.0

Determine from the above information the *analytic* expression for the boundary line through A–B, for the boundary line through B–C, and for the boundary line through C–D; the results shall be given on the form: $V = a + b \cdot pH$ (volt). In calculations, ideal solutions and that $[Zn^{++}] = [HZnO_2^-] = 1 \cdot 10^{-6}\,\text{mol}/\ell$ shall be assumed!

Literature

The following literature and supplementary readings are recommended for further reference on the subjects covered in Chapter 6.

References

In Chapter 6, a number of references to the literature have been made; the following is a complete list of citations of the references used.

- Ewing, G.W.: *Instrumental Methods of Chemical Analysis*, McGraw-Hill Book Compagny, New York 1969.
- Pourbaix, M.: *Atlas d'Équilibres Électrochimiques*, Gauthier-Villars & Co[ie], Paris 1963.
- Beckris, J.O'M, Bonciocat, N. & Gutma, F.: *An Introduction to Electrochemical Science*, Wykeham Publications Ltd., London 1974.
- Bates, R.G. & Bowel, V.E.: *Journal of Research*, National Bureau of Standards No. 53, 1954.

Supplementary literature

- Rieger, Philip Henri: *Electrochemistry*, Springer Verlag, 1994.

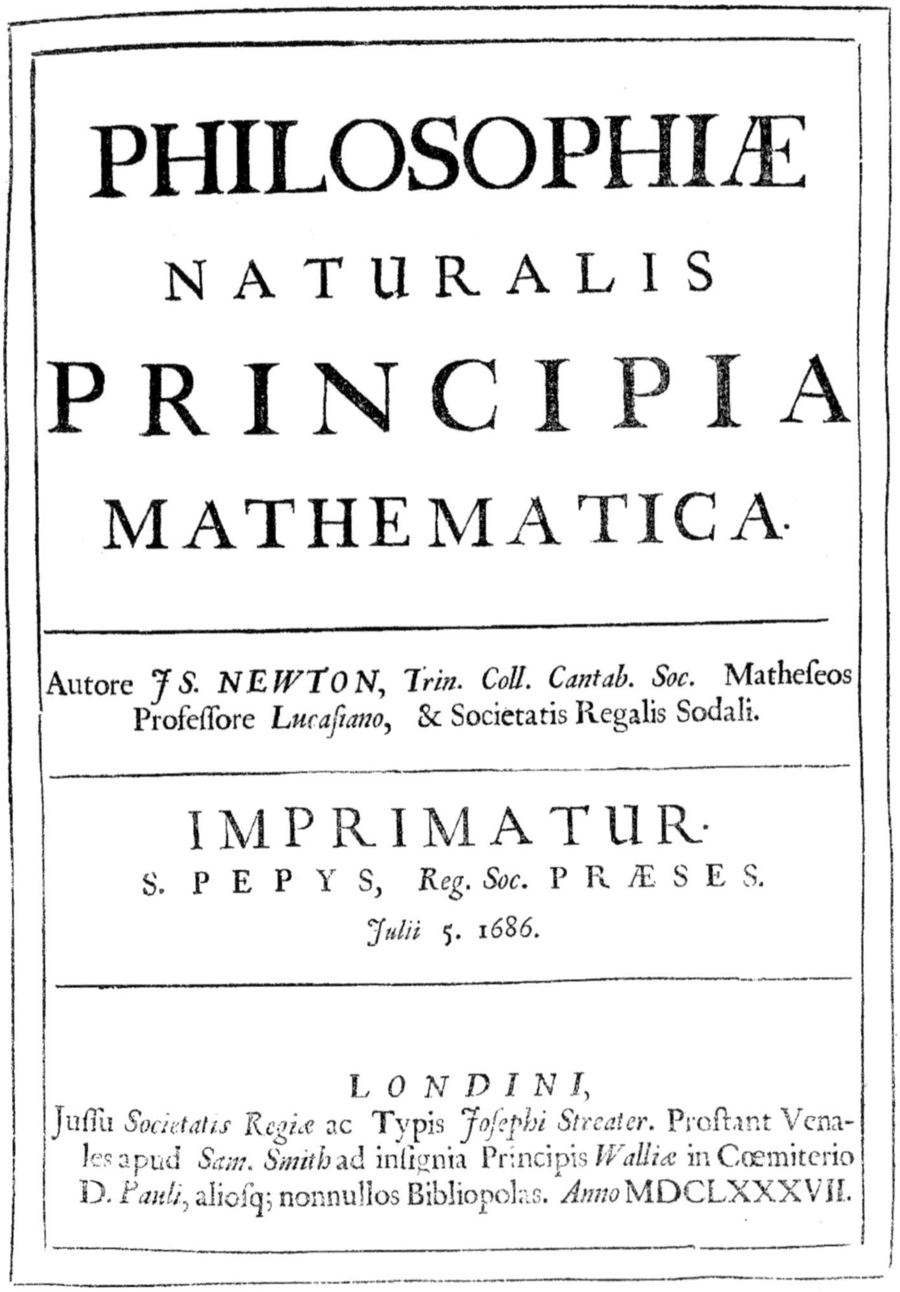

PHILOSOPHIÆ
NATURALIS
PRINCIPIA
MATHEMATICA.

Autore *J S. NEWTON*, *Trin. Coll. Cantab. Soc.* Matheſeos Profeſſore *Lucaſiano*, & Societatis Regalis Sodali.

IMPRIMATUR.
S. PEPYS, *Reg. Soc.* PRÆSES.
Julii 5. 1686.

LONDINI,
Juſſu *Societatis Regiæ* ac Typis *Joſephi Streater*. Proſtant Venales apud *Sam. Smith* ad inſignia Principis *Walliæ* in Cœmiterio D. *Pauli*, alioſq; nonnullos Bibliopolas. *Anno* MDCLXXXVII.

Title page from Newton's famous landmark work "Principia"; the work, which contains three volumes, was published in the years 1686-87.

APPENDIX A

Mathematical appendix

The following appendix gives a brief summary of selected physical and mathematical **concepts** and **methods**. The subjects are:

- **Thermodynamic derivations** that support and expand the theoretical sections in the text (Maxwell's relations, Debye Hückel's law, etc.).
- **Mathematical concepts**, which are fundamental for the understanding of physical or chemical definitions and derivations in the text, but which due to their length would make it harder to get an overview of the text (Linear regression, Exact differential, etc.).
- **Numerical methods**, which are suitable for solving practical calculation problems, but which if not included here will have to be looked up in textbooks or notes every time they are used (Newton-Rapson iteration, Cramer's formula, etc.).

An overview of some works of reference and textbooks that may be useful to consult is given at the end of each of the individual sections.

Dimensional analysis is an efficient tool for solving technical problems. Therefore, a short introduction of this tool is provided here and is supplemented with practical examples.

Isaac Newton (1642-1727)

Mathematician and physicist, professor at University of Cambridge.

Contents

1 Numerical calculations

Normally the results of technical calculations are imbued with a certain **uncertainty**, which is due to the uncertainties inherent in the input data used. Therefore, in a calculated result, the numerical information should be adjusted to the accuracy that can be achieved using the input data. For this part of the calculation work, one can lean on certain simple rules for uncertainty calculation.

Input data

Input data for technical calculations can be **exact** quantities or **uncertain** quantities. Normally, exact quantities specify a **number** of units (e.g. **12** protons, **60** seconds), or they express a physical or mathematical **definition** (e.g. **101325** Pa/atm or the value of $\boldsymbol{\pi}$). Exact quantities are *not* encumbered with uncertainty. In contrast, uncertain quantities are **measured** physical quantities or **rounded** numerical values encumbered with more or less uncertainty (e.g. R = $8.31441 \pm 0.00026\,\mathrm{J/mol\,K}$, $\pi \simeq 3.14$).

Specification of uncertainty

The uncertainty on a measured physical quantity depends on the measuring equipment used and how the measurement is performed, and also on the number of measurements. Conventionally, the result of a measurement is given so that the uncertainty can be seen directly or indirectly from the numerical value. When the actual **standard deviation** s_x for a measured quantity x is known, the result is, for example, given on the form:

$$x \pm s_x \qquad \text{example : } 3.74\,m \pm 0.02\,m \tag{a}$$

If no measured standard deviation is available, the numerical value should be given to as many **significant digits** that the uncertainty is confined to the last digit given. In this case – and in case no other information is available – the last digit of the number will be given the uncertainty ± 1. In this case, a given physical quantity x is to be read in the following way:

$$\text{statement : } 37.8\,^{\circ}\mathrm{C} \qquad \text{to be read as: } 37.8 \pm 0.1^{\circ}\mathrm{C} \tag{b}$$

Note that the indirect statement of uncertainty in the expression (b) is not very well defined. In both of the above cases, the number of **significant digits** has been determined so that the uncertainty concerns the **last digit** in the result. Generally, for estimates of uncertainty, an accuracy in the estimate of uncertainty better than 10 - 20 % is seldom necessary; therefore, the uncertainty is given with a maximum of two significant digits.

Recapitulation (c)

Measured or calculated physical quantities are given to exactly as many **significant digits** that the uncertainty is confined to the **last** or **next-to-last** digit in the result.

Significant digits

As explained above, the uncertainty of a physical quantity can indirectly be stated through the number of **significant digits**.

Significant digits (d)

The number of **significant digits** in a number c are the number of digits in the number exclusive of leading or trailing zero that are given with the *sole* purpose of specifying the position of the decimal separator.

Generally, the number of significant digits in a stated number can be seen unambiguously from the numerical data; only where the integer has a trailing zero is the number of significant digits not uniquely established in the numerical data given. The concept of significant digits according to (d) can best be illustrated by some examples.

Example no.	Number c	Significant digits in bold	Number of significant digits
1	3.700	**3.700**	4
2	370.0	**370.0**	4
3	3700	**37**00?, **370**0?, **3700** ?	2, 3 or 4?
4	0.00370	0.0**370**	3
5	0.00003	0.0000**3**	1

Example no. 3 illustrates that an integer with one or more trailing 0 digits does not unambiguously establish the number of significant digits. In such cases, "*scientific notation*" that establishes the number of significant digits in all cases should always be used. Example 3 in the table is stated as follows: $3.7 \cdot 10^3$, $3.70 \cdot 10^3$, and $3.700 \cdot 10^3$, respectively, corresponding to the number of significant digits being 2, 3, or 4.

The fact that a number is specified with many decimals does *not* necessarily imply high accuracy - which could easily be believed when using a pocket calculator. From example 5, it should be noted that a number with 5 decimals and one significant digit carries an uncertainty of around 30 % of the numerical value.

Mathematical uncertainty calculation

In a calculation of a composite expression, the uncertainty in the individual terms will each contribute to the uncertainty of the final result. A brief description explaining how the uncertainty of the result of a composite calculation can be determined from the uncertainties in the individual terms is given in the following.

Consider a composite calculation expression of the form

$$y = y(a,\ b,\ c,\ \cdots)$$

where a known uncertainty Δa, Δb, Δc, $\cdots$ can be ascribed to the individual terms of the expression: a, b, c, $\cdots$. Then, the uncertainty Δy of the calculated result y in accordance with the mathematical theory of uncertainty (Coleman, H.W. & Steele, W.G.: Experimentation and Uncertainty Analysis for Engineers) is determined by

$$\Delta y = \sqrt{(\frac{\partial y}{\partial a} \cdot \Delta a)^2 + (\frac{\partial y}{\partial b} \cdot \Delta b)^2 + (\frac{\partial y}{\partial c} \cdot \Delta c)^2 + \cdots} \tag{e}$$

Use of the propagation of uncertainty (e) can be illustrated by some simple special cases:

Linear expression:

$$y = 2a^2 + 5b - 3c^{-1}$$

$$\frac{\partial y}{\partial a} = 4a;\ \frac{\partial y}{\partial b} = 5;\ \frac{\partial y}{\partial c} = 3c^{-2}$$

$$\Delta y = \sqrt{(4a \cdot \Delta a)^2 + (5 \cdot \Delta b)^2 + (3c^{-2} \cdot \Delta c)^2}$$

Product of factors:

$$y = a^2 \cdot b \cdot c$$

$$\frac{\partial y}{\partial a} = 2abc;\ \frac{\partial y}{\partial b} = a^2 c;\ \frac{\partial y}{\partial c} = a^2 b$$

$$\Delta y = \sqrt{(2abc \cdot \Delta a)^2 + (a^2 c \cdot \Delta b)^2 + (a^2 b \cdot \Delta c)^2}$$

The procedure outlined above for calculating the uncertainty Δy of a calculated result can be applied to all conventional calculation expressions. In practical calculations, however, it is only necessary to perform a detailed uncertainty calculation in exceptional cases; normally, the uncertainty of a calculated numerical value can be determined with sufficient accuracy by simpler methods for rough estimates that are provided in the following section.

The mathematical uncertainty calculation for a composite calculation expression can be summarized in the following procedure:

Uncertainty calculation, composite expression (f)

For a composite calculation expression $y = y(a, b, c, \cdots)$ where the terms have known, mutually independent uncertainties Δa, Δb, Δc, $\cdots$, the uncertainty Δy on the result y is determined by the following procedure:

(1) Determine the **partial derivatives** $\frac{\partial y}{\partial a}, \frac{\partial y}{\partial b}, \frac{\partial y}{\partial c}, \cdots$ by differentiation.

(2) Calculate the **numerical value** of the partial derivatives by **insertion.**

(3) Calculate the **uncertainty** Δy on the result as:

$$\Delta y = \sqrt{(\frac{\partial y}{\partial a} \cdot \Delta a)^2 + (\frac{\partial y}{\partial b} \cdot \Delta b)^2 + (\frac{\partial y}{\partial c} \cdot \Delta c)^2 + \cdots}$$

(4) **Round off** the result y so that the uncertainty is put on the last **significant digit**.

Approximate uncertainty calculation

In practical calculations the number of significant digits in the result can normally be established with sufficient accuracy using the following simple rules.

Addition/subtraction: $y = a + b$; given : $\Delta a \gg \Delta b$

$$\Delta y = \sqrt{(1 \cdot \Delta a)^2 + (1 \cdot \Delta b)^2} \simeq \Delta a$$

Rule no. 1: *By addition or subtraction of terms the absolute uncertainty Δy on the result y is approximately equal to the absolute uncertainty on the most uncertain element a.*

- The result is given with the same decimal accuracy as used for the most uncertain element a.
- Example: 3.**4** + 1.23861 = 4.**6**

Multiplication: $y = a \cdot b$; given : $\frac{\Delta a}{a} \gg \frac{\Delta b}{b}$

$$\frac{\Delta y}{y} = \frac{\sqrt{(b \cdot \Delta a)^2 + (a \cdot \Delta b)^2}}{a \cdot b} \simeq \frac{\Delta a}{a}$$

Rule no. 2: *By multiplication of terms, the relative uncertainty $\Delta y/y$ on the result y is approximately equal to the relative uncertainty on the most uncertain factor.*

- The result y is stated with the same number of significant digits as used for the most uncertain factor
- Example: **3.4** · 1.23861 = **4.2**

Logarithmic function: $y = \ln(a)$; given : Δa

$$\Delta y = \sqrt{(\frac{1}{a} \cdot \Delta a)^2} = \frac{\Delta a}{a}$$

Rule no. 3: By *logarithmic calculation* $\ln(a)$, *the absolute uncertainty* Δy *on the result* y *is equal to the relative uncertainty on the argument* a.

- The result y is stated with the same number of decimals as the number of significant digits in the argument a.
- Example: $\ln(\mathbf{25.3}) = 3.\mathbf{231}$

Exponential function: $y = \exp(a);\quad \text{given}: \Delta a$

$$\frac{\Delta y}{y} = \frac{\sqrt{(\exp(a)\cdot\Delta a)^2}}{\exp(a)} = \Delta a$$

Rule no. 4: *By calculation of the exponential expression* $\exp(a)$ *the relative uncertainty* $\Delta y/y$ *of the result* y *is equal to the absolute uncertainty on the argument* a.

- The result y is stated with $(n+1)$ significant digits, where n is the number of decimals in the argument a.
- Example: $\exp(3.\mathbf{462}) = \mathbf{31.88}$

Exponential function: $y = a^n;\quad \text{given}: \Delta a;\quad n \text{ exact}$

$$\frac{\Delta y}{y} = \frac{\sqrt{(na^{n-1}\cdot\Delta a)^2}}{a^n} = |n|\cdot\frac{\Delta a}{a}$$

Rule no. 5: *When, for example,* a *is raised to the nth power, the relative uncertainty* $\Delta y/y$ *of the result becomes n times the relative uncertainty in the argument* a.

- Example: $(3.4 \pm \mathbf{2}\ \%)^3 = 39.304 \pm \mathbf{6}\ \% = 39$

These simple rules for approximate uncertainty calculation have been summarized in the following overview table

Rule no.	Relation	Given	Result
1	$y = a + b$	$\Delta a \gg \Delta b$	$\Delta y \simeq \Delta a$
2	$y = a \cdot b$	$\frac{\Delta a}{a} \gg \frac{\Delta b}{b}$	$\frac{\Delta y}{y} \simeq \frac{\Delta a}{a}$
3	$y = \ln(a)$	Δa	$\Delta y = \frac{\Delta a}{a}$
4	$y = \exp(a)$	Δa	$\frac{\Delta y}{y} = \Delta a$
5	$y = a^n$	Δa	$\frac{\Delta y}{y} = \lvert n\rvert \cdot \frac{\Delta a}{a}$

Combining these rules for approximate uncertainty calculation, it can be ensured that a calculation result expresses the accuracy on which the input data are based. This prevents, for example, that the density of a mortar sample is stated as 2.29583 g/cm^3 when the value has been calculated from a measured quantity of 5.51 g and a measured volume of 2.4 cm^3.

For most calculations, an approximate uncertainty calculation is adequate. On the other hand, for a problem where the very uncertainty on the result is significant, a complete mathematical uncertainty calculation should be performed following the procedure (f).

■ Thermodynamic calculation of the ion product and pH of water

Pure water $H_2O(\ell)$ always contains modest concentrations of **hydrogen ions** $H^+(aq)$ and **hydroxide ions** $OH^-(aq)$. These ions are produced by **dissociation** of water molecules according to the equation:

$$H_2O(\ell) \longrightarrow H^+(aq) + OH^-(aq) \qquad (g)$$

By the dissociation equilibrium (g), a concentration of hydrogen ions $[H^+]$ will be established, and thus the **pH value** of the water can be calculated.

Problem. From thermodynamic data for the standard free energy of the components, calculate the following equilibrium quantities for the dissociation reaction (g) at 25.0°C:

- The thermodynamic equilibrium constant K_a
- The pH value of the water

The results should be stated on such a form that they reflect the uncertainty that is due to the input data used.

Conditions. In *Physical Chemistry*, P.W. Atkins specifies the following thermodynamic standard values for $H_2O(\ell)$, $H^+(aq)$ and $OH^-(aq)$:

$$G^{\ominus}(H_2O) = -237.13\,kJ/mol; \quad G^{\ominus}(H^+) = 0\,kJ/mol$$
$$G^{\ominus}(OH^-) = -157.24\,kJ/mol$$

The Danish Standards Association's *Standards for SI units* (in Danish: Dansk Standardiseringsråd *Standarder for SI-enheder*) specifies the following value for the gas constant R:

$$R = 8.31441 \pm 0.00026\,J/molK.$$

In the calculations, the following assumptions are made:

- At the concentration concerned, hydrogen ions $H^+(aq)$ and hydroxide ions $OH^-(aq)$ form an **ideal solution** in water.
- The uncertainties of the standard values $G^{\ominus}(H_2O)$ and $G^{\ominus}(OH^-)$ provided are both taken as 0.01 kJ/mol, since no uncertainty value has been specified. The standard value $G^{\ominus}(H^+)$ is exact because we are dealing with a definition.
- The uncertainty with respect to the temperature $\theta = 25.0\,°C$ is assumed to be 0, because we are dealing with a chosen value. Since $T_0 = 273.15\,K$ is an exact value, the uncertainty of the thermodynamic temperature $T = 298.15\,K$ becomes 0.

Solution. *Calculation with detailed uncertainty determination*

In the case of equilibrium, the dissociation (g) is reversible, and $\Delta_r G = 0$ for the reaction. This entails that the following thermodynamic equilibrium condition is fulfilled

$$K_a = \frac{a(H^+) \cdot a(OH^-)}{a(H_2O)} = \exp\left(-\frac{\Delta_r G_T^{\ominus}}{RT}\right) \qquad (h)$$

The standard free reaction energy $\Delta_r G_T^{\ominus}$ for the reaction is determined by

$$\Delta_r G_T^{\ominus} = (G_T^{\ominus}(H^+) + G_T^{\ominus}(OH^-)) - G_T^{\ominus}(H_2O) \qquad (i)$$

$$\Delta_r G_{298}^{\ominus} = (0 + (-157.24)) - (-237.13)\,kJ/mol \qquad = \mathbf{79.89\,kJ/mol}$$

The uncertainty $\Delta(\Delta G^{\ominus})$ is determined by the procedure (f).

$$\frac{\partial \Delta_r G^{\ominus}}{\partial G^{\ominus}(H^+)} = 1; \quad \frac{\partial \Delta_r G^{\ominus}}{\partial G^{\ominus}(OH^-)} = 1; \quad \frac{\partial \Delta_r G^{\ominus}}{\partial G^{\ominus}(H_2O)} = -1;$$

$$\Delta(\Delta_r G^{\ominus}) = \sqrt{(1 \cdot 0)^2 + (1 \cdot 0.01)^2 + (-1 \cdot 0.01)^2} \qquad = \mathbf{0.014\,kJ/mol}$$

The thermodynamic equilibrium constant K_a at $T = 298.15$ K is calculated from (h)

$$K_a = \exp\left(-\frac{\Delta_r G_T^{\ominus}}{RT}\right) = \exp\left(-\frac{79890}{8.31441 \cdot 298.15}\right) \qquad = \mathbf{1.00876 \cdot 10^{-14}}$$

The uncertainty on the equilibrium constant K_a is then calculated by the procedure (f); it should be remembered that there are only two uncertain quantities in the expression:

$\Delta_r G^\ominus_{298}$ and R. For calculation of numerical values, the value for K_a found above can be inserted into the exponential function

$$\frac{\partial K_a}{\partial \Delta_r G^\ominus} = -\frac{1}{RT}\cdot \exp(-\frac{\Delta_r G^\ominus_T}{RT}) = -\frac{1.00876\cdot 10^{-14}}{8.31441\cdot 298.15} = 4.1\cdot 10^{-18}$$

$$\frac{\partial K_a}{\partial R} = -\frac{\Delta_r G_T}{R^2 T}\cdot \exp(-\frac{\Delta_r G^\ominus_T}{RT}) = -\frac{79890\cdot 1.00876\cdot 10^{-14}}{8.31441^2\cdot 298.15} = 3.9\cdot 10^{-14}$$

The uncertainty $\Delta(K_a)$ on the thermodynamic equilibrium constant is now determined using item (3) in the procedure (f); note the change to the unit J/mol for $\Delta_r G^\ominus$

$$\Delta(K_a) = \sqrt{(4.1\cdot 10^{-18}\cdot 14)^2 + (3.9\cdot 10^{-14}\cdot 0.00026)^2} = \mathbf{0.0058\cdot 10^{-14}}$$

When the pH value is calculated, it should be noted that $a(\mathrm{H^+}) = a(\mathrm{OH^-})$ in this solution. Since the activity of the water is $a(\mathrm{H_2O}) = 1$, the equilibrium relation (h) can be directly rewritten as follows

$$K_a = \frac{a(\mathrm{H^+})\cdot a(\mathrm{OH^-})}{a(\mathrm{H_2O})} = \frac{a(\mathrm{H^+})^2}{1} \quad\Rightarrow\quad a(\mathrm{H^+}) = \sqrt{K_a} \qquad \text{and thus}$$

$$\mathrm{pH} \overset{\text{def}}{=} -\log(a(\mathrm{H^+})) = -\log(\sqrt{K_a}\,) = -\log(\sqrt{1.00876\cdot 10^{-14}}\,) = \mathbf{6.998}$$

The uncertainty Δ (pH) on the calculated pH value is found using the procedure (f)

$$\frac{\partial \mathrm{pH}}{\partial K_a} = -\frac{1}{2\cdot \ln(10)\cdot \mathrm{K_a}} = \frac{1}{2\cdot 0.217\cdot 1.00876\cdot 10^{-14}} = \mathbf{-2.3\cdot 10^{14}}$$

$$\Delta(\mathrm{pH}) = \sqrt{(-2.3\cdot 10^{14}\cdot 0.0058\cdot 10^{-14})^2} = \mathbf{0.013}$$

The uncertainty determined is rounded to one decimal and the number of significant digits are rounded so that the uncertainty applies to the digit stated last. Thus, the calculations lead to the following result

> **Result**
>
> Thermodynamic equilibrium constant : $K_a = 1.009\cdot 10^{-14} \pm 0.006\cdot 10^{-14}$
>
> pH value of pure water : $\mathrm{pH} = 7.00 \pm 0.01$

Solution. *Calculation with approximate uncertainty determination*

$\Delta_r G^\ominus_{298}$ is determined with the same number of decimals as used for specifying $G^\ominus$ values (**Rule 1**)

$$\Delta_r G^\ominus_{298} = (0 + (-157.24)) - (-237.13)\,\mathrm{kJ/mol} = \mathbf{79.89\,kJ/mol}$$

The relative uncertainty on $\Delta_r G^\ominus$ is $\sim 1\cdot 10^{-4}$ and the relative uncertainty on the gas constant R is $\sim 3\cdot 10^{-5}$; according to **Rule 2** and **Rule 5**, respectively, $\Delta_r G^\ominus/RT$ is given with four significant digits corresponding to a relative uncertainty of $\sim 1\cdot 10^{-4}$

$$\frac{\Delta_r G^\ominus}{RT} = \frac{79890}{8.31441\cdot 298.15} = \mathbf{32.23}$$

The equilibrium constant is calculated and specified according to **Rule 4** with three significant digits; the pH value is calculated according to **Rule 3** with 3 decimals

$$K_a = \exp(-32.23) = \mathbf{1.01\cdot 10^{-14}}$$

$$\mathrm{pH} \overset{\text{def}}{=} -\log(a(\mathrm{H^+})) = -\log(\sqrt{K_a}\,) = -\tfrac{1}{2}\cdot\log(1.01\cdot 10^{-14}) = \mathbf{6.998}$$

Thus, the calculations performed with approximate uncertainty assessment have lead to the following result

> **Result**
>
> Thermodynamic equilibrium constant : $K_a = 1.01\cdot 10^{-14}$
>
> pH value of pure water : $\mathrm{pH} = 6.998$

Discussion. By accurate **electrochemical** measurements of the ion product of water at 25 °C the following was found: $K_a = 1.0083 \cdot 10^{-14}$ (Duncan A. MacInnes: *The Principles of Electrochemistry*). Within the uncertainty interval established, the thermodynamically calculated value for K_a corresponds very well to this:

Mathematical uncertainty calculation: $1.003 \cdot 10^{-14} < K_a < 1.015 \cdot 10^{-14}$

Approximate uncertainty calculation: $1.00 \cdot 10^{-14} < K_a < 1.02 \cdot 10^{-14}$

As seen from the example, the mathematical uncertainty calculation by the procedure (f) involves quite an amount of extra calculation work. An actual mathematical uncertainty calculation, therefore, is typically performed only in tasks where it is important to know the magnitude of the uncertainty. This applies, for example, to processing and tabulation of measurement results.

By contrast, an approximate uncertainty calculation can be performed during the process of calculation itself without increasing the amount of calculation work to a noticeable degree. Therefore, these simple estimates of the numbers are a natural part of any technical calculation.

Literature

Coleman, H.W. & Steele, W.G.: *Experimentation and Uncertainty Analysis for Engineers*, 2nd ed, Wiley-Interscience, New York.

Gordus, A.A.: *Analytical Chemistry*, Schaum's Outline Series in Science, McGraw-Hill Book Company, New York 1985

Kreyszig, E.: *Advanced Engineering Mathematics*, John Wiley and Sons, Inc., New York 1972

2 Dimensional analysis

Dimensional analysis is a very efficient tool for solving technical-physical problems. The following brief description is meant as an introduction to this tool and - hopefully - an inspiration for the reader to seek a complete description of the method in the existing literature. It is beyond the scope of this appendix to give an actual interpretation of the theory of the dimensional analysis.

Fields of application

Dimensional analysis is characterized by an extraordinarily wide range of applications - from the routine control of a calculation result to fundamental phenomena in theoretical physics. As examples of application of dimensional analysis to solution of practical engineering tasks, the following can be mentioned

- Checking of formulas by analysis of their dimensional homogeneity.
- Conversion of design constants and formulas from one unit system to another.
- Support of experimental investigations by choice of suitable test parameters and reduction of the number of parameters that need to be investigated.
- Planning and analysis of model tests, including formulation of conditions of similitude.
- Choice of suitable forms for illustration of physical laws or test results.
- Derivation of physical laws that cannot be described by general mathematical methods.

Limitations

Dimensional analysis can be used to supplement, but certainly not replace insight into physical laws. As will be seen from the following, choosing a set of physical parameters that can give an unambiguous description of an actual physical phenomenon requires a good deal of intuition and insight. It goes without saying that a result cannot be correct if it is calculated from erroneous physical assumptions; therefore, the initial choice of parameters is an important part of any dimensional analysis. For the same reason, the result of a dimensional analysis should always be verified by tests or be substantiated by a critical numerical evaluation.

Physical dimension

The international system of SI units is based on the following seven **base quantities**: *length*, *mass*, *time*, *electric current*, *temperature*, *amount of substance*, and *luminosity*. For these seven base quantities, the corresponding **base dimensions** are formally symbolized by L, M, T, I, Θ, N, and J.

For each base quantity, a practical unit of measurement - the so-called **base unit** - has been defined. Thus, the base unit for *length* is **metre** and the base unit for *mass* is **kilogram**. The SI unit system is based on the following base system of quantities, dimensions, and units.

Base quantity	Base unit	Symbol	Dimension
length	metre	m	L
mass	kilogram	kg	M
time	second	s	T
electric current	ampere	A	I
thermodynamic temperature	kelvin	K	Θ
amount of substance	mol	mol	N
luminosity	candela	cd	J

Based on these seven base units of the SI system, **derived units** for *all* other physical quantities can be formed. In this connection, any other physical quantity Q appears as an exponential product of the base quantities forming the unit system. Therefore, any derived physical quantity Q can be given a **dimension** *dim* Q determined by

$$\mathit{dim}\ \mathrm{Q} = \mathrm{L}^{\alpha}\,\mathrm{M}^{\beta}\,\mathrm{T}^{\gamma}\,\mathrm{I}^{\delta}\,\Theta^{\varepsilon}\,\mathrm{N}^{\zeta}\,\mathrm{J}^{\eta} \qquad \text{(a)}$$

where the exponents $(\alpha,\ \beta,\ \gamma,\ \cdots)$ are called **dimensional exponents**. One or more of these exponents can be zero; if all of the dimensional exponents in a physical quantity are zero, we have

$$\mathit{dim}\ \mathrm{Q} = \mathrm{L}^{0}\,\mathrm{M}^{0}\,\mathrm{T}^{0}\,\mathrm{I}^{0}\,\Theta^{0}\,\mathrm{N}^{0}\,\mathrm{J}^{0} = 1 \qquad \text{(b)}$$

and the quantity in this case is said to be **non-dimensional**. The following table gives examples of *dim* Q for some frequently used derived SI quantities.

Quantity	Unit	Symbol	Dimension
force	newton	N	$\mathrm{L\,M\,T^{-2}}$
energy	joule	J	$\mathrm{L^2\,M\,T^{-2}}$
effect	watt	W	$\mathrm{L^2\,M\,T^{-3}}$
pressure	pascal	Pa	$\mathrm{L^{-1}\,M\,T^{-2}}$
electric charge	coulomb	C	$\mathrm{T\,I}$
potential diff.	volt	V	$\mathrm{L^2\,M\,T^{-3}\,I^{-1}}$
electric resistance	ohm	Ω	$\mathrm{L^2\,M\,T^{-3}\,I^{-2}}$

Various collections of tables contain more comprehensive overviews of the dimension of SI quantities; in addition, the dimension of any physical quantity can in a simple way be derived from the equation of definition for the quantity considered.

Types of dimension

In dimensional analysis it is important to distinguish between the following four types of quantity that occur in physical relations

- *Dimensional variables*: Are typically dependent or independent test variables that are given a physical dimension. Example: pressure p Pa with the dimension ($L^{-1}\,M\,T^{-2}$) and volume V m^3 with the dimension (L^3).
- *Dimensional constants*: Many physical relations include constants that can be given a physical dimension. The numerical value of these constants can *only* be changed if a different unit system is applied. Example: The gravitational constant $G = 6.6720 \cdot 10^{-11}$ Nm^2/kg^2 with the dimension ($L^3\,M^{-1}T^{-2}$) and the gas constant $R = 8.31441$ J/mol K with the dimension ($L^2M\,T^{-2}\,\Theta^{-1}\,N^{-1}$).
- *Non-dimensional variables*: These denote the relation between physical variables that are non-dimensional according to condition (b) in the above. Example: Reynolds' number $Re = v\varrho\ell/\eta$ (1) and the Fourier number $Fo = \lambda\tau/\varrho c\delta^2$ (1).
- *Non-dimensional constants*: These denote pure numerical quantities that are incorporated in a physical relation. Example: 2π, $\frac{1}{2}$.

When drawing up parameter lists (described in the following), it is important to introduce all dimensional constants incorporated in the problem considered; it is notorious that these quantities are easily neglected, since they do not occur as "active" test variables.

Dimensional-homogeneous equation

In the dimensional analysis, any equation containing physical quantities is assumed to be **homogeneous** with regard to physical dimensions. A physical equation can normally be written as the sum of two or more terms

$$Q_1 + Q_2 + \cdots = 0 \tag{c}$$

Such equations are dimensional-homogeneous if the terms have the *same* physical dimension, i.e.

$$dim\,Q_1 = dim\,Q_2 = \cdots \tag{d}$$

According to eqn. (a), this means that the dimension (L) has the same exponent in all terms, that the dimension (M) has the same exponent in all terms, etc.

Assume, for example, an object falling freely under the influence of a field of gravity g. According to the time of falling t, the velocity is v and the distance travelled is s

$$v = gt; \qquad dim(v) = (L\,T^{-1}) \qquad dim(gt) = (L\,T^{-2})(T) = (L\,T^{-1}) \tag{e}$$

$$s = \tfrac{1}{2}gt^2; \qquad dim(s) = (L) \qquad dim(gt^2) = (L\,T^{-2})(T^2) = (L) \tag{f}$$

Both of these equations are *numerically correct* and *homogeneous* with regard to dimensions. Contrary to this, the equation

$$v + s = gt + \tfrac{1}{2}gt^2 \tag{g}$$

is dimensionally *inhomogeneous*, since the dimensions ($L\,T^{-1}$) and (L) occur in the same expression. Nevertheless, eqn. (g) is *numerically correct*: for any t, the numerical values of the left-hand side and the right-hand side will coincide. This last example should be borne in mind when using dimensional analysis.

The Buckingham pi theorem

The basic principle of dimensional analysis is as follows: -*a given set of dimensional variables describing a physical phenomenon is rewritten to a reduced set of non-dimensional variables.*

From this - apparently simple - principle, a very efficient and versatile tool appears. At the early 20th century, the principle was developed to different analytical procedures by, among others, *Lord Rayleigh*, *Buckingham*, and *Bridgman*. In the following we shall use the so-called **Buckingham pi theorem**, because the formulation of this theorem most clearly illustrates the underlying principle of the analysis.

Assume a physical phenomenon that can only be fully described in one way, i.e. by specification of p parameters: $\{P_1, P_2, P_3, \cdots, P_p\}$ through a dimensional-homogeneous expression of the form

$$f(P_1, P_2, P_3, \cdots, P_p) = 0 \tag{h}$$

and assume that among the p parameters there are d dimensionally independent parameters, i.e. parameters that cannot be formed by combining the others. Now, the Buckingham pi theorem predicates that any physical expression of this form can be reduced to a relation between $p - d$ independent non-dimensional parameters

$$\Phi(\Pi_1, \Pi_2, \cdots, \Pi_{p-d}) = 0 \tag{i}$$

In other words, the original number of parameters can be reduced from p to $p - d$. The number of **dimensionally independent** parameters will normally be equal to the number of base dimensions in the actual set of parameters, however, exceptions may occur.

Using the above-mentioned free fall, eqn. (e), as an example, the phenomenon is described by $p = 3$ parameters, and the corresponding formal functional expression is: $f(v, g, t) = v - gt = 0$. The number of dimensionally independent parameters is $d = 2$; the parameters $t(\mathrm{T})$ and $v(\mathrm{LT}^{-1})$ are in this case independent with regard to dimensions. According to the Buckingham pi theorem, $f(v, g, t)$ can be reduced to $p-d = 3-2 = 1$, i.e. one non-dimensional parameter. The parameter we will actually find using the theorem is: $\Pi = (v/gt)$.

In conclusion,

The Buckingham pi theorem (j)

Any dimensional-homogeneous expression of the form

$$f(P_1, P_2, P_3, \cdots P_p) = 0$$

which with p independent **dimensional** parameters fully describes a physical phenomenon, can be rewritten to an expression of the form

$$\Phi(\Pi_1, \Pi_2, \cdots, \Pi_{p-d}) = 0$$

with $p - d$ independent **non-dimensional** parameters, where d denotes the number of **dimensionally independent** parameters in the expression.

The Buckingham pi theorem was formulated in 1914; the final proof of its universal validity was presented by *Bridgman* in 1922 (Bridgman, P.W.: *Dimensional Analysis*).

Determination of pi parameters

Use of the Buckingham pi theorem may at first seem a bit incalculable when described formally; however, as will be shown later through examples, the practical manipulation of parameters is simple. In any case, it is recommended that the formal description be compared with one or more of the calculated examples.

Assume a full set of parameters $\{P_1, P_2, \cdots, P_p\}$ with p dimensional parameters that describe a physical phenomenon. We want to rewrite this set of P parameters to a set of non-dimensional Π parameters according to the Buckingham theorem (j).

First, it is noted that any non-dimensional Π parameter is a product of selected P parameters raised to a suitable power e. This condition can be generally formulated by

$$dim(\Pi) = dim(P_1^{e_1} \cdot P_2^{e_2} \cdot P_3^{e_3} \cdot \cdots \cdot P_p^{e_p}) = 1 \tag{k}$$

The dimensional P parameters themselves are formed by a combination of the base units (L, M, T,$\cdots$) of the unit system; after involution, therefore, the dimensions of the parameters are determined by

$$dim(P_1^{e_1}) = \mathrm{L}^{\alpha_1 e_1} \cdot \mathrm{M}^{\beta_1 e_1} \cdot \mathrm{T}^{\gamma_1 e_1} \cdot \ldots \tag{1}$$
$$dim(P_2^{e_2}) = \mathrm{L}^{\alpha_2 e_2} \cdot \mathrm{M}^{\beta_2 e_2} \cdot \mathrm{T}^{\gamma_2 e_2} \cdot \ldots \qquad etc.$$

A non-dimensional Π parameter results when the sum of the dimensional exponents is zero for the base dimensions included in the equation system; the exponents $\{e_1, e_2, e_3, \cdots\}$, therefore, must satisfy the following **linear** and **homogeneous** equation system

$$\begin{array}{ll} \mathrm{L}: & \alpha_1 e_1 + \alpha_2 e_2 + \alpha_3 e_3 + \cdots + \alpha_p e_p = 0 \\ \mathrm{M}: & \beta_1 e_1 + \beta_2 e_2 + \beta_3 e_3 + \cdots + \beta_p e_p = 0 \\ \mathrm{T}: & \gamma_1 e_1 + \gamma_2 e_2 + \gamma_3 e_3 + \cdots + \gamma_p e_p = 0 \\ \vdots & \\ \mathrm{J}: & \eta_1 e_1 + \eta_2 e_2 + \eta_3 e_3 + \cdots + \eta_p e_p = 0 \end{array} \tag{m}$$

In this equation system, the first column is determined by the dimension and exponent of the P_1 parameter, and the second column is determined by the the P_2 parameter, etc. It is noted that the number of **actual** equations correspond to just the number of **base dimensions** that are incorporated in the actual set of P parameters.

For minor tasks with a few P parameters it is possible in a simple manner to determine the targeted non-dimensional Π parameters by solving the equation system (m); the known **dimensional exponents** $(\alpha, \beta, \gamma, \cdots)$ are introduced into the expression (1), whereupon the targeted exponents $(\mathrm{e}_1, \mathrm{e}_2, \mathrm{e}_3 \cdots)$ are determined.

Previously, reference was made to the number d of **dimensionally independent** parameters in the physical relation considered. When the homogeneous equation system (m) is solved, the ranking of the equation system ϱ will correspond to the number of dimensionally independent P parameters.

Normally, the rank ϱ of the equation system (m) is less than the number p of parameters, i.e. $\varrho < p$. Then the equation system formally contains a $(p - \varrho)$ infinity of solutions. In the present case, however, we are only interested in determining a set of **particular solutions** to the equation system. These are obtained by successively giving one of the $(p - \varrho)$ surplus exponents the value 1, and the others the value 0. Thus a $(p - \varrho)$ set of inhomogeneous linear equation systems is obtained, each of which determining one Π parameter. Thus, the final solution becomes

$$\Phi(\Pi_1, \Pi_2, \cdots, \Pi_{p-\varrho}) = 0 \tag{n}$$

If the number of P parameters exceeds 3 or 4, it is an advantage to solve the equation by using simple matrix operations. Furthermore, going over one or more examples and comparing the practical calculations with the formal description of solution method described above is recommendable.

Rewriting of pi parameters

When solving the homogeneous, linear equation system (m), one or more non-dimensional Π parameters of the function in eqn. (n) are established. However, it should be borne in mind that the dimensional analysis *cannot* determine the nature of the function $\Phi(\ )$.

In practice, it is often necessary to be able to rewrite the derived Π parameters for specific purposes. The following rules can be used for this

Rewriting of pi parameters (o)

Rule 1: An arbitrary Π parameter in eqn. (n) can be written as a function of the remaining parameters
example: $\Pi_2 = \Phi'(\Pi_1, \Pi_3, \cdots)$

Rule 2: Any Π parameter of the function $\Phi(\ )$ can be replaced by an arbitrary function $f(\Pi)$ of the parameter
example: $\Phi'(\Pi_1, k \cdot \sqrt{\Pi_2}, \Pi_3, \cdots) = 0$

Rule 3: Any Π parameter can be multiplied or divided by any of the other parameters
example: $\Phi'(\Pi_1, \Pi_2/\Pi_1, \Pi_3, \cdots) = 0$

By a combination of these rules it is in many cases possible to obtain a more suitable set of parameters than that directly obtained by solving the equation system (m). Especially - where this is possible - one should by relevant rewriting seek generally used non-dimensional quantities such as the *Reynolds number* **Re**, the *Fourier number* **Fo**, etc. through description of systems where these variables are traditionally used.

Synthetic base units

This brief introduction to the methods of dimensional analysis is based on the seven base units of the SI unit system with corresponding dimensions: (L,M,T,I,Θ,N,J). Most of the general analyses can be performed on the basis of this set of standardized base units.

The principle of dimensional analysis, however, is *not* a priori related to any unit system in particular. Therefore, one is free to form other, *matched* systems of units - which in the following are denoted **synthetic base units**. Thus, in certain situations, more detailed information of a physical system can be obtained. Some of the most frequently used synthetic base units, therefore, will be briefly mentioned in the following. Furthermore, it is recommended that the literature on this subject be consulted.

• *Vectorial base units*: All of the base units of the SI system are **scalar quantities**. A number of physical problems can be analysed in far more detail when directional length dimensions (L_x, L_y, L_z) related to a normal rectilinear xyz coordinate system are used.

By breaking down length dimensions into **vectors** one achieves, among other things, that the base unit of **force** - in accordance with the general perception - becomes a vector of the dimension ($L_x MT^{-2}$), for example. Further, it will apply that physical quantities such as **work** and **force moment** can be identified as different quantities through a dimensional analysis. With the usual SI dimensions, work and force moment have the *same* dimension: (L^2MT^{-2}); using vectorial base units, work, for example, gets the dimension: ($L_x^2MT^{-2}$), while force moment may get the dimension: ($L_xL_yMT^{-2}$).

• *Inertial and gravitational mass*: In physical relations the base unit of **mass** occurs with two fundamentally different meanings. Mass can specify a **quantity** of a substance; in this meaning, the quantity is, for example, included in the definition of density ϱ with the dimension: ($L^{-3}M$). Mass can also be included as a **proportionality factor** between force and acceleration in *Newton's 2nd law*, and in this context, **inertial mass** is meant.

In the base dimensions of the SI system, the two meanings of the concept of mass are not distinguished from one another, and they are given the same dimension (M). In the dimensional analysis, considerable advantages can often be gained by introducing a new base unit for the quantity of a substance with the dimension (M_μ) and a new base unit for inertial mass with the dimension (M_i).

• *Unit for temperature*: The SI unit system defines a base unit for temperature: **kelvin** with the dimension (Θ). The physical origin of this definition is the thermodynamic temperature scale where the temperature unit is defined as being proportional to the kinetic mean energy of the molecules in a system of matter. Therefore, for dimensional analysis it is admissible to use the same unit for temperature quantities as for energy, if this is expedient. In this case, the dimension for temperature is: (L^2MT^{-2}).

Calculation examples

Doing a dimensional analysis can in many ways be compared to playing a game of chess. It is fairly simple to learn the workmanship - the rules - but it requires a good deal of experience, routine and intuition to transform these rules to results of substance.

The following examples illustrate different aspects of a dimensional analysis process. The examples have been chosen so that they illustrate the main principles in an analysis and show the practical application of the rules developed.

However, it shall be emphasized that a versatile tool such as the dimensional analysis can only be learnt in one way, namely through training and independent work with solution of practical tasks. May these few examples serve as inspiration!

■ Non-steady, convective cooling of concrete cross-sections

Often, when planning casting of concrete it is necessary to evaluate the heating or cooling process for a concrete cross-section. For such evaluations, different diagram templates and analytical and numerical calculation methods can be found in the literature.

In the following example we shall investigate *how* the result of a dimensional analysis of a cooling process is obtained and compare the result with the exact analytical solution to the same problem; this will reveal the pros and cons of dimensional analysis.

The example illustrates the main principles of a dimensional analysis together with application of all significant methods and rules. Therefore, rather detailed calculations have been performed in the example so it may serve as a paradigm for the reader's own exercises.

Problem. Assume a plane, large and homogeneous concrete wall of the thickness 2δ, which in the initial state has the temperature $\theta_0 > 0\,^\circ\mathrm{C}$ all over the cross-section. At the time $\tau = 0$, the wall is placed in surroundings at a constant temperature $\theta_u = 0\,^\circ\mathrm{C}$. The heat is transferred between the wall surface and the surroundings by **convection**.

Examine, what fundamental parameters are critical for the cooling process of the cross-section considered and propose a suitable formulation for illustrating this process!

Conditions. First, we will try to draw up a list of the **dimensional** parameters that fully describe an actual cooling process. In support of this first exploration it can be useful to split up the main problem into a number of subproblems.

• *What is the total amount of heat that should be removed from the cross-section?* The total content of heat in the cross-section is proportional to the concrete volume considered; for 1 m^2 of the wall, this volume is proportional to the **characteristic dimension** δ of the wall. The heat content per unit volume is ϱc, where ϱ is the concrete **density** and c denotes the **mass-specific heat capacity** of the concrete.

In the following, we use the volumetric heat capacity ϱc. Furthermore, the heat content is proportional to the original temperature difference θ_0 between concrete and its surroundings. This indicates the following P parameters: $(\delta,\ \varrho c,\ \theta_0)$.

• *What parameters are critical for the heat transport to the surface of the wall?* At a given state of cooling, the transport rate is proportional to the **coefficient of thermal conductivity** λ of the concrete, proportional to the characteristic temperature difference $\Delta\theta$, and inversely proportional to the characteristic dimension δ. This indicates the following P parameters: $(\Delta\theta,\ \lambda,\ \delta)$.

• *What parameters are critical for the heat transfer from a wall surface to its surroundings?* At a given state of cooling, the heat transfer is proportional to the **convective heat transfer coefficient** α and proportional to the characteristic temperature difference $\Delta\theta$. This indicates the following P parameters: $(\alpha,\ \Delta\theta)$.

• *What parameters are necessary for characterizing a cooling process?* A cooling process can be described by specifying a characteristic degree of cooling $\Delta\theta/\theta_0$ as a function of the time of cooling τ. This indicates the following P parameters: $(\Delta\theta,\ \theta_0,\ \tau)$.

Based on these considerations, we can assume that the following dimensional parameters are necessary for describing a cooling process: $(\delta,\ \varrho c,\ \theta_0,\ \lambda,\ \Delta\theta,\ \alpha,\ \tau)$. Comparing this with the **Buckingham pi theorem** (j), we have:

$$f(P_1,\ P_2,\ \cdots,\ P_7) = f(\delta,\ \varrho c,\ \theta_0,\ \lambda,\ \Delta\theta,\ \alpha,\ \tau) = 0 \qquad \text{(p)}$$

In this expression, a total of seven dimensional variables are included; using the Buckingham pi theorem, we shall now find out what non-dimensional parameters are included in the expression.

Solution. To determine the non-dimensional Π parameters, the homogeneous equation system according to eqn. (m) for the seven parameters is drawn up. The dimensional exponents $(\alpha,\ \beta,\ \cdots)$ of this equation system can be directly seen from the dimensions of the P parameters, which are

$$\begin{aligned} &dim(\delta) = \mathrm{L}^1 && dim(\lambda) = \mathrm{L}^1\mathrm{M}^1\mathrm{T}^{-3}\Theta^{-1} \\ &dim(\varrho c) = \mathrm{L}^{-1}\mathrm{M}^1\mathrm{T}^{-2}\Theta^{-1} && dim(\Delta\theta) = \Theta^1 \\ &dim(\tau) = \mathrm{T}^1 && dim(\alpha) = \mathrm{M}^1\mathrm{T}^{-3}\Theta^{-1} \\ &dim(\theta) = \Theta^1 && \end{aligned}$$

For clarity, it is suitable to arrange and set up the parameter dimensions in the following **dimensional matrix**

	independent					dependent				
Dimension	P_1	P_2	P_3	P_4		P_5	P_6	P_7		
L	0	0	−1	1		1	0	0		
M	0	0	1	0		1	0	1		
T	0	1	−2	0		−3	0	−3		
Θ	1	0	−1	0		−1	1	−1		
Exponent:	e_1	e_2	e_3	e_4		e_5	e_6	e_7		
Parameter:	θ_0	τ	ϱc	δ		λ	$\Delta\theta$	α		

By formation of the dimensional matrix, the maximum number of **dimensionally independent** parameters are drawn and written in the left-side block, and the remaining $(p-\varrho)$ dependent parameters are written in the right-side block. Due to the subsequent calculations, parameters with simple dimensions should be preferred as independent parameters. In the dimensional matrix shown above, $(\theta_0,\ \tau,\ \varrho c,\ \delta)$ have been chosen as independent parameters, because these parameters have especially simple, mutually independent dimensions.

Now, an unknown exponent $(e_1,\ e_2,\ \cdots)$ belongs to each of the parameters $(P_1,\ P_2,\ \cdots)$ in the dimensional matrix; this is seen by comparing the dimensional matrix with the equation system (m). The rank ϱ of the equation system is 4. The equations contain 7 unknown exponents. In the mathematical sense, this means that there is a three infinity of solutions to the equation system. Among these, three particular solutions are determined in the following way.

First, we normalize P_5 corresponding to the coefficient of thermal conductivity λ. This is achieved by giving the corresponding exponent e_5 the value of 1 and taking the remaining exponents e_6 and e_7 as zero. Thus, the following **inhomogeneous** equation system for determination of Π_1 is obtained

$$\begin{cases} -1e_3 + 1e_4 = -1 \\ 1e_3 = -1 \\ 1e_2 - 2e_3 = 3 \\ 1e_1 - 1e_3 = 1 \end{cases} \quad \text{from which} \quad \begin{pmatrix} e_1 \\ e_2 \\ e_3 \\ e_4 \end{pmatrix} = \begin{pmatrix} 0 \\ 1 \\ -1 \\ -2 \end{pmatrix} \tag{q}$$

The set of solutions found determines the non-dimensional Π_1, cf. the expression (k)

$$\Pi_1 = P_1^{e_1} \cdot P_2^{e_2} \cdot P_3^{e_3} \cdot P_4^{e_4} \cdot P_5^1 = \theta_0^0 \cdot \tau^1 \cdot (\varrho c)^{-1} \cdot \delta^{-2} \cdot \lambda^1 = \frac{\lambda\tau}{\varrho c \delta^2} \tag{r}$$

Similarly, the non-dimensional groups Π_2 and Π_3 are determined by successively giving e_6 and e_7 the value of 1. However, with the actual number of parameters, it is simpler to perform this sequence of calculations by a matrix operation. Denoting the matrix of independent exponents $\mathbf{U}$ and the matrix with dependent exponents $\mathbf{A}$, i.e.

$$\mathbf{U} = \begin{pmatrix} 0 & 0 & -1 & 1 \\ 0 & 0 & 1 & 0 \\ 0 & 1 & -2 & 0 \\ 1 & 0 & -1 & 0 \end{pmatrix} \quad ; \quad \mathbf{A} = \begin{pmatrix} 1 & 0 & 0 \\ 1 & 0 & 1 \\ -3 & 0 & -3 \\ -1 & 1 & -1 \end{pmatrix} \tag{s}$$

we obtain the solution to the three inhomogeneous equation systems by the following simple matrix operation

$$\mathbf{U}^{-1}(-\mathbf{A}) = \begin{pmatrix} 0 & 1 & 0 & 1 \\ 0 & 2 & 1 & 0 \\ 0 & 1 & 0 & 0 \\ 1 & 1 & 0 & 0 \end{pmatrix} \begin{pmatrix} -1 & 0 & 0 \\ -1 & 0 & -1 \\ 3 & 0 & 3 \\ 1 & -1 & 1 \end{pmatrix} = \begin{pmatrix} 0 & -1 & 0 \\ 1 & 0 & 1 \\ -1 & 0 & -1 \\ -2 & 0 & -1 \end{pmatrix} \tag{t}$$

whereupon the three columns in the calculated matrix contain the desired, particular sets of solutions (e_1, e_2, e_3, e_4); for example, the solution {0, 1, −1, −2} determined earlier is retrieved as the set of figures in the first column.

We can now write the three Π parameters by the procedure used in eqn. (r). The following non-dimensional groups are found by insertion

$$\Phi(\Pi_1, \Pi_2, \Pi_3) = \Phi\left(\frac{\lambda\tau}{\varrho c\delta^2}, \frac{\Delta\theta}{\theta_0}, \frac{\alpha\tau}{\varrho c\delta}\right) = 0 \tag{u}$$

A cooling process is most appropriately described by stating the relative cooling $\Delta\theta/\theta_0$ as a function of a time-dependent parameter. Using **Rule no. 1** for converting pi parameters, we get

$$\Pi_2 = \Phi(\Pi_1, \Pi_3) \qquad \Rightarrow \qquad \frac{\Delta\theta}{\theta_0} = \Phi\left(\frac{\lambda\tau}{\varrho c\delta^2}, \frac{\alpha\tau}{\varrho c\delta}\right)$$

Correspondingly, the parameter Π_3 can be made independent of time τ by division by Π_1 according to **Rule no. 2**

$$\Pi_2 = \Phi(\Pi_1, \Pi_3/\Pi_1) \qquad \Rightarrow \qquad \frac{\Delta\theta}{\theta_0} = \Phi\left(\frac{\lambda\tau}{\varrho c\delta^2}, \frac{\alpha\delta}{\lambda}\right)$$

Through dimensional analysis, we have thus arrived at the following statement on a convective cooling process for a plane wall:

- The relative degree of cooling ($\Delta\theta/\theta_0$) can be fully described as a function of only two independent non-dimensional variables

$$\frac{\Delta\theta}{\theta_0} = \Phi\left(\frac{\lambda\tau}{\varrho c\delta^2}, \frac{\alpha\delta}{\lambda}\right) \tag{v}$$

- It is expedient to illustrate a cooling process in a coordinate system with **ordinate** ($\Delta\theta/\theta_0$) and **abscissa** ($\lambda\tau/\varrho c\delta^2$). Then, for each fixed value of ($\alpha\delta/\lambda$), only one cooling process is found.

Discussion. In the calculation example we have reduced the number of independent variables from six in the original expression (p) to only two in eqn. (v). In experimental investigations this entails a considerable reduction of the extent of tests. Make a somewhat crude assumption that the phenomenon described shall be investigated for

10 different values of each independent variable. The number of individual measurements would then be reduced by the factor $10^6/10^2 = 10000$. In other words, it can be a good investment to perform a dimensional analysis *before* launching out on major experimental investigations.

Through the dimensional analysis we have obtained two of the most fundamental parameters in the theory of convective heat balance, namely:

$$\text{The Fourier number}: \quad \mathbf{Fo} = \frac{\lambda\tau}{\varrho c \delta^2}; \qquad \text{The Biot number}: \quad \mathbf{Bi} = \frac{\alpha\delta}{\lambda}$$

The Fourier number **Fo** is a generalized, non-dimensional time variable, and the Biot number **Bi** is a non-dimensional relation between the heat transfer resistance (δ/λ) in wall cross-sections and the convective transition resistance $(1/\alpha)$ at the wall surface.

By the analysis, however, we are stuck with an unknown function of the two non-dimensional quantities: $\Phi(\mathbf{Fo}, \mathbf{Bi})$ - and this path leads no further. In conclusion, however, it is interesting to compare the result with the exact mathematical solution to the problem. For the temperature development at the middle of cross-sections we obtain the following through a not quite simple analysis

$$\frac{\Delta\theta}{\theta} = \Phi(\mathbf{Fo}, \mathbf{Bi}) = \sum_{n=1}^{\infty} \frac{2\sin(\mu_n)}{\mu_n + \sin(\mu_n)\cos(\mu_n)} \cdot \exp(-\mu_n^2 \cdot \mathbf{Fo})$$

where: μ_n denotes the nth eigenvalue in the equation: $\mu_n = \mathbf{Bi} \cdot \cot(\mu_n)$.

In other words, the unknown function $\Phi(\mathbf{Fo},\mathbf{Bi})$ takes the form of a Fourier series; application of this expression entails a rather extensive numerical calculation involving, for example, determination of the eigenvalues μ_n. The graphical illustration of $\Phi(\mathbf{Fo},\mathbf{Bi})$ is also found in the so-called **Temperature Response Diagram.**

■ Capillary cohesion in particle systems

We know by experience that a sand castle should be built of sand with a suitable moisture content. Dry sand just trickles down into a conical heap due to gravitation. The internal **cohesion** making moist sand dimensionally stable is based on the **surface tension** of the water. At certain moisture contents, the pore water is located especially around contact points between the individual sand particles. Here, the surface tension forms **menisci** - concave liquid surfaces - that bring about underpressure in the pore water; this underpressure maintains the particles at the contact points already formed.

This kind of capillary cohesion is vital for many of the phenomena dealt with in the science of materials. Mention can be made of, for example, cracking due to **plastic shrinkage** of concrete, **stability** and **workability** of ceramic materials during processing, **drying shrinkage** of hardened concrete, and **load-carrying capacity** of soil.

However, one can easily underestimate or ignore the significance of capillary cohesion, because the effect is dimensionally dependent! In a heap of stones, the effect is completely insignificant, in a heap of sand, the effect has just become noticeable; however, if we consider particles of microscopic dimension, the capillary cohesion forces constitute a quite dominating property parameter. The following example shows how a dimensional analysis can be used to evaluate the dimensional dependence of the effect.

Problem. Assume a system of particles with the characteristic dimension δ. The moisture content u of the particle system is given; the capillary cohesion forces between the particles cause the system to show a characteristic cohesion strength of p.

State qualitatively how the cohesion strength p must be expected to depend on the particle dimension δ in geometrically similar systems with the same moisture content u!

Conditions. The particle system is assumed to be influenced by capillary cohesion forces only; the effect of gravitational forces and electric surface forces is ignored. The surface tension of the water is denoted σ. The moisture content of the system is assumed to be constant, and particle system and menisci are assumed to be geometrically similar.

The determining parameters of the system are: (p, δ, σ); according to the **Buckingham pi theorem** (j), we can then write the formal function

$$f(P_1, P_2, \cdots) = f(p, \delta, \sigma) = 0$$

with a total of three dimensional parameters.

Solution. The non-dimensional Π parameters are determined from the **homogeneous** equation system for the actual parameters corresponding to eqn. (m). The dimensional exponents $(\alpha, \beta, \cdots)$ are seen from the physical dimension of the parameters, namely

$$dim(p) = \mathrm{L}^{-1}\mathrm{MT}^{-2}; \qquad dim(\sigma) = \mathrm{MT}^{-2}; \qquad dim(\delta) = L$$

Here, δ and σ can be taken as dimensionally independent parameters and thus form the following **dimensional matrix**

	independent				dependent				
Dimension	P_1	P_2			P_3				
L	1	0			−1				
M	0	1			1				
T	0	−2			−2				
Exponent:	e_1	e_2			e_3				
Parameter:	δ	σ			p				

In this case, there are three equations, but only two dimensionally independent parameters. This corresponds to the rank of the equation system being $\varrho = 2$, and it is seen that the dimensional matrix can be solved by simple elimination of the last row. The particular solution is obtained by giving e_3 the value 1, whereby

$$\begin{cases} 1e_1 = 1 \\ 1e_2 = -1 \\ -2e_2 = 2 \end{cases} \quad \text{from which} \quad \begin{pmatrix} e_1 \\ e_2 \\ e_3 \end{pmatrix} = \begin{pmatrix} 1 \\ -1 \\ 1 \end{pmatrix}$$

The non-dimensional parameter Π can now be determined from eqn. (k) and then introduced into the Buckingham Φ function

$$\Pi_1 = P_1^{e_1} \cdot P_2^{e_2} \cdot P_3^1 = \delta^1 \cdot \sigma^{-1} \cdot p^1 = \frac{\delta p}{\sigma} \qquad \Rightarrow \quad \Phi(\Pi_1) = \Phi\left(\frac{\delta p}{\sigma}\right) = 0$$

Rewriting the above expression according to **Rule no. 1**, we obtain a qualitative expression of the dimensional dependence of the cohesion force

$$\Pi_1 = \left(\frac{\delta p}{\sigma}\right) = \text{constant} \qquad \Rightarrow \quad p = \text{constant} \cdot \left(\frac{\sigma}{\delta}\right)$$

The dimensional analysis shows that in geometrically similar systems, the cohesion force p of the particle system is proportional to the surface tension σ and inversely proportional to the particle dimension δ.

Discussion. An indicative estimate of the dimensional dependency of the capillary cohesion force can be obtained by comparing the characteristic dimension δ for some frequently occurring particle systems.

Particle system	δ (m)	δ	$p/p(sand)$ *
Coarse aggregates	$\sim 10^{-2}$	~ 1 cm	~ 0.1
Concrete sand	$\sim 10^{-3}$	~ 1 mm	~ 1
Portland cement	$\sim 10^{-5}$	~ 10 μm	~ 100
Silica fume	$\sim 10^{-7}$	~ 100 nm	~ 10000

* *Besides the capillary cohesion in practical systems, electric surface forces will occur, which may play a significant role for small particle dimensions.*

The formal procedure for determination of Π parameters has been used in this example. In simple cases, the non-dimensional groups can often be determined by experiments, i.e. by investigating possible parameter combinations. For example, it can be verified rather quickly that the parameters p, σ, δ can only be combined in one way into a non-dimensional product, namely on the form: $\delta p/\sigma$. When dealing with larger problems, however, one should always set up a dimensional matrix and determine the Π parameters in a systematic way.

Literature

Bridgman, P.W.: *Dimensional Analysis*, Yale University Press, New Haven 1963.

Focken, C.M.: *Dimensional Methods and their Applications*, Edward Arnold & Co., London 1953.

Huntley, H.E.: *Dimensional Analysis*, Dover Publications, Inc. , New York 1967.

Kaye, G.W.C. and Laby, T.H.: *Tables of Physical and Chemical Constants*, 5th ed., Longman, Harlow, 1992.

3 Newton–Raphson iteration

In physical-chemical calculations, **algebraic** or **transcendental** equations of the following form are often encountered

$$f(x) = 0 \tag{a}$$

If an equation of the form (a) cannot be solved explicitly in x by the usual operations, numerical solution methods must be used. For those equations of the above-mentioned form that fulfil the following special conditions,

- the derivative $f'(x) = \dfrac{df(x)}{dx}$ has a simple form (b)
- the root can be approximated by a simple estimate (c)

the **Newton–Raphson** iteration method is especially suitable for numerical calculation of the root value. Equations fulfilling these terms occur, for example, when calculations are made using the **van der Waals equation**; in this case, a first estimate for the targeted root is obtained using the **ideal gas law**.

The calculation rule for root determination according to the Newton–Raphson method is

Newton–Raphson iteration

(1) The derivative $f'(x) = \dfrac{df(x)}{dx}$ is determined by differentiation.

(2) A reasonable first estimate x_0 for the desired root is determined.

(3) The root is approximated by the desired accuracy using the algorithm successively:

$$x_{i+1} = x_i - \frac{f(x_i)}{f'(x_i)} \qquad i = 0, 1, \cdots n$$

where x_{i+1} for each further step in the calculation represents the next approximation to the desired root.

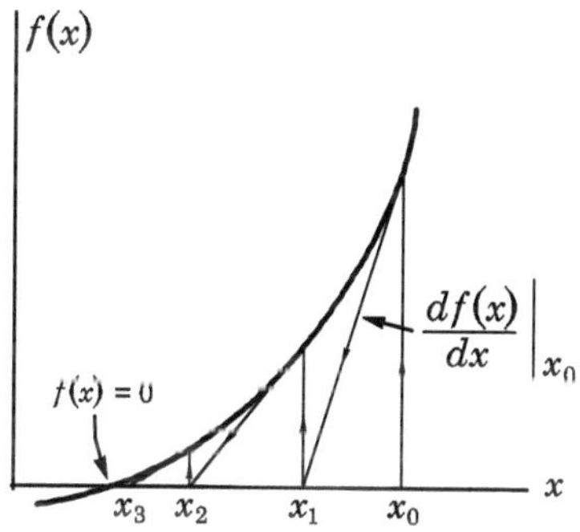

Figure A.1. Root determination for $f(x)$ by Newton-Raphson iteration.

With a suitable first estimate x_0, the Newton–Raphson iteration method will in most cases converge very quickly. When applied to programming, the correction $|\, x_{i+1} - x_i \,|$ can be compared with a requirement made for accuracy ε: $|\, x_{i+1} - x_i \,| > \varepsilon$ continue iteration

$|\, x_{i+1} - x_i \,| \leq \varepsilon$ discontinue iteration; root $= x_{i+1}$

The principle of iteration according to the Newton-Raphson method can be seen from the figure; as seen from the graph, the successively calculated values x_{i+1} specify the point of intersection between the axis $y = 0$ and the tangent to the curve $f(x)$ at the point x_i.

Under the given conditions, the method is suitable for programming; as will be seen from the following example, the iteration can also be applied to simple hand calculation.

■ Calculation of amount of substance, the van der Waals equation

A pressure tank of the volume V = 4 ℓ has been filled with carbon dioxide $CO_2(g)$ at 20 °C and a pressure of p = 28.5 atm, corresponding to $2.89 \cdot 10^6$ Pa. The **amount of substance** $n(CO_2)$ in the tank shall be calculated using the van der Waals equation

$$(p + a \cdot (\frac{n}{V})^2) \cdot (V - n \cdot b) = n \cdot R\,T$$

With table values for the van der Waals constants a and b, this leads to the following algebraic equation in the amount of substance n:

$$f(n) = n^3 - 93.75 \cdot n^2 + 2.627 \cdot 10^3 \cdot n - 1.191 \cdot 10^4 = 0$$

The differential coefficient $f'(n)$ is determined:

$$f'(n) = 3 \cdot n^2 - 187.5 \cdot n + 2.627 \cdot 10^3$$

A first estimate, n_0, for the root in the equation is obtained by using the ideal gas law

$$n_0 = \frac{p \cdot V}{R \cdot T} = \frac{2.89 \cdot 10^6 \text{Pa} \cdot 4 \cdot 10^{-3}\text{m}^3}{8.314\text{J/mol} \cdot \text{K} \cdot 293.15\text{K}} = \qquad \mathbf{4.74\,mol}$$

When successively applying the Newton–Raphson algorithm (3) in the above box and using the first estimate $n_0 = 4.74$ mol, we find

$$n_1 = n_0 - \frac{f(n_0)}{f'(n_0)}$$

$$n_1 = 4.74 - \frac{4.74^3 - 93.75 \cdot 4.74^2 + 2.627 \cdot 10^3 \cdot 4.74 - 1.191 \cdot 10^4}{3 \cdot 4.74^2 - 187.5 \cdot 4.74 + 2.627 \cdot 10^3} = \qquad 5.547\,\text{mol}$$

Repeating the operation, we find the following convergence

$$n_2 = 5.547 - \frac{f(5.547)}{f'(5.547)} = 5.578; \qquad n_3 = 5.578 - \frac{f(5.578)}{f'(5.578)} = 5.578$$

In this case, the root n = **5.578 mol** is determined with four significant digits in the second calculation.

The above set of figures refer to *Example 1.4:* **Accelerated testing of concrete – carbonation properties** in chapter 1.

Literature

Margenau, H. & Murphy, G.M.: *Mathematics of Physics and Chemistry*, D.van Nostrand Company Inc., New York 1955.

4 Cramer's formula

Linear equation systems of the form $\mathbf{Ax} = \mathbf{b}$ are conventional in practical calculation problems. For solution of such equation systems by pocket calculator, one can benefit from **Cramer's formula** in case of two or three unknowns. The advantage in this solution method is, for example, that keying and calculation can be performed in a simple and systematic manner.

Considering, for example, an equation system consisting of two equations with two unknowns (x_1, x_2), the matrix notation $\mathbf{Ax} = \mathbf{b}$ corresponds to the equation system

$$\begin{pmatrix} a_{11} & a_{12} \\ a_{21} & a_{22} \end{pmatrix} \begin{pmatrix} x_1 \\ x_2 \end{pmatrix} = \begin{pmatrix} b_1 \\ b_2 \end{pmatrix} \quad \text{or :} \quad \begin{cases} a_{11} \cdot x_1 + a_{12} \cdot x_2 = b_1 \\ a_{21} \cdot x_1 + a_{22} \cdot x_2 = b_2 \end{cases} \qquad \text{(a)}$$

To solve this equation system, we can, for example, multiply the first equation by (a_{22}), multiply the second equation by $(-a_{12})$, and then add together the two equations thus found

$$(a_{11} \cdot a_{22} - a_{21} \cdot a_{12}) \cdot x_1 = (b_1 \cdot a_{22} - b_2 \cdot a_{12}) \qquad \text{(b)}$$

By this operation we have eliminated x_2. Correspondingly, multiplying the first equation by $(-a_{21})$ and the second equation by (a_{11}), we obtain the following by addition of the equations thus found

$$(a_{11} \cdot a_{22} - a_{21} \cdot a_{12}) \cdot x_2 = (b_2 \cdot a_{11} - b_1 \cdot a_{21}) \qquad \text{(c)}$$

Similarly, we have now eliminated x_1. If the quantity $(a_{11} \cdot a_{22} - a_{21} \cdot a_{12}) \neq 0$, the unknowns (x_1, x_2) are determined by

$$x_1 = \frac{b_1 \cdot a_{22} - b_2 \cdot a_{12}}{a_{11} \cdot a_{22} - a_{21} \cdot a_{12}}; \qquad x_2 = \frac{b_2 \cdot a_{11} - b_1 \cdot a_{21}}{a_{11} \cdot a_{22} - a_{21} \cdot a_{12}} \qquad \text{(d)}$$

In this solution, the denominator is seen to be det $\mathbf{A}$, i.e. the **determinant** of the coefficient matrix $\mathbf{A}$. The numerator of the solution to x_1 is the determinant of $\mathbf{A}$, if the first column in $\mathbf{A}$ is replaced by the elements in $\mathbf{b}$. Correspondingly, the numerator in the solution to x_2 is the determinant of $\mathbf{A}$, if the second column is replaced by the elements in $\mathbf{b}$. For $b_1 = b_2 = 0$, eqn. (a) constitutes a **homogeneous** equation system and has at least the solution $x_1 = x_2 = 0$. If b_1 and/or $b_2 \neq 0$, eqn. (a) constitutes an **inhomogeneous** equation system; if det $\mathbf{A} \neq 0$, the solution to this inhomogeneous equation system is eqn. (d). For two equations with two unknowns, the solution to the inhomogeneous equation system is given by **Cramer's formula**

Cramer's formula, two unknowns

Determination of solution to the linear equation system $\mathbf{Ax} = \mathbf{b}$

$$\begin{pmatrix} a_{11} & a_{12} \\ a_{21} & a_{22} \end{pmatrix} \begin{pmatrix} x_1 \\ x_2 \end{pmatrix} = \begin{pmatrix} b_1 \\ b_2 \end{pmatrix}$$

(1) Calculate $\mathrm{D} = \det \mathbf{A} = \begin{vmatrix} a_{11} & a_{12} \\ a_{21} & a_{22} \end{vmatrix} = a_{11} \cdot a_{22} - a_{21} \cdot a_{12}$

If $\mathrm{D} = 0$, there is no solution; if $\mathrm{D} \neq 0$, the following should be done:

(2) Calculate: $\mathrm{D}_1 = \begin{vmatrix} b_1 & a_{12} \\ b_2 & a_{22} \end{vmatrix}$ and $\mathrm{D}_2 = \begin{vmatrix} a_{11} & b_1 \\ a_{21} & b_2 \end{vmatrix}$

(3) Solution to equation system: $x_1 = \mathrm{D}_1/\mathrm{D}$; $x_2 = \mathrm{D}_2/\mathrm{D}$

Cramer's formula can be generalized for all $n \geq 2$. However, as a solution method, the formula is less suitable for $n \geq 3$. For three equations with three unknowns (x_1, x_2, x_3), the solution is determined by

Cramer's formula, three unknowns

Determination of solution to the linear equation system $\mathbf{Ax} = \mathbf{b}$

$$\begin{pmatrix} a_{11} & a_{12} & a_{13} \\ a_{21} & a_{22} & a_{23} \\ a_{31} & a_{32} & a_{33} \end{pmatrix} \begin{pmatrix} x_1 \\ x_2 \\ x_3 \end{pmatrix} = \begin{pmatrix} b_1 \\ b_2 \\ b_3 \end{pmatrix}$$

(1) Calculate $\mathrm{D} = \det \mathbf{A} = \begin{vmatrix} a_{11} & a_{12} & a_{13} \\ a_{21} & a_{22} & a_{23} \\ a_{31} & a_{32} & a_{33} \end{vmatrix}$

If $\mathrm{D} = 0$, there is no solution; if $\mathrm{D} \neq 0$, the following is calculated:

(2) $\mathrm{D}_1 = \begin{vmatrix} b_1 & a_{12} & a_{13} \\ b_2 & a_{22} & a_{23} \\ b_3 & a_{32} & a_{33} \end{vmatrix}$; $\mathrm{D}_2 = \begin{vmatrix} a_{11} & b_1 & a_{13} \\ a_{21} & b_2 & a_{23} \\ a_{31} & b_3 & a_{33} \end{vmatrix}$; $\mathrm{D}_3 = \begin{vmatrix} a_{11} & a_{12} & b_1 \\ a_{21} & a_{22} & b_2 \\ a_{31} & a_{32} & b_3 \end{vmatrix}$

(3) Solution to equation system: $x_1 = \mathrm{D}_1/\mathrm{D}$; $x_2 = \mathrm{D}_2/\mathrm{D}$; $x_3 = \mathrm{D}_3/\mathrm{D}$

■ Solution of linear equation system

The following linear equation system of the form $\mathbf{Ax} = \mathbf{b}$ is given:

$$\begin{pmatrix} 1.3 & 4.1 \\ 3.8 & 10.5 \end{pmatrix} \begin{pmatrix} x_1 \\ x_2 \end{pmatrix} = \begin{pmatrix} 32.32 \\ 86.16 \end{pmatrix}$$

Determine the unknowns (x_1, x_2) using **Cramer's formula!**

Solution. The calculations are made using the procedure for two unknowns

(1) $\mathrm{D} = \det \mathbf{A} = \begin{vmatrix} 1.3 & 4.1 \\ 3.8 & 10.5 \end{vmatrix} = -1.93 \neq 0 \quad \Rightarrow$ solution according to (2) in the first box above

(2) $\mathrm{D}_1 = \begin{vmatrix} 32.32 & 4.1 \\ 86.16 & 10.5 \end{vmatrix} = -13.896; \; \mathrm{D}_2 = \begin{vmatrix} 1.3 & 32.32 \\ 3.8 & 86.16 \end{vmatrix} = -10.808$

(3) $x_1 = \mathrm{D}_1/\mathrm{D} = -13.896/(-1.93) = 7.2 \qquad x_2 = \mathrm{D}_2/\mathrm{D} = -10.808/(-1.93) = 5.6$

We thus get the solution: $(x_1, x_2) = (\mathbf{7.2},\ \mathbf{5.6})$

As mentioned, Cramer's formula is only suitable for solving equation systems with two or three unknowns; the advantage using this solution method is the clear systematics when keying on a pocket calculator. To solve linear equation systems with more unknowns than three, we can, for example, use **Gauss elimination**.

Literature

Kreyszig, E.: *Advanced Engineering Mathematics*, John Wiley & Sons Inc., New York 1972.

5 Linear regression

Processing of experimental results is an important task in all kinds of material testing. During investigations aiming at further developments, the analysis of a test material can make considerable demands on the persons doing the tests. It may be that a theory needs to be verified, or that physical parameters must be determined, or that new calculation methods need to be developed. For such tasks, **linear regression** is a vital calculation tool.

Assume that for n values of an **independent** variable $X(X_1, X_2, \cdots, X_k)$, corresponding values of a **dependent** variable Y are measured. The independent variable Y is assumed to be able to express a linear combination of X

$$Y = a_0 + a_1 X_1 + a_2 X_2 + \cdots + a_k X_k \; + \varepsilon \qquad (\varepsilon = \text{residual}) \tag{a}$$

From this test material, the coefficients $(a_0, a_1, a_2, \cdots, a_k)$ are determined so that the **S**um of **S**quared **D**eviations (SSD) assumes its minimum value

$$SSD = \sum_{i=1}^{n} (Y_i - (a_0 + a_1 X_{i.1} + a_2 X_{i.2} + \cdots + a_k X_{i.k}))^2 = \sum_{i=1}^{n} (\varepsilon_i)^2 \tag{b}$$

The deviation ε_i between estimated value and measured value is called the **residual**. In the following we shall address the calculation approach to solve this problem in the form of a simple example.

Assume that n values of the dependent variable Y have been found by tests and that these are expected to be described by a linear combination of the form: $a_0 + a_1 X_1 + a_2 X_2$, where X_1, X_2 are independent variables. Let Y denote the 14 days' strength of a concrete – which is a dependent variable – and let X_1, X_2 be the w/c ratio or air content, respectively, of the concrete – which are independent test variables. The n test results now represent a linear equation system of the form

$$\begin{array}{ccccccccc} Y_1 & = & a_0 & + & a_1X_{1.1} & + & a_2X_{1.2} & + & \varepsilon_1 \\ Y_2 & = & a_0 & + & a_1X_{2.1} & + & a_2X_{2.2} & + & \varepsilon_2 \\ Y_3 & = & a_0 & + & a_1X_{3.1} & + & a_2X_{3.2} & + & \varepsilon_3 \\ \vdots & & \vdots & & \vdots & & \vdots & & \vdots \\ Y_n & = & a_0 & + & a_1X_{n.1} & + & a_2X_{n.2} & + & \varepsilon_n \end{array} \qquad \text{(c)}$$

For $n > 3$, this equation system will be a **redundant** system. The coefficients (a_0, a_1, a_2) will now be determined so that the sum of squared deviations SSD according to eqn. (b) assumes its minimum value.

Before solving this problem it should be emphasized that the equation system (c) can be expressed more clearly by the matrix form: $\mathbf{Y} = \mathbf{Xk} + \boldsymbol{\varepsilon}$, where

$$\mathbf{Y} = \begin{pmatrix} Y_1 \\ Y_2 \\ Y_3 \\ \vdots \\ Y_n \end{pmatrix}; \quad \mathbf{X} = \begin{pmatrix} 1 & X_{1.1} & X_{1.2} \\ 1 & X_{2.1} & X_{2.2} \\ 1 & X_{3.1} & X_{3.2} \\ \vdots & \vdots & \vdots \\ 1 & X_{n.1} & X_{n.2} \end{pmatrix}; \quad \mathbf{k} = \begin{pmatrix} a_0 \\ a_1 \\ a_2 \end{pmatrix}; \quad \boldsymbol{\varepsilon} = \begin{pmatrix} \varepsilon_1 \\ \varepsilon_2 \\ \varepsilon_3 \\ \vdots \\ \varepsilon_n \end{pmatrix}$$

We shall now determine the coefficients (a_0, a_1, a_2) so that the sum of squared deviations (b) is at its minimum. To satisfy this requirement, the coefficients (a_0, a_1, a_2) shall fulfil the following three conditions

$$\frac{\partial}{\partial a_0}(\sum_{i=1}^{n}(Y_i - (a_0 + a_1X_{i.1} + a_2X_{i.2}))^2) = 0$$

$$\frac{\partial}{\partial a_1}(\sum_{i=1}^{n}(Y_i - (a_0 + a_1X_{i.1} + a_2X_{i.2}))^2) = 0 \qquad \text{(d)}$$

$$\frac{\partial}{\partial a_2}(\sum_{i=1}^{n}(Y_i - (a_0 + a_1X_{i.1} + a_2X_{i.2}))^2) = 0$$

By conventional differentiation and rearranging of terms, the equation system (d) transforms into the so-called **normal equation** for the equations (c)

$$\begin{pmatrix} n & \sum X_1 & \sum X_2 \\ \sum X_1 & \sum X_1^2 & \sum X_1X_2 \\ \sum X_2 & \sum X_1X_2 & \sum X_2^2 \end{pmatrix} \begin{pmatrix} a_0 \\ a_1 \\ a_2 \end{pmatrix} = \begin{pmatrix} \sum Y \\ \sum YX_1 \\ \sum YX_2 \end{pmatrix} \qquad \text{(e)}$$

where it is inherent that the summation Σ includes all n test sets. The normal equation formed will be a regular, square matrix.

Comparing the normal equation (e) with the original equation system on matrix form, the normal equation can be expressed by

$$\mathbf{X}^{\mathrm{T}}\mathbf{Xk} = \mathbf{X}^{\mathrm{T}}\mathbf{Y} \qquad \text{(f)}$$

where $\mathbf{X}^{\mathrm{T}}$ denotes the **transposed** matrix $\mathbf{X}$. The equation system (f) determines (a_0, a_1, a_2) so that the requirement for minimum of the sum of squared deviations SSD in eqn. (c) is fulfilled. In other cases than that of singularity, i.e. $\det(\mathbf{X}^{\mathrm{T}}\mathbf{X}) \neq 0$, the formal solution to this equation system can be directly written; using matrix notation as introduced, the solution is

$$\mathbf{k} = (\mathbf{X}^{\mathrm{T}}\mathbf{X})^{-1}(\mathbf{X}^{\mathrm{T}}\mathbf{Y}) \qquad \text{(g)}$$

Here we have developed the solution to an example with the independent variable $X(X_1, X_2)$; the solution (g), however, is general regardless of the number k

of independent variables $X(X_1, \cdots, X_k)$ in a problem presented. Thus, in conclusion we have the following procedure for determination of the coefficients $\mathbf{k} = \{a_0, a_1, \cdots, a_k\}$ by linear regression

Linear regression, determination of coefficients

In a test material with n sets of numbers the **dependent variable** Y is to be expressed by a linear combination of the **independent** variable X using

$$Y = a_0 + a_1X_1 + a_2X_2 + \cdots + a_kX_k + \varepsilon \qquad (n \geq k+1)$$

so that $SSD = \sum(\varepsilon^2)$ assumes its minimum value.

(1) Write the matrices:

$$\mathbf{Y} = \begin{pmatrix} Y_1 \\ Y_2 \\ \vdots \\ Y_n \end{pmatrix}; \quad \mathbf{X} = \begin{pmatrix} 1 & X_{1.1} & X_{1.2} & \cdots & X_{1.k} \\ 1 & X_{2.1} & X_{2.2} & \cdots & X_{2.k} \\ \vdots & \vdots & \vdots & \ddots & \vdots \\ 1 & X_{n.1} & X_{n.2} & \cdots & X_{n.k} \end{pmatrix}; \quad \mathbf{k} = \begin{pmatrix} a_0 \\ a_1 \\ \vdots \\ a_k \end{pmatrix}$$

(2) Form the equation system:

$$\mathbf{X}^{\mathrm{T}}\mathbf{X}\mathbf{k} = \mathbf{X}^{\mathrm{T}}\mathbf{Y}$$

(3) Determine the desired coefficients $\mathbf{k} = \{a_0, a_1, \cdots, a_k\}$ by calculation:

$$\mathbf{k} = (\mathbf{X}^{\mathrm{T}}\mathbf{X})^{-1}(\mathbf{X}^{\mathrm{T}}\mathbf{Y})$$

or by simple solution of the linear equation system under item (2) above.

For solution of practical problems, the independent variables $X_1 \cdots X_k$ may be mutually correlated. For example, this is the case if two or more of the independent variables are functions of a common parameter. Thermodynamic relations, for example, often have the form:

$$Y = a_0 + a_1 \cdot f_1(\mathrm{T}) + a_2 \cdot f_2(\mathrm{T}) + \cdots + a_k \cdot f_k(\mathrm{T})$$

where T denotes the thermodynamic temperature K. The specified procedure for linear regression, for example, can also be used in cases of partial correlation between the independent variables. Reference is made to special literature concerning the theory of partial correlation.

The matrix algebra used for calculation of the coefficient matrix **k** is particularly favourable for programming linear regression. If the number of independent variables X_k exceed three or four, it is in many cases better to solve the equation system (3) by **Gauss elimination** instead of inverting the matrix $\mathbf{X}^{\mathrm{T}}\mathbf{X}$ in the traditional way.

■ Influence of the force-fibre angle on the compressive strength of wood

The compressive strength of wood depends on the angle γ between **force direction** and **fibre direction**. When wood is loaded in the fibre direction, i.e. with $\gamma = 0$, its compressive strength is $\sigma_\gamma = \sigma_0$. When wood is loaded at right angles to the fibres, its compressive strength is $\sigma_\gamma = \sigma_{90}$. Generally, the strength in the transverse direction ($\gamma = 90^o$) is considerably lower than the strength in the fibre direction ($\gamma = 0^o$). Therefore, when dimensioning wood structures, it is necessary to take into account the influence of the force-fibre angle on the strength.

A frequently used expression for the dependence of the compressive strength on force-fibre angle γ is the so-called **sine formula**

$$\sigma_\gamma = \sigma_0 - (\sigma_0 - \sigma_{90}) \cdot \sin(\gamma)$$

where σ_0 and σ_{90} denote the compressive strength in the fibre direction ($\gamma = 0^o$) and the compressive strength at right angles to the fibre direction ($\gamma = 90^o$), respectively, and σ_γ denotes the compressive strength for the angle γ between force direction and fibre direction.

Problem. For a quantity of spruce, the following measurement results have been found for the compressive strength σ_γ for different angles γ between force and fibre direction:

γ	0	0	18	33	62	90	90	^{o}
σ_γ	48.6	49.5	36.1	25.6	11.2	6.2	6.8	MPa

Based on these results, determine the constants (σ_0, σ_{90}) in the sine formula so that the sum of squared deviations SSD between test data and values calculated by the sine formula is at a minimum.

Solution. We determine the constants by linear regression using the following substitutions:

$$Y_i = \sigma_{\gamma.i} \quad a_0 = \sigma_0 \quad a_1 = -(\sigma_0 - \sigma_{90}) \quad X_{i.1} = \sin(\gamma_i)$$

Using (1) of the procedure, the following matrices can be written:

$$\mathbf{Y} = \begin{pmatrix} 48.6 \\ 49.5 \\ 36.1 \\ 25.6 \\ 11.2 \\ 6.2 \\ 6.8 \end{pmatrix}; \quad \mathbf{X} = \begin{pmatrix} 1 & 0.00 \\ 1 & 0.00 \\ 1 & 0.31 \\ 1 & 0.54 \\ 1 & 0.88 \\ 1 & 1.00 \\ 1 & 1.00 \end{pmatrix}; \quad \mathbf{k} = \begin{pmatrix} \sigma_0 \\ -(\sigma_0 - \sigma_{90}) \end{pmatrix}$$

Next, according to (2) in the procedure, $\mathbf{X}$ and $\mathbf{Y}$ are multiplied by the transposed matrix $\mathbf{X}^{\mathrm{T}}$. By this operation, the **normal equation** of the form (e) is found

$$(\mathbf{X}^{\mathrm{T}}\mathbf{X})\mathbf{k} = \mathbf{X}^{\mathrm{T}}\mathbf{Y}$$

$$\begin{pmatrix} 7.00 & 3.73 \\ 3.73 & 3.16 \end{pmatrix} \mathbf{k} = \begin{pmatrix} 184.00 \\ 47.87 \end{pmatrix}$$

Item (3) in the procedure is used to determine the reciprocal matrix $(\mathbf{X}^{\mathrm{T}}\mathbf{X})^{-1}$ from

$$(\mathbf{X}^{\mathrm{T}}\mathbf{X})^{-1} = (7.00 \cdot 3.16 - 3.73^2)^{-1} \cdot \begin{pmatrix} 3.16 & -3.73 \\ -3.73 & 7.00 \end{pmatrix} = \begin{pmatrix} 0.385 & -0.454 \\ -0.454 & 0.853 \end{pmatrix}$$

$$\text{of which . } \mathbf{k} = \begin{pmatrix} \sigma_0 \\ -(\sigma_0 - \sigma_{90}) \end{pmatrix} = \begin{pmatrix} 0.385 & -0.454 \\ -0.454 & 0.853 \end{pmatrix} \begin{pmatrix} 184.00 \\ 47.87 \end{pmatrix} = \begin{pmatrix} 49.1 \\ -42.7 \end{pmatrix}$$

Thus, the solution is: $\sigma_0 = \mathbf{49.1\ MPa}$; $\sigma_{90} = 49.1 - 42.7 = \mathbf{6.4\ MPa}$

$$\textbf{Sine formula}: \ \sigma_\gamma = 49.1 - 47.2 \cdot \sin(\gamma)\ \text{MPa}$$

Literature

Edwards, A.L.: *Linear Regression and Correlation*, W.H.Freeman and Company, New York 1984.

Edwards, A.L.: *Multiple Regression and the Analysis of Variance and Covariance*, W.H.Freeman and Company, San Francisco, 1979.

6 Exact differential

In thermodynamic calculations, the value of a function $u(x, y)$ often has to be determined for two states (x_1, y_1) and (x_2, y_2) by integrating an expression of the form

$$du(x, y) = M(x, y)\,dx + N(x, y)\,dy \tag{a}$$

between the limits u_1 and u_2. The integrals occurring in the process are of the form

$$\int_{x_1}^{x_2} M(x, y)\,dx \qquad \text{and} \qquad \int_{y_1}^{y_2} N(x, y)\,dy \tag{b}$$

Integrals of the form (b) have no meaning, unless y can be eliminated in $M(x, y)$ by the relation $y = f(x)$, and correspondingly that x can be eliminated in $N(x, y)$ by the relation $x = g(y)$. This elimination corresponds to uniquely defining an integration path in the xy-plane; the integral (b), therefore, is called a **line integral**. Generally, many different line integrals can be defined from u_1 to u_2, each rendering different values.

If (a) is an **exact differential** of $u(x, y)$, the line integral from u_1 to u_2 has special properties. The differential (a) is said to be an exact differential of $u(x, y)$, if it applies all over the area of definition that

$$du(x, y) = (\frac{\partial u(x, y)}{\partial x})_y\,dx + (\frac{\partial u(x, y)}{\partial y})_x\,dy \tag{c}$$

This condition is fulfilled, and (a) is an exact differential, if it applies that

$$\frac{\partial M(x, y)}{\partial y} = \frac{\partial u(x, y)^2}{\partial x \partial y} = \frac{\partial N(x, y)}{\partial x}) \tag{c}$$

where the order of succession of the differentiation can be chosen arbitrarily. The expression (c) can be used to investigate whether a differential of the form (a) is exact. The special properties of an exact differential have been summarized in the following.

Line integral of exact differential (d)

(1) $du(x, y) = M(x, y)\,dx + N(x, y)\,dy$ is an **exact differential** if

$$(\frac{\partial M(x, y)}{\partial y})_x = (\frac{\partial N(x, y)}{\partial x})_y$$

(2) For an exact differential $du(x, y)$ it applies that

$$\int_{(1)}^{(2)} du(x, y) = u(x_2, y_2) - u(x_1, y_1)$$

i.e. the value of the line integral is **independent** of the integration path chosen.

(3) If $du(x, y)$ is an exact differential, integration along a closed curve will give the value 0:

$$\oint du(x, y) = 0$$

In thermodynamics, exact differentials occur in the form of so-called **state variables** and **state functions**; the special properties (2) and (3) for these differentials are fundamental in thermodynamic analysis.

A simple conception of the notion of "exact differential" can be obtained from the following down-to-earth example: In a hilly terrain, two points (1) and (2) are delimited. A variable $h(x, y)$ denoting the height level of the terrain as a function of the xy coordinate is analogous to what is designated a **state variable** in thermodynamics. The differential $dh(x, y)$ is an exact differential because it is evident that

$$\int_{(1)}^{(2)} dh(x,y) = \int_{(1)}^{(2)} (\frac{\partial h(x,y)}{\partial x})_y dx + \int_{(1)}^{(2)} (\frac{\partial h(x,y)}{\partial y})_x dy = h(2) - h(1)$$

Regardless of the integration path chosen from point (1) to point (2), the increase of height level $h(2) - h(1)$ remains the same. A non-exact differential can in this connection be represented by the differential ds of the path length from point (1) to point (2). Evidently, the resulting path length will be determined by the integration path chosen, i.e. by the route one follows in the terrain.

■ Change of state, ideal gas

In thermodynamics, exact and non-exact differentials occur in a number of cases. In the following, we consider a system consisting of n mol of an ideal gas. The state of the system is described by the **ideal gas law**

$$pV = nRT \qquad \text{(e)}$$

During an infinitesimal change of state, the gas volume is increased by dV; in the process, the work dW is done on the system. Check whether dV and/or dW are exact differentials according to condition (1)!

Solution. From the ideal gas law (e), it follows that $V = f(p, T)$; the differential dV can thus be expressed on the form

$$dV = (\frac{\partial V}{\partial T})_p dT + (\frac{\partial V}{\partial p})_T dp \qquad \text{(f)}$$

Using the gas law (e), the partial differentials can be introduced and we obtain

$$dV = (\frac{nR}{p})\, dT + (-\frac{nRT}{p^2})\, dp = M(p,T) \cdot dT + N(p,T) \cdot dp \qquad \text{(g)}$$

To decide whether dV is an exact differential, the relation (d) is applied to this differential

$$\frac{\partial}{\partial p}(\frac{nR}{p})_T = -\frac{nR}{p^2}; \qquad \frac{\partial}{\partial T}(-\frac{nRT}{p^2})_p = -\frac{nR}{p^2}$$

Therefore, the differential dV is an exact differential and in accordance with experience, the gas volume V is a **state function** in a thermodynamic sense.

Considering the volume work $dW = -p \cdot dV$ exerted on the system by the surroundings during the process, we have, based on the above

$$dW = -p \cdot dV = -nR\, dT + (\frac{nRT}{p})\, dp$$

Using the condition (d), it is now tested whether dW is an exact differential

$$\frac{\partial}{\partial p}(-nR)_T = 0; \qquad \frac{\partial}{\partial T}(\frac{nRT}{p})_p = \frac{nR}{p} \neq 0$$

Thus, the differential dW does not fulfil the condition for being an exact differential. Accordingly, the work $W_{1.2}$ during a change of state is dependent on the process path; in a thermodynamic sense, the work W is therefore not a state variable. In the main text therefore, the symbol δW is used instead of dW for the differential of the thermodynamic work W.

Literature

Kreyszig, E.: *Advanced Engineering Mathematics*, John Wiley and Sons, Inc., New York 1972.

Margenau, H. & Murphy,G.M.: *The Mathematics of Physics and Chemistry*, D. van Nostrand Company Inc., New York 1955.

7 Gradient field

A **scalar field** $\Phi(x, y, z)$ denotes a volume in space in which each point (x, y, z) can be given a scalar value $\Phi(x, y, z)$. As a simple, but illustrative example of a scalar field, the temperature distribution $\theta(x, y, z)$ in the atmosphere at a given time can be mentioned.

A **vector field** $\mathbf{f}(x, y, z)$ denotes a volume in space in which any point (x, y, z) can be given a vector value $\mathbf{f}(x, y, z)$. A vector field can be illustrated by the wind velocity distribution $\mathbf{v}(x, y, z)$ in the atmosphere at a given time.

The **gradient** of a continuous and differentiable scalar function $\Phi(x, y, z)$ is defined as follows

Gradient of scalar function (a)

$$\mathbf{grad}\,\Phi = \frac{\partial \Phi}{\partial x}\mathbf{i} + \frac{\partial \Phi}{\partial y}\mathbf{j} + \frac{\partial \Phi}{\partial z}\mathbf{k}$$

where: **i**, **j**, **k** denote the unit vectors in the principal directions x, y, z of the coordinate system.

As seen from this definition, **grad** Φ of a scalar function $\Phi(x, y, z)$ itself denotes a vector field, i.e. the so-called **gradient field** belonging to the scalar field.

In the literature, **grad** Φ is often expressed by application of the symbolic **differential operator** $\boldsymbol{\nabla}$ ("*nabla*") defined by

$$\boldsymbol{\nabla} = \frac{\partial}{\partial x}\mathbf{i} + \frac{\partial}{\partial y}\mathbf{j} + \frac{\partial}{\partial z}\mathbf{k} \tag{b}$$

Thus, we have the following conventional form of notation for the gradient of a scalar function:

Notation for gradient of scalar function (c)

$$\mathbf{grad}\,\Phi = \boldsymbol{\nabla}\Phi = \left(\frac{\partial}{\partial x}\mathbf{i} + \frac{\partial}{\partial y}\mathbf{j} + \frac{\partial}{\partial z}\mathbf{k}\right)\Phi = \frac{\partial \Phi}{\partial x}\mathbf{i} + \frac{\partial \Phi}{\partial y}\mathbf{j} + \frac{\partial \Phi}{\partial z}\mathbf{k}$$

Due to the use of the del operator, the gradient of Φ can be expressed in a shorter form. Again, it should be borne in mind that *nabla* $\boldsymbol{\nabla}$ is a **linear differential operator** that forms a vector function by operating on a scalar function; thus, $\boldsymbol{\nabla}$ should not in itself be taken as a vector.

In physics, the gradient of scalar functions occur in a number of cases; therefore, we shall briefly draw attention to some important properties of this quantity. Assume a continuous and differentiable scalar function $\Phi(x, y, z)$ in a volume in space. Through the volume, curve $C(s)$ runs through a point Q. The **unit tangent vector T** to the curve in Q can then be expressed by

$$\mathbf{T} = \frac{dx}{ds}\mathbf{i} + \frac{dy}{ds}\mathbf{j} + \frac{dz}{ds}\mathbf{k} \tag{c}$$

Expressing by conventional differentiation the differential of the scalar field at the curve point Q, we obtain

$$\frac{d\Phi}{ds} = \left(\frac{\partial \Phi}{\partial x}\right)\frac{dx}{ds} + \left(\frac{\partial \Phi}{\partial y}\right)\frac{dy}{ds} + \left(\frac{\partial \Phi}{\partial z}\right)\frac{dz}{ds} \tag{d}$$

Comparing the expression (d) with (a) and (c), we find that the differential of Φ at the point Q can be expressed as the **scalar product** of **grad** Φ and the unit tangent vector $\mathbf{T}$ at this point; in other words, we have that

$$\frac{d\Phi}{ds} = \mathbf{grad}\,\Phi \cdot \mathbf{T} = \nabla\Phi \cdot \mathbf{T} \qquad \text{(e)}$$

The differential of a scalar field Φ at a point $Q(x, y, z)$ in a direction given by the unit vector $\mathbf{T}$ is equal to the scalar product of **grad** Φ and $\mathbf{T}$ at the point Q.

The scalar product $\nabla\Phi \cdot \mathbf{T}$ in eqn. (e) can be expressed by the angle θ between the gradient vector and the unit vector $\mathbf{T}$ at the point Q

$$\frac{d\Phi}{ds} = \nabla\Phi \cdot \mathbf{T} = |\,\nabla\Phi\,| \cdot |\,\mathbf{T}\,| \cdot \cos(\theta) \qquad \text{(f)}$$

From this it is seen that $\cos(\theta)$, and thus the differential assume their largest value for $\theta = 0$, i.e. when $\mathbf{T}$ has the same direction as **grad** Φ

$$\max(\frac{d\Phi}{ds}) = |\,\mathbf{grad}\,\Phi\,| = |\,\nabla\Phi\,| \qquad \text{(g)}$$

Correspondingly, from eqn. (f) it is seen that the differential assumes the value 0 for $\cos(\theta) = 0$, i.e. for all directions perpendicular to the gradient vector at the point Q. Mathematically, this means that universally, the gradient is **normal** to the **level** surface where the scalar field $\Phi(x, y, z)$ is constant. In conclusion, we thus have

Gradient vector properties (h)

At a point $Q(x, y, z)$, belonging to a continuous, differentiable scalar field $\Phi(x, y, z)$, the following applies:

(1) The gradient vector **grad** Φ denotes the **direction**, in which the directional differential $\frac{d\Phi}{ds}$ assumes its largest value.

(2) The numerical value of the gradient vector $|$ **grad** Φ $|$ denotes the **magnitude** of the directional differential of the scalar field $\max(\frac{d\Phi}{ds})$.

(3) The gradient vector **grad** Φ is **normal** to the level surface through the point $Q(x, y, z)$, given by $\Phi(x, y, z) = \textit{constant}$.

The above-mentioned properties of a scalar field are, for example, utilized by the definition of a **potential field** in physics. From the expression (e) it is seen that the curve integral from a point $A(x, y, z)$ to a point $B(x, y, z)$ in the field is determined by

$$\int_A^B d\Phi = \int_A^B \mathbf{grad}\,\Phi \cdot d\mathbf{s} = \Phi(B) - \Phi(A) \qquad \text{(i)}$$

Thus, the value of the curve integral is equal to the difference between the value of the scalar field at point B and point A independently of the integration path chosen. From this it is seen that the curve integral along a closed curve $ABCA$ is 0.

Integral of gradient field (j)

$$\int_A^B \mathbf{grad}\,\Phi{\cdot}d\mathbf{s} = \Phi(B) - \Phi(A); \qquad \oint \mathbf{grad}\,\Phi{\cdot}d\mathbf{s} = 0$$

A number of the **force fields** occurring in physics can be described as gradients of a scalar field $\Phi(x, y, z)$. With negative sign, the scalar field denotes the so-called **potential field** belonging to the actual force field. From the above, it is seen that the work with a given displacement in this kind of force field is independent of the path chosen in the field; these kinds of force fields, therefore, will always be **conservative** fields.

■ Calculation of gradient and curve integral

A continuous and differentiable scalar field $\Phi(x, y, z)$ is given by:

$$\Phi(x, y, z) = 2x^2 + 3y^2 + z^2$$

Determine the following quantities for the given scalar field:

a) Direction and magnitude of the maximum scalar field gradient in $Q(2, 1, 3)$.

b) The directional differential in Q determined by $\mathbf{a} = \mathbf{i} - 2\mathbf{k}$.

c) The curve integral of the gradient field from $Q(2, 1, 3)$ to $P(3, 4, 1)$.

Solution.

a) For point $Q(2, 1, 3)$ it applies that

$$\mathbf{grad}\,\Phi = 4x\,\mathbf{i} + 6y\,\mathbf{j} + 2z\,\mathbf{k} = \qquad 8\,\mathbf{i} + 6\,\mathbf{j} + 6\,\mathbf{k}$$

$$|\,\mathbf{grad}\,\Phi\,| = \sqrt{8^2 + 6^2 + 6^2} = \sqrt{136} = \qquad 2\sqrt{34}$$

Thus, the maximum gradient of the scalar field is $2\sqrt{34}$ in the direction determined by the gradient vector $\mathbf{grad}\,\Phi = 8\mathbf{i} + 6\mathbf{j} + 6\mathbf{k}$ at the point $Q(2, 1, 3)$.

b) For $|\,\mathbf{a}\,| = \sqrt{1^2 + 2^2} = \sqrt{5}$, the direction of the unit vector $\mathbf{T}$ in the $\mathbf{a}$ direction is

$$\mathbf{T} = \frac{\mathbf{a}}{|\,\mathbf{a}\,|} = \frac{1}{\sqrt{5}}\mathbf{i} - \frac{2}{\sqrt{5}}\mathbf{k}$$

The directional differential at the point Q through the vector $\mathbf{a}$ can thus be determined from eqn. (e)

$$\frac{d\Phi}{ds} = \mathbf{grad}\,\Phi{\cdot}\mathbf{T} = (8\,\mathbf{i} + 6\,\mathbf{j} + 6\,\mathbf{k}){\cdot}(\frac{1}{\sqrt{5}}\mathbf{i} - \frac{2}{\sqrt{5}}\mathbf{k}) = \qquad \mathbf{4\sqrt{5}}$$

c) From the expression (j), the value of the integral can now be determined

$$\Phi(Q) = (2 \cdot 2^2 + 3 \cdot 1^2 + 1 \cdot 3^2) = 76; \qquad \Phi(P) = (2 \cdot 3^2 + 3 \cdot 4^2 + 1 \cdot 1^2) = 67$$

$$\int_Q^P \mathbf{grad}\,\Phi{\cdot}d\mathbf{s} = \Phi(P) - \Phi(Q) = 67 - 76 = \qquad \mathbf{-9}$$

Literature

Hsu, Hwei P.: *Applied Vector Analysis*, Harcourt Brace Jovanovich, San Diego, 1964.

Kreyszig, E.: *Advanced Engineering Mathematics*, John Wiley and Sons, Inc., New York 1972.

8 Maxwell's relations

The thermodynamic concepts formulated in chapters 2 to 6 mainly addressed the **first** and **second law**. The use of these laws has been illustrated by a number of examples from engineering practice. We have shown how subject areas such as equilibrium in systems of matter, electrochemical phenomena, and thermal effects can be analysed based on the definitions of these laws.

The following brief outline explains how to develop a set of more general relations of thermodynamic systems based on the laws of thermodynamics. Knowledge of these relations is useful in connection with experimental investigations aimed at determining thermodynamic data for materials based on laboratory measurements.

State quantities

The general and practical applicability characterizing the laws of thermodynamics is especially based on two conditions: the analysis is based on **state variables** and **state functions**. The names specify that these quantities depend on the **state** of the system only and that they are *independent* of how this state was achieved.

The idea is clear: Assume a change of state connecting two **states of equilibrium** (1) and (2). If only one **reversible process** connecting these two states can be identified, the change of all state quantities can be calculated. The calculated change will then be the same as for another *arbitrary* process connecting these states. In other words, the use of state quantities makes analysis of the result of **irreversible processes** possible by replacing these by imaginary, reversible processes by the same terminal states.

The state quantities used in thermodynamics can be divided into two classes:

- *State variables* such as **volume** V, **pressure** p, and **temperature** T all of which are well-known *measurable* physical quantities.
- *State functions* such as **internal energy** U, **enthalpy** H, **entropy** S, and the **Gibbs free energy** G expressing mathematical mutual relations between state quantities; thus, state functions are physical quantities that are measurable only in exceptional cases.

Using mathematics, we shall now see how to make a change of variables in state functions and thus simplify these functions or rewrite them to a form that is measurable.

Fundamental equations

First, we compare the **first law** of thermodynamics eqn. (3.2) and the **second law** of thermodynamics eqn. (4.12); in these laws, two fundamental state functions are defined, namely **internal energy** U and **entropy** S

$$\textbf{1st Law}: \quad dU \stackrel{\text{def}}{=} \delta Q + \delta W \qquad \textbf{2nd Law}: \quad dS \stackrel{\text{def}}{=} \frac{\delta Q_{rev}}{T} \tag{a}$$

For any **reversible process** where *only* **volume work** occurs, it will apply that: $\delta W = -pdV$, and that: $\delta Q = \delta Q_{rev} = TdS$. Under these circumstances the differential of the internal energy U can thus be expressed on the following form

$$dU = TdS - pdV \tag{b}$$

Under the *same* conditions, the differential of **enthalpy** H, defined in eqn. (3.11), can be expressed by

$$dH = dU + d(pV) = TdS - pdV + pdV + Vdp = TdS + Vdp \tag{c}$$

For the state functions the **Gibbs free energy** G eqn. (5.1) and the **Helmholtz free energy** A eqn. (5.2), we obtain under the same conditions:

$$dG = dH - d(TS) = TdS + Vdp - TdS - SdT = Vdp - SdT \tag{d}$$

$$dA = dU - d(TS) = TdS - pdV - TdS - SdT = -pdV - SdT \tag{e}$$

In the expression (d) we recognize the *Gibbs-Duhem equation* (eqn. (5.14)) that was previously derived and was used for describing, for example, the influence of pressure and temperature on phase equilibrium in single-component systems.

The relations (b)...(e) are called the **fundamental equations** of thermodynamics. Derivation of these relations had to be limited to reversible processes during which only volume work occurred. However, since internal energy U, enthalpy H, the Gibbs free energy G, and the Helmholtz free energy A are all **state functions**, the relations derived apply to *any* change - reversible as well as irreversible - connecting states of equilibrium in a system. Thus, in conclusion we have

Fundamental equations of thermodynamics (f)

For **changes of state** in a thermodynamic system where *only* **volume work** $\delta W = -pdV$ occurs, the following applies

$$dU = TdS - pdV \qquad U = U(S,V) \qquad (1)$$

$$dH = TdS + Vdp \qquad H = H(S,p) \qquad (2)$$

$$dG = -SdT + Vdp \qquad G = G(T,p) \qquad (3)$$

$$dA = -SdT - pdV \qquad A = A(T,V) \qquad (4)$$

Note that the fundamental equations (f) express a change of **internal energy** dU when the **entropy** dS and the **volume** dV are changed. It is said that (S,V) are the "natural" variables of the internal energy $U = U(S,V)$, because dU has a particularly simple relation to dS and dV. Correspondingly, (S,p) are the natural variables for **enthalpy** $H = H(S,p)$, and (T,p) are the natural variables for the **Gibbs free energy** $G = G(T,p)$.

Maxwell's relations

The fundamental equations (f) denote **exact differentials** of the thermodynamic state functions U, H, G, and A (see: *Mathematical appendix, subchapter A6 on Exact differential*); for example, the differential of the **internal energy** $U(S,V)$ means that

$$dU = \left(\frac{\partial U}{\partial S}\right)_V dS + \left(\frac{\partial U}{\partial V}\right)_S dV \quad \Leftrightarrow \quad dU = TdS - pdV \qquad (g)$$

Comparing the differential with the fundamental equation we see that **thermodynamic temperature** T and **pressure** p have the following relation to the internal energy $U(S,V)$

$$T = \left(\frac{\partial U}{\partial S}\right)_V \qquad p = -\left(\frac{\partial U}{\partial V}\right)_S \qquad (h)$$

In other words, if the quantities $(\frac{\partial U}{\partial S})_V$ or $(\frac{\partial U}{\partial V})_S$ occur in a thermodynamic relation, we can replace these complex quantities by the **temperature** T and **pressure** p of the system, which are simple *measurable* physical quantities. Correspondingly, by comparing the exact differential of **enthalpy** $H(S,p)$ with the fundamental equation (f.2), we obtain

$$dH = \left(\frac{\partial H}{\partial S}\right)_p dS + \left(\frac{\partial H}{\partial p}\right)_S dp \quad \Leftrightarrow \quad dH = TdS + Vdp \qquad (i)$$

Thus, the **temperature** T and **volume** V of the system have the following relation to enthalpy $H(S,p)$

$$T = \left(\frac{\partial H}{\partial S}\right)_p \qquad V = \left(\frac{\partial H}{\partial p}\right)_S \qquad (j)$$

Once more, we see that complex quantities such as $(\frac{\partial H}{\partial S})_p$ and $(\frac{\partial H}{\partial p})_S$ can be replaced by simple, measurable physical quantities, namely **temperature** (T) and

volume (V). Using the same procedure, we see from the differential of $G(T,p)$ and $A(T,V)$ that

$$S = -\left(\frac{\partial G}{\partial T}\right)_p \quad V = \left(\frac{\partial G}{\partial p}\right)_T \qquad S = -\left(\frac{\partial A}{\partial T}\right)_V \quad p = -\left(\frac{\partial A}{\partial V}\right)_T \tag{k}$$

Consider again the differential (g) of the **internal energy** $U(S,V)$. By partial differentiation of the coefficient to dS and utilizing that the sequence of the differentiation is immaterial (see: *Mathematical appendix, subchapter A6*) we can rewrite as follows

$$\left(\frac{\partial T}{\partial V}\right)_S = \left(\frac{\partial}{\partial V}\left(\frac{\partial U}{\partial S}\right)_V\right)_S = \left(\frac{\partial}{\partial S}\left(\frac{\partial U}{\partial V}\right)_S\right)_V = \left(\frac{\partial(-p)}{\partial S}\right)_V \tag{l}$$

where we in the last rewriting have utilized the expression (h) previously derived for pressure: $p = -(\frac{\partial U}{\partial V})_S$. Thus, by rewriting we obtain the following relation between **temperature** T and **pressure** p

$$\left(\frac{\partial T}{\partial V}\right)_S = -\left(\frac{\partial p}{\partial S}\right)_V \tag{m}$$

Correspondingly, for the differential (i) of **enthalpy** $H(S,p)$, we obtain by partial differentiation of the coefficient to dS that

$$\left(\frac{\partial T}{\partial p}\right)_S = \left(\frac{\partial}{\partial p}\left(\frac{\partial H}{\partial S}\right)_p\right)_S = \left(\frac{\partial}{\partial S}\left(\frac{\partial H}{\partial p}\right)_S\right)_p = \left(\frac{\partial V}{\partial S}\right)_p \tag{n}$$

where, once more, in the last rewriting we use the expression for volume (j) derived above: $V = (\frac{\partial H}{\partial p})_S$. Thus, we have a corresponding relation between **temperature** T and **volume** V

$$\left(\frac{\partial T}{\partial p}\right)_S = \left(\frac{\partial V}{\partial S}\right)_p \tag{o}$$

Using the same procedure, we can form relations between T and p based on the differential of the G function and relations between T and V based on the differential of the A function. Summarizing these derivations, we have the so-called **Maxwell relations** between thermodynamic state quantities.

Maxwell's relations (p)

For a **change of state** in a thermodynamic system where *only* **volume work** $\delta W = -pdV$ occurs, the following relations apply

$$T = \left(\frac{\partial U}{\partial S}\right)_V \qquad p = -\left(\frac{\partial U}{\partial V}\right)_S \qquad \left(\frac{\partial T}{\partial V}\right)_S = -\left(\frac{\partial p}{\partial S}\right)_V \tag{1}$$

$$T = \left(\frac{\partial H}{\partial S}\right)_p \qquad V = \left(\frac{\partial H}{\partial p}\right)_S \qquad \left(\frac{\partial T}{\partial p}\right)_S = \left(\frac{\partial V}{\partial S}\right)_p \tag{2}$$

$$S = -\left(\frac{\partial G}{\partial T}\right)_p \qquad V = \left(\frac{\partial G}{\partial p}\right)_T \qquad -\left(\frac{\partial S}{\partial p}\right)_T = \left(\frac{\partial V}{\partial T}\right)_p \tag{3}$$

$$S = -\left(\frac{\partial A}{\partial T}\right)_V \qquad p = -\left(\frac{\partial A}{\partial V}\right)_T \qquad \left(\frac{\partial S}{\partial V}\right)_T = \left(\frac{\partial p}{\partial T}\right)_V \tag{4}$$

At first, **Maxwell's relations** and the foregoing mathematical exercises may seem somewhat inaccessible. However, as shown in the following examples, the aim of these relations is to *rewrite* and *simplify* thermodynamic functions. Before considering the practical application of fundamental equations and Maxwell's relations, however, it is expedient to summarize some of the definitions previously introduced.

Substance parameters

In chapter 3, we introduced the thermodynamic definition of heat capacity c_V at **constant volume**, eqn. (3.10), and heat capacity c_p at **constant pressure**, eqn. (3.16).

Isochoric and isobaric heat capacity (q)

$$c_V \stackrel{\text{def}}{=} \left(\frac{\partial U}{\partial T}\right)_V \qquad \text{heat capacity, constant } V \quad (\text{J/mol K}) \tag{1}$$

$$c_p \stackrel{\text{def}}{=} \left(\frac{\partial H}{\partial T}\right)_p \qquad \text{heat capacity, constant } p \quad (\text{J/mol K}) \tag{2}$$

Also, in chapter 1, we defined the **coefficient of volume expansion** α and **compressibility** κ of a substance. These substance parameters are defined by

Coefficient of expansion, compressibility, and elastic modulus (r)

$$\alpha \stackrel{\text{def}}{=} \frac{1}{V}\left(\frac{\partial V}{\partial T}\right)_p \qquad \text{coefficient of volume expansion} \quad (\text{K}^{-1}) \tag{1}$$

$$\beta \stackrel{\text{def}}{=} \frac{1}{L}\left(\frac{\partial L}{\partial T}\right)_p \qquad \text{coefficient of linear expansion} \quad (\text{K}^{-1}) \tag{2}$$

$$\kappa \stackrel{\text{def}}{=} -\frac{1}{V}\left(\frac{\partial V}{\partial p}\right)_T \qquad \text{compressibility} \quad (\text{Pa}^{-1}) \tag{3}$$

$$E \stackrel{\text{def}}{=} \frac{L}{A}\left(\frac{\partial F}{\partial L}\right)_T \qquad \text{elastic modulus} \quad (\text{N/m}^2) \tag{4}$$

The substance parameters: c_V, c_p, α, β, κ, and E are measurable, physical quantities; the numerical values of these quantities can be found in most tables.

The examples calculated in the following show how to solve practical problems by combining **fundamental equations** (f) and **Maxwell's relations** (p) with table data for substance parameters. However, please note that this brief explanation is only an introduction to the subject of *Thermodynamic relations*. In the literature list, reference has been made to literature in which the subject has been thoroughly explained.

■ Joule's law for ideal gas

In chapter 3.5 we have formulated **Joule's law**, eqn. (3.9), for an ideal gas: The **internal energy** $U = U(T)$ of an **ideal gas** only depends on the gas **temperature** T and is *independent* of the gas **volume** V and **pressure** p.

This important rule for ideal gases was originally formulated based on "Joule's experiments" (1845). Joule proved by laboratory experiments that the temperature T of an ideal gas remains *unchanged*, when the gas expands into a vacuum. In its original form, Joule's law was given as an empirical law. We now try to look into the validity of this statement using the fundamental equations (f) and Maxwell's relations (p).

Joule's law, eqn. (3.19) states that $(\frac{\partial U}{\partial V})_T = 0$, i.e. the **internal energy** U of an ideal gas does *not* change during an **isothermal** change of volume! For a constant temperature T, the fundamental equation (f.1) can be written on the form

$$(\partial U)_T = T(\partial S)_T - p(\partial V)_T \qquad \Rightarrow \qquad \left(\frac{\partial U}{\partial V}\right)_T = T\left(\frac{\partial S}{\partial V}\right)_T - p \tag{s}$$

However, according to Maxwell's relation: $(\frac{\partial S}{\partial V})_T = (\frac{\partial p}{\partial T})_V$; this is inserted into eqn. (s)

$$\left(\frac{\partial U}{\partial V}\right)_T = T\left(\frac{\partial p}{\partial T}\right)_V - p \tag{t}$$

$p = RT/V$ for an ideal gas; introducing the differential of this: $(\frac{\partial p}{\partial T})_V = R/V$ into eqn. (s), we obtain

$$\left(\frac{\partial U}{\partial V}\right)_T = T \cdot \frac{R}{V} - p = p - p = 0 \quad \Rightarrow \quad \text{“Joule's law”} \tag{u}$$

Thus, the ideal gas is seen to satisfy Joule's law. On the other hand, if we had considered a **real gas** determined by the **van der Waals equation** (1.11), the pressure p and the differential $(\frac{\partial p}{\partial T})_V$ would have assumed the form

$$p = \frac{nRT}{V-nb} - a\left(\frac{n}{V}\right)^2; \qquad \left(\frac{\partial p}{\partial T}\right)_V = \frac{nR}{V-nb} \tag{v}$$

Introducing these values into eqn. (t), we obtain the corresponding expression for $(\frac{\partial U}{\partial V})_T$ in the real gas

$$\left(\frac{\partial U}{\partial V}\right)_T = \frac{nRT}{V-nb} - \frac{nRT}{V-nb} + a\left(\frac{n}{V}\right)^2 = a\left(\frac{n}{V}\right)^2 > 0 \tag{x}$$

Therefore, for all real gases, the internal energy $U = U(T, V)$ is *increased* by an **isothermal** expansion. This indicates that the potential energy between the gas particles grows when they are removed from one another. During a free **adiabatic** expansion, molecular-kinetic energy is spontaneously transformed to potential energy; for real gases, this is seen by the fact that the gas temperature decreases during adiabatic expansion into a vacuum.

■ Influence of pressure on the entropy of condensed substances

In chapter 4.7, it was assumed that the influence of pressure on the entropy content of solid and liquid substances can normally be disregarded. For ideal gases, the entropy change with the pressure p is determined by eqn. (4.22). We shall now estimate this assumption based on **Maxwell's relations.**

The dependence on pressure of entropy at constant temperature T can be estimated from the differential: $(\frac{\partial S}{\partial p})_T$. In Maxwell's relations, this differential occurs under eqn. (p.3) in the Maxwell relations box above

$$-\left(\frac{\partial S}{\partial p}\right)_T = \left(\frac{\partial V}{\partial T}\right)_p \tag{y}$$

The right-hand side of this expression is seen to contain the differential $(\frac{\partial V}{\partial T})_p$ that is included in the definition equation (r.1) for the coefficient of volume expansion α of substances; therefore, the relation (y) can be rewritten to the simpler form

$$\left(\frac{\partial S}{\partial p}\right)_T = -\alpha V \tag{z}$$

Most solid and liquid substances have a *positive* coefficient of volume expansion α. The expression drawn up indicates that the entropy S *decreases* with increasing pressure p. This indicates that under increased pressure, substances tend towards a state of increased molecular order. Exceptions from the rule that $\alpha > 0$ can, for example, be seen for water. In the temperature range $0 - 4\,^\circ$C it is well-known that an anomaly exists for water with *negative* coefficient of volume expansion α so that the maximum density of water occurs at $4\,^\circ$C (see the table in Appendix B on Water and water vapour).

The following table denotes $(\frac{\partial S}{\partial p})_T$ for some selected substances calculated from the relation (z). The given table data refer to $25\,^\circ$C at the pressure $p = p^\ominus = 1$ atm.

Substance	V (m^3/mol)	α (K^{-1})	S_{298} (J/mol K)	$(\frac{\partial S}{\partial p})_T$ (J/mol K · atm)
$H_2O(\ell)$	$18.1 \cdot 10^{-6}$	$2.1 \cdot 10^{-4}$	69.9	$-3.8 \cdot 10^{-4}$
Fe(s)	$7.1 \cdot 10^{-6}$	$3.5 \cdot 10^{-5}$	27.3	$-2.5 \cdot 10^{-4}$
Al(s)	$10.0 \cdot 10^{-6}$	$6.9 \cdot 10^{-5}$	28.3	$-7.0 \cdot 10^{-4}$
$H_2(g)$	$24.5 \cdot 10^{-3}$	$3.4 \cdot 10^{-3}$	130.6	-8.4

As seen from these table values, the influence of pressure on the entropy S is negligible for solid and liquid substances if no extreme pressures occur. For ideal gases, $(\frac{\partial S}{\partial p})_T$ is typically 10^4 times the value for solid and liquid substances.

■ Temperature change for adiabatic compression of substances

In Chapter 3.7, the **adiabatic equations** for adiabatic compression of an **ideal gas** were derived; for example, the following relation between temperature T and pressure p was established: $p^{1-\gamma}T^{\gamma} = constant$, where $\gamma = c_p/c_v$. Through an example, it will be illustrated in the following how **fundamental equations** and **Maxwell's relations** can be used in a simple way to derive a *general* expression for $T = T(p)$ by adiabatic compression of solid substances, liquids and gases.

In a **reversible, adiabatic** process: $\delta Q_{rev} = TdS = 0$. This condition, therefore, corresponds to a process with constant S ("isentropic"). Thus, the differential we are looking for is: $(\frac{\partial T}{\partial p})_S$. In the Maxwell relations derived in eqn. (p), it is seen that the differential can be expressed by the natural variables (S, V) of the internal energy U

$$\left(\frac{\partial T}{\partial p}\right)_S = \left(\frac{\partial V}{\partial S}\right)_p \tag{æ}$$

We now break down the differential: $(\frac{\partial V}{\partial S})_p$ into known physical quantities using the following rewriting

$$\left(\frac{\partial V}{\partial S}\right)_p = \left(\frac{\partial V}{\partial T}\right)_p \left(\frac{\partial T}{\partial H}\right)_p \left(\frac{\partial H}{\partial S}\right)_p = (\alpha V)\cdot\left(\frac{1}{c_p}\right)\cdot(T) \tag{ø}$$

where we have applied the relations (r.1), (q.2), and (p.2) in the order of succession mentioned. Thus, we can express the differential considered: $(\frac{\partial T}{\partial p})_S$ for adiabatic change of pressure by simple, measurable substance parameters

$$\left(\frac{\partial T}{\partial p}\right)_S = \frac{\alpha V T}{c_p} \tag{å}$$

This expression is universal for **solid substances**, for **liquids**, and for **real** and **ideal gases**. We test the expression on an ideal gas, where: $\alpha = 1/T$, $V = RT/p$, and $\gamma = c_p/c_V$

$$\left(\frac{\partial T}{\partial p}\right)_S = \frac{\alpha V T}{c_p} = \frac{RT}{p\,c_p} \quad \Rightarrow \quad d\ln(T) = \frac{R}{c_p}\cdot d\ln(p) = \frac{\gamma - 1}{\gamma}\cdot d\ln(p) \tag{aa}$$

which by integration leads to the adiabatic equation previously derived: $p^{1-\gamma}T^{\gamma} =$ *constant*. The following table indicates $(\frac{\partial T}{\partial p})_S$ for some selected substances calculated from the relation (å). The table values indicated refer to 25 °C at the pressure $p = p^{\ominus} = 1$ atm.

Substance	V (m^3/mol)	α (K^{-1})	c_p (J/mol K)	$(\frac{\partial T}{\partial p})_S$ (K/atm)
$H_2O(\ell)$	$18.1 \cdot 10^{-6}$	$2.1 \cdot 10^{-4}$	75.3	0.0015
Fe(s)	$7.1 \cdot 10^{-6}$	$3.5 \cdot 10^{-5}$	25.1	0.0003
Al(s)	$10.0 \cdot 10^{-6}$	$6.9 \cdot 10^{-5}$	24.3	0.0009
H_2(g)	$24.5 \cdot 10^{-3}$	$3.4 \cdot 10^{-3}$	28.8	87.4

Since most solid and liquid substances have a *positive* coefficient of volume expansion α, an increased pressure according to eqn. (å) will result in a temperature rise. As seen from the table, this temperature rise is modest for condensed phases, unless extreme changes of pressure occur. For the special case of liquid water in the temperature range 0 − 4 °C, an increased pressure will result in a temperature decrease due to the anomalous, negative α value in this area.

■ Pressure increase in the case of heating with delayed expansion

In the design of structures it is important to predict and take into consideration thermally induced differential movements. Although the coefficient of thermal expansion for solid and liquid substances may seem rather modest, damaging imposed stresses can arise if a structure is not able to absorb the movements due to temperature changes.

A typical example is heating of a liquid-filled steel tank in which *no* expansion relief has been ensured. The coefficient of volume expansion α is 10 − 30 times larger for liquids than for steel; during heating, therefore, expansion of the liquid is prevented and a pressure is built up in the tank. In these circumstances, even the most modest temperature changes can make the tank burst.

In conclusion, we shall estimate the increase in pressure arising in a substance when heated at constant volume. With the notation used here, this corresponds to determining the numerical value for the differential: $(\frac{\partial p}{\partial T})_V$. Based on a formal state equation of a substance: $f(p, V, T) = 0$, we can express the volume as a function of temperature and pressure: $V = V(T, p)$. Then the differential of V is

$$dV = \left(\frac{\partial V}{\partial T}\right)_p dT + \left(\frac{\partial V}{\partial p}\right)_T dp \tag{ab}$$

It is seen that dV is an exact differential, because: $(\frac{\partial}{\partial p}(\frac{\partial V}{\partial T})_p)_T = (\frac{\partial}{\partial T}(\frac{\partial V}{\partial p})_T)_p$. Introducing the assumption of constant volume V, we obtain

$$0 = \left(\frac{\partial V}{\partial T}\right)_p (\partial T)_V + \left(\frac{\partial V}{\partial p}\right)_T (\partial p)_V \quad \Rightarrow \quad \left(\frac{\partial V}{\partial T}\right)_p + \left(\frac{\partial V}{\partial p}\right)_T \left(\frac{\partial p}{\partial T}\right)_V = 0 \qquad \text{(ac)}$$

The differential: $(\frac{\partial p}{\partial T})_V$ is now included in the expression drawn up. It should be noted that the other two differentials are included in the definition of the coefficient of expansion of the substance α (r.1) and the compressibility κ (r.3), respectively. We can therefore express the desired quantity by these measurable substance parameters

$$\left(\frac{\partial p}{\partial T}\right)_V = -\left(\frac{\partial V}{\partial T}\right)_p \left(\frac{\partial p}{\partial V}\right)_T = -(\alpha V)\cdot\left(-\frac{1}{\kappa V}\right) = \frac{\alpha}{\kappa} \qquad \text{(ad)}$$

From the above it is seen that $(\frac{\partial p}{\partial T})_V$ can be expressed by a simple relationship between the coefficient of volume expansion α and its compressibility κ. In the following table, this quantity has been calculated based on eqn. (ad) for some selected substances. The calculations have been performed for $\theta = 25\,^{\circ}\mathrm{C}$ based on the expression $p = p^{\ominus} = 1$ atm. Hydrogen $H_2(g)$ is assumed to be an ideal gas so that $\kappa = -\frac{1}{V}(\frac{\partial V}{\partial p})_T = \frac{1}{p}$.

Substance	κ (Pa^{-1})	α (K^{-1})	$(\frac{\partial p}{\partial T})_V$ (atm/K)
$H_2O(\ell)$	$4.57 \cdot 10^{-10}$	$2.1 \cdot 10^{-4}$	4.3
Fe(s)	$5.91 \cdot 10^{-12}$	$3.5 \cdot 10^{-5}$	58.4
Al(s)	$1.32 \cdot 10^{-11}$	$6.9 \cdot 10^{-5}$	51.6
$H_2(g)$	$9.87 \cdot 10^{-6}$	$3.4 \cdot 10^{-3}$	0.0034

As seen from this table, there are considerable compressive stresses at volume-constant heating of solid and liquid substances. For steel, there is a pressure increase of $\simeq 60$ atm per K. This pressure increase is approximately 10^4 times that obtained by isochoric heating of an ideal gas.

Literature

Margenau, H. & Murphy, G.M.: *Mathematics of Physics and Chemistry*, D. van Nostrand Compagny Inc., New York 1955.

Abbott, M.M. & vanNess, H.C.: *Thermodynamics*, Schaum Outline Series, McGraw-Hill Inc., New York 1976.

Atkins, P.W.: *Physical Chemistry*, Oxford University Press, 1986.

9 Debye-Hückel's law

In the calculation of the activity of dissolved substances, ideal solutions were assumed throughout the chapters 5 and 6. A substance of the concentration c (mol/ℓ) in an **ideal solution** will have the activity $a = c/c^{\ominus}$, where $c^{\ominus} = 1$ mol/ℓ.

For dissolved, *neutral* substances, it is normally an acceptable approximation to assume ideality in concentrations less than 0.1–1.0 mol/ℓ. For ions in solutions, however, the assumption only applies to highly diluted solutions; even at concentrations of 0.001–0.01 mol/ℓ, ionic solutions can deviate noticeably from ideality. In the following, we shall briefly explain how it is possible in moderate concentrations of dissolved and dissociated salts to correct for this deviation from ideality.

Ions in solutions

By mutual influence, electrically *neutral* atoms and molecules can form **induced dipoles**; between these induced dipoles, weak, attractive **van der Waals forces** are acting. The effective distance for these forces is a few times the diameter of the atoms. The potential bond energy $\Phi(r)$, between induced dipoles can be mathematically expressed by the **Lennard-Jones potential** (1.12). Van der Waals' forces are *short-range forces*; the potential bond energy is inversely proportional to the effective distance of the 6th order, i.e. $\Phi(r) \propto 1/r^6$.

When a salt is dissolved in water, it dissociates into electrically charged ions that are mutually influenced by strong electrostatic **Coulomb forces**. Positive cations *attract* negatively charged anions and *repel* positively charged cations and vice versa. According to the **Coulomb law**, the potential energy between two electrically charged particles is inversely proportional to the effective distance in the first order, i.e. $\Phi(r) \propto 1/r$. Therefore, the effective distance of Coulomb forces between ions is considerably larger than that of van der Waals' forces between neutral atoms and molecules.

Consider an aqueous solution of sodium chloride NaCl that is fully dissociated into positively charged cations (Na^+) and negatively charged anions (Cl^-)

$$\mathrm{NaCl(s)} \quad \rightarrow \quad \mathrm{Na^+(aq) + Cl^-(aq)} \tag{a}$$

In the solution, the charged ions are mutually influenced by electrostatic **Coulomb forces**; therefore, the ions are *not* homogeneously distributed in the solution.

All ions tend to surround themselves by reversely charged ions ("counter ions") and to withdraw from ions of the same charge. As a result, the potential energy of the ions, and thus their readiness to enter into chemical reactions, is reduced. Macroscopically, this is seen by the ion activity a being lower than for a corresponding, ideal solution: $a < c/c^{\ominus}$. In calculations, this deviation from ideality is taken into account by introducing a special correction factor: the activity coefficient γ.

Activity coefficient

The activity a of a substance included in a non-ideal solution can be expressed as follows:

$$a = \gamma \cdot a_{ideal} = \gamma \cdot \left(\frac{c}{c^{\ominus}}\right) \tag{b}$$

The **activity coefficient** γ is a measure of the deviation from ideality. For ideal solutions, the activity coefficient is $\gamma = 1$ so that the calculated expressions introduced above still apply. For low and moderate ion concentrations, $a < a_{ideal}$ and the activity coefficient is $\gamma < 1$. With increased ion concentration, the activity a can again increase and then, in highly concentrated ion solutions, it can exceed a_{ideal} corresponding to $\gamma > 1$. However, the theory of concentrated ion solutions is beyond the scope of this overview.

Debye-Hückel's law

Based on a theoretical analysis of the electrostatic **Coulomb forces** acting between ions in an aqueous solution, *P. Debye* and *E. Hückel* formulated in the 1920s the following expression for calculation of the activity coefficient γ for a dissolved ion with the charge z

Debye-Hückel's law (c)

In an aqueous solution, the **activity coefficient** γ of an **ion** with the **charge** z is determined by

$$\log_{10}(\gamma) \simeq -\frac{A \cdot z^2 \cdot \sqrt{I}}{1 + \sqrt{I}}$$

where: $I = \frac{1}{2} \cdot \Sigma c_i z_i^2$ mol/ℓ (ion strength of the solution I)

$A \simeq 0.51$ (constant for aqueous solutions)

$z =$ ion charge

The Debye-Hückel theory expands the concentration area where aqueous solutions of ions can be described by calculations. However, it is emphasized once again that this theory does *not* apply to more concentrated ion solutions; therefore, it is not advisable to use this expression for ion strengths I that exceed 0.1–0.2 mol/ℓ. The procedure for use of the Debye-Hückel law is illustrated by a simple example in the following.

■ Ion strength and activity coefficient in saline solution

An aqueous solution contains 0.0050 mol of sodium sulphate Na_2SO_4 and 0.0100 mol of sodium chloride NaCl per litre. The dissolved salts are assumed to be completely dissociated.

(1) Determine the activities $a(Na^+)$, $a(SO_4^{--})$ and $a(Cl^-)$, if the solution is ideal!

(2) Determine the activities $a(Na^+)$, $a(SO_4^{--})$ and $a(Cl^-)$ if corrections for the deviation from ideality using the Debye-Hückel law are made!

Solution. *QUESTION (1)* Na_2SO_4 and NaCl dissociate according to the following equations

$$Na_2SO_4(s) \rightarrow 2Na^+(aq) + SO_4^{--}(aq) \qquad (d)$$

$$NaCl(s) \rightarrow Na^+(aq) + Cl^-(aq) \qquad (e)$$

Thus, we have the following molar concentrations of the individual ions in the solution

$$[Na^+] = 0.0200\,\text{mol}/\ell; \qquad [SO_4^{--}] = 0.0050\,\text{mol}/\ell; \qquad [Cl^-] = 0.0100\,\text{mol}/\ell$$

For ideal solutions, $a = c/c^{\ominus} = c$ (1 mol/ℓ); thus, we can directly write the actual activities for an ideal solution

$$a(Na^+) = \mathbf{0.0200}; \qquad a(SO_4^{--}) = \mathbf{0.0050}; \qquad a(Cl^-) = \mathbf{0.0100}$$

QUESTION (2) The ion strength I (mol/ℓ) is calculated from the Debye-Hückel law (c)

$$I = \tfrac{1}{2} \cdot \Sigma c_i z_i^2 = \tfrac{1}{2} \cdot (0.0200 \cdot 1^2 + 0.0050 \cdot 2^2 + 0.0100 \cdot 1^2) \qquad = \mathbf{0.0500}$$

Next, $\log_{10}(\gamma)$ and γ are determined for the three ions in the solution. For Na^+ we obtain using the Debye-Hückel law (e)

$$\log_{10}(\gamma(Na^+)) = -\frac{0.51 \cdot 1^2 \cdot \sqrt{0.0500}}{1 + \sqrt{0.0500}} = -0.093; \qquad \gamma(Na^+) = 10^{-0.093} = \mathbf{0.81}$$

In a corresponding way, $\log_{10}(\gamma)$, γ for SO_4^{--}, and Cl^- are determined using the Debye-Hückel law (e)

$$\log_{10}(\gamma(SO_4^{--})) = -\frac{0.51 \cdot 2^2 \cdot \sqrt{0.0500}}{1 + \sqrt{0.0500}} = -0.37; \qquad \gamma(SO_4^{--}) = 10^{-0.37} = \mathbf{0.42}$$

$$\log_{10}(\gamma(Cl^-)) = -\frac{0.51 \cdot 1^2 \cdot \sqrt{0.0500}}{1 + \sqrt{0.0500}} = -0.093; \qquad \gamma(Cl^-) = 10^{-0.093} = \mathbf{0.81}$$

When the coefficients of activity for Na^+, SO_4^{--} and Cl^- are known, the activities can now be determined using the relation (b)

$$a(Na^+) = \gamma(Na^+) \cdot \frac{c}{c^{\ominus}} = 0.81 \cdot 0.0200 \qquad = \mathbf{0.016}$$

$$a(SO_4^{--}) = \gamma(SO_4^{--}) \cdot \frac{c}{c^{\ominus}} = 0.42 \cdot 0.0050 \qquad = \mathbf{0.0021}$$

$$a(Cl^-) = \gamma(Cl^-) \cdot \frac{c}{c^{\ominus}} = 0.81 \cdot 0.0100 \qquad = \mathbf{0.0081}$$

The result shows that the activities of the dissolved ions at the actual ion strength $I \simeq 0.05$ deviate considerably from the ideal activity $c/c^{\ominus}$. Furthermore, it is noted that the monovalent ions Na^+ and Cl^- deviate less from ideality than does the divalent ion SO_4^{--}.

Literature

Atkins, P.W.: *Physical Chemistry*, Oxford University Press, 1986.

A glass blower from the mid 19th century; a double-acting pair of bellows supply air to a nozzle that through a large alcohol flame is directed towards the glass to be formed.

APPENDIX B

Tables

This appendix contains a selection of tables of **physical constants** and **substance data** that are frequently needed in calculations in the science of construction materials.

The table collection is meant as a support for the theoretical treatment of the book, covering the special types of problems occurring within construction materials. It is well known that standard tables only sparingly deal with subjects such as **cement chemistry**, **cement hydration**, and **corrosion phenomena**. Therefore, numerical data have been gathered and presented to facilitate calculations with respect to these special subjects. For calculations of moisture, a rather extensive numerical material has been selected and prepared, including properties of water and water vapour in the temperature range from −15 °C to +195 °C.

It goes without saying that **thermochemical data** form an important part of the collection of tables in this book. Tables of thermochemical data have been divided into three main groups: Thermochemical data (1) dealing with **inorganic compounds** thermochemical data (2) especially dealing with **cement-chemical compounds**, and thermochemical data (3) containing a smaller selection of **organic compounds**. In the selection of thermochemical data, the aim has been to further practical engineering applications to the science of construction materials.

L. Boltzmann (1844-1906)

German professor of physics, who, among other things, is known for his contributions to statistical thermodynamics.

Contents

Physical constants

Planck constant	$h = 6.626176 \cdot 10^{-34}$	J s
Gravitational constant	$G = 6.6720 \cdot 10^{-11}$	$\mathrm{N\,m^2/kg^2}$
Velocity of light in vacuum	$c = 2.99792458 \cdot 10^{8}$	m/s
Boltzmann constant	$k = 1.380662 \cdot 10^{-23}$	J/K
Gas constant	$R = 8.31441$	J/(mol K)
Elementary charge	$e = 1.6021892 \cdot 10^{-19}$	C
Faraday constant	$\mathcal{F} = 9.648456 \cdot 10^{4}$	C/mol
Avogadro constant	$\mathcal{N} = 6.022045 \cdot 10^{23}$	$\mathrm{mol^{-1}}$
Atomic mass constant	$m_u = 1.6605655 \cdot 10^{-27}$	kg
Electron mass	$m_e = 0.9109534 \cdot 10^{-30}$	kg
Proton mass	$m_p = 1.6726485 \cdot 10^{-27}$	kg
Neutron mass	$m_n = 1.6749543 \cdot 10^{-27}$	kg
Stefan-Boltzmann constant	$\sigma = 5.67032 \cdot 10^{-8}$	$\mathrm{W/(m^2\,K^4)}$
Wien displacement constant	$w = 2.8978 \cdot 10^{-3}$	m K
Permittivity in vacuum	$\varepsilon_0 = 8.8541878 \cdot 10^{-12}$	$\mathcal{F}$/m
Coulomb constant	$k_c = 8.987552 \cdot 10^{9}$	$\mathrm{N\,m^2/C^2}$
Permeability in vacuum	$\mu_0 = 4\pi \cdot 10^{-7}$	H/m
Rydberg constant	$R_\infty = 1.097373177 \cdot 10^{7}$	$\mathrm{m^{-1}}$
Standard pressure	$p^\ominus = 101325$	Pa
Standard gravity	$\mathrm{g} = 9.80665$	N/kg

Ref.: DS "Standards for SI units Handbook 3",1. ed. 1985 (in Danish).

Elements

The table contains the following key data for the elements in the periodic system: Z: **atomic number**; M: **molar mass**; ϱ: **density**; θ_m: **melting point**; θ_b **boiling point**, ΔH_{mh}: **melting heat**. All data refer to the standard state 101325 Pa at 20 °C. Radioactive isotopes are marked with (*); substances that sublime are marked with (^{s}).

Element		Z	M $\left(\frac{\mathrm{g}}{\mathrm{mol}}\right)$	ϱ $\left(\frac{\mathrm{g}}{\mathrm{cm^3}}\right)$	θ_m (°C)	θ_b (°C)	ΔH_{mh} $\left(\frac{\mathrm{kJ}}{\mathrm{mol}}\right)$
Actinium	Ac(s)*	89	227	10.07	1050	3200	10.06
Aluminium	Al(s)	13	26.982	2.699	660	2467	10.67
Americium	Am(s)*	95	243	13.67	9940	2607	–
Antimony	Sb(s)	51	121.75	6.691	631	1750	19.83
Argon	Ar(g)	18	39.948	0.0016	−189	−186	1.18
Arsenic	As(s)	33	74.922	5.73	817	^{s}613	–
Astatine	At(s)*	85	210	–	302	337	–
Barium	Ba(s)	56	137.34	3.51	725	1640	7.66
Berkelium	Bk(s)*	97	247	14	–	–	–
Beryllium	Be(s)	4	9.012	1.848	1278	2970	–

Element		Z	M $\left(\frac{g}{mol}\right)$	ϱ $\left(\frac{g}{cm^3}\right)$	θ_m (°C)	θ_b (°C)	ΔH_{mh} $\left(\frac{kJ}{mol}\right)$
Bismuth	Bi(s)	83	208.981	9.747	271	1560	11
Boron	B(s)	5	10.81	2.34	2300	^{s}2550	22.18
Bromine	$Br_2(\ell)$	35	159.808	3.12	−7	59	10.54
Cadmium	Cd(s)	48	112.40	8.65	321	765	6.07
Caesium	Cs(s)	55	132.905	1.873	28	669	2.13
Calcium	Ca(s)	20	40.08	1.55	839	1484	–
Californium	Cf(s)*	98	251	–	–	–	–
Carbon	C(s)	6	12.011	–	3550	–	–
Cerium	Ce(s)	58	140.12	6.657	799	3426	9.20
Chlorine	$Cl_2(g)$	17	70.906	0.0032	−101	−35	6.41
Chromium	Cr(s)	24	51.996	7.19	1857	2672	13.82
Cobalt	Co(s)	27	58.933	8.0	1495	2870	15.23
Copper	Cu(s)	29	63.546	8.96	1083	2567	13.05
Curium	Cm(s)*	96	247	13.51	1340	–	–
Dysprosium	Dy(s)	66	162.50	8.550	1412	2562	–
Einsteinium	Es()*	99	254	–	–	–	–
Erbium	Er(s)	68	167.26	9.066	1529	2863	17.15
Europium	Eu(s)	63	151.96	5.243	822	1597	10.46
Fermium	Fm(s)*	100	257	–	–	–	–
Flourine	$F_2(g)$	9	37.997	0.0017	−220	−188	5.10
Francium	Fr(s)*	87	223	–	27	677	–
Gadolinium	Gd(s)	64	157.25	7.900	1313	3266	15.48
Gallium	Ga(s)	31	69.737	5.904	30	2430	5.59
Germanium	Ge(s)	32	72.59	5.323	937	2830	31.8
Gold	Au(s)	79	196.967	18.88	1064	3080	12.36
Hafnium	Hf(s)	72	178.49	13.31	2227	4602	21.76
Helium	He(g)	2	4.003	0.0002	−272	−270	0.02
Holmium	Ho(s)	67	164.930	8.795	1474	2695	17.15
Hydrogen	$H_2(g)$	1	2.016	0.0001	−259	−253	0.11
Indium	In(s)	49	114.82	7.31	157	2080	3.26
Iodine	$I_2(s)$	53	253.809	4.93	114	184	0.02
Iridium	Ir(s)	77	192.22	22.42	2410	4130	26.36
Iron	Fe(s)	26	55.847	7.874	1535	2750	15.36
Krypton	Kr(g)	36	83.80	0.0037	−157	−152	1.64
Lanthanum	La(s)	57	138.906	6.145	921	3457	11.30
Lawrencium	Lr(s)*	103	257	–	–	–	–
Lead	Pb(s)	82	207.2	11.35	328	1740	4.77
Lithium	Li(s)	3	6.941	0.534	181	1342	3.02
Lutetium	Lu(s)	71	174.967	9.840	1663	3395	–
Magnesium	Mg(s)	12	24.305	1.738	649	1090	8.95
Manganese	Mn(s)	25	54.938	7.21	1244	1962	14.64
Mendelevium	Md(s)*	101	257	–	–	–	–
Mercury	$Hg(\ell)$	80	200.59	13.546	−39	357	2.30
Molybdenum	Mo(s)	42	95.94	10.22	2617	4612	27.61
Neodymium	Nd(s)	60	144.24	6.80	1021	3068	10.88

Element		Z	M $\left(\frac{g}{mol}\right)$	ϱ $\left(\frac{g}{cm^3}\right)$	θ_m (°C)	θ_b (°C)	ΔH_{mh} $\left(\frac{kJ}{mol}\right)$
Neon	Ne(g)	10	20.179	0.0009	−249	−246	0.34
Neptunium	Np(s)	93	237.048	20.25	640	3902	–
Nickel	Ni(s)	28	58.71	8.902	1453	2732	17.61
Niobium	Nb(s)	41	92.906	8.57	2468	4742	26.78
Nitrogen	N_2(g)	7	28.013	0.0013	−210	−196	0.72
Nobelium	No(s)*	102	–	–	–	–	–
Osmium	Os(s)	76	190.2	22.57	3045	5027	29.29
Oxygen	O_2(g)	8	32.000	0.0014	−218	−183	0.44
Palladium	Pd(s)	46	106.4	12.02	1554	2970	16.74
Phosphorus	P_4(s)	15	123.895	1.82	44	280	–
Platinum	Pt(s)	78	195.09	21.45	1772	3827	19.66
Plutonium	Pu(s)*	94	239.13	19.84	641	3232	–
Polonium	Po(s)*	84	209	9.32	254	962	–
Potassium	K(s)	19	39.098	0.862	63	760	2.32
Praseodymium	Pr(s)	59	140.908	6.773	931	3512	10.04
Promethium	Pm(s)*	61	–	7.22	1168	2460	–
Protactinium	Pa(s)*	91	231.036	–	–	–	–
Radium	Ra(s)*	88	226.025	∼ 5	700	1140	8.37
Radon	Rn(g)*	86	272	0.0097	−71	−62	2.9
Rhenium	Re(s)	75	186.21	21.02	3180	5627	33.05
Rhodium	Rh(s)	45	102.906	12.41	1966	3727	21.76
Rubidium	Rb(s)	37	85.468	1.532	39	686	2.34
Ruthenium	Ru(s)	44	101.41	12.41	2310	3900	25.52
Samarium	Sm(s)	62	150.4	7.52	1077	1791	11.09
Scandium	Sc(s)	21	44.956	2.989	1541	2831	16.11
Selenium	Se(s)	34	78.96	4.79	217	685	5.44
Silicon	Si(s)	14	28.086	2.33	1410	2355	46.44
Silver	Ag(s)	47	107.868	10.50	962	2212	11.30
Sodium	Na(s)	11	22.990	0.971	98	883	2.60
Strontium	Sr(s)	38	87.62	2.54	769	1384	9.20
Sulphur	S_8(s)	16	256.48	2.07	113	445	11.28
Tantalum	Ta(s)	73	180.948	16.654	2996	5425	31.38
Technetium	Tc(s)*	43	98.906	11.50	2172	4877	–
Tellurium	Te(s)	52	127.60	6.24	450	990	17.49
Terbium	Tb(s)	65	158.925	8.229	1356	3123	–
Thallium	Tl(s)	81	204.37	11.85	304	1457	4.27
Thorium	Th(s)*	90	232.038	11.72	1750	4790	15.65
Thulium	Tm(s)	69	168.934	9.321	1545	1947	–
Tin	Sn(s)	50	118.69	5.75	232	2270	7.20
Titanium	Ti(s)	22	47.90	4.54	1660	3287	15.48
Tungsten	W(s)	74	183.85	19.3	3410	5660	35.22
Uranium	U(s)	92	238.029	18.95	1132	3818	15.48
Vanadium	V(s)	23	50.942	6.11	1890	3380	17.57
Xenon	Xe(g)	54	131.30	0.0059	−112	−107	2.30
Ytterbium	Yb(s)	70	173.04	6.965	819	1194	–

Element		Z	M $\left(\frac{g}{mol}\right)$	ϱ $\left(\frac{g}{cm^3}\right)$	θ_m (°C)	θ_b (°C)	ΔH_{mh} $\left(\frac{kJ}{mol}\right)$
Yttrium	Y(s)	39	88.906	4.469	1522	3338	17.15
Zinc	Zn(s)	30	65.38	7.133	420	907	7.38
Zirconium	Zr(s)	40	91.22	6.506	1852	4377	16.74

Ref.: "Handbook of Chemistry and Physics", 64th ed., CRC Press Inc., Boca Raton, 1984.
Andersen, Jespersgaard & Østergaard: "Databog, fysik og kemi", F & K forlaget,1984 (in Danish).

Van der Waals constants

For a number of selected substances, the table specifies the constants a and b in the **van der Waals equation** (1.11) for **real gases**

$$\left(p + a\left(\frac{n}{V}\right)^2\right) \cdot (V - nb) = nRT \tag{1.11}$$

Further, the table contains **critical temperature** θ_c (°C) and **critical pressure** p_c (atm) calculated from the relation (1.18): $T_c = 8a/(27\,b\,R)$ (K); $p_c = a/(27\,b^2)$ (Pa).

Substance	Formula	a $\left(\frac{m^6 Pa}{mol^2}\right)$	b $\left(\frac{m^3}{mol}\right)$	θ_c (°C)	p_c (atm)
Acetone	$(CH_3)_2CO$	1.4094	$0.0994 \cdot 10^{-3}$	232	52
Acetylene	C_2H_2	0.4448	$0.0514 \cdot 10^{-3}$	35	61
Ammonium	NH_3	0.4225	$0.0371 \cdot 10^{-3}$	133	112
Argon	Ar	0.1363	$0.0322 \cdot 10^{-3}$	−122	48
Benzene	C_6H_6	1.8239	$0.1154 \cdot 10^{-3}$	290	50
n-Butane	C_4H_{10}	1.4662	$0.1226 \cdot 10^{-3}$	153	36
Carbon dioxide	CO_2	0.3640	$0.0427 \cdot 10^{-3}$	31	73
Carbon disulphide	CS_2	1.1774	$0.0769 \cdot 10^{-3}$	272	73
Carbon monoxide	CO	0.1505	$0.0399 \cdot 10^{-3}$	−139	35
Carbon tetrachloride	CCl_4	2.0660	$0.1383 \cdot 10^{-3}$	259	39
Chlorine	Cl_2	0.6579	$0.0562 \cdot 10^{-3}$	144	76
Chloroform	$CHCl_3$	1.5371	$0.1022 \cdot 10^{-3}$	263	54
Ethane	C_2H_6	0.5562	$0.0638 \cdot 10^{-3}$	38	50
Ethanol	C_2H_5OH	1.2179	$0.0841 \cdot 10^{-3}$	243	63
Ethylene	C_2H_4	0.4530	$0.0571 \cdot 10^{-3}$	10	51
Helium	He	0.0035	$0.0237 \cdot 10^{-3}$	−268	2
Hydrogen	H_2	0.0248	$0.0266 \cdot 10^{-3}$	−240	13
Hydrogen chloride	HCl	0.3716	$0.0408 \cdot 10^{-3}$	51	82
Krypton	Kr	0.2349	$0.0398 \cdot 10^{-3}$	−63	54
Mercury	Hg	0.8200	$0.0170 \cdot 10^{-3}$	1446	1037
Methane	CH_4	0.2283	$0.0428 \cdot 10^{-3}$	−83	46
Methanol	CH_3OH	0.9656	$0.0670 \cdot 10^{-3}$	240	79
Neon	Ne	0.0213	$0.0171 \cdot 10^{-3}$	−229	27
Nitrogen	N_2	0.1408	$0.0391 \cdot 10^{-3}$	−145	34
Nitrogen dioxide	NO_2	0.5354	$0.0442 \cdot 10^{-3}$	159	100

Substance	Formula	a $\left(\frac{m^6 Pa}{mol^2}\right)$	b $\left(\frac{m^3}{mol}\right)$	θ_c (°C)	p_c (atm)
Nitrogen oxide	NO	0.1358	$0.0279 \cdot 10^{-3}$	−100	64
Oxygen	O_2	0.1378	$0.0318 \cdot 10^{-3}$	−119	50
n-Pentane	C_5H_{12}	1.9262	$0.1460 \cdot 10^{-3}$	197	33
Propane	C_3H_8	0.8779	$0.0845 \cdot 10^{-3}$	97	45
Propylene	C_2H_6	0.8490	$0.0827 \cdot 10^{-3}$	93	45
Sulphur dioxide	SO_2	0.6803	$0.0564 \cdot 10^{-3}$	157	78
Toluene	$C_6H_6CH_3$	2.4379	$0.1463 \cdot 10^{-3}$	321	42
Water vapour	H_2O	0.5536	$0.0305 \cdot 10^{-3}$	374	218
Xenon	Xe	0.4250	$0.0511 \cdot 10^{-3}$	59	23
Xylene	$C_6H_4(CH_3)_2$	3.0762	$0.1772 \cdot 10^{-3}$	346	36

Ref.: "Handbook of Chemistry and Physics", 64th ed., CRC Press Inc., Boca Raton, 1984.

Thermochemical data (1)

inorganic compounds

The table specifies **molar mass** M and **thermodynamic standard values** $H^\ominus_{298}$, $G^\ominus_{298}$, $S^\ominus_{298}$, and c_p for selected elements, compounds, and ions. The **states** of the substances are given by: (s): solid, (ℓ): liquid, (g): gaseous, and (aq) for ions and gases in aqueous solution. **Standard state:** $p^\ominus = 101325\,\text{Pa}$, and for dissolved substances: $c^\ominus = 1\,\text{mol}/\ell$. **Reference temperature:** $T = 298.15\,\text{K}$ (25 °C).

	Substance		M $\left(\frac{g}{mol}\right)$	$H^\ominus_{298}$ $\left(\frac{kJ}{mol}\right)$	$G^\ominus_{298}$ $\left(\frac{kJ}{mol}\right)$	$S^\ominus_{298}$ $\left(\frac{J}{mol\,K}\right)$	c_p $\left(\frac{J}{mol\,K}\right)$
ALUMINIUM	**Al**	(s)	26.98	0.0	0.0	28.3	24.3
	Al	(g)	26.98	329.7	289.1	164.6	21.4
	Al^{+++}	(aq)	–	−531.4	−485.3	−321.7	–
	$AlCl_3$	(s)	133.34	−705.6	−630.0	109.3	91.1
	$AlCl_3 \cdot 6H_2O$	(s)	241.43	−2691.6	−2260.9	318.0	296.2
	AlF_3	(s)	83.98	−1510.4	−1431.4	66.5	75.1
	AlF_3	(g)	83.98	−1209.3	−1192.7	276.7	62.5
	$Al(NO_3)_3 \cdot 6H_2O$	(s)	321.09	−2850.5	−2203.9	467.8	433.0
	AlO_2^-	(aq)	58.98	−918.8	−823.0	−20.9	–
	α-Al_2O_3	(s)	101.96	−1675.7	−1582.3	50.9	79.0
	γ-Al_2O_3	(s)	101.96	−1656.9	−1563.9	52.3	82.7
	$Al(OH)_3$ (amorphous)	(s)	78.00	−1276.1	−1138.7	71.1	93.1
	$Al_2O_3 \cdot H_2O$ (boehmite)	(s)	119.98	−1980.7	−1831.4	96.9	131.3
	$Al_2O_3 \cdot 3H_2O$ (gibbsite)	(s)	156.01	−2586.5	−2310.1	136.9	183.5
	$Al_2O_3 \cdot SiO_2$	(s)	162.05	−2590.3	−2442.9	93.8	122.8
	$3Al_2O_3 \cdot 2SiO_2$	(s)	426.05	−6820.5	−6443.0	274.9	325.4
	$Al_2O_3 \cdot 2SiO_2 \cdot 2H_2O$	(s)	258.16	−4119.6	−3799.4	205.0	246.1
	$Al_2(SO_4)_3$	(s)	342.15	−3440.8	−3099.6	239.3	259.4
ANTIMONY	**Sb**	(s)	121.75	0.0	0.0	45.5	25.2

Substance		M $\left(\frac{g}{mol}\right)$	$H^\ominus_{298}$ $\left(\frac{kJ}{mol}\right)$	$G^\ominus_{298}$ $\left(\frac{kJ}{mol}\right)$	$S^\ominus_{298}$ $\left(\frac{J}{mol\,K}\right)$	c_p $\left(\frac{J}{mol\,K}\right)$	
Sb	(g)	121.75	246.6	224.4	180.3	20.8	ARGON
$SbCl_3$	(s)	228.11	−381.2	−322.4	183.3	110.5	
Sb_2O_3	(s)	291.50	−720.3	−634.3	110.5	111.8	
Sb_2S_3	(s)	339.70	−141.8	−140.3	182.2	119.8	
Ar	(g)	39.95	0.0	0.0	154.8	20.8	
As	(s)	74.92	0.0	0.0	35.7	24.7	ARSENIC
$AsCl_3$	(s)	181.28	−305.0	−259.1	216.3	133.5	
As_2O_3	(s)	197.84	−654.8	−576.6	117.0	97.0	
As_2O_5	(s)	229.84	−924.9	−782.1	105.4	116.5	
As_2O_5	(s)	246.04	−167.4	−166.2	163.6	116.5	
As_4S_4	(s)	427.95	−138.1	−133.0	254.0	188.2	
Ba	(s)	137.33	0.0	0.0	62.4	28.1	BARIUM
Ba	(g)	137.33	182.0	149.9	170.2	20.8	
Ba^{++}	(aq)	–	−537.6	−560.7	9.6	–	
$BaCO_3$	(s)	197.34	−1216.3	−1137.7	112.1	85.4	
$BaCl_2$	(s)	208.23	−858.6	−810.3	123.7	75.1	
$BaCrO_4$	(s)	253.32	−1446.0	−1345.3	158.6	120.2	
BaF_2	(s)	175.32	−1208.8	−1158.4	96.4	72.2	
$Ba(NO_3)_2$	(s)	261.34	−992.1	−796.6	213.8	151.4	
BaO	(s)	153.33	−553.5	−525.3	70.4	42.3	
BaO_2	(s)	169.33	−634.3	−582.3	93.1	67.3	
$Ba(OH)_2$	(s)	171.34	−946.3	−859.5	107.1	101.6	
$BaO \cdot 2SiO_2$	(s)	273.50	−2548.1	−2411.0	153.1	134.2	
$2BaO \cdot SiO_2$	(s)	366.74	−2287.8	−2175.1	176.1	134.9	
$2BaO \cdot 3SiO_2$	(s)	486.91	−4184.8	−3963.1	258.2	224.6	
BaS	(s)	169.39	−460.2	−455.4	78.2	49.4	
$BaSO_4$	(s)	233.39	−1473.2	−1362.1	132.2	102.2	
Be	(s)	9.01	0.0	0.0	9.4	16.4	BERYLLIUM
Be^{++}	(aq)	–	−382.8	−379.7	−129.7	–	
$BeCl_2$	(s)	79.92	−496.2	−449.5	75.8	62.4	
BeF_2	(s)	47.01	−1026.8	−979.4	53.4	51.8	
BeO	(s)	25.01	−608.4	−579.1	13.8	25.6	
$Be(OH)_2$	(s)	40.03	−902.9	−815.9	53.6	65.7	
$2BeO \cdot SiO_2$	(s)	110.1	−2145.6	−2031.2	64.4	93.5	
$BeSO_4$	(s)	105.08	−1200.8	−1089.4	78.0	86.0	
$BeSO_4 \cdot 2H_2O$	(s)	141.02	−1823.0	−1597.8	163.2	152.8	
Bi	(s)	208.98	0.0	0.0	56.7	25.6	BISMUTH
Bi	(g)	208.98	209.6	170.8	187.0	20.8	
$BiCl_3$	(s)	315.34	−379.1	−315.1	177.0	100.4	
Bi_2O_3	(s)	465.96	−573.9	−493.5	151.5	113.5	
Bi_2S_3	(s)	514.16	−143.1	−140.3	200.4	122.1	
$Bi_2(SO_4)_3$	(s)	706.15	−2544.3	−2208.0	312.5	278.8	
α-**B**	(s)	10.81	0.0	0.0	5.8	11.3	BORON
B (amorphous)	(s)	10.81	48.9	42.7	26.6	11.3	
B	(g)	10.81	560.0	516.0	153.4	20.8	

	Substance		M $\left(\frac{\mathrm{g}}{\mathrm{mol}}\right)$	$H^{\ominus}_{298}$ $\left(\frac{\mathrm{kJ}}{\mathrm{mol}}\right)$	$G^{\ominus}_{298}$ $\left(\frac{\mathrm{kJ}}{\mathrm{mol}}\right)$	$S^{\ominus}_{298}$ $\left(\frac{\mathrm{J}}{\mathrm{mol\,K}}\right)$	c_p $\left(\frac{\mathrm{J}}{\mathrm{mol\,K}}\right)$
(Boron)	B_4C	(s)	55.26	−71.1	−70.6	27.1	53.1
	BCl_3	(s)	117.17	−403.0	−388.0	290.2	62.4
	BF_3	(g)	67.81	−1136.6	−1119.9	254.1	50.0
	BN	(s)	24.82	−254.4	−228.5	14.8	19.7
	B_2O_3	(s)	69.62	−1271.9	−1192.8	54.0	62.6
	B_2O_3 (amorphous)	(s)	69.62	−1253.4	−1181.5	78.5	62.8
	$B(OH)_3^-$	(aq)	78.84	−1344.0	−1153.3	102.5	–
	H_3BO_3	(s)	61.83	−1094.0	−968.5	88.7	81.4
Cadmium	**Cd**	(s)	112.41	0.0	0.0	51.8	25.9
	Cd	(g)	112.41	111.8	77.2	167.7	20.8
	Cd^{++}	(aq)	–	−75.9	−77.6	−73.2	–
	$CdCO_3$	(s)	172.42	−670.5	−751.9	92.5	82.4
	$CdCl_2$	(s)	183.32	−391.5	−343.9	115.3	74.6
	CdO	(s)	128.41	−259.0	−229.3	54.8	43.6
	$Cd(OH)_2$	(s)	146.43	−560.7	−473.8	96.0	–
	CdS	(s)	144.48	−149.4	−143.7	64.9	48.7
	$CdSO_4$	(s)	208.48	−933.3	−822.6	123.0	99.6
Caesium	**Cs**	(s)	132.91	0.0	0.0	85.1	32.2
	Cs	(g)	132.91	76.5	49.5	175.6	20.8
	Cs_2	(g)	265.81	106.3	72.3	284.1	38.1
	Cs_2CO_3	(s)	325.82	−1147.3	−1064.0	204.5	123.8
	CsCl	(s)	168.36	−442.8	−414.4	101.2	52.4
	CsO	(g)	148.91	62.8	42.5	255.5	36.5
	CsO_2	(s)	164.90	−286.2	−242.0	142.3	79.1
	Cs_2O_3	(s)	313.81	−520.1	−446.2	230.1	125.5
	CsOH	(s)	149.91	−416.7	−370.7	98.7	67.9
	Cs_2SO_4	(s)	361.87	−1443.0	−1323.5	211.9	135.1
Calcium	**Ca**	(s)	40.08	0.0	0.0	41.4	25.3
	Ca	(g)	40.08	178.2	144.4	154.9	20.8
	Ca^{++}	(aq)	–	−542.8	−553.6	−53.1	–
	$CaBr_2$	(s)	199.89	−683.2	−664.2	129.7	75.0
	CaB_4O_7	(s)	195.32	−3360.3	−3167.0	134.7	157.9
	CaC_2	(s)	64.10	−59.8	−64.9	70.0	62.7
	$CaCO_3$ (calcite)	(s)	100.09	−1206.9	−1128.8	92.9	83.5
	$CaCO_3$ (aragonite)	(s)	100.09	−1207.1	−1127.8	88.7	82.3
	$CaMg(CO_3)_2$ (dolomite)	(s)	184.40	−2326.3	−2163.6	155.2	157.5
	$CaCl_2$	(s)	110.98	−795.8	−748.1	104.6	72.9
	CaF_2	(s)	78.08	−1225.9	−1173.5	68.6	68.6
	$CaHPO_4$	(s)	136.06	−1814.4	−1681.2	111.4	110.0
	$CaHPO_4 \cdot 2H_2O$	(s)	172.09	−2403.7	−2154.7	189.5	196.9
	CaI_2	(s)	293.89	−536.8	−533.1	145.3	77.2
	$Ca(NO_3)_2$	(s)	164.09	−938.4	−743.1	193.3	149.4
	$Ca(NO_3)_2 \cdot 2H_2O$	(s)	200.12	−1540.8	−1229.0	269.4	231.5
	$Ca(NO_3)_2 \cdot 3H_2O$	(s)	218.13	−1838.0	−1471.6	319.2	267.1
	$Ca(NO_3)_2 \cdot 4H_2O$	(s)	236.15	−2132.3	−1713.1	375.3	300.5

Substance		M $\left(\frac{g}{mol}\right)$	$H^{\ominus}_{298}$ $\left(\frac{kJ}{mol}\right)$	$G^{\ominus}_{298}$ $\left(\frac{kJ}{mol}\right)$	$S^{\ominus}_{298}$ $\left(\frac{J}{mol\,K}\right)$	c_p $\left(\frac{J}{mol\,K}\right)$	
CaO	(s)	56.08	−635.1	−603.5	38.1	42.1	(CALCIUM)
$Ca(OH)_2$	(s)	74.09	−986.1	−898.5	83.4	87.5	
$Ca_2P_2O_7$	(s)	254.10	−3333.4	−3126.5	189.2	187.8	
$Ca_3(PO_4)_2$	(s)	318.18	−4120.8	−3885.0	236.0	227.8	
CaS	(s)	72.14	−473.2	−468.2	56.6	47.4	
$CaSO_3$	(s)	120.14	−1159.4	−1076.0	101.4	91.7	
$CaSO_3 \cdot \frac{1}{2}H_2O$	(s)	129.15	−1311.7	−1199.4	121.3	112.4	
$CaSO_4$	(s)	136.14	−1434.1	−1321.7	106.7	99.6	
$CaSO_4 \cdot \frac{1}{2}H_2O$	(s)	145.15	−1576.7	−1436.6	130.5	118.0	
$CaSO_4 \cdot 2H_2O$ (gypsum)	(s)	172.17	−2022.6	−1797.2	194.1	186.0	
C (graphite)	(s)	12.01	0.0	0.0	5.7	8.5	CARBON
C (diamond)	(s)	12.01	1.9	2.9	2.4	6.1	
CCl_4	(ℓ)	153.82	−132.8	−62.5	216.2	133.9	
CCl_4	(g)	153.82	−100.4	−58.2	310.2	83.4	
CO	(g)	28.01	−110.5	−137.2	197.7	29.1	
CO_2	(g)	44.01	−393.5	−394.4	213.8	37.1	
CO_2	(aq)	44.01	−413.8	−386.0	117.6	–	
CO_3^{--}	(aq)	60.01	−677.1	527.9	−56.9	–	
HCO_3^-	(aq)	61.02	−692.0	−586.8	91.2	–	
H_2CO_3	(aq)	62.03	−699.6	−623.2	187.4	–	
$COCl_2$	(g)	98.92	−220.1	−205.9	283.9	57.7	
CNS^-	(aq)	58.08	76.4	92.7	144.3	−40.2	
$\mathbf{Cl_2}$	(g)	70.91	0.0	0.0	223.1	33.9	CHLORINE
Cl_2	(aq)	70.91	−23.4	6.9	121.3	–	
Cl_2O	(g)	86.91	80.3	79.9	266.1	45.4	
ClO_2	(g)	67.45	102.5	120.5	256.7	42.0	
HCl	(g)	36.46	−92.3	−95.3	186.8	29.1	
HCl	(aq)	36.46	−167.2	−131.3	56.5	−136.4	
Cl^-	(aq)	–	−167.2	−131.3	56.5	−136.4	
ClO^-	(aq)	51.45	−107.1	−36.8	41.8	–	
ClO_3^-	(aq)	83.45	−99.2	−3.3	−162.3	–	
ClO_4^-	(aq)	99.45	−12.9	−8.6	182.0	–	
Cr	(s)	52.00	0.0	0.0	23.6	23.3	CHROMIUM
$CrCl_2$	(s)	122.90	−395.4	−356.2	115.3	71.2	
$CrCl_3$	(s)	158.35	−556.5	−486.3	123.0	91.8	
CrN	(s)	66.00	−117.2	−92.7	37.4	52.3	
Cr_2N	(s)	118.00	−125.5	−102.2	64.9	66.0	
Cr_2O_3	(s)	151.99	−1139.7	1058.1	81.2	120.4	
CrO_4^{--}	(aq)	115.99	−881.1	−727.8	50.2	–	
$Cr_2O_7^{--}$	(aq)	215.99	−1490.3	−1301.2	261.9	–	
$HCrO_4^-$	(aq)	117.00	−878.2	−764.8	184.1	–	
$Cr_2(SO_4)_3$	(s)	392.18	−2949.7	−2617.1	258.8	281.4	
Co	(s)	58.93	0.0	0.0	30.0	24.8	COBALT
Co	(g)	58.93	428.4	383.9	179.5	23.0	
Co^{++}	(aq)	58.93	−58.2	−54.4	−113.0	–	

Element	Substance		M $\left(\frac{g}{mol}\right)$	$H^{\ominus}_{298}$ $\left(\frac{kJ}{mol}\right)$	$G^{\ominus}_{298}$ $\left(\frac{kJ}{mol}\right)$	$S^{\ominus}_{298}$ $\left(\frac{J}{mol\,K}\right)$	c_p $\left(\frac{J}{mol\,K}\right)$
(Cobalt)	Co^{+++}	(aq)	58.93	92.0	133.9	−305.4	–
	$CoCO_3$	(s)	118.94	−713.0	−636.8	87.9	80.1
	$CoCl_2$	(s)	129.84	−312.5	−269.7	109.3	78.5
	$CoCl_2 \cdot 2H_2O$	(s)	165.87	−923.0	−764.8	188.3	–
	$CoCl_2 \cdot 6H_2O$	(s)	237.93	−2115.4	−1725.5	343.1	–
	Co_3N	(s)	190.81	8.4	34.4	98.7	91.9
	CoO	(s)	74.93	−237.9	−214.2	53.0	55.1
	$CoSO_4$	(s)	155.00	−888.3	−782.4	117.4	103.2
	$CoSO_4 \cdot 6H_2O$	(s)	263.09	−2683.6	−2235.7	367.6	353.4
	Co_3O_4	(s)	240.80	−910.0	−794.9	114.3	123.0
	$Co(OH)_2$	(s)	92.95	−539.7	−454.2	79.0	97.1
	CoS_2	(s)	123.07	−153.1	−145.6	69.0	68.2
Copper	**Cu**	(s)	63.55	0.0	0.0	33.2	24.4
	Cu	(g)	63.55	337.6	297.9	166.4	20.8
	Cu^{+}	(aq)	63.55	71.7	50.0	40.6	–
	Cu^{++}	(aq)	63.55	64.8	65.5	−99.6	–
	$CuCl$	(s)	99.00	−155.6	−138.7	87.7	52.5
	$CuCl_2$	(s)	134.45	−218.0	−173.8	108.0	71.9
	$CuCl_2 \cdot 2H_2O$	(s)	170.48	−821.3	−656.1	167.4	–
	$Cu(NH_3)_4^{++}$	(aq)	131.66	−348.5	−111.3	273.6	–
	CuO	(s)	79.55	−156.1	−128.3	42.6	42.2
	Cu_2O	(s)	143.09	−170.1	−147.9	92.3	62.5
	$Cu(OH)_2$	(s)	97.56	−443.1	−359.0	87.0	87.9
	CuS	(s)	95.61	−53.1	−53.5	66.5	47.8
	Cu_2S	(s)	159.16	−81.2	−86.5	116.2	76.9
	$CuSO_4$	(s)	159.61	−771.4	−662.1	109.0	98.8
	$CuSO_4 \cdot H_2O$	(s)	177.63	−1085.8	−918.0	146.0	134.0
	$CuSO_4 \cdot 3H_2O$	(s)	213.66	−1684.3	−1399.9	221.3	205.0
	$CuCO_3$	(s)	123.56	−595.0	−518.0	87.9	–
	$CuCO_3 \cdot Cu(OH)_2$	(s)	221.10	−1051.4	−893.7	186.2	–
Fluorine	$\mathbf{F_2}$	(g)	38.00	0.0	0.0	202.8	31.3
	F	(g)	19.00	78.9	61.8	158.8	22.7
	F^{-}	(aq)	19.00	−332.6	−278.8	−13.8	−106.7
	HF	(g)	20.01	−271.1	−273.2	173.7	29.1
	HF_2^{-}	(aq)	39.01	−649.9	−578.1	92.5	–
Gold	**Au**	(s)	196.97	0.0	0.0	47.5	25.3
	Au	(g)	196.97	368.2	328.5	180.5	20.8
	AuCl	(s)	232.42	−36.4	−14.6	85.9	48.7
	$AuCl_3$	(s)	303.33	−117.6	−47.8	148.1	94.8
	$AuCl_4^{-}$	(aq)	338.78	−322.2	−235.2	266.9	–
Helium	**He**	(g)	4.00	0.0	0.0	126.1	20.8
	He	(aq)	4.00	−1.7	19.2	55.6	–
Hydrogen	$\mathbf{H_2}$	(g)	2.02	0.0	0.0	130.7	28.8
	H	(g)	1.01	218.0	203.3	114.7	20.8
	H^{+}	(aq)	1.01	0.0	0.0	0.0	0.0

Substance		M $\left(\frac{g}{mol}\right)$	$H^\ominus_{298}$ $\left(\frac{kJ}{mol}\right)$	$G^\ominus_{298}$ $\left(\frac{kJ}{mol}\right)$	$S^\ominus_{298}$ $\left(\frac{J}{mol\,K}\right)$	c_p $\left(\frac{J}{mol\,K}\right)$	
H_2	(aq)	2.02	−4.2	17.6	57.3	–	(HYDROGEN)
H_3BO_3	(s)	61.83	−1094.0	−968.5	88.7	81.4	
HBr	(g)	80.91	−36.4	−53.4	198.7	29.1	
HCN	(g)	27.03	135.1	124.7	201.8	35.9	
HCl	(g)	36.46	−92.3	−95.3	186.9	29.1	
HCl	(aq)	36.46	−167.2	−131.3	56.5	−136.4	
H_2O	(ℓ)	18.02	−285.8	−237.2	69.9	75.3	
H_2O	(g)	18.02	−241.8	−228.6	188.7	33.6	
OH^-	(aq)	17.01	−230.0	−157.3	−10.8	−148.5	
H_3O^+	(aq)	19.03	−285.8	−237.2	69.9	75.3	
H_2O_2	(ℓ)	34.02	−187.8	−120.3	109.6	89.1	
H_2O_2	(g)	34.02	−136.1	−105.4	233.0	43.1	
H_2O_2	(aq)	34.02	−191.2	−134.1	143.9	–	
I_2	(s)	253.81	0.0	0.0	116.1	54.4	IODINE
I_2	(g)	253.81	62.4	19.3	260.7	36.9	
I_2	(aq)	253.81	141.0	16.4	137.2	–	
I^-	(aq)	126.91	−55.2	−51.6	111.3	−142.3	
IO^-	(aq)	142.91	−107.5	−38.5	5.4	–	
IO_3^-	(aq)	174.90	−221.3	−128.0	118.4	–	
HI	(g)	127.91	26.4	1.6	206.6	29.2	
Ir	(s)	192.22	0.0	0.0	35.5	25.0	IRIDIUM
Ir	(g)	192.22	669.4	622.3	193.6	20.8	
$IrCl_3$	(s)	298.58	−245.6	−169.5	114.9	85.8	
IrO_2	(s)	224.22	−242.7	−188.4	58.6	60.0	
IrS_2	(s)	256.35	−133.1	−123.9	69.0	65.9	
Fe	(s)	55.85	0.0	0.0	27.3	25.0	IRON
Fe	(g)	55.85	415.5	369.8	180.5	25.7	
Fe^{++}	(aq)	55.85	−89.1	−78.9	−137.6	–	
Fe^{+++}	(aq)	55.85	−48.5	−4.6	−315.9	–	
Fe_3C (cementite)	(s)	179.55	25.1	20.0	104.6	105.9	
$FeCO_3$	(s)	115.86	−740.6	−666.7	92.9	82.1	
$FeCl_2$	(s)	126.75	−341.8	−302.3	117.9	76.7	
$FeCl_3$	(s)	162.21	−399.4	−333.9	142.3	96.6	
Fe_4N	(s)	237.40	−11.1	3.7	155.3	122.6	
FeO	(s)	71.85	−272.0	−251.4	60.8	49.9	
Fe_2O_3 (hematite)	(s)	159.69	−824.2	−742.3	87.4	103.9	
Fe_3O_4 (magnetite)	(s)	231.54	−1118.4	−1015.2	146.1	150.3	
$Fe(OH)_2$	(s)	89.86	−569.0	487.0	88.0	97.0	
$Fe(OH)_3$	(s)	106.87	−823.0	−696.5	106.7	101.5	
$Fe_2O_3 \cdot H_2O$	(s)	177.71	−1118.0	−975.9	118.8	175.7	
FeS	(s)	87.91	−101.7	−102.0	60.3	50.5	
FeS_2 (pyrite)	(s)	119.98	−171.5	−160.1	52.9	62.1	
$FeSO_4$	(s)	151.91	−928.8	−824.9	121.0	100.5	
$FeSO_4 \cdot 7H_2O$	(s)	278.02	−3014.6	−2510.3	409.2	394.5	
$Fe_2(SO_4)_3$		399.89	−2583.0	−2262.8	307.5	265.0	

	Substance		M $\left(\frac{\text{g}}{\text{mol}}\right)$	$H^{\ominus}_{298}$ $\left(\frac{\text{kJ}}{\text{mol}}\right)$	$G^{\ominus}_{298}$ $\left(\frac{\text{kJ}}{\text{mol}}\right)$	$S^{\ominus}_{298}$ $\left(\frac{\text{J}}{\text{mol K}}\right)$	c_p $\left(\frac{\text{J}}{\text{mol K}}\right)$
	FeSi	(s)	83.93	−78.9	−78.4	44.7	45.2
	$FeSi_2$	(s)	112.02	−81.2	−78.4	55.6	64.2
KRYPTON	**Kr**	(g)	83.80	0.0	0.0	164.1	20.8
	Kr	(aq)	83.80	−15.5	15.1	61.5	–
LEAD	**Pb**	(s)	207.20	0.0	0.0	64.8	26.8
	Pb	(g)	207.20	195.2	162.2	175.4	20.8
	Pb^{++}	(aq)	–	−1.7	−24.4	10.5	–
	$Pb(BO_2)_2$	(s)	292.82	−1556.4	−1450.2	130.5	107.6
	$PbCO_3$	(s)	267.21	−699.1	−625.4	131.0	87.4
	$PbCl_2$	(s)	278.11	−359.4	−314.1	136.0	77.1
	PbI_2	(s)	461.01	−175.4	−173.6	174.8	77.6
	PbO (yellow)	(s)	223.20	−218.1	−188.6	68.7	45.8
	PbO (red)	(s)	223.20	−219.4	−189.3	66.3	45.8
	PbO_2	(s)	239.20	−274.5	−215.4	71.8	61.1
	Pb_3O_4	(s)	685.60	−718.7	−601.6	212.0	154.9
	PbS	(s)	239.27	−98.6	−97.0	91.3	49.4
	$PbSiO_3$	(s)	283.28	−1145.0	−1061.1	109.9	90.1
	$PbSO_4$	(s)	303.26	−923.1	−816.2	148.5	86.4
LITHIUM	**Li**	(s)	6.94	0.0	0.0	29.1	24.6
	Li	(g)	6.94	159.3	126.6	138.8	20.8
	Li^+	(aq)	6.94	−278.5	−293.3	13.4	68.6
	Li_2CO_3	(s)	73.89	−1216.0	−1132.1	90.2	96.2
	LiCl	(s)	42.39	−408.3	−384.0	59.3	48.0
	$LiClO_4$	(s)	106.39	−380.7	−253.9	125.5	105.0
	$LiNO_3$	(s)	68.95	−483.1	−381.2	90.0	–
	$LiNO_3 \cdot 3H_2O$	(s)	122.99	−1374.4	−1103.7	223.4	–
	Li_2O	(s)	29.88	−597.9	−561.2	37.6	54.1
	LiOH	(s)	23.95	−484.9	−439.0	42.8	49.7
	Li_2SO_4	(s)	109.94	−1436.5	−1321.8	115.1	117.6
	$Li_2SO_4 \cdot H_2O$	(s)	127.96	−1735.5	−1565.7	163.6	151.1
MAGNESIUM	**Mg**	(s)	24.31	0.0	0.0	32.7	24.9
	Mg	(g)	24.31	146.4	111.9	148.6	20.8
	Mg^{++}	(aq)	24.31	−466.9	−454.8	−138.1	–
	MgC_2	(s)	48.33	87.9	84.8	54.4	56.2
	$MgCO_3$	(s)	84.31	−1095.8	−1012.2	65.7	75.5
	$MgCl_2$	(s)	95.21	−641.6	−592.1	89.6	71.7
	$Mg(NO_3)_2$	(s)	148.32	−790.7	−589.2	164.0	141.9
	$Mg(NO_3)_2 \cdot 6H_2O$	(s)	256.41	−2613.3	−2080.7	451.9	–
	MgO	(s)	40.30	−601.2	−568.9	26.9	37.1
	$Mg(OH)_2$	(s)	58.32	−924.7	−833.6	63.2	77.2
	MgS	(s)	56.37	−345.7	−341.4	50.3	45.6
	$MgSO_4$	(s)	120.37	−1284.9	−1170.6	91.6	96.2
	$MgSO_4 \cdot 6H_2O$	(s)	228.47	−3087.0	−2632.2	348.1	348.1
	$MgSO_4 \cdot 7H_2O$	(s)	246.48	−3388.7	−2871.9	372.4	–
MANGANESE	**Mn**	(s)	54.94	0.0	0.0	32.0	26.3

Substance		M $\left(\frac{g}{mol}\right)$	$H^{\ominus}_{298}$ $\left(\frac{kJ}{mol}\right)$	$G^{\ominus}_{298}$ $\left(\frac{kJ}{mol}\right)$	$S^{\ominus}_{298}$ $\left(\frac{J}{mol\,K}\right)$	c_p $\left(\frac{J}{mol\,K}\right)$	
Mn	(g)	54.94	283.3	241.0	173.7	20.8	(MANGANESE)
Mn^{++}	(aq)	54.94	−220.7	−228.0	−73.6	50.2	
$MnCO_3$	(s)	114.95	−894.1	−816.7	85.8	81.5	
$MnCl_2$	(s)	125.84	−481.3	−440.5	118.2	73.0	
$MnCl_2 \cdot H_2O$	(s)	143.86	−789.9	−696.2	174.1	–	
$MnCl_2 \cdot 2H_2O$	(s)	161.87	−1092.0	−942.2	218.8	–	
$MnCl_2 \cdot 4H_2O$	(s)	197.91	−1687.4	−1423.8	303.3	–	
MnO	(s)	70.94	−385.2	−362.9	59.7	44.1	
MnO_2	(s)	86.94	−520.0	−465.1	53.0	54.4	
Mn_2O_3	(s)	157.87	−959.0	−881.1	110.5	99.0	
Mn_3O_4	(s)	228.81	−1387.8	−1383.2	155.6	140.5	
$Mn(OH)_2$	(s)	88.95	−695.4	−615.0	99.2	–	
MnS	(s)	87.00	−214.2	−218.4	78.2	49.9	
MnS_2	(s)	119.07	−223.8	−225.0	99.9	70.1	
$MnSO_4$	(s)	151.00	−1065.3	−957.2	112.1	100.2	
Hg	(ℓ)	200.59	0.0	0.0	75.9	28.0	MERCURY
Hg	(g)	200.59	61.4	31.9	175.0	20.8	
Hg^{++}	(aq)	200.59	171.1	164.4	−32.2	–	
Hg_2^{++}	(aq)	401.18	172.4	153.6	84.5	–	
$HgCl_2$	(s)	271.50	−230.1	−184.1	144.5	73.9	
Hg_2Cl_2	(s)	472.09	−264.9	−210.6	192.5	101.9	
Hg_2CO_3	(s)	461.19	−553.5	−468.2	179.9	–	
HgO (red)	(s)	216.59	−90.8	−58.5	70.3	44.1	
HgO (yellow)	(s)	216.59	−90.5	−58.4	71.1	–	
HgS (red)	(s)	232.65	−58.2	−50.6	82.4	48.4	
HgS (yellow)	(s)	232.65	−53.6	−47.7	88.3	–	
$HgSO_4$	(s)	296.65	−707.5	−594.8	140.2	102.3	
Hg_2SO_4	(s)	497.24	−743.1	−625.9	200.8	132.0	
Mo	(s)	95.94	0.0	0.0	28.6	23.9	MOLYBDENUM
Mo	(g)	95.94	658.5	612.8	182.0	20.8	
MoO_4^{--}	(aq)	159.94	−997.9	−836.4	27.2	–	
MoC	(s)	107.95	28.5	−29.1	36.7	30.7	
MoC_2	(s)	203.89	−53.1	−54.0	65.8	60.2	
MoO_2	(s)	127.94	−588.9	−533.0	46.3	56.0	
MoO_3	(s)	143.94	−745.1	−688.0	77.7	75.0	
MoS_2	(s)	160.07	−276.1	−267.2	62.6	63.5	
Ne	(g)	20.18	0.0	0.0	146.3	20.8	NEON
Ne	(aq)	20.18	−4.6	19.2	66.1	–	
Ni	(s)	58.69	0.0	0.0	29.9	26.1	NICKEL
Ni	(g)	58.69	430.1	384.7	182.2	23.4	
Ni^{++}	(aq)	58.69	−54.0	−45.6	−128.9	–	
Ni_3C	(s)	188.08	67.4	64.2	106.3	106.7	
$NiCO_3$	(s)	118.70	−694.5	−617.9	86.2	86.2	
$NiCl_2$	(s)	129.60	−305.3	−259.1	98.0	71.7	
$NiCl_2 \cdot 2H_2O$	(s)	165.65	−922.2	−760.2	175.7	–	

	Substance		M $\left(\frac{g}{mol}\right)$	$H^{\ominus}_{298}$ $\left(\frac{kJ}{mol}\right)$	$G^{\ominus}_{298}$ $\left(\frac{kJ}{mol}\right)$	$S^{\ominus}_{298}$ $\left(\frac{J}{mol\,K}\right)$	c_p $\left(\frac{J}{mol\,K}\right)$
(NICKEL)	$NiCl_2 \cdot 4H_2O$	(s)	201.68	−1516.7	−1235.1	242.7	–
	$NiCl_2 \cdot 6H_2O$	(s)	237.71	−2103.2	−1713.5	344.3	–
	$Ni(OH)_2$	(s)	92.73	−529.7	−447.3	87.9	–
	NiO	(s)	74.69	−239.7	−211.5	38.0	44.3
	NiS	(s)	90.76	−87.9	−85.2	53.0	47.1
	$NiSO_4$	(s)	154.75	−872.9	−759.5	92.0	138.0
NITROGEN	$\mathbf{N_2}$	(g)	28.01	0.0	0.0	191.6	29.1
	NH_3	(g)	17.03	−45.9	−16.4	192.8	35.7
	NH_3	(aq)	17.03	−80.3	−26.6	111.3	–
	NH_4^+	(aq)	18.04	−132.5	−79.4	113.4	79.9
	NH_4Cl	(s)	53.49	−314.6	−203.1	95.0	86.6
	NH_4NO_3	(s)	80.04	−365.6	−184.0	151.1	139.3
	HNO_3	(ℓ)	63.01	−174.1	−80.8	155.6	111.3
	NO_3^-	(aq)	62.00	−207.4	−111.3	146.4	−86.6
	HNO_2	(aq)	47.01	−119.2	−50.6	135.6	–
	NO_2^-	(aq)	46.01	−104.6	−32.2	123.0	−97.5
	NO	(g)	30.01	90.3	86.6	210.8	29.8
	NO_2	(g)	46.01	33.1	51.3	240.0	36.7
	N_2O	(g)	44.01	82.0	104.2	220.0	38.8
	N_2O_4	(g)	92.01	9.1	97.8	304.4	77.6
	N_2O_4	(ℓ)	92.01	−19.6	97.5	209.2	142.5
OXYGEN	$\mathbf{O_2}$	(g)	32.00	0.0	0.0	205.1	29.4
	O_2	(aq)	32.00	−11.7	16.3	110.9	–
	O_3 (ozone)	(g)	48.00	142.7	163.2	238.8	39.2
	H_2O	(ℓ)	18.02	−285.8	−237.2	69.9	75.3
	H_2O	(g)	18.02	−241.8	−228.6	188.7	33.6
	H_3O^+	(aq)	19.03	−285.8	−237.2	69.9	75.3
	OH^-	(aq)	17.01	−230.0	−157.3	−10.8	−148.5
	H_2O_2	(ℓ)	34.02	−187.8	−120.3	109.6	89.1
	H_2O_2	(g)	34.02	−136.1	−105.4	233.0	43.1
	H_2O_2	(aq)	34.02	−191.2	−134.1	143.9	–
PHOSPHORUS	**P** (white)	(s)	30.97	0.0	0.0	41.1	23.8
	P (red)	(s)	30.97	−17.5	−12.0	22.9	21.2
	PCl_3	(g)	137.33	−288.6	−269.5	311.7	71.6
	PCl_5	(g)	208.24	−374.9	−304.9	364.1	111.6
	PH_3	(g)	34.00	5.6	13.5	210.3	37.1
	H_3PO_4	(s)	98.00	−1279.0	−1119.2	110.5	106.1
PLATINUM	**Pt**	(s)	195.08	0.0	0.0	41.6	25.9
	Pt	(g)	195.08	564.5	519.9	192.4	25.5
	$PtCl_2$	(s)	265.99	−106.7	−93.2	219.6	75.4
	$PtCl_3$	(s)	301.44	−168.2	−129.6	246.9	121.3
	$PtCl_4$	(s)	336.89	−229.3	−163.7	267.9	125.5
	PtS	(s)	227.15	−83.1	−77.5	55.1	43.4
	PtS_2	(s)	259.22	−110.5	−101.2	74.7	65.9
POTASSIUM	**K**	(s)	39.10	0.0	0.0	64.7	29.3

Substance		M $\left(\frac{g}{mol}\right)$	$H^\ominus_{298}$ $\left(\frac{kJ}{mol}\right)$	$G^\ominus_{298}$ $\left(\frac{kJ}{mol}\right)$	$S^\ominus_{298}$ $\left(\frac{J}{mol\,K}\right)$	c_p $\left(\frac{J}{mol\,K}\right)$	
K	(g)	39.10	89.0	60.5	160.3	20.8	(POTASSIUM)
K^+	(aq)	39.10	−252.4	−283.3	102.5	21.8	
$KAl(SO_4)_2$	(s)	258.21	−2470.2	−2239.7	204.6	193.0	
$KAl(SO_4)_2 \cdot 3H_2O$	(s)	312.25	−3381.1	−2974.6	314.0	314.3	
$KAl(SO_4)_2 \cdot 12H_2O$	(s)	474.39	−6061.8	−5140.7	687.4	651.9	
K_2CO_3	(s)	138.21	−1150.2	−1064.5	155.5	114.2	
KCl	(s)	74.55	−436.7	−408.8	82.6	51.7	
$KClO_4$	(s)	138.55	−430.1	−300.3	151.0	112.4	
K_2CrO_4	(s)	194.19	−1403.7	−1295.4	200.0	146.0	
KH_2PO_4	(s)	136.09	−1568.3	−1415.7	134.9	116.6	
K_2HPO_4	(s)	174.18	−1775.8	−1636.5	179.1	141.3	
K_3PO_4	(s)	212.27	−1988.2	−1858.9	211.7	164.9	
KI	(s)	166.00	−327.9	−323.0	106.4	52.8	
KNO_3	(s)	101.10	−494.6	−394.7	133.1	96.4	
K_2O	(s)	94.20	−361.5	−322.8	102.0	74.4	
KOH	(s)	56.02	−424.7	−378.9	78.9	64.9	
K_2S	(s)	110.26	−376.6	−362.7	115.1	74.7	
Si	(s)	28.09	0.0	0.0	18.8	20.0	SILICON
SiC (carborundum)	(s)	40.10	−73.2	−70.9	16.6	27.0	
K_2SO_3	(s)	158.26	−1126.8	−1038.0	171.5	123.4	
K_2SO_4	(s)	174.26	−1437.8	−1319.7	175.6	131.2	
$SiCl_4$	(g)	169.90	−662.7	−622.8	330.9	90.3	
SiH_4 (silane)	(g)	32.12	34.3	56.8	204.7	42.8	
Si_2H_6 (disilane)	(g)	62.22	80.3	127.1	272.7	79.1	
SiO_2 (quartz)	(s)	60.08	−910.9	−856.4	41.5	44.4	
H_2SiO_3	(s)	78.10	−1188.7	−1092.4	133.9	–	
Ag	(s)	107.87	0.0	0.0	42.7	25.4	SILVER
Ag	(g)	107.87	284.1	245.2	173.0	20.8	
Ag^+	(aq)	107.87	105.6	77.1	72.7	21.8	
$Ag(NH_3)_2^+$	(aq)	141.93	−111.3	−17.2	245.2	–	
Ag_2CO_3	(s)	275.75	−505.8	−436.8	167.4	111.6	
$AgCl$	(s)	143.32	−127.1	−109.8	96.2	52.9	
$AgNO_3$	(s)	169.87	−124.4	−33.3	140.6	93.1	
Ag_2O	(s)	231.74	−31.0	−11.2	121.3	65.9	
Ag_2S	(s)	247.80	−32.6	−40.5	144.0	76.2	
Ag_2SO_4	(s)	311.80	−715.9	−618.3	200.4	131.5	
Na	(s)	22.99	0.0	0.0	51.5	28.2	SODIUM
Na	(g)	22.99	107.3	76.8	153.7	20.8	
Na^+	(aq)	22.99	−240.1	−261.9	59.0	46.4	
Na_2CO_3	(s)	106.00	−1130.8	−1048.0	138.8	111.0	
$Na_2CO_3 \cdot H_2O$	(s)	124.00	−1431.3	−1285.4	168.1	–	
$Na_2CO_3 \cdot 7H_2O$	(s)	232.10	−3200.0	−2714.6	426.8	–	
$Na_2CO_3 \cdot 10H_2O$	(s)	286.14	−4081.3	−3428.2	564.0	550.3	
$NaHCO_3$	(s)	84.01	−950.8	−852.9	101.7	87.6	
$NaCl$	(s)	58.44	−411.1	−384.0	72.1	50.5	

	Substance		M $\left(\frac{\text{g}}{\text{mol}}\right)$	$H^{\ominus}_{298}$ $\left(\frac{\text{kJ}}{\text{mol}}\right)$	$G^{\ominus}_{298}$ $\left(\frac{\text{kJ}}{\text{mol}}\right)$	$S^{\ominus}_{298}$ $\left(\frac{\text{J}}{\text{mol K}}\right)$	c_p $\left(\frac{\text{J}}{\text{mol K}}\right)$
(SODIUM)	$NaClO_4$	(s)	122.44	−382.8	−254.2	142.2	111.3
	NaH	(s)	24.00	−56.4	−33.6	40.0	36.4
	$NaNO_2$	(s)	69.00	−359.0	−290.1	121.3	69.0
	$NaNO_3$	(s)	85.00	−468.0	−367.0	116.3	93.0
	Na_2O	(s)	61.98	−418.0	−379.1	75.0	68.9
	NaOH	(s)	40.00	−425.9	−379.7	64.4	59.6
	Na_3PO_4	(s)	163.94	−1917.4	−1788.6	173.8	150.0
	Na_2S	(s)	78.05	−366.1	−354.6	96.2	82.8
	Na_2SO_3	(s)	126.04	−1100.8	−1012.3	145.9	120.2
	Na_2SO_4	(s)	142.04	−1387.8	−1269.8	149.6	128.2
	$Na_2SO_4 \cdot 10H_2O$	(s)	322.20	−4327.3	−3647.4	592.0	–
SULPHUR	**S** (rhombic)	(s)	32.07	0.0	0.0	32.1	22.8
	S (monoclinic)	(s)	32.07	0.3	0.1	32.9	24.7
	S_8	(g)	256.53	100.4	48.6	430.3	155.8
	SO_2	(g)	64.07	−296.8	−300.1	248.2	39.9
	SO_3	(g)	80.06	−395.8	−371.0	256.8	50.7
	S_2^{--}	(aq)	64.13	30.1	79.5	28.5	–
	SO_3^{--}	(aq)	80.06	−635.5	−486.6	−29.3	–
	SO_4^{--}	(aq)	96.06	−909.3	−744.6	20.1	−292.9
	H_2S	(g)	34.08	−20.6	−33.6	205.7	34.2
	H_2S	(aq)	34.08	−39.7	−27.9	121.3	–
	HSO_3	(aq)	81.07	−626.2	−527.8	139.7	–
	HSO_4^-	(aq)	97.07	−887.3	−756.0	131.8	−83.7
	H_2SO_4	(aq)	98.08	−909.3	−744.6	20.1	−292.9
TIN	**Sn** (white)	(s)	118.71	0.0	0.0	51.2	27.0
	Sn (grey)	(s)	118.71	−2.0	0.1	44.3	25.8
	Sn	(g)	118.71	301.2	266.3	168.5	21.3
	Sn^{++}	(aq)	118.71	−8.8	−27.2	−16.7	–
	Sn^{++++}	(aq)	118.71	30.5	2.5	−117.2	–
	$SnCl_2$	(s)	189.62	−328.0	−286.2	134.1	78.0
	$SnCl_4$	(ℓ)	260.52	−511.3	−440.1	258.7	165.3
	SnO	(s)	134.71	−285.8	−256.8	56.5	47.8
	SnO_2	(s)	150.71	−580.8	−520.0	52.3	52.6
	SnS	(s)	150.78	−107.9	−106.1	77.0	49.3
	SnS_2	(s)	182.84	−153.6	−145.2	87.4	70.1
	Sn_2S_3	(s)	333.62	−263.6	−253.4	164.4	120.1
	$SnSO_4$	(s)	214.77	−887.0	−781.2	138.6	150.6
	$Sn(SO_4)_2$	(s)	310.84	−1648.5	−1414.1	149.8	284.5
TITANIUM	**Ti**	(s)	47.88	0.0	0.0	30.8	25.1
	Ti	(g)	47.88	473.6	429.0	180.3	24.4
	TiC	(s)	59.89	−184.5	−180.8	24.2	33.8
	$TiCl_2$	(s)	118.79	−515.5	−465.8	87.4	69.8
	$TiCl_3$	(s)	154.24	−721.7	−654.5	139.7	97.1
	$TiCl_4$	(s)	189.69	−804.2	−737.2	252.4	145.2
	TiH_2	(s)	49.90	−144.3	−105.1	29.7	30.3

Substance		M $\left(\frac{g}{mol}\right)$	$H^\ominus_{298}$ $\left(\frac{kJ}{mol}\right)$	$G^\ominus_{298}$ $\left(\frac{kJ}{mol}\right)$	$S^\ominus_{298}$ $\left(\frac{J}{mol\,K}\right)$	c_p $\left(\frac{J}{mol\,K}\right)$	
TiN	(s)	61.89	−337.9	−309.2	30.3	37.1	(TITANIUM)
TiO	(s)	63.88	−542.7	−513.3	34.8	40.0	
TiO_2 (rutile)	(s)	79.88	−944.7	−889.4	50.3	55.1	
Ti_2O_3	(s)	143.76	−1520.9	−1433.8	77.2	95.8	
TiS	(s)	79.95	−272.0	−270.1	56.5	48.1	
TiS_2	(s)	112.01	−407.1	−402.2	78.4	67.9	
W	(s)	183.85	0.0	0.0	32.7	24.3	TUNGSTEN
W	(g)	183.85	851.0	808.9	174.0	21.3	
WC	(s)	195.86	−40.2	−38.4	32.4	35.4	
W_2C	(s)	379.71	−26.4	−21.9	56.1	76.6	
WO_2	(s)	215.85	−589.7	−533.9	50.5	55.7	
WO_3	(s)	231.85	−842.9	−764.1	75.9	72.8	
WS_2	(s)	247.98	−259.4	−249.9	64.9	63.5	
V	(s)	50.94	0.0	0.0	28.9	24.9	VANADIUM
V	(g)	50.94	512.2	468.5	182.3	26.0	
V_2C	(s)	113.89	−117.2	−113.4	51.1	55.9	
VN	(s)	64.95	−217.2	−191.1	37.3	38.0	
VO	(s)	66.94	−431.8	−404.2	39.0	45.5	
VO_2	(s)	82.94	−232.6	−241.9	265.3	43.9	
V_2O_3	(s)	149.88	−1218.8	−1139.1	98.1	105.0	
VO_2^{++}	(aq)	66.94	−486.6	−446.4	−133.9	–	
HVO_4^{--}	(aq)	115.95	−1159.0	−974.9	16.7	–	
$H_2VO_4^-$	(aq)	116.96	−1174.0	−1020.9	121.3	–	
Xe	(g)	131.29	0.0	0.0	169.7	20.8	XENON
Xe	(aq)	131.29	−17.6	13.4	65.7	–	
Zn	(s)	65.39	0.0	0.0	41.6	25.4	ZINC
Zn	(g)	65.39	130.4	94.8	161.0	20.8	
Zn^{++}	(aq)	65.39	−153.9	−147.0	−112.1	46.0	
$ZnCO_3$	(s)	125.40	−812.8	−731.5	82.4	80.1	
$ZnCl_2$	(s)	136.30	−415.1	−369.4	111.5	71.3	
Zn_3N_2	(s)	224.18	−22.6	39.3	108.8	109.3	
ZnO	(s)	81.39	−350.5	−320.5	43.6	41.1	
$Zn(OH)_2$	(s)	99.39	−643.2	−555.1	81.6	72.4	
$Zn_3(PO_4)_2$	(s)	386.11	−2899.5	−2663.8	237.0	234.1	
ZnS	(s)	97.45	−191.8	−190.1	68.0	45.9	
$ZnSO_4$	(s)	161.45	−982.8	−871.4	110.5	99.1	
$ZnSO_4 \cdot H_2O$	(s)	179.47	−1301.5	−1131.1	145.5	153.6	
$ZnSO_4 \cdot 2H_2O$	(s)	197.48	−1596.0	−1370.0	192.5	198.7	
$ZnSO_4 \cdot 6H_2O$	(s)	269.55	−2779.0	−2323.5	355.9	358.0	
$ZnSO_4 \cdot 7H_2O$	(s)	287.56	−3078.5	−2563.3	388.7	379.2	
Zr	(s)	91.22	0.0	0.0	38.9	25.2	ZIRCONIUM
ZrC	(s)	103.24	−196.6	−193.3	33.3	37.9	
ZrS_2	(s)	155.36	−577.4	−570.0	78.2	68.8	

Ref.: "Handbook of Chemistry and Physics", 64th ed., CRC Press Inc., Boca Raton, 1984.
Barin, I.: "Thermochemical Data of Pure Substances", VCH, Basel 1989.

Thermochemical data (2)

cement-chemical compounds

The table specifies **thermodynamic standard values** $H^{\ominus}_{298}$, $G^{\ominus}_{298}$, $S^{\ominus}_{298}$, and c_p for selected substances included in cement-chemical calculations. The **state** of the substances is given by: (s): solid, (ℓ): liquid, (g): gaseous, and (aq) for ions and gases in aqueous solution. **Standard state:** $p^{\ominus} = 101325\,\text{Pa}$, and for dissolved substances: $c^{\ominus} = 1\,\text{mol}/\ell$. **Reference temperature:** $T = 298.15\,\text{K}$ (25 °C).

Substance		$H^{\ominus}_{298}$ $\left(\frac{\text{kJ}}{\text{mol}}\right)$	$G^{\ominus}_{298}$ $\left(\frac{\text{kJ}}{\text{mol}}\right)$	$S^{\ominus}_{298}$ $\left(\frac{\text{J}}{\text{mol K}}\right)$	c_p $\left(\frac{\text{J}}{\text{mol K}}\right)$
Al	(s)	0.0	0.0	28.3	24.3
Al	(g)	329.7	289.1	164.6	21.4
Al^{+++}	(aq)	−531.4	−485.3	−321.7	–
$AlCl_3$	(s)	−705.6	−630.0	109.3	91.1
$AlCl_3 \cdot 6H_2O$	(s)	−2691.6	−2260.9	−318.0	296.2
AlO_2^-	(aq)	−919.8	−823.0	−20.9	–
α-Al_2O_3	(s)	−1675.7	−1582.3	50.9	79.0
γ-Al_2O_3	(s)	−1656.9	−1563.9	52.3	82.7
δ-Al_2O_3	(s)	−1666.5	−1573.0	50.6	81.8
κ-Al_2O_3	(s)	−1666.5	−1573.8	53.6	80.7
AlOCl	(s)	−793.3	−737.2	54.4	56.9
$Al(OH)_3$	(s)	−1276.1	−1138.7	71.1	93.1
$Al_2O_3 \cdot H_2O$ (diaspore)	(s)	−1999.1	−1842.0	70.7	106.2
$Al_2O_3 \cdot H_2O$ (boehmite)	(s)	−1980.7	−1831.4	96.9	131.3
$Al_2O_3 \cdot 3H_3O$ (gibbsite)	(s)	−2586.5	−2310.1	136.9	183.5
$2Al_2O_3 \cdot 2MgO \cdot 5SiO_2$	(s)	−9161.7	−8651.3	407.1	452.3
$Al_2O_3 \cdot SiO_2$ (kyanite)	(s)	−2594.1	−2443.9	84.5	121.8
$Al_2O_3 \cdot SiO_2$ (andalusite)	(s)	−2590.3	−2443.0	93.8	122.8
$Al_2O_3 \cdot SiO_2$ (sillimanite)	(s)	−2587.8	−2441.1	96.1	122.2
$3Al_2O_3 \cdot 2SiO_2$ (mullite)	(s)	−6820.5	−6443.0	274.9	325.4
$Al_2O_3 \cdot 2SiO_2 \cdot 2H_2O$ (kaolinite)	(s)	−4119.6	−3799.4	205.0	246.1
$Al_2O_3 \cdot 2SiO_2 \cdot 2H_2O$ (dickite)	(s)	−4118.3	−3795.8	197.1	239.5
$Al_2O_3 \cdot 2SiO_2 \cdot 2H_2O$ (halloysite)	(s)	−4101.2	−3780.6	203.3	246.3
$Al_2(SO_4)_3$	(s)	−3440.8	−3099.6	239.3	259.4
$Al_2(SO_4)_3 \cdot 6H_2O$	(s)	−5311.7	−4622.6	469.0	492.9
CO_2	(g)	−393.5	−394.4	213.8	37.1
CO_2	(aq)	−413.8	−386.0	117.6	–
CO_3^{--}	(aq)	−677.1	−527.9	−56.9	–
HCO_3^-	(aq)	−692.0	−586.8	91.2	–
H_2CO_3	(aq)	−699.6	−623.2	187.4	–
Ca	(s)	0.0	0.0	41.4	25.3
Ca	(g)	178.2	144.4	154.9	20.8
Ca^{++}	(aq)	−542.8	−553.6	−53.1	–
$CaCO_3$ (calcite)	(s)	−1206.9	−1128.8	92.9	83.5
$CaCO_3$ (aragonite)	(s)	−1207.1	−1127.8	88.7	82.3

Substance		$H^{\ominus}_{298}$ $\left(\frac{kJ}{mol}\right)$	$G^{\ominus}_{298}$ $\left(\frac{kJ}{mol}\right)$	$S^{\ominus}_{298}$ $\left(\frac{J}{mol\,K}\right)$	c_p $\left(\frac{J}{mol\,K}\right)$
$CaMg(CO_3)_2$ (dolomite)	(s)	−2326.3	−2163.6	155.2	157.5
$CaCl_2$	(s)	−795.8	−748.1	104.6	72.9
CaO	(s)	−635.1	−603.5	38.1	42.1
$CaO \cdot Al_2O_3$	(s)	−2326.3	−2208.8	114.2	120.8
$CaO \cdot 2Al_2O_3$	(s)	−4025.8	−3818.7	177.8	198.2
$2CaO \cdot Al_2O_3$	(s)	−2958.0	−2801.4	127.1	164.3
$3CaO \cdot Al_2O_3$	(s)	−3587.8	−3411.8	205.7	209.7
$12CaO \cdot 7Al_2O_3$	(s)	−19430.0	−18466.7	1046.8	1084.8
$3CaO \cdot Al_2O_3 \cdot 3CaSO_4 \cdot 31H_2O^{\dagger}$	(s)	−17208.1	−14886.9	1688.6	–
$3CaO \cdot Al_2O_3 \cdot CaSO_4 \cdot 12H_2O^{\dagger}$	(s)	−8718.6	−7717.3	719.2	–
$3CaO \cdot Al_2O_3 \cdot 6H_2O$	(s)	−5548.0	−5005.6	376.6	432.6
$2CaO \cdot Al_2O_3 \cdot 8H_2O^{\dagger}$	(s)	−5404.1	−4780.4	414.4	–
$4CaO \cdot Al_2O_3 \cdot 13H_2O^{\dagger}$	(s)	−8300.8	−7321.3	686.5	–
$4CaO \cdot Al_2O_3 \cdot 19H_2O^{\dagger}$	(s)	−10084.1	−8757.1	920.9	–
$CaO \cdot Al_2O_3 \cdot 2SiO_2$	(s)	−4227.9	−4002.2	199.3	211.3
$2CaO \cdot Al_2O_3 \cdot SiO_2$ (gehlenite)	(s)	−3981.5	−3782.9	210.0	205.4
$3CaO \cdot Al_2O_3 \cdot 3SiO_2$ (grossular)	(s)	−6646.2	−6280.4	241.4	323.1
$CaO \cdot Al_2O_3 \cdot 2SiO_2 \cdot 2H_2O$	(s)	−4858.5	−4505.2	237.7	233.0
$CaO \cdot Fe_2O_3$	(s)	−1520.3	−1412.7	145.7	150.2
$2CaO \cdot Fe_2O_3$	(s)	−2139.3	−2001.7	188.8	190.1
$Ca(OH)_2$	(s)	−986.1	−898.5	83.4	87.5
$CaO \cdot MgO$	(s)	−1243.9	−1180.4	66.3	79.6
$CaO \cdot MgO \cdot SiO_2$ (monticellite)	(s)	−2263.1	−2145.7	109.6	123.2
$CaO \cdot MgO \cdot 2SiO_2$ (diopside)	(s)	−3206.2	−3032.0	142.9	156.1
$2CaO \cdot MgO \cdot 2SiO_2$ (akermanite)	(s)	−3877.2	−3679.8	209.2	212.0
$3CaO \cdot MgO \cdot 2SiO_2$ (merwinite)	(s)	−4567.7	−4340.5	253.1	252.3
$2CaO \cdot 5MgO \cdot 8SiO_2 \cdot H_2O$	(s)	−12360.0	−11632.4	548.9	655.5
$CaO \cdot SiO_2$ (wollastonite)	(s)	−1634.9	−1549.7	81.9	85.3
$CaO \cdot SiO_2$ (psd. wollast.)	(s)	−1628.4	−1544.7	87.4	86.5
β-$2CaO \cdot SiO_2$ (larnite)	(s)	−2304.8	−2190.3	127.7	128.8
γ-$2CaO \cdot SiO_2$ (olivine)	(s)	−2315.2	−2198.6	120.8	126.7
$3CaO \cdot SiO_2$ (tricalciumsilicate)	(s)	−2929.2	−2783.9	168.6	171.9
$3CaO \cdot 2SiO_2$ (rankinite)	(s)	−3961.0	−3761.5	210.8	214.4
$CaO \cdot 2SiO_2 \cdot 2H_2O$	(s)	−3138.0	−2873.4	171.1	162.1
$2CaO \cdot SiO_2 \cdot \frac{7}{6}H_2O$	(s)	−2665.2	−2479.8	160.7	164.8
$2CaO \cdot 3SiO_2 \cdot 2\frac{1}{2}H_2O$	(s)	−4920.4	−4541.3	271.5	295.2
$3CaO \cdot 2SiO_2 \cdot 3H_2O$	(s)	−4782.3	−4404.4	312.1	328.4
$4CaO \cdot 3SiO_2 \cdot 1\frac{1}{2}H_2O$	(s)	−6020.8	−5642.9	330.3	353.8
$5CaO \cdot 6SiO_2 \cdot 3H_2O$	(s)	−9928.6	−9257.7	513.2	595.8
$5CaO \cdot 6SiO_2 \cdot 5\frac{1}{2}H_2O$	(s)	−10686.8	−9871.3	611.5	698.6
$5CaO \cdot 6SiO_2 \cdot 10\frac{1}{2}H_2O$	(s)	−12179.6	−11075.0	808.1	889.9
$6CaO \cdot 6SiO_2 \cdot H_2O$	(s)	−10024.9	−944.8	507.5	548.3
$CaSO_4$	(s)	−1434.1	−1321.7	106.7	99.6
$CaSO_4 \cdot \frac{1}{2}H_2O$ (hemihydrate)	(s)	−1576.7	−1436.6	130.5	118.0
$CaSO_4 \cdot 2H_2O$ (dihydrate, gypsum)	(s)	−2022.6	−1797.2	194.1	186.0
SO_4^{--}	(aq)	−909.3	−744.6	20.1	−292.9

Substance		$H^{\ominus}_{298}$ $\left(\frac{kJ}{mol}\right)$	$G^{\ominus}_{298}$ $\left(\frac{kJ}{mol}\right)$	$S^{\ominus}_{298}$ $\left(\frac{J}{mol\,K}\right)$	c_p $\left(\frac{J}{mol\,K}\right)$
Fe^{++}	(aq)	−89.1	−78.9	−137.6	–
Fe^{+++}	(aq)	−48.5	−4.6	−315.9	–
$FeCO_2$	(s)	−740.6	−666.7	92.9	82.1
FeO	(s)	−272.0	−251.4	60.8	49.9
Fe_2O_3 (hematite)	(s)	−824.2	−742.3	87.4	103.9
Fe_3O_4 (magnetite)	(s)	−1118.4	−1015.2	146.1	150.7
$FeO \cdot Al_2O_3$	(s)	−1995.3	−1879.7	106.3	123.5
$FeOCl$	(s)	−377.0	−329.1	80.8	77.0
$Fe(OH)_2$	(s)	−569.0	−487.0	88.0	97.0
$Fe(OH)_3$	(s)	−823.0	−696.5	106.7	101.5
$Fe_2O_3 \cdot H_2O$ (goethite)	(s)	−1118.0	−975.9	118.8	175.7
$FeO \cdot SiO_2$	(s)	−1195.0	−1117.5	93.9	89.5
$2FeO \cdot SiO_2$	(s)	−1479.9	−1379.0	145.2	132.9
H^+	(aq)	0.0	0.0	0.0	0.0
H_2	(aq)	−4.2	17.6	57.3	–
H_2O	(ℓ)	−285.8	−237.2	69.9	75.3
H_2O	(g)	−241.8	−228.6	188.7	33.6
OH^-	(aq)	−230.0	−157.3	−10.8	−148.5
K	(s)	0.0	0.0	64.7	29.3
K^+	(aq)	−252.4	−283.3	102.5	21.8
$KAl(SO_4)_2$	(s)	−2470.2	−2239.7	204.6	193.0
$KAl(SO_4)_2 \cdot 3H_2O$	(s)	−3381.1	−2974.6	314.0	314.3
$KAl(SO_4)_2 \cdot 12H_2O$	(s)	−6061.8	−5140.7	687.4	651.9
K_2CO_3	(s)	−1150.2	−1064.5	155.5	114.2
K_2O	(s)	−361.5	−322.8	102.0	74.4
$KAlSiO_4$ (kaliophilite)	(s)	−2121.3	−2005.3	133.1	119.8
$KAlSi_2O_6$ (leucite)	(s)	−3034.2	−2871.4	200.0	164.1
$KAlSi_3O_8$ (microcline)	(s)	−3968.1	−3742.8	214.2	202.4
$KAlSi_3O_8$ (adularia)	(s)	−3954.1	−3734.7	234.3	190.5
$KAlSi_3O_8$ (sanidine)	(s)	−3959.7	−3739.9	232.9	204.5
Mg	(s)	0.0	0.0	32.7	24.9
$MgCO_3$	(s)	−1095.8	−1012.2	65.7	75.5
MgO	(s)	−601.2	−568.9	26.9	37.1
$MgO \cdot Al_2O_3$	(s)	−2299.9	−2175.0	80.6	116.2
$Mg(OH)_2$	(s)	−924.7	−833.6	63.2	77.2
$Mg(OH)Cl$	(s)	−799.6	−731.5	83.8	74.1
$MgO \cdot SiO_2$	(s)	−1548.9	−1462.0	67.8	81.9
$2MgO \cdot SiO_2$	(s)	−2176.9	−2057.9	95.1	118.7
Na	(s)	0.0	0.0	51.5	28.2
$NaCO_3$	(s)	−1130.8	−1048.0	138.8	111.0
$NaCl$	(s)	−411.1	−384.0	72.1	50.5
$NaHCO_3$	(s)	−950.8	−852.9	101.7	87.6
Na_2O	(s)	−418.0	−379.1	75.0	68.9
$NaAlO_2$	(s)	−1133.2	−1069.2	70.4	73.5
$NaAlSiO_4$ (nepheline)	(s)	−2094.7	−1980.0	124.3	115.8

Substance		$H^{\ominus}_{298}$ $\left(\frac{\text{kJ}}{\text{mol}}\right)$	$G^{\ominus}_{298}$ $\left(\frac{\text{kJ}}{\text{mol}}\right)$	$S^{\ominus}_{298}$ $\left(\frac{\text{J}}{\text{mol K}}\right)$	c_p $\left(\frac{\text{J}}{\text{mol K}}\right)$
$NaAlSi_2O_6$ (jadeite)	(s)	−3032.8	−2854.1	133.5	159.9
$NaAlSi_3O_8$ (albite)	(s)	−3937.0	−3713.5	207.4	205.1
$NaOH$	(s)	−425.9	−379.7	64.4	59.6
$Na_2Si_2O_5$	(s)	−2470.1	−2324.2	164.1	157.0
Na_4SiO_4	(s)	−2106.6	−1975.7	195.6	184.7
$Na_6Si_2O_7$	(s)	−3632.0	−3407.0	309.6	306.2
Na_2SO_4	(s)	−1387.8	−1269.8	149.6	128.2
O_2	(g)	0.0	0.0	205.1	29.4
O_2	(aq)	−11.7	16.3	110.9	–
Si	(s)	0.0	0.0	18.8	20.0
SiO_2 (quartz)	(s)	−910.9	−856.7	41.8	44.4
SiO_2 (cristobalite)	(s)	−909.5	−855.5	42.7	44.2
SiO_2 (trimydite)	(s)	−909.1	−855.3	43.5	44.6
SiO_2 (amorphous)	(s)	−903.5	−850.7	46.9	44.4
H_2SiO_3	(s)	−1188.7	−1092.4	133.9	–

Ref.: "Handbook of Chemistry and Physics", 64th ed., CRC Press Inc., Boca Raton, 1984.
Barin, I.: "Thermochemical Data of Pure Substances", VCH, Basel 1989.
Mcedlov–Petrosjan: "Thermodynamik der Silikate", VEB Verlag Bauwesen, Berlin 1966.

†) Mcedlov–Petrosjan 1966; these data for calcium aluminate hydrates are uncertain.

Thermochemical data (3)

organic compounds

The table specifies **thermodynamic standard values** $H^{\ominus}_{298}$, $G^{\ominus}_{298}$, $S^{\ominus}_{298}$, and c_p for some selected organic substances. The **state** is given by: (s): solid, (ℓ): liquid, (g): gaseous and (aq) for ions and gases in aqueous solution. **Standard state**: $p^{\ominus} = 101325\,\text{Pa}$, and for dissolved substances: $c^{\ominus} = 1\,\text{mol}/\ell$. **Reference temperature**: $T = 298.15\,\text{K}$ (25 °C).

Substance		$H^{\ominus}_{298}$ $\left(\frac{\text{kJ}}{\text{mol}}\right)$	$G^{\ominus}_{298}$ $\left(\frac{\text{kJ}}{\text{mol}}\right)$	$S^{\ominus}_{298}$ $\left(\frac{\text{J}}{\text{mol K}}\right)$	c_p $\left(\frac{\text{J}}{\text{mol K}}\right)$
C (graphite)	(s)	0.00	0.00	5.74	8.53
CO_2	(g)	−393.51	−394.36	213.74	37.11
CO_2	(aq)	−413.8	386.0	117.6	–
CH_4 (methane)	(g)	−74.81	−50.72	186.26	35.31
CH_3 (methylene)	(g)	145.69	147.92	194.2	38.70
C_2H_2 (ethyne)	(g)	226.73	209.20	200.94	43.93
C_2H_4 (ethene)	(g)	52.26	68.15	219.56	43.56
C_2H_6 (ethane)	(g)	−84.68	−32.82	229.60	52.63
C_3H_6 (propene)	(g)	20.42	62.78	267.04	63.89
C_3H_6 (cyclopropane)	(g)	53.30	104.45	237.55	55.94
C_3H_8 (propane)	(g)	−103.85	−23.49	269.91	73.5
C_4H_8 (1–butene)	(g)	−0.13	71.39	305.71	85.65
C_4H_8 (cis–2–butene)	(g)	−6.99	65.95	300.94	78.91
C_4H_{10} (butane)	(g)	−126.15	−17.03	310.23	97.45
C_5H_{12} (pentane)	(g)	−146.44	−8.20	348.40	120.2

Substance		$H^{\ominus}_{298}$ $\left(\frac{kJ}{mol}\right)$	$G^{\ominus}_{298}$ $\left(\frac{kJ}{mol}\right)$	$S^{\ominus}_{298}$ $\left(\frac{J}{mol\,K}\right)$	c_p $\left(\frac{J}{mol\,K}\right)$
C_6H_6 (benzene)	(ℓ)	49.0	124.3	173.3	136.1
C_6H_6 (benzene)	(g)	82.93	129.72	269.31	81.67
C_6H_{12} (cyclohexane)	(ℓ)	−37.3	6.4	204.3	156.5
$C_6H_5CH_3$ (methylbenzene)	(g)	50.0	122.0	320.7	103.6
C_7H_{16} (heptane)	(ℓ)	−224.4	1.0	328.6	224.3
C_8H_{18} (octane)	(ℓ)	−249.9	6.4	361.1	–
CH_3OH (methanol)	(ℓ)	−238.66	−166.27	126.8	81.6
CH_3OH (methanol)	(g)	−200.66	−161.96	239.81	43.89
C_2H_5OH (ethanol)	(ℓ)	−277.69	−174.78	160.7	111.46
C_2H_5OH (ethanol)	(g)	−235.10	−168.49	282.70	65.44
C_6H_5OH (phenol)	(s)	−165.0	−50.9	146.0	–
$HCOOH$ (methane acid)	(ℓ)	−424.72	−361.35	128.95	99.04
CH_3COOH (ethane acid)	(ℓ)	−484.5	−389.9	159.8	124.3
CH_3COOH (ethane acid)	(aq)	−485.76	−396.46	178.7	–
CH_3COO^- (ethanoate ion)	(aq)	−486.01	−369.31	86.6	−6.3
C_6H_5COOH (benzoic acid)	(s)	−385.1	−245.3	167.6	146.8
$HCHO$ (methanal)	(g)	−108.57	−102.53	218.77	35.40
CH_3CHO (ethanal)	(ℓ)	−192.30	−128.12	160.2	–
CH_3CHO (ethanal)	(g)	−166.19	−128.86	250.3	57.3
CH_3COCH_3 (propanone)	(ℓ)	−248.1	−155.4	200.4	124.7

Ref.: Atkins, P.W.: "Physical Chemistry", Third ed., Oxford University Press, Oxford 1986.

Molar heat capacity

The table specifies for selected **elements** and **compounds** the molar heat capacity c_p (J/mol K) as a function of the thermodynamic temperature T (K). The temperature dependence of the heat capacity is given on the form

$$c_p = a + b \cdot T + d \cdot T^{-2}$$

Furthermore, the table specifies the range of the given functional expression as well as $c_p(298)$ calculated using the given constants.

Substance		a $\left(\frac{J}{mol\,K}\right)$	$b \cdot 10^3$ $\left(\frac{J}{mol\,K^2}\right)$	$d \cdot 10^{-5}$ $\left(\frac{J\,K}{mol}\right)$	Range (K)	$c_p(298)$ $\left(\frac{J}{mol\,K}\right)$
Al	(s)	20.08	13.47	–	273 - 931	24.1
$AlCl_3$	(s)	55.44	117.15	–	273 - 465	90.4
Al_2O_3	(s)	92.38	37.53	−21.86	273 - 1973	79.0
$Al_2 \cdot SiO_2$ (sillimanite)	(s)	170.67	19.93	−41.54	273 - 1573	129.9
$Al_2O_3 \cdot SiO_2$ (andalusite)	(s)	183.93	8.05	−45.44	273 - 1573	135.2
$3Al_2O_3 \cdot 2SiO_2$ (mullite)	(s)	249.58	280.33	–	273 - 576	333.1
Sb	(s)	23.05	7.45	–	273 - 903	25.3
$SbCl_3$	(s)	43.10	213.80	–	273 - 346	106.8
Sb_2O_3	(s)	79.91	71.55	–	273 - 929	101.2
Sb_2S_3	(s)	101.25	55.23	–	273 - 821	117.7

Substance		a $\left(\frac{J}{mol\,K}\right)$	$b \cdot 10^3$ $\left(\frac{J}{mol\,K^2}\right)$	$d \cdot 10^{-5}$ $\left(\frac{J\,K}{mol}\right)$	Range (K)	$c_p(298)$ $\left(\frac{J}{mol\,K}\right)$
As	(s)	21.63	9.79	–	273 - 1168	24.5
As_2O_3	(s)	35.02	203.34	–	273 - 548	95.6
$BaCl_2$	(s)	71.13	13.97	–	273 - 1198	75.3
$BaCO_3$	(s)	72.22	54.81	–	273 - 1083	88.6
$BaSO_4$	(s)	89.33	58.99	–	273 - 1323	106.9
Be	(s)	19.66	6.51	−5.06	273 - 1173	15.9
BeO	(s)	36.36	15.27	−13.10	273 - 1175	26.2
BiS	(s)	22.51	10.88	–	273 - 544	25.8
Bi_2O_3	(s)	97.36	46.23	–	273 - 777	111.1
B	(s)	6.44	18.41	–	273 - 1174	11.9
B_2O_3	(s)	21.51	133.99	–	273 - 513	61.4
Cd	(s)	22.84	10.32	–	273 - 594	25.9
CdO	(s)	40.38	8.70	–	273 - 2086	43.0
CdS	(s)	53.97	3.77	–	273 - 1273	55.1
Ca	(s)	22.22	13.93	–	273 - 673	26.4
$CaCl_2$	(s)	70.71	16.15	–	273 - 1055	75.5
$CaCO_3$	(s)	82.34	49.75	−12.87	273 - 1033	82.7
CaF_2	(s)	61.50	15.90	–	273 - 1651	66.2
CaO	(s)	41.84	20.25	−4.52	273 - 1173	42.8
$CaO \cdot Al_2O_3 \cdot 2SiO_2$	(s)	264.14	62.76	−64.31	273 - 1673	210.5
$CaO \cdot MgO \cdot 2SiO_2$	(s)	227.86	24.04	−62.76	273 - 1573	164.4
$CaO \cdot SiO_2$ (wollastonite)	(s)	116.94	8.60	−31.20	273 - 1573	84.4
$CaO \cdot SiO_2$ (psd. wollast.)	(s)	106.61	17.29	−20.42	273 - 1673	88.8
$CaSO_4$	(s)	77.49	91.92	−5.56	273 - 1373	97.5
C (graphite)	(s)	11.18	10.95	−4.89	273 - 1373	8.9
C (diamond)	(s)	9.05	12.80	−5.45	273 - 1313	6.7
CO_2	(g)	43.26	11.46	−8.18	273 - 1200	37.5
Cs	(s)	8.20	76.15	–	273 - 301	30.9
CsCl	(s)	48.95	12.93	–	273 - 752	52.8
Cl_2	(g)	34.64	2.34	–	273 - 2000	35.3
Cr	(s)	20.25	12.34	–	273 - 1823	23.9
Cr_2O_3	(s)	108.78	16.74		273 - 2263	113.8
Co	(s)	21.42	13.95	–	273 - 1763	25.6
Cu	(s)	22.76	6.12		273 - 1357	24.6
CuO	(s)	45.48	14.96	−6.30	273 - 810	42.9
CuS	(s)	44.35	11.05	–	273 - 1273	47.6
Cu_2S	(s)	39.25	130.54	–	273 - 376	78.2
F_2	(g)	27.20	4.18	–	300 3000	28.4
Au	(s)	23.47	6.02	–	273 - 1336	25.3
H_2	(g)	27.70	3.39	–	273 - 2500	28.7
HCl	(g)	28.03	3.51	–	273 - 2000	29.1
H_2S	(g)	30.12	15.06	–	300 - 600	34.6
Fe	(s)	17.28	26.69	–	273 - 1041	25.2
Fe_3C (cementite)	(s)	105.31	9.33	–	273 - 1173	108.1
FeO	(s)	52.80	6.24	−3.19	273 - 1173	51.1

Substance		a $\left(\frac{J}{mol\,K}\right)$	$b \cdot 10^3$ $\left(\frac{J}{mol\,K^2}\right)$	$d \cdot 10^{-5}$ $\left(\frac{J\,K}{mol}\right)$	Range (K)	$c_p(298)$ $\left(\frac{J}{mol\,K}\right)$
Fe_2O_3 (hematite)	(s)	103.43	67.11	−17.72	273 - 1097	103.5
Fe_3O_4 (magnetite)	(s)	172.26	78.74	−40.98	273 - 1065	149.6
FeS	(s)	8.49	163.18	–	273 - 411	57.1
FeS_2 (pyrite)	(s)	44.77	55.90	–	273 - 773	61.4
FeSi	(s)	44.10	19.16	–	273 - 903	49.8
Pb	(s)	24.14	8.45	–	273 - 600	26.7
$PbCl_2$	(s)	66.44	34.94	–	273 - 771	76.9
PbO	(s)	43.22	13.31	–	273 - 544	47.2
PbO_2	(s)	53.14	32.64	–	273 - ?	62.9
PbS	(s)	44.48	16.78	–	273 - 873	49.5
Li	(s)	2.85	75.31	–	273 - 459	25.3
LiCl	(s)	46.02	14.18	–	273 - 887	50.2
$LiNO_3$	(s)	38.37	150.62	–	273 - 523	83.3
Mg	(s)	25.94	5.56	−2.84	273 - 923	24.4
$MgCl_2$	(s)	72.38	15.77	–	273 - 991	77.1
MgO	(s)	45.44	5.01	−8.73	273 - 2073	37.1
Mn	(s)	15.73	31.25	–	273 - 1108	25.0
$MnCl_2$	(s)	67.78	21.76	–	273 - 923	74.3
$HgCl_2$	(s)	64.02	43.10	–	273 - 553	76.9
HgS	(s)	45.61	15.27	–	273 - 853	50.2
Mo	(s)	23.81	7.87	−2.10	273 - 1773	23.8
MoO_3	(s)	63.18	50.63	–	273 - 1068	78.3
MoS_2	(s)	82.42	13.18	–	273 - 729	86.3
Ni	(s)	17.82	26.78	–	273 - 626	25.8
NiO	(s)	47.28	9.00	–	273 - 1273	50.0
NiS	(s)	38.70	26.78	–	273 - 597	46.7
$\mathbf{N_2}$	(g)	27.20	4.18	–	300 - 3000	28.4
NH_3	(g)	28.03	26.36	–	300 - 800	35.9
NH_4Cl	(s)	41.00	153.97	–	273 - 457	86.9
NO	(g)	33.68	0.97	−6.54	300 - 5000	26.6
$\mathbf{O_2}$	(g)	34.60	1.08	−7.85	300 - 5000	26.1
P (red)	(s)	0.88	75.31	–	273 - 472	23.3
Pt	(s)	24.77	4.85	–	273 - 1873	26.2
K	(s)	21.92	23.24	–	273 - 336	28.8
Si	(s)	24.02	2.58	−4.23	273 - 1174	20.0
SiC	(s)	37.20	12.18	−11.88	273 - 1629	27.5
SiO_2 (quartz)	(s)	45.48	36.45	−10.09	273 - 848	45.0
SiO_2 (cristobalite)	(s)	15.27	100.40	–	273 - 523	45.2
SiO_2 (amorphous)	(s)	53.61	18.70	−12.64	273 - 1973	45.0
S (rhombic)	(s)	15.19	26.78	–	273 - 368	23.2
S (monoclinic)	(s)	18.33	18.40	–	368 - 392	23.8
Sn	(s)	21.13	20.08	–	273 - 504	27.0
Zn	(s)	21.97	11.30	–	273 - 692	25.3
ZnO	(s)	47.70	6.07	−7.63	273 - 1573	40.9

Ref.: Perry, R.H.: "Chemical Engineers' Handbook", McGraw–Hill Kogakusha, Ltd., 1973.

Surface tension

pure substances and solutions

The table specifies **surface tension** σ (mN/m) for **water** as well as for selected aqueous **solutions** and **pure substances**. The unit (mN/m) corresponds numerically to the unit (dyn/cm) that often occurs in old tables.

1. Surface tension of water as a function of temperature.

θ	(°C)	−8	−5	0	5	10
σ	(mN/m)	77.0	76.4	75.6	74.9	74.22
θ	(°C)	15	20	25	30	40
σ	(mN/m)	73.49	72.75	71.97	71.18	69.56
θ	(°C)	50	60	70	80	100
σ	(mN/m)	67.91	66.18	64.4	62.6	58.9

2. Surface tension of aqueous solutions at constant temperature.

HCl	weight-%	1.78	3.52	6.78	16.97	35.29
(20 °C)	σ (mN/m)	72.55	72.45	72.25	71.75	65.75
HNO_3	weight-%	4.21	8.64	14.99	34.87	–
(20 °C)	σ (mN/m)	72.15	71.65	70.95	68.75	–
NaOH	weight-%	2.72	5.66	16.66	30.56	35.90
(18 °C)	σ (mN/m)	74.35	75.85	83.05	96.05	101.05
NaCl	weight-%	0.58	2.84	5.43	14.92	25.92
(20 °C)	σ (mN/m)	72.92	73.75	74.39	77.65	82.55
$MgCl_2$	weight-%	0.94	4.55	8.69	16.00	25.44
(20 °C)	σ (mN/m)	73.07	74.00	75.75	79.15	85.75
$Al_2(SO_4)_3$	weight-%	2.54	4.06	9.40	19.32	25.50
(25 °C)	σ (mN/m)	72.32	72.92	73.51	76.06	79.73
$MgSO_4$	weight-%	1.19	5.68	10.75	19.41	24.53
(20 °C)	σ (mN/m)	73.01	73.78	74.85	77.35	79.25
$NaCO_3$	weight-%	2.58	5.03	9.59	13.72	–
(20 °C)	σ (mN/m)	73.45	74.05	75.45	76.75	–
$NaNO_3$	weight-%	0.85	4.08	7.84	29.82	47.06
(20 °C)	σ (mN/m)	72.87	73.75	73.95	78.35	87.05
Methanol	vol-%	7.50	10.00	25.00	50.00	60.00
(20 °C)	σ (mN/m)	60.90	59.04	46.38	35.31	27.26
Ethanol	vol-%	5.00	10.00	24.00	48.00	60.00
(40 °C)	σ (mN/m)	54.92	48.25	35.50	28.93	26.18

3. Surface tension σ (mN/m) of pure substances at 20 °C.

Acetaldehyde	C_2H_4O	21.2	Acetic acid	$C_2H_4O_2$	27.8
Acetone	C_3H_6O	23.70	Aniline	C_6H_7N	42.9
Benzene	C_6H_6	28.85	*n*-Butyl alcohol	$C_4H_{10}O$	24.6
Chloroform	$CHCl_3$	27.14	Ethanol	C_2H_6O	22.75
Ethyl ether	$C_4H_{10}O$	17.01	Glycerol	$C_3H_8O_3$	63.4
Methanol	CH_4O	22.61	Phenol	C_6H_6O	40.9

Ref.: "Handbook of Chemistry and Physics", 64th ed., CRC Press Inc., Boca Raton, 1984.

Electrochemical standard potential

reduction potentials

The table specifies the **standard potential** $V^{\ominus}$ (volt) for **reduction** of selected metals. The reaction equations are given alphabetically.

Ag^{+}	$+\ e^-$	→	Ag	0.80	Hg_2^{++}	$+\ 2e^-$	→	2Hg	0.80
Al^{+++}	$+\ 3e^-$	→	Al	−1.66	K^{+}	$+\ e^-$	→	K	−2.93
Au^{+}	$+\ e^-$	→	Au	1.69	Li^{+}	$+\ e^-$	→	Li	−3.04
Au^{+++}	$+\ 3e^-$	→	Au	1.50	Mg^{++}	$+\ 2e^-$	→	Mg	−2.37
Be^{++}	$+\ 2e^-$	→	Be	−1.85	Mn^{++}	$+\ 2e^-$	→	Mn	−1.19
Ca^{++}	$+\ 2e^-$	→	Ca	−2.87	Mo^{+++}	$+\ 3e^-$	→	Mo	−0.20
Cd^{++}	$+\ 2e^-$	→	Cd	−0.40	Na^{+}	$+\ e^-$	→	Na	−2.71
Ce^{+++}	$+\ 3e^-$	→	Ce	−2.48	Pb^{++}	$+\ 2e^-$	→	Pb	−0.13
Co^{++}	$+\ 2e^-$	→	Co	−0.28	Pt^{++}	$+\ 2e^-$	→	Pt	1.12
Cr^{++}	$+\ 2e^-$	→	Cr	−0.91	Sn^{++}	$+\ 2e^-$	→	Sn	−0.14
Cr^{+++}	$+\ 3e^-$	→	Cr	−0.74	Te^{++++}	$+\ 4e^-$	→	Te	0.57
Cu^{+}	$+\ e^-$	→	Cu	0.52	Ti^{++}	$+\ 2e^-$	→	Ti	−1.63
Cu^{++}	$+\ 2e^-$	→	Cu	0.34	Tl^{+}	$+\ e^-$	→	Tl	−0.34
Fe^{++}	$+\ 2e^-$	→	Fe	−0.45	V^{++}	$+\ 2e^-$	→	V	−1.18
Fe^{+++}	$+\ 3e^-$	→	Fe	−0.04	Zn^{++}	$+\ 2e^-$	→	Zn	−0.76

Ref.: "Handbook of Chemistry and Physics", 64th ed., CRC Press Inc., Boca Raton, 1984.

Water and water vapour

vapour pressure table etc.

In the temperature range, 0 – 195 °C, the table specifies the following data for water vapour $H_2O(g)$: partial pressure p_s (Pa) of **saturated water vapour**, the ratio $p_s/p^{\ominus}$, **absolute moisture content** u (g/m^3) in moisture saturated atmospheric air; for water $H_2O(\ell)$, the table denotes: The **density** ϱ_ℓ (kg/m^3) of water, the **specific volume** v_ℓ (m^3/kg) of water, and the **evaporation enthalpy** ΔH_e of water given in (kJ/kg).

1. Water vapour and water, temperature range 0 °C – 99 °C.

0 °C

Temperature	p_s	$p_s/p^{\ominus}$	u	ϱ_ℓ	$10^3 \cdot v_\ell$	ΔH_e
°C	(Pa)	(–)	$\left(\frac{g}{m^3}\right)$	$\left(\frac{kg}{m^3}\right)$	$\left(\frac{m^3}{kg}\right)$	$\left(\frac{kJ}{kg}\right)$
0	611.28	0.00604	4.86	999.78	1.00022	2500.5
1	657.16	0.00649	5.20	999.85	1.00015	2498.2
2	706.05	0.00697	5.56	999.90	1.00010	2495.8
3	758.13	0.00748	5.95	999.93	1.00007	2493.4
4	813.59	0.00803	6.36	999.95	1.00005	2491.1
5	872.60	0.00861	6.80	999.94	1.00006	2488.7
6	935.37	0.00923	7.27	999.92	1.00008	2486.3
7	1002.09	0.00989	7.76	999.89	1.00011	2484.0
8	1072.97	0.01059	8.28	999.84	1.00016	2481.6
9	1148.25	0.01133	8.82	999.77	1.00023	2479.3

Temperature °C	p_s (Pa)	$p_s/p^{\ominus}$ (−)	u $\left(\frac{\text{g}}{\text{m}^3}\right)$	ϱ_ℓ $\left(\frac{\text{kg}}{\text{m}^3}\right)$	$10^3 \cdot v_\ell$ $\left(\frac{\text{m}^3}{\text{kg}}\right)$	ΔH_e $\left(\frac{\text{kJ}}{\text{kg}}\right)$	
10	1228.1	0.01212	9.41	999.69	1.00031	2476.9	10 °C
11	1312.9	0.01296	10.02	999.60	1.00040	2474.5	
12	1402.7	0.01384	10.67	999.49	1.00051	2472.2	
13	1497.9	0.01478	11.35	999.37	1.00063	2469.9	
14	1598.8	0.01578	12.08	999.24	1.00076	2467.5	
15	1705.6	0.01683	12.84	999.09	1.00091	2465.1	
16	1818.5	0.01795	13.64	998.93	1.00107	2462.8	
17	1938.0	0.01913	14.49	998.76	1.00124	2460.4	
18	2064.4	0.02037	15.38	998.58	1.00142	2458.0	
19	2197.9	0.02169	16.32	998.39	1.00161	2455.7	
20	2338.8	0.02308	17.31	998.19	1.00182	2453.4	20 °C
21	2487.7	0.02455	18.35	997.97	1.00203	2451.0	
22	2644.7	0.02610	19.44	997.75	1.00226	2448.6	
23	2810.4	0.02774	20.59	997.52	1.00249	2446.2	
24	2985.0	0.02946	21.80	997.27	1.00274	2443.9	
25	3169.1	0.03128	23.07	997.02	1.00299	2441.6	
26	3362.9	0.03319	24.40	996.75	1.00326	2439.2	
27	3567.0	0.03520	25.79	996.48	1.00353	2436.8	
28	3781.8	0.03732	27.26	996.20	1.00381	2434.4	
29	4007.8	0.03955	28.79	995.91	1.00411	2432.0	
30	4245.5	0.04190	30.40	995.61	1.00441	2429.6	30 °C
31	4495.3	0.04437	32.08	995.30	1.00472	2427.3	
32	4757.8	0.04696	33.85	994.99	1.00504	2425.0	
33	5033.5	0.04968	35.70	994.66	1.00537	2422.6	
34	5322.9	0.05253	37.63	994.33	1.00570	2420.2	
35	5626.7	0.05553	39.65	993.99	1.00605	2417.8	
36	5945.4	0.05868	41.76	993.64	1.00640	2415.4	
37	6279.5	0.06197	43.97	993.28	1.00676	2413.1	
38	6629.8	0.06543	46.28	992.92	1.00713	2410.7	
39	6996.9	0.06905	48.69	992.55	1.00751	2408.3	
40	7381.4	0.07285	51.21	992.17	1.00789	2405.9	40 °C
41	7784.0	0.07682	53.83	991.78	1.00829	2403.5	
42	8205.4	0.08098	56.57	991.39	1.00869	2401.0	
43	8646.4	0.08533	59.43	990.99	1.00909	2398.7	
44	9107.6	0.08989	62.41	990.58	1.00951	2396.3	
45	9589.8	0.09464	65.52	990.17	1.00993	2393.9	
46	10093.8	0.09962	68.75	989.74	1.01036	2391.5	
47	10620.5	0.10482	72.12	989.32	1.01080	2389.1	
48	11170.6	0.11025	75.63	988.88	1.01124	2386.6	
49	11744.9	0.11591	79.28	988.44	1.01170	2384.3	
50	12344	0.12183	83.08	987.99	1.01215	2381.9	50 °C
51	12970	0.12800	87.03	987.54	1.01262	2379.5	
52	13623	0.13445	91.14	987.08	1.01309	2377.0	
53	14303	0.14116	95.41	986.61	1.01357	2374.6	
54	15012	0.14816	99.85	986.13	1.01406	2372.2	

	Temperature	p_s	$p_s/p^\ominus$	u	ϱ_ℓ	$10^3 \cdot v_\ell$	ΔH_e
	°C	(Pa)	(–)	$\left(\frac{\text{g}}{\text{m}^3}\right)$	$\left(\frac{\text{kg}}{\text{m}^3}\right)$	$\left(\frac{\text{m}^3}{\text{kg}}\right)$	$\left(\frac{\text{kJ}}{\text{kg}}\right)$
	55	15752	0.15546	104.46	985.65	1.01455	2369.8
	56	16522	0.16306	109.25	985.17	1.01505	2367.4
	57	17324	0.17097	114.23	984.68	1.01556	2364.9
	58	18159	0.17922	119.39	984.18	1.01608	2362.5
	59	19028	0.18779	124.75	983.67	1.01660	2360.0
60 °C	60	19932	0.19671	130.30	983.16	1.01712	2357.7
	61	20873	0.20600	136.07	982.65	1.01766	2355.2
	62	21851	0.21565	142.04	982.13	1.01820	2352.8
	63	22868	0.22569	148.24	981.60	1.01875	2350.3
	64	23925	0.23612	154.65	981.07	1.01930	2347.9
	65	25022	0.24695	161.30	980.53	1.01986	2345.4
	66	26163	0.25821	168.19	979.98	1.02043	2342.9
	67	27347	0.26989	175.32	979.43	1.02100	2340.5
	68	28576	0.28202	182.69	978.88	1.02158	2338.1
	69	29852	0.29462	190.33	978.32	1.02216	2335.6
70 °C	70	31176	0.30768	198.23	977.75	1.02276	2333.1
	71	32549	0.32123	206.40	977.18	1.02336	2330.6
	72	33972	0.33528	214.85	976.60	1.02396	2328.1
	73	35448	0.34984	223.58	976.02	1.02457	2325.6
	74	36978	0.36494	232.61	975.42	1.02519	2323.1
	75	38563	0.38059	241.94	974.84	1.02581	2320.6
	76	40205	0.39679	251.58	974.24	1.02644	2318.2
	77	41905	0.41357	261.53	973.64	1.02708	2315.7
	78	43665	0.43094	271.80	973.03	1.02772	2313.2
	79	45487	0.44892	282.41	972.41	1.02837	2310.7
80 °C	80	47373	0.46754	293.36	971.79	1.02902	2308.1
	81	49324	0.48679	304.65	971.17	1.02969	2305.6
	82	51342	0.50671	316.31	970.54	1.03035	2303.1
	83	53428	0.52729	328, 32	969.91	1.03103	2300.6
	84	55585	0.54854	340.72	969.27	1.03171	2298.0
	85	57815	0.57059	353.49	968.62	1.03239	2295.5
	86	60119	0.59333	366.66	967.98	1.03308	2293.0
	87	62499	0.61682	380.23	967.32	1.03378	2290.4
	88	64958	0.64109	394.20	966.66	1.03449	2287.9
	89	67496	0.66613	408.60	966.00	1.03520	2285.3
90 °C	90	70117	0.69200	423.43	965.33	1.03591	2282.7
	91	72823	0.71871	438.70	964.66	1.03664	2280.2
	92	75614	0.74625	454.41	963.98	1.03736	2277.6
	93	78495	0.77469	470.58	963.30	1.03810	2274.9
	94	81465	0.80400	487.23	962.61	1.03884	2272.3
	95	84529	0.83424	504.3	961.92	1.03959	2269.7
	96	87688	0.86541	522.0	961.22	1.04034	2267.2
	97	90945	0.89756	540.1	960.52	1.04110	2264.6
	98	94301	0.93068	558.7	959.82	1.04186	2261.9
	99	97759	0.96481	577.8	959.11	1.04264	2259.3

2. Water and water vapour, temperature range 100 °C – 195 °C.

Temperature °C	p_s (Pa)	$p_s/p^\ominus$ (–)	u $\left(\frac{\text{g}}{\text{m}^3}\right)$	ϱ_ℓ $\left(\frac{\text{kg}}{\text{m}^3}\right)$	$10^3 \cdot v_\ell$ $\left(\frac{\text{m}^3}{\text{kg}}\right)$	ΔH_e $\left(\frac{\text{kJ}}{\text{kg}}\right)$
100	101320	1.0000	597.5	958.39	1.04341	2256.6
105	120790	1.1921	704.2	954.75	1.04739	2243.4
110	143240	1.4137	826.2	951.00	1.05153	2230.0
115	169020	1.6681	964.3	947.13	1.05582	2216.3
120	198480	1.9588	1120.8	943.16	1.06027	2202.4
125	232010	2.2898	1297.2	939.07	1.06488	2188.3
130	270020	2.6649	1495.4	934.88	1.06965	2174.0
135	312930	3.0884	1717.2	930.59	1.07459	2159.4
140	361190	3.5647	1964.7	926.18	1.07970	2144.6
145	415290	4.0986	2240.0	921.67	1.08498	2129.5
150	475720	4.6950	2545.4	917.06	1.09044	2114.1
155	542990	5.3589	2883.4	912.33	1.09609	2098.4
160	617660	6.0958	3256.4	907.50	1.10193	2082.4
165	700290	6.9113	3667.0	902.56	1.10796	2065.9
170	791470	7.8112	4118.1	897.51	1.11420	2049.2
175	891800	8.8014	4612.7	892.34	1.12065	2032.1
180	1001900	9.8880	5154	887.06	1.12732	2014.6
185	1122500	11.0782	5745	881.67	1.13422	1996.6
190	1254200	12.3780	6390	876.15	1.14136	1978.2
195	1397600	13.7932	7091	870.51	1.14875	1959.5

Ref.: Haar, L., Gallagher, J.S., & Kell, G.S., : "NBS/NRC Wasserdampftafeln", Springer–Verlag, Berlin 1988.

100 °C

150 °C

3. Water vapour over ice and supercooled water, –1 °C to – 15 °C.

Ice °C	p_s (Pa)	$p_s/p^\ominus$ (–)	Water °C	p_s (Pa)	$p_s/p^\ominus$ (–)
– 1	562.2	0.005549	– 1	567.7	0.005603
– 2	517.3	0.005105	– 2	527.4	0.005205
– 3	475.7	0.004695	– 3	489.7	0.004833
– 4	437.3	0.004316	– 4	454.6	0.004487
– 5	401.7	0.003964	– 5	421.7	0.004162
– 6	368.6	0.003638	– 6	390.8	0.003857
– 7	338.2	0.003338	– 7	362.0	0.003572
– 8	310.1	0.003061	– 8	335.2	0.003308
– 9	284.1	0.002804	– 9	310.1	0.003061
– 10	260.0	0.002566	– 10	286.5	0.002828
– 11	238.0	0.002349	– 11	264.9	0.002614
– 12	217.6	0.002147	– 12	244.5	0.002413
– 13	198.7	0.001961	– 13	225.4	0.002225
– 14	181.5	0.001791	– 14	208.0	0.002053
– 15	165.5	0.001633	– 15	191.5	0.001889

Ref.: "Handbook of Chemistry and Physics", 64th ed., CRC Press Inc., Boca Raton, 1984.

–1 °C

–10 °C

Set-up from the 18th century used for exact plotting of the boiling point of water during calibration of laboratory thermometers.

APPENDIX C

Solutions to check-up questions and exercises

This appendix contains a systematically arranged list of solutions to all **check-up questions** and **exercises** in the book. The solutions are set up consecutively with reference to **chapter** and **section**. For each individual section, the number of the exercise is boxed; for example: [5.] gives the solution to exercise no. 5 in the section concerned. The results given are only to a minor degree followed by explanations with regard to method and intermediary results.

Anders Celsius (1701-1744)

Swedish astronomer and professor at the University of Uppsala; known as the inventor of the Celsius temperature scale.

Contents

Solutions to check-up questions and exercises

In the following section, solutions to the check-up questions and exercises posed in Chapters 1-6 are given.

1. Systems of matter

1.1 Atoms

1. protium ^1_1H; deuterium ^2_1H; tritium ^3_1H. **2.** $^{37}_{17}Cl$: A = 37, N = 20, Z = 17; $^{56}_{26}Fe$: A = 56, N = 30, Z = 26; ^7_3Li : A = 7, N = 4, Z = 3. **3.** $4.5 \cdot 10^{22}$ electrons. **4.** $^{98}_{42}Mo$; $^{40}_{20}Ca$. **5.** A_1 and A_3 are isotopes of the same element.

1.2 Relative atomic mass

1. $A_r = 107.869$. **2.** $m = 4.4805 \cdot 10^{-26}$ kg. **3.** $A_r = 63.929$. **4.** $A_r = 12$; $M = 1.9927 \cdot 10^{-26}$ kg. **5.** ^1_1H : 99.986 %; ^2_1H : 0.014 %.

1.3 Relative molecular mass

1. $BaCl_2$ (crystal); SiO_2 (crystal); H_2O (crystal); in these cases we are dealing with extensive, organized crystal structures. **2.** $M_r = 12$. **3.** $CaSO_4 \cdot 2H_2O$: $M_r = 172.18$ $3CaO \cdot Al_2O_3$: $M_r = 270.20$. **4.** $N = 3.34 \cdot 10^{22}$ water molecules. **5.** $N = 5.15 \cdot 10^{21}$ Na^+ ions.

1.4 Amount of substance - the mole

1. a) 1 mole of atoms; b) 6 moles of protons; c) 6 moles of neutrons; d) 12 moles of nucleons; e) 6 moles of electrons. **2.** a) 0.5 mole of Fe atoms and 1 mole of S atoms; b) 4.17 moles of Fe atoms and 8.33 moles of S atoms. **3.** 35.8 moles of electrons. **4.** $9.6 \cdot 10^8$ tons, corresponding to the total world production of cement for approximately 10 years. **5.** 111.2 moles of electrons.

1.5 Molar mass

1. "Oxygen" can be O, O_2 and O_3; "ferric oxide" can be FeO, Fe_2O_3 and F_3O_4; "stannic chloride" can be SnCl and $SnCl_2$ **2.** $M = 18.016 \cdot 10^{-3}$ kg/mol = 18.016 g/mol. **3.** 45.44 mol O atoms. **4.** a) oxygen O_2; b) same number of O atoms in both cases. **5.** Hematite Fe_2O_3 is formed.

1.6 Mixture of substances

1. $x(Pb) = 0.46$; $x(Sn) = 0.54$. **2.** $Fe_2O_3 : w_{Fe} = 0.70$; $Fe_3O_4 : w_{Fe} = 0.72$. **3.** $[CaCl_2] = 0.180\,mol/\ell$; $[Cl^-] = 0.360\,mol/\ell$. **4.** $[H_2O] = 1.28\,mol/m^3$; $\varrho(H_2O(g)) = 0.0231\,kg/m^3$. **5.** $[H_2SO_4] = 6.75\,mol/\ell$; $\tilde{m} = 9.412\,mol/kg$.

1.7 The ideal gas law

1. a) $V(H_2) = 11.93\,m^3$; b) $V(CO_2) = 0.5466\,m^3$. **2.** $\varrho = 1.204\,kg/m^3$. **3.** $A_r(N) = 14.01$. **4.** $N = 1.94 \cdot 10^{18}$ molecules. **5.** $\alpha = 3.66 \cdot 10^{-3}\,K^{-1}$.

1.8 Ideal gas mixture

1. 278.7 g $O_2(g)$. **2.** $p = 100$ kPa. **3.** Partial pressure of $H_2O(g)$: 3396 Pa. **4.** 58.2 kg of water vapour. **5.** $\kappa = 9.87 \cdot 10^{-6}\,Pa^{-1}$.

1.9 Real gases

1. The intermolecular forces are much weaker between He(g), than between $H_2O(g)$! **2.** 34 Å at 1 atm; 16 Å at 10 atm; 11 Å at 30 atm. **3.** 0.9 Pa. **4.** As gas: $1.23 \cdot 10^5$ Pa; as ideal gas: $1.24 \cdot 10^5$ Pa. **5.** 0.07 % $\simeq$ 0.1 %.

1.10 Intermolecular forces

1. Dipole bond: 0.1-1.0 kJ/mol; hydrogen bond: 20-30 kJ/mol; primary bonds: 100-1000 kJ/mol. **2.** Both of the OH groups are hydrogen bound to the O atom on the two other water molecules so that a three-dimensional network of hydrogen bound water molecules is formed. **3.** ca. 9.233 Å. **4.** 1 atm: $\bar{r} = 34.4$ Å, $\Phi(\bar{r}) = -0.0049$ J/mol; 10 atm: $\bar{r} = 16.0$ Å, $\Phi(\bar{r}) = -0.49$ J/mol; 50 atm: $\bar{r} = 9.3$ Å, $\Phi(\bar{r}) = -12.2$ J/mol. **5.** By reduction we get: $r^* = r_0/\sqrt[6]{2} \Rightarrow r^* = 2.56$ Å for He(g), and $r^* = 3.70$ Å for N_2(g).

1.11 Critical temperature

1. CO_2(g) can liquefy because its critical temperature is higher than room temperature; N_2(g) and O_2(g) cannot liquefy at room temperature. **2.** (1.16): $-RT/(\vartheta - b)^2 + 2a/\vartheta^3 = 0$; (1.17): $2RT/(\vartheta - b)^3 - 6a/\vartheta^4 = 0$. **3.** $p_c \simeq 1042$ atm; $T_c \simeq 1450\,°C$. **4.** $a = 0.438$ m^{6}Pa/mol^2; $b = 57.9 \cdot 10^{-6}$ m^3/mol. **5.** Calculated for butane: $T_c \simeq 426$ K $= 153\,°C$ is above room temperature, and $p_c \simeq 36$ atm; therefore, butane can liquefy and be delivered as liquid gas.

1.12 SI units

1. $5.4 \cdot 10^{-10}$ s $= 0.54$ ns; 0.023 GJ $= 23 \cdot 10^6$ J. **2.** [kg/Pa m s] = [s]. **3.** $c = 8.55 \cdot 10^{-4}$ mol/ℓ $= 855\,\mu$mol/ℓ. **4.** $1.02 \cdot 10^{-2}$ eV. **5.** d $= 10^5 = 10^4$ nm $= 10^{-2}$ mm $= 10^{-5}$ m.

Exercises, chapter 1

1.1 $M = 68.119$ g/mol. **1.2** $n = 4.38$ mol. **1.3** 55.846. **1.4** Fe_2O_3: w(Fe) $= 0.698$; x(Fe) $= 0.40$; Fe_3O_4: w(Fe) $= 0.724$; x(Fe) $= 0.43$. **1.5** 582 g of water vapour.

1.6 Fe_3C : w(Fe) = 6.7 %; x(Fe) = 0.25. **1.7** $[Cl^-] = 0.98$ mol/ℓ. **1.8** $\varrho = 260$ kg/m^3; $\lambda = 0.049$ W/m K. **1.9** $R = 62.36$ ℓ · Torr/mol · K; $p = 876$ Torr. **1.10** a) $RH = 52$ %; b) $p = 316.9$ Pa; c) $u = 6.91$ g/m^3.

1.11 Ideal gas: 5.8 kg of N_2(g); real gas: 6.1 kg of N_2(g). **1.12** 1230 A. **1.13** a) Propane: $a = 0.8779$ m^{6}Pa/mol^2; b $= 84.45 \cdot 10^{-6}$ m^3/mol; butane: $a = 1.466$ m^6 Pa/mol^2; $b = 122.6 \cdot 10^{-6}$ m^3/mol; b) propane: $T_c = 370$ K $= 97\,°C$; $p_c = 4.56 \cdot 10^6$ Pa $= 45$ atm; butane: $T_c = 426$ K $= 153\,°C$; $p_c = 3.61 \cdot 10^6$ Pa $= 36$ atm; c) Both propane and butane can liquefy at room temperature, because $T_c >$ room temperature. **1.14** $p_2 = 0.121 \cdot 10^6$ Pa $\simeq 1.20$ atm; $\Delta p = 0.020 \cdot 10^6$ Pa $\simeq 0.20$ atm. **1.15** $\Delta V = -118.1$ mℓ; $\Delta V/V_0 \cdot 100\,\% = 10.6$ %.

1.16 Increase in solid substance volume: 272 %. **1.17** a) 233.8 g NaCl; b) 20.3 wt-% NaCl; c) $\widetilde{m} = 4.37$ mol/kg of solvent. **1.18** Ca^{++} and Cl^- ions carry $+1.74 \cdot 10^5$ C and $-1.74 \cdot 10^5$ C, respectively. 1 A shall run for $1.74 \cdot 10^5$ s $\simeq 48$ hours! **1.19** a) 34.2 Å; b) 129 Å; 1.8 %.

2. Thermodynamic concepts

2.1 Thermodynamic system

1. Closed system: can exchange energy, but not matter **2.** Isolated system: can neither exchange energy nor matter. **3.** Closed system: can exchange energy, but not matter. **4.** Open system: exchanges both energy and matter. **5.** Open system: has exchanged both energy and matter.

2.2 Description of state

1. 2, e.g. (p, T), because then V is given and the state is fully described. **2.** Pressure p, temperature T, volume V, mass m, density ϱ, mole fraction x(NaCl), mole fraction $x(H_2O)$, molar concentration [NaCl], molality $\widetilde{m}$(NaCl). **3.** No! - At 20 °C, the partial pressure of saturated water vapour is 2338.8 Pa > 1705 Pa, i.e. the water will evaporate spontaneously. **4.** Either: all of the $Ca(OH)_2$ is dissolved homogeneously in the water. Or: a homogeneous, saturated solution of $Ca(OH)_2$ is formed with the surplus, solid $Ca(OH)_2$ as precipitate. **5.** Cooling of water from 20 °C to 0 °C is a state of non-equilibrium; transformation of water into ice at 0 °C is a state of non-equilibrium; cooling of ice from 0 °C to −18 °C is a state of non-equilibrium; ice at −18 °C is a state of equilibrium.

2.3 Thermodynamic variables

1. Intensive: density ϱ, viscosity η, compressibility κ, mole fraction x. Extensive: amount of substance n, mass m, volume V. **2.** Intensive, because the value per m is independent of the magnitude of the system. **3.** $(p + a/\vartheta^2)(\vartheta - b) = RT$, where $\vartheta = V/n$ denotes the molar gas volume. **4.** By molar mass M (kg/mol). **5.** Intensive: pressure p, temperature T, density ϱ, mole fraction $x(O_2)$, molar concentration $c(N_2)$. Extensive: volume V, total mass m, mass $m(O_2)$, amount of substance $n(N_2)$, amount of substance $n(O_2)$.

2.4 Temperature

1. 70.5 °F = 21.4 °C = 294.5 K; −63.5 °F = −53.1 °C = 220.1 K; 896 °F = 480 °C = 753 K. **2.** 0.00 °C = 32.00 °F = 273.15 K; 25.0 °C = 77.0 °F = 298.2 K; 100 °C = 212 °F = 373 K. **3.** −40 °C = −40 °F. **4.** $24 \cdot 10^{-6}\,°C^{-1} = 24 \cdot 10^{-6}\,K^{-1} = 13.3 \cdot 10^{-6}\,°F^{-1}$. **5.** $T_R = \Theta(°F) + 459.67$; $\Theta(°F) = T_R - 459.67$.

2.5 Work

1. $W_{1.2} = mg \cdot (h_2 - h_1) = 9.81$ kJ. **2.** $W_{1.2} = 2.46$ kJ. **3.** $W_{1.2} = 512.3$ kJ. **4.** $W = -2500$ J. **5.** $W_{1.2} = 1.94$ J.

2.6 Heat

1. Because of the low coefficient of thermal expansion of solid substances, the volume work $\delta W = -pdV$ during heating of a solid substance is negligible compared to the heat $\delta Q = cdT$. **2.** Mole-specific heat capacity (J/mol K) divided by molar mass M (kg/mol) gives the mass-specific heat capacity (J/kg K). **3.** Mole-specific: c = 25.1 J/mol K; mass-specific: c = 0.449 kJ/kg K. **4.** $1.29 \cdot 10^5$ kJ **5.** 359.7 kJ.

2.7 Thermodynamic process

1. a) Yes! - An isolated system can neither exchange heat nor work, i.e. the system is adiabatic ($Q = 0$). b) No! - An adiabatic system ($Q = 0$) can, for example, exchange work with its surroundings and thus, it is not necessarily an isolated system! **2.** a) During reversible melting, the temperature of ice and water deviates infinitesimally dT from the melting point of ice T_{mp}, and the system is in temperature equilibrium. b) During irreversible melting, the water temperature is $> T_{mp}$, and the system is not in temperature equilibrium. **3.** The process is irreversible because the system is not in solution equilibrium with NaCl - the concentration $c(NaCl)$ is not identical over the entire system. **4.** a) In a VT diagram, the line: $V = k \cdot T$ is seen, i.e. a straight line through the origo. b) A pV diagram shows a straight line parallel with the V axis. c) A pT diagram shows a straight line parallel with the T axis. **5.** $W = -3326$ J.

Exercises, chapter 2

2.1 Intensive variables: modulus of elasticity E, viscosity η, molar mass M, coefficient of thermal expansion α. — Extensive variables: mass m, electric charge Q, cement content C, elongation $\Delta\ell$. — $\varepsilon = \Delta\ell/\ell$ where the strain ε is an intensive variable; $\varrho = m/V$ where the density ϱ is an intensive variable. **2.2** 1) 57.5 °F = 14.2 °C; 2) 385.2 K = 233.5 °F; 3) 33.5 R = −426.2 °F = −254.5 °C 4) 1052 °F = 840 K; 5) 852.4 K = 1534 R; 6) 4.2 K = −269.0 °C. **2.3** c = 0.501 kJ/kg K. **2.4** θ_m = 418 °C; $\alpha = 32.4 \cdot 10^{-6}$ K; λ = 197 W/m K. **2.5** C = 14.0 kJ/kg; Q_h = 791 kJ; Q = 474 kJ/kg of cement.

2.6 $W_{1.2} = 21$ J. **2.7** $W_{1.2} = 70.2$ kJ. **2.8** $(p + a \cdot c_m^2)(1/c_m - b) = RT$; $p = 1.03 \cdot 10^5$ Pa. **2.9** $M = 18.03$ g/mol; $c_m = 0.522\,\text{mol/m}^3$. **2.10** $W = (V_b - V_a) \cdot Q = 12.0$ J.

2.11 $W_{1.2} = 304$ J. **2.12** C = 0.582 kJ/K; c = 0.554 kJ/kg K. **2.13** $\Delta\theta_a$ = 64.2 °C. **2.14** Surface work: W = 103 J. **2.15** System: C_p = 20.9 kJ/K; Mole-specific: c_p = 75.3 J/mol K; Mass-specific: c_p = 4.18 J/g K

2.16 Q = 165 kJ **2.17** Molar mass: M = 29.0 g/mol; $W_{1.2} = -78.3$ J; $p_2 = 84.1 \cdot 10^3$ Pa. **2.18** Electrical work: $W = 64.3 \cdot 10^6$ J = 17.9 kWh. **2.19** a) C = 44.5 J/mol K; b) Q = 569 kJ; c) Q = 407 kJ. **2.20** Electrical work: $W = 14.6 \cdot 10^6$ J = 4.05 kWh.

3. First law

3.1 Energy

1. As increase in potential energy between atoms. **2.** Chemical energy is transformed into molecular kinetic energy. **3.** a) $\Delta V = nc_V\Delta T = 4000\,\text{J}$. b) The system is supplied with molecular kinetic energy. **4.** The energy of an ideal gas is molecular kinetic energy only. – The energy of a real gas is molecular kinetic energy + potential energy between gas particles **5.** a) In an isolated system, the total energy E_{total} is constant. b) The potential energy E_p decreases. c) The molecular kinetic energy E_k increases, i.e. the temperature in the system increases.

3.2 First law

1. The isolated system has no exchange of energy or substance with its surroundings, and therefore: $\Delta U = Q + W = 0$. **2.** $\Delta U = Q + W = 1025 - 625 = 400\,\text{kJ}$; note that W done on the system is assumed to be positive. **3.** $Q = nc_p\Delta T = 416\,\text{J}$; $W = -p\Delta V = -166\,\text{J}$; $\Delta U = Q + W = 250\,\text{J}$. **4.** $Q = -840\,\text{kJ}$; $W = 0$ (isochoric); $\Delta U = q + W = -840\,\text{kJ}$. **5.** $\Delta U = \frac{1}{2}F_{\text{max}} \cdot \Delta\ell_{\text{max}} = 39\,\text{J}$.

3.3 Internal energy U

1. $\Delta U = 379.5\,\text{J/mol}$. **2.** a) No! – It will be unclear to what the amount of substance mol refers! b) Yes! – Per mol of NaCl. c) Yes! – Per mol of NaCl. **3.** $\Delta U = 0$, because the system is isolated. **4.** $W = \Delta U - Q = -1800\,\text{kJ}$, i.e. the system has done work on its surroundings. **5.** $\Delta U = 3510\,\text{J}$; $\Delta U = 190\,\text{J/mol}$; $C_V = 468\,\text{J/K}$; $c_V = 25.3\,\text{J/mol K}$.

3.4 Enthalpy II

1. $H - U = pV$: a) 2479 J; b) 1.8 J; c) 0.7 J; **2.** $\Delta H = 1000\,\text{J}$; $\Delta H = 18.02\,\text{J/mol K}$. **3.** If (pV) = constant! **4.** $Q = 0$ (adiabatic) $\Rightarrow \Delta H = 0$ in an isobaric process. **5.** $\Delta H = 4660\,\text{J}$; $\Delta H = 252\,\text{J/mol}$; $C_p = 621\,\text{J/K}$; $c_p = 33.6\,\text{J/mol K}$.

3.5 Ideal gas

1. $\Delta U = \Delta H = 0$ for constant T. **2.** $\Delta H = 52.0\,\text{J/mol}$; $c_V = 12.5\,\text{J/mol K}$. **3.** $c_p = 27.1\,\text{J/mol K}$; $c_V = 18.8\,\text{J/mol K}$; $\Delta H = 54.2\,\text{J/mol}$. **4.** $\Delta H - \Delta U = \Delta(pV) = \Delta(nRT) = 99.8\,\text{J}$. **5.** $Q = -3436\,\text{J}$.

3.6 Isothermal change of state

1. In an isolated system, $Q = 0$ and $W = 0$; this means that: $\Delta U = Q + W = 0 = U(T)$ cf. eqn. (3.23), so that the temperature T = constant and $\Delta T = 0$; $\Delta U = Q + W = 0$; $\Delta H = \Delta H(T) = 0$ because the process is isothermal. **2.** $Q = -8447\,\text{J}$; $W = 8447\,\text{J}$; $\Delta U = 0\,\text{J}$; $\Delta H = 0\,\text{J}$ **3.** $Q = -5000\,\text{J}$; $W = 5000\,\text{J}$; $\Delta U = 0\,\text{J}$ (isothermal); $\Delta H = 0\,\text{J}$ (isothermal). **4.** Ideal gas: $p_1V_1 = p_2V_2$ = constant in an isothermal process; $W_{1.2} = nRT \cdot \ln(p_2/p_1)$; $Q_{1.2} = -nRT \cdot \ln(p_2/p_1)$. **5.** No! – Since $\Delta U = Q + W = 0 + W = 0$ in an adiabatic system, the state cannot be changed isothermally!

3.7 Adiabatic change of state

1. $\gamma = c_p/c_V$; $c_p - c_V = R \Rightarrow$ a) $\gamma = c_p/(c_p - R) = 1.33$; b) $\gamma = (c_V + R)/c_p = 1.67$; c) $\gamma = c_p/(c_p - R) = 1.40$ ($R = 1.986\,\text{cal/mol°C}$). **2.** $T_2 = 109\,\text{K}$ ($-164\,°\text{C}$); $V_2 = 3.98\,\text{m}^3$. **3.** $C_V = 775\,\text{J/K}$. **4.** $\theta_2 = 789\,°\text{C}$. **5.** $\Delta U = -1495\,\text{J/mol}$; $W = -1495\,\text{J/mol}$.

3.8 Thermochemical equation

1. a) No! – It may, for example, be $CO_2(g)$ and $CO_2(aq)$ rendering different values of ΔH. b) No! – It may, for example, be $H_2O(\ell)$ and $H_2O(g)$ rendering different values of ΔH. **2.** a) Exothermic! b) Exothermic! c) Endothermic! d) Exothermic! **3.** a) $\Delta H = -45050\,\text{J/mol}$; b) $\Delta H = -6026\,\text{J/mol}$; c) $\Delta H = -51076\,\text{J/mol}$. **4.** Thermochemical equation: $\frac{3}{2}Fe_2O_3(s) + 3Al(s) \rightarrow 3Fe(s) + \frac{3}{2}Al_2O_3(s)$; $\Delta H = -1276.6\,\text{kJ/mol}$. **5.** ΔH will be larger! – Adding the process: $H_2O(\ell) \rightarrow H_2O(g)$ with $\Delta H = 44\,\text{kJ/mol}$ to the given reaction equation, we obtain the new reaction!

3.9 Standard enthalpy

1. a) −824.2 kJ/mol; b) −394.4 kJ/mol; c) −233.7 kJ/mol; d) 0.98 kJ/mol. **2.** Assuming a state of ideal gas, H is independent of the partial pressure: a) −241.3 kJ/mol; b) −242.3 kJ/mol. **3.** a) −285.8 kJ/mol. b) Because $V \cdot \Delta p = 16\,\text{J} = 0.016\,\text{kJ/mol}$, H is unchanged within the given, significant digits: −285.8 kJ/mol. c) 0.006 %. **4.** At the standard state and 298.15 K, $H_2(g)$ is the stable form of hydrogen; this is seen from $H^\ominus_{298} = 0$! – The dissociation of $H_2(g)$ at: $H_2(g) \rightarrow 2H(g)$ is strongly endothermic, i.e. energy must be supplied to break the bond H–H in the hydrogen molecule. **5.** The enthalpy content (a lower potential energy) of $H_2O(\ell)$ is lower than $H_2O(g)$ because the molecular space in the liquid state corresponds to a minimum of potential energy, cf. figure 1.20. – To transform (ℓ) into (g), the potential energy of the molecules must be increased; this corresponds to the evaporation heat supplied.

3.10 Reaction enthalpy

1. 44.8 kJ/mol at 5 °C; 43.0 kJ/mol at 50 °C; 41.1 kJ/mol at 95 °C. **2.** $\Delta_r H_{303} = -17.4\,\text{kJ/mol}$. **3.** $\Delta(\Delta_r H) = \Delta c_p \cdot (373 - 293) = -3.34\,\text{kJ/mol}$. **4.** $\Delta_r H = -241.8\,\text{kJ/mol} \Rightarrow Q_{\text{developed}} = -\Delta_r H = 241.8\,\text{kJ/mol}$. **5.** $\Delta_r H = 25.7\,\text{kJ/mol} \Rightarrow$ endothermic process!

Exercises, chapter 3

3.1 $Q_{1.2} = 9158\,\text{J}$; $W_{1.2} = -3658\,\text{J}$; $\Delta U = 5500\,\text{J}$. **3.2** $T_2 = 977\,\text{K}$ (704 °C), i.e. $\Delta T = 684\,\text{K}$. **3.3** $\Delta_r H^\ominus_{298} = -725.7\,\text{kJ/mol}$ (exothermic). $Q_{\text{developed}} = 725.7\,\text{kJ}$. **3.4** $H = -14916.8\,\text{kJ}$. **3.5** $\Delta H = 120.6\,\text{kJ}$.

3.6 a) $\Delta U = 300\,\text{J}$; b) $\Delta H = 482\,\text{J}$; c) $c_P = 22.0\,\text{J/mol K}$; d) $c_V = 13.7\,\text{J/mol K}$. **3.7** a) $\Delta U = -427\,\text{J}$; b) $c_V = 12.7\,\text{J/mol K}$; c) $c_p = 21.0\,\text{J/mol K}$. **3.8** a) $p_2 = 84438\,\text{Pa}$; b) $Q_{1.2} = 92.4\,\text{J}$; c) $W_{1.2} = -92.4\,\text{J}$. **3.9** a) $\theta_2 = 145.9\,°\text{C}$; b) $V_2 = 27.8\,\ell$; c) $\Delta U = 3223\,\text{J}$. **3.10** a) $H^\ominus_{338} = -282.8\,\text{kJ/mol}$; b) $H^\ominus_{273} = -1689.4\,\text{kJ/mol}$; c) $H^\ominus_{773} = -889.7\,\text{kJ/mol}$; d) $H^\ominus_{423} = 3.4\,\text{kJ/mol}$.

3.11 a) −1675.7 kJ/mol; b) −65.2 kJ/mol; c) −12.7 kJ/mol; d) −17.6 kJ/mol. **3.12** a) $\Delta_r H^\ominus_{1100} = 174.9\,\text{kJ/mol}$; b) $\Delta_r H^\ominus_{664} = 58.8\,\text{kJ/mol}$; c) $\Delta_r H^\ominus_{800} = 103.3\,\text{kJ/mol}$. **3.13** $\Delta_r H^\ominus_{298} = -851.5\,\text{kJ/mol}$ (exothermic). **3.14** $\Delta_r H^\ominus_{298} = -824.2\,\text{kJ/mol}$; $Q = 76\,\text{kJ}$. **3.15** 1614 J/g is developed.

3.16 C_2H_2: $48.3 \cdot 10^3$ kJ/kg; $52.2 \cdot 10^3$ kJ/m^3; H_2: $119.8 \cdot 10^3$ kJ/kg; $10.1 \cdot 10^3$ kJ/m^3. **3.17** 100 °C: (1) −1201 kJ/mol; (2) −1199 kJ/mol deviation 0.2 %. 500 °C: (1) −1169 kJ/mol; (2) −1152 kJ/mol deviation 1.5 %. 1000 °C: −1128 kJ/mol (2) −1089 kJ/mol deviation 3.5 %. **3.18** $W_{1.2} = 21\,\text{J}$ $\Delta U = Q + W = 0 + 21 = 21\,\text{J}$. **3.19** a) $\Delta_r H^\ominus_{298} = -28.3\,\text{kJ/mol}$. b) 235 J/g SiO_2. **3.20** $W_{1.2} = -17.8\,\text{kJ}$ $\Delta_r H_T = 175.0\,\text{kJ/mol}$.

4. Second law

4.1 Introduction

1. The solution process is irreversible (spontaneous) because the system in not in thermodynamic equilibrium during the process! **2.** No! – A process would require exchange of energy with the surroundings, and this is not possible for an isolated system. **3.** The first law only takes into account whether the total energy is constant and not whether a given process is spontaneous, reversible or impossible! **4.** No! – But all forces shall be balanced: when there is equilibrium in a graduated cylinder with water, the hydrostatic pressure is, for example, larger at the bottom of the cylinder than directly below the surface of the water! **5.** At 40 °C, the saturated vapour pressure of water is $p_s = 7381.4\,\text{Pa}$; then the following occurs: a) reversible evaporation for $p = p_s$ b) irreversible (spontaneous) evaporation for $p < p_s$, c) evaporation is impossible for $p > p_s$.

4.2 The Carnot cycle

1. The efficiency $\eta \rightarrow 1$ a) when the temperature in the hot reservoir $T_1 \rightarrow \infty$, and b) when the temperature in the cold reservoir approaches absolute thermodynamic zero $T_3 \rightarrow 0$. **2.** $\eta = 1 - 283/923 = 0.69$. **3.** A reduction of the temperature in the cold reservoir by 10 °C gives the highest efficiency! **4.** No! – This would mean that $\eta = 1 - T/T > 0$! **5.** $T_1 = 718\,\text{K}$ (445 °C).

4.3 Second law

[1.] $T = (Q_{1.2}/\Delta S)_{rev} = 300\,K$ [2.] In the adiabatic system, $\delta Q = 0$, from which it follows that $dS_{system} \geq \delta Q/T = 0$, cf. the Clausius inequality. [3.] Since Q and W are zero for an isolated system, the process is spontaneous, i.e. irreversible! [4.] According to eqn. (3.27), $Q = -W$, i.e. $\Delta S = Q_{rev}/T = -4000/298 = -13.4\,J/K$. [5.] During the adiabatic processes, $Q = 0$, and thus $\Delta S = 0$ for $T_1 \to T_3$ and for $T_3 \to T_1$, i.e. the process lines are parallel with the T axis. During the isothermal processes, T is constant; the isothermal process lines $\Delta S = Q/T$ are therefore parallel with the S axis.

4.4 Temperature dependence of entropy

[1.] $\Delta S = nc_p \cdot \ln(T_2/T_1) = -13.9\,J/K$. [2.] $\Delta S_{sy} = 289.5\,J/K$; $\Delta S = 1.74\,J/mol\,K$. [3.] a) $\Delta S = 8.10\,J/K$; b) $\Delta S = 0.16\,J/K$. [4.] $\Delta S = nc_V \cdot \ln(T_2/T_1) = -3.36\,J/K$. [5.] The water is cooled from 305.47 K to 293.50 K; $\theta_1 = 32.3\,°C$, $\theta_2 = 20.4\,°C$.

4.5 Change of entropy, ideal gas

[1.] $\Delta S = -nR \cdot \ln(p_2/p_1) = +75.4\,J/K$. [2.] $\Delta S = -R \cdot \ln(p_2/p_1) = +11.5\,J/mol\,K$. [3.] $\Delta S = +316.7\,J/K$. [4.] $\Delta S = R \cdot \ln(V_2/V_1) = -5.8\,J/mol\,K$. [5.] The gas pressure has been increased!

4.6 Entropy change by phase transformation

[1.] $\Delta S = 147.6\,J/mol\,K$. [2.] $\Delta_r H$(sublimation) $= 51058\,J/mol$ (Hess's law) $\Rightarrow \Delta S = 186.9\,J/mol\,K$. [3.] Heat of evaporation: $\Delta H = 40673\,J/mol = 2257\,J/g$ [4.] $\Delta S = -7.4\,J/mol\,K$ [5.] A decrease in system entropy indicates that a higher degree of order arises at the molecular level!

4.7 Standard entropy

[1.] a) $S(Ca(OH)_2) = 94.4\,J/mol\,K$; $S(CaO) = 43.4\,J/mol\,K$. b) $S(Ca(OH)_2) = 71.8\,J/mol\,K$; $S(CaO) = 32.5\,J/mol\,K$. [2.] a) $S(O_2) = 213.2\,J/mol\,K$; b) $S(O_2) = 265.0\,J/mol\,K$. [3.] $S(Ar) = 167.0\,J/mol\,K$. [4.] S(water vapour) $= 215.6\,J/mol\,K$. [5.] $\Delta S = 93.7\,J/mol\,K$.

4.8 Reaction entropy

[1.] $\Delta_r S = -200.8\,J/mol\,K$. [2.] $\Delta_r S_{298} = -172.1\,J/mol\,K$. [3.] $\Delta_r S_{1073} = 153.5\,J/mol\,K$. [4.] $\Delta_r S$ is increased with decreasing CO_2 pressure because the products formed have an increased content of entropy (increased disorder)! [5.] A positive value of $\Delta_r S$ indicates a higher degree of molecular disorder of the products formed than was the case for the original reactants!

4.9 Chemical equilibrium

[1.] $\Delta S_{univ} = 0$ (reversible process); $\Delta S_{system} = 2441\,J/K$; $\Delta S_{surr} = -2441\,J/K$. [2.] $\Delta S_{surr} = 0$ (adiabatic process); $\Delta S_{system} = nc_p \cdot \ln(T_2/T_1) = 14.2\,J/K$; $\Delta S_{univ} = \Delta S_{surr} + \Delta S_{system} = 14.2\,J/K$. The increase in the entropy of the universe is > 0; this shows that the process is irreversible! [3.] $\Delta S_{univ} = 4.3\,J/mol\,K > 0 \Rightarrow$ spontaneous! [4.] We consider the transformation: Sn(white) $\to$ Sn(grey); for this process: $\Delta S_{univ} = \Delta S_{system} - \Delta H/T = -0.1\,J/mol\,K$; the process is impossible because $\Delta S_{univ} < 0$, i.e. white tin is the stable form. This is also seen from the fact that $H^{\ominus}_{298} = 0$ for this modification. [5.] $p = 0.27\,Pa$.

4.10 The concept of entropy

[1.] The probability is: $(\frac{1}{52}) \cdot (\frac{1}{51}) \cdot \ldots \cdot (\frac{1}{2}) \cdot (\frac{1}{1}) = \frac{1}{52!} \simeq 1.24 \cdot 10^{-68}$. [2.] $k = R/N = 1.38 \cdot 10^{-23}\,J/K = 3.30 \cdot 10^{-24}\,cal/K$. [3.] a) $\Omega \simeq \exp(1.1 \cdot 10^{23})$ b) $\Omega \simeq \exp(1.7 \cdot 10^{23})$. The difference indicates a more organized and regularly built crystalline structure of diamonds! [4.] Example 1: Mercury Hg $S(\ell) = 76.0\,J/mol\,K < S(g) = 174.8\,J/mol\,K$. Example 2: Water H_2O $S(\ell) = 69.9\,J/mol\,K < S(g) = 188.7\,J/mol\,K$. Example 3: Ethanol CH_3CH_2OH $S(\ell) = 160.7\,J/mol\,K < S(g) = 282.6\,J/mol\,K$. [5.] $\Phi \simeq \exp(-0.001/k) \simeq \exp(-7.2 \cdot 10^{19})$.

Exercises, chapter 4

[4.1] $\eta = 51.4\,\%$; $Q_3 = 946\,kJ$. [4.2] $W_{1.2} = -8000\,J$; $\Delta S = 5.5\,J/mol\,K$. [4.3] $\Delta S = 3890\,J/K$. [4.4] $\Delta S = \int (c_p(T)/T)dT = 22.8\,J/mol\,K$. [4.5] $\Delta S = nR \cdot \ln(V_2/V_1) = 2.09\,J/K$.

4.6 $\Delta S = 127.2\,\text{J/mol K}$. **4.7** $Q_{1.2} = 14.8 \cdot 10^3\,\text{J}$; $W_{1.2} = -14.8 \cdot 10^3\,\text{J}$; $\Delta S = 45.8\,\text{J/K}$. **4.8** $S(\text{CaCO}_3) = 122.1\,\text{J/mol K}$; $S(\text{H}_2) = 165.5\,\text{J/mol K}$; $S(\text{H}_2\text{O}(\ell)) = 88.8\,\text{J/mol K}$; $S(\text{H}_2\text{O(g)}) = 206.5\,\text{J/mol K}$. **4.9** $S(\ell) = 80.0\,\text{J/mol K}$. **4.10** H_2O at 0 °C: $S(\ell) = 63.3\,\text{J/mol K}$; $S(\text{s}) = 41.3\,\text{J/mol K}$; $S(\text{g}) = 228.4\,\text{J/mol K}$.

4.11 $\Delta_r S_{473} = -289\,\text{J/mol K}$. **4.12** $p_s \simeq 3221\,\text{Pa}$ (deviates by 1.6 % from table value). **4.13** $\Delta S_{\text{univ}} = +9.2\,\text{J/K}$; since there is an increase in the entropy of the universe, the process is irreversible! **4.14** $Q_{\text{tot}} = 115092\,\text{J} \Rightarrow Q = 23.0\,\text{J/g} = 4.77\,\text{kJ/mol}$. **4.15** $\theta_k \simeq 17.46\,°\text{C}$.

4.16 $p \simeq 9.6\,\text{atm}$. **4.17** $T_{\text{trans}} \simeq 600\,\text{K}$ (327 °C). **4.18** $p_c = 2372\,\text{Pa}$; $p_a = 1816\,\text{Pa}$; the concrete will desiccate, since $p_c > p_a$!

5. Calculations of equilibrium

5.1 The Gibbs free energy

1. $\Delta G = \Delta H - T\Delta S = 876\,\text{J/mol} > 0$, i.e. the process is "impossible", cf. eqn. (5.7). **2.** $\Delta G = \Delta H - T\Delta S = -26\,\text{J/mol} < 0$, i.e. the process can proceed spontaneously, cf. eqn. (5.7). **3.** $\Delta G = -0.1\,\text{kJ/mol} < 0$, i.e. the transformation can proceed spontaneously; therefore, S(rhombic) is the stable form at 25 °C! **4.** $p_s = 3181\,\text{Pa}$ which deviates by approximately 0.4 % from the table value. **5.** ΔG => 0; the process cannot proceed spontaneously!

5.2 The Clapeyron equation

1. $\theta = -0.55\,°\text{C}$. **2.** For: $(\ell) \to (\text{s})$, we have $\Delta H < 0$, and since $V_s < V_\ell$, we have $\Delta V < 0$, and thus $dp/dT = \Delta H/(T\Delta V) > 0$; this means that an increased pressure increases the melting point temperature! **3.** $\Delta p \simeq 346\,\text{atm}$. **4.** At the triple point there is phase equilibrium between ice, water and water vapour in a closed system (see figure 2.10). The pressure in this system is the partial pressure of saturated water vapour at approximately 0 °C, i.e. $p \simeq 610\,\text{Pa}$. The reference point 0 °C is defined by equilibrium between ice and water at the atmospheric pressure of $p = 101325\,\text{Pa}$. The difference in pressure $\Delta p = 610 - 101325 = -100715\,\text{Pa}$ results in an increase of the freezing point $\Delta\theta \simeq 0.01\,°\text{C}$. Therefore, we find that the triple-point temperature: $\theta_{\text{trip}} = 0.01\,°\text{C}$, corresponding to 273.16 K. **5.** The freezing point is increased by 0.40 °C, i.e. $\theta_{\text{m}} = -38.45\,°\text{C}$.

5.3 The Clausius-Clapeyron equation

1. $\Delta H = 44715\,\text{J/mol} = 2481\,\text{J/g}$. **2.** $p_s \simeq 133.4 \cdot 10^3\,\text{Pa} = 1.32\,\text{atm}$. **3.** $p_s = 455\,\text{Pa}$ (table value: 454.6 Pa). **4.** $p_s = 4247\,\text{Pa}$ (table value: 4245.5 Pa). **5.** $dp/dT = p \cdot \Delta H/(RT^2) = 1900\,\text{Pa/K}$.

5.4 Activity

1. $a = 1.7 \cdot 10^{-4}$; $G = -283.4\,\text{kJ/mol}$. **2.** The precipitate of $CaCO_2$(s) is a pure substance with $a = 1$; therefore, we have that: $G_{298} = G^\ominus_{298} = -1128.8\,\text{kJ/mol}$. **3.** Partial pressure of water vapour: $p = RH \cdot p_s = 2535.3\,\text{Pa} \Rightarrow a = p/p^\ominus = 2.50 \cdot 10^{-2}$; $G = -237.7\,\text{kJ/mol}$. **4.** $a = p/p^\ominus$; with data from table 1.4, we get: $a(\text{O}_2) = 0.2095$; $a(\text{N}_2) = 0.7809$; $a(\text{CO}_2) = 0.0003$, and thus: $G(\text{O}_2) = -3.9\,\text{kJ/mol}$; $G(\text{N}_2) = -0.6\,\text{kJ/mol}$; $G(\text{CO}_2) = -414.5\,\text{kJ/mol}$. **5.** $a(\text{C}_2\text{H}_5\text{OH}) = c/c^\ominus = 0.050$; $G(\text{C}_2\text{H}_5\text{OH}) = -189.4\,\text{kJ/mol}$; $a(\text{H}_2\text{O}) = x/x^\ominus \simeq 1.0$; $G(\text{H}_2\text{O}) = -237.2\,\text{kJ/mol}$.

5.5 Thermodynamic equilibrium constant

1. At 30 Pa: $K_a = 2.96 \cdot 10^{-4}$; at 3000 Pa: $K_a = 2.96 \cdot 10^{-2}$; at 1 atm: $K_a = 1.0$. **2.** $p \simeq p^\ominus \cdot \exp(-200) \simeq 0$; the process aims at formation of hematite Fe_2O_3(s). **3.** $\Delta_r G^\ominus_{298} = -17700\,\text{J/mol}$; $p = 868\,\text{Pa}$; $RH = 27\,\%$. **4.** $\Delta_r G^\ominus_{298} = 56800\,\text{J/mol}$; $K_a = 1.1 \cdot 10^{-10}$; $[\text{BaSO}_4] = 1.1 \cdot 10^{-5}\,\text{mol}/\ell = 2.5\,\text{mg}/\ell$. **5.** $\Delta_r G^\ominus_{298} = 79900\,\text{J/mol}$; $K_a = 1.00 \cdot 10^{-14}$; $[\text{H}^+] = 1.0 \cdot 10^{-7}\,\text{mol}/\ell$; pH = 7.0.

5.6 Temperature dependence of equilibrium

1. $\Delta H^\ominus = 43.3\,\text{kJ/mol}$; $\Delta S^\ominus = 116.4\,\text{J/mol K}$. **2.** The decreasing solubility shows that $\ln(K_a)$ diminishes with increasing temperature, i.e. $d\ln(K_a)/dT = \Delta H/RT^2 < 0$ from which it follows that $\Delta H < 0$. The process is therefore exothermic (heat-releasing).

3. $\Delta_r H^\ominus_{298} = 56.5\,\mathrm{kJ/mol}$. **4.** For processes that only include pure substances, all activities $a = 1$, i.e. $K_a = 1$, and $\ln(K_a) = 0$. Thus we have that: $\Delta H^\ominus - T \cdot \Delta S^\ominus = \Delta H - T \cdot \Delta S = \Delta G = 0$ at equilibrium! **5.** $\Delta_r H^\ominus = 49.6\,\mathrm{kJ/mol}$; $\Delta_r S^\ominus = 135.7\,\mathrm{J/mol\,K}$.

Exercises, chapter 5

5.1 315 K (42 °C). **5.2** a) $\Delta H > 0 \Rightarrow$ Endothermic; b) $\Delta G = \Delta H - T\Delta S > 0 \Rightarrow$ impossible process at all T. **5.3** The equilibrium form has the lowest $G^\ominus_{298}$! **5.4** $\Delta G(\alpha \to \beta) = \Delta H - T\Delta S = 0 \Rightarrow T_{eq} = \Delta H/\Delta S = 279.3\,\mathrm{K} = 6.1\,°\mathrm{C}$. **5.5** $\Rightarrow V_A > V_B$(molar vol.) $\Rightarrow dG_A = V_A dp > dG_B = V_B dp \Rightarrow$ raised $p \Rightarrow$ lowered T_m.

5.6 The Clausius-Clapeyron equation: $\Rightarrow \Delta H = -44.44\,\mathrm{kJ/mol} = -2466\,\mathrm{kJ/kg}$. **5.7** a) $G_{mol} = -3990\,\mathrm{J/mol}$; b) $a = \exp((G - G^\ominus)/RT) = 0.20$; c) $p(N_2) = 20265\,\mathrm{Pa}$. **5.8** a) $a(Ca^{++}) = 4.32 \cdot 10^{-3}$; $a(OH^-) = 8.64 \cdot 10^{-3}$; b) $G(Ca^{++}) = -567.1\,\mathrm{kJ/mol}$; $G(OH^-) = -169.9\,\mathrm{kJ/mol}$. **5.9** $a = x/x^\ominus = 0.996$. **5.10** $K_a = 0.0728$; $\Delta G^\ominus = -6816\,\mathrm{J/mol}$.

5.11 $\Delta_r G^\ominus_{298} = 1.04\,\mathrm{kJ/mol}$; red PbO(s) is the stable modification at 25 °C! **5.12** $\Delta_r G^\ominus_{298} = 1081\,\mathrm{J/mol}$; calcite is the stable modification at 25 °C! **5.13** Maximum freezing pressure: $p \simeq 267\,\mathrm{atm}$. **5.14** 30 °C : $p_s \simeq 4229\,\mathrm{Pa}$; 40 °C : $p_s \simeq 7365\,\mathrm{Pa}$. **5.15** $a = 0.036$; $G = -237.9\,\mathrm{kJ/mol}$.

5.16 1) At $p = 22000$ Pa, $\Delta_r G_{298} = -3153.2\,\mathrm{kJ/mol}$; 2) At $p = 1$ Pa, $\Delta_r G_{298} = -3078.8\,\mathrm{kJ/mol}$: Since $\Delta G < 0$ at $p = 1\,\mathrm{Pa}$, the oxidation will be spontaneous although the partial pressure has been reduced! **5.17** $K_a = 1.009 \cdot 10^{-14}$; pH = 7.00. **5.18** $\Delta_r G_{298} = -0.227\,\mathrm{kJ/mol}$; Yes! – $\Delta G < 0$ means that the process is spontaneous and that the concrete will desiccate! **5.19** $K_a = 5.1 \cdot 10^{-17}$; solubility $2.3 \cdot 10^{-4}\,\mathrm{g/\ell}$. **5.20** $RH = 62.8\,\%$.

6. Electrochemistry

6.1 Electric current and charge

1. $Q = I \cdot t = 1.44 \cdot 10^5\,\mathrm{C}$. **2.** $I = Q/t = 13.3\,\mathrm{A}$. **3.** $N = I \cdot t/e = 3.1 \cdot 10^{19}$ electrons per second. **4.** $I = n \cdot e \cdot \mathcal{F}/t = 2 \cdot (m/M) \cdot \mathcal{F}/t = 123\,\mathrm{A}$. **5.** $Q = n(\mathrm{Na^+}) \cdot \mathcal{F} = n(\mathrm{NaCl}) \cdot \mathcal{F} = 3.3 \cdot 10^5\,\mathrm{C}$.

6.2 Electric potential

1. $W_{1.2} = (V_b - V_a) \cdot dQ = -60\,\mathrm{J}$. **2.** $\Delta V = (V_b - V_a) = W_{a.b}/dQ = -55\,\mathrm{volt}$. **3.** $1\,\mathrm{eV} = 1.602 \cdot 10^{-19}\,\mathrm{J}$. **4.** $|\mathbf{F}| = |\mathbf{E}| \cdot Q = 1.602 \cdot 10^{-16}\,\mathrm{N}$. **5.** $W = \Delta V \cdot Q = \Delta V \cdot I \cdot t = 1.8 \cdot 10^5$ J per hour.

6.3 Electric conductivity

1. $N = C/e = I \cdot t/e = 3.1 \cdot 10^{19}$ electrons. **2.** $Q = 3 \cdot n \cdot \mathcal{F} = 1.5 \cdot 10^6\,\mathrm{C}$. **3.** $\sigma = 30 \cdot 10^6\,\Omega^{-1}\mathrm{m}^{-1}$ **4.** $G = \sigma \cdot A/L = 4.6\,\Omega^{-1}$; $R = 1/G = 2.2 \cdot 10^{-1}\,\Omega$. **5.** $\sigma = (I/\Delta V) \cdot (L/A) = 0.2\,\Omega^{-1}\mathrm{m}^{-1}$; $\rho = 1/\sigma = 5\,\Omega\mathrm{m}$.

6.4 Electrochemical reaction

1. For the galvanic cell, $\Delta W < 0$, i.e. the cell does electrical work on its surroundings; for the electrolytic cell, $W > 0$, and the surroundings do work on the cell. **2.** a) $Fe^{++}(aq) + 2e^- \to Fe(s)$; b) $Mn(s) \to Mn^{++++}(aq) + 4e^-$; c) $Cu^{++}(aq) + e^- \to Cu^+(aq)$; d) $2H^+(aq) + 2e^- \to H_2(g)$. **3.** a) Reduction of Mg^{++}; b) oxidation of Cr^{++}; c) reduction of H^+. **4.** a) OH^- anion; b) Zn^{++} cation; c) NH_4^+ cation; d) SO_4^{--} anion; e) H^+ cation; f) NO_3^- anion; g) O^{--} anion; h) Li^+ cation. **5.** $Q = 3 \cdot n(\mathrm{Al^{+++}}) \cdot \mathcal{F} = 1.07 \cdot 10^7\,\mathrm{C}$.

6.5 Electrochemical potential

1. $\Delta V(Zn\|Ag) = \Delta V(Zn\|Cu) - \Delta V(Ag\|Cu) = +1.56\,\mathrm{volt}$. **2.** Cu has less tendency to oxidize than Zn; copper will therefore occur more often than zinc as cathode in a corrosive environment. Since oxidation, and thus metal decay, occurs during the anode process, Cu has a higher resistance against corrosion than Zn. **3.** $Zn(s) + 2Ag^+ \to Zn^{++}(aq) + 2Ag(s)$; during the process, Zn is oxidized while Ag^+ is reduced. **4.** $Zn(s) + Fe^{++}(aq) \to Zn^{++}(aq) + Fe(s)$; during the process, Zn is oxidized and Fe^{++}

is reduced. **5.** The anode has negative potential in relation to the cathode; cations such as Zn^{++} and Cu^{++} will therefore be influenced by an electric field leading them to the anode; correspondingly, SO^{--} will be influenced by a force directed towards the cathode.

6.6 The Nernst equation

1. Cathode reaction: $Fe^{++}(aq) + 2e^{-} \rightarrow Fe(s)$; anode reaction: $Zn(s) \rightarrow Zn^{++}(aq) + 2e^{-}$; oxidized state ($ox$) : Fe^{++}, Zn^{++} reduced state (red) : Fe(s), Zn(s) **2.** According to the Gibbs-Duhem equation, the Gibbs free energy is changed with the pressure as: $dG = Vdp$. The change is negligible for condensed phases with little molar volume, whereas the free energy of gas phases is strongly dependent on the large molar volume. The electrochemical potential is sensitive to pressure if the redox reaction includes gases, e.g. $H_2(g)$; if all phases are condensed, the potential can be assumed to be independent of pressure, unless extreme changes of pressure occur. **3.** $|Ag||Zn|$ redox reaction: $Zn(s) + 2Ag^{+}(aq) \rightarrow Zn^{++}(aq) + 2Ag$; (z = 2); $\Delta G^{\ominus}_{298} = -301.2\,kJ/mol$; $\Delta V^{\ominus} = -\Delta G^{\ominus}/(z \cdot \mathcal{F}) = +1.56\,volt$. **4.** $|Zn||Pb|$ redox reaction: $Zn(s) + Pb^{++}(aq) \rightarrow Zn^{++}(aq) + Pb(s)$; (z = 2); $\Delta G^{\ominus}_{298} = -122.6\,kJ/mol$; $\Delta V^{\ominus} = +0.64\,volt$; For $|Pb||Zn|$, $\Delta V^{\ominus} = -0.64\,volt$. **5.** $Q = 2 \cdot n(\mathrm{Fe}) \cdot \mathcal{F} = 3.46 \cdot 10^{5}\,C$.

6.7 Temperature dependence of the potential

1. (z=1): 96500 C/mol; (z=2): 193000 C/mol; (z=3): 289500 C/mol. **2.** $\Delta V^{\ominus} = -\Delta G^{\ominus}/(z \cdot \mathcal{F}) = +0.78\,volt$. **3.** $\Delta V = -\Delta G/(z \cdot \mathcal{F}) \Rightarrow \Delta G = -\Delta V \cdot z \cdot \mathcal{F} = -73.4\,kJ/mol$. **4.** $(\partial V^{\ominus}/\partial T)_p = \Delta_r S^{\ominus}/(z \cdot \mathcal{F}) = 0.6 \cdot 10^{-4}\,volt/K$. **5.** $\Delta S = (\partial V/\partial T) \cdot z \cdot \mathcal{F} \simeq (\Delta V/\Delta T) \cdot z \cdot \mathcal{F} = -32.8\,J/mol\,K$.

6.8 Notation rules

1. a) $Na(s) \rightarrow Na^{+}(aq) + e^{-}$ (oxidation of Na(s)); b) $Cl^{-}(aq) \rightarrow Cl(aq) + e^{-}$ (oxidation of Cl^{-}); c) $Cu^{+}(aq) \rightarrow Cu^{++}(aq) + e^{-}$ (oxidation of Cu^{+}). **2.** a) $Ag^{+}(aq) + e^{-} \rightarrow Ag(s)$; b) $Fe^{+++}(aq) + e^{-} \rightarrow Fe^{++}(aq)$; c) $Cu^{++}(aq) + e^{-} \rightarrow Cu^{+}(aq)$ **3.** Anode reactions are described by: a) $Ca(s)\,|\,Ca^{++}(aq)$; b) $(Pt)\,H_2(aq)\,|\,2H^{+}(aq)$; c) $Sn(s)\,|\,Sn^{++}(aq)$. **4.** Cathode reactions are described by: a) $Fe^{+++}(aq)\,|\,Fe(s)$; b) $\frac{1}{2}O_2(aq), H_2O(\ell)\,|\,2OH^{-}(aq)\,(Pt)$; c) $Sn^{++++}(aq)\,|\,Sn^{++}(aq)\,(Pt)$. **5.** A corrosion cell is formed as: $Zn(s)\,|\,Zn^{++}(aq)\,||\,\frac{1}{2}O_2(aq), H_2O(\ell)|2OH^{-}\,(Fe)$.

6.9 Standard potential

1. Since $V^{\ominus}(SHE) < V^{\ominus}(Cu)$, the Cu electrode will form a positive cathode, cf. eqn. (6.50). **2.** The standard potential $V^{\ominus}$ is calculated for a reduction process: $Cu^{+}(aq) + e^{-} \rightarrow Cu(s)$; $\Delta G^{\ominus}_{298} = -50$ kJ/mol; (z=1); $V^{\ominus} = -\Delta G^{\ominus}/(z \cdot \mathcal{F}) = +0.52$ volt. **3.** $V^{\ominus} = 0.80 - (-2.34) = +3.14$ volt. **4.** $\Delta G^{\ominus}_{298} = +78.9$ kJ/mol; (z=2); $V^{\ominus} = -0.41$ volt; $V = V^{\ominus} - (RT/(z \cdot \mathcal{F})) \cdot \ln(K_a) = -0.59$ volt. **5.** Cell reaction: $Fe(s) + 2Ag^{+}(aq) \rightarrow Fe^{++}(aq) + 2Ag(s)$; $\Delta G^{\ominus}_{298} = -233.1$ volt; (z=2); $\Delta V^{\ominus} = +1.21$ volt.

6.10 Passivation

1. Reduction process: $\Delta G^{\ominus}_{298} = 485.3$ kJ/mol; (z=3); $V^{\ominus} = -\Delta G^{\ominus}/(z \cdot \mathcal{F}) = -1.68$ volt. **2.** Reduction process: $\Delta G^{\ominus}_{298} = -77.1$ kJ/mol; (z=1); $V^{\ominus} = -\Delta G^{\ominus}/(z \cdot \mathcal{F}) = +0.80$ volt. **3.** Reduction process: $\Delta G^{\ominus}_{298} = +147.0$ kJ/mol; (z=2); $V^{\ominus} = -0.76$ volt; $V = V^{\ominus} - (RT/(z \cdot \mathcal{F})) \cdot \ln(K_a) = -0.82$ volt. **4.** Reduction process: $\Delta G^{\ominus}_{298} = 127.1$ kJ/mol; (z=2); $V^{\ominus} = -\Delta G^{\ominus}/(z \cdot \mathcal{F}) = +0.66$ volt; $K_a = 1 \cdot 10^{63}$; $V = V^{\ominus} - (RT/(z \cdot \mathcal{F})) \cdot \ln(K_a) = -1.20$ volt. **5.** If a metal rod is used, two unknown electrodes are introduced and their potential contribution shall be included in the cell potential; this is avoided by a conductive salt bridge that at the same time prevents the two different electrolytes from being mixed.

Exercises, chapter 6

6.1 k1: $1.602 \cdot 10^{-19}$ J/eV; k2: $6.242 \cdot 10^{18}$ eV/J. **6.2** Integrate from $t = 0$ to 40: $Q = \int I(t)dt = 10.8 \cdot 10^{3}$ C. **6.3** Per hour: 78.7 mol Cu(s) $\sim$ 157.4 mol e; $I = Q/t = 4.22 \cdot 10^{3}$ A. **6.4** a) $\Delta V = R \cdot I = 40$ volt; b) $Q = \int I(t)dt = 7200$; C c) $W = \Delta V \cdot Q = 288$ kJ. **6.5** a) $R = \Delta V/I = 120\,\Omega$; b) $Q = \int I(t)dt = 60$ C; c) $W = (V_b - V_a) \cdot Q = 720$ J.

6.6 a) $G = 1/R = 1.11\,\Omega^{-1}$; b) $\sigma = G \cdot L/A = 3.53 \cdot 10^{7}\,\Omega^{-1}m^{-1}$; c) $\rho = 1/\sigma = 2.83 \cdot 10^{-8}\,\Omega m$. **6.7** $W = (V_b - V_a) \cdot Q = -12\,eV = -1.92 \cdot 10^{-18}$ J. **6.8** a)

$Q = 2 \cdot n \cdot \mathcal{F} = 1.64 \cdot 10^6$ C; b) $I = Q/t = 456$ A. **6.9** $G = 1/R = 3.08 \cdot 10^{-2}\,\Omega^{-1}$; a) $\sigma = G \cdot L/A = 9.23 \cdot 10^{-2}\Omega^{-1}\mathrm{m}^{-1}$; b) $\rho = 1/\sigma = 10.8\,\Omega\mathrm{m}$. **6.10** 33.6 kg Al(s) per hour.

6.11 $K_a = 300$; $\Delta V^\ominus = 0$ volt; (z=2); $\Delta V = \Delta V^\ominus - (RT/(z\mathcal{F}))\cdot\ln(K_a) = -0.073$ volt; $\Delta G^\ominus_{298} = -\Delta V \cdot (z\mathcal{F}) = +14.1$ kJ/mol. **6.12** a) $\Delta V^\ominus = +0.76$ volt; $(\partial\Delta V^\ominus/\partial T)_p = -1.2 \cdot 10^{-4}$ volt/K; b) $\Delta V^\ominus = +0.35$ volt; $(\partial\Delta V^\ominus/\partial T)_p = +5.8 \cdot 10^{-5}$ volt/K; c) $\Delta V^\ominus = +0.81$ volt; $(\partial\Delta V^\ominus/T)_p = -1.5 \cdot 10^{-3}$ volt/K. **6.13** a) $\Delta V^\ominus = +0.35$ volt; $\Delta V = +0.32$ volt; b) $\Delta V^\ominus = +0.24$ volt; $\Delta V = +0.30$ volt. **6.14** $V = V^\ominus - (RT/(z\mathcal{F})) \cdot \ln(1/(c/c^\ominus)) = -0.76 + 0.013 \cdot \ln(c/c^\ominus)$ volt; $c = 10^{-6}$mol/ℓ: $V = -0.94$ volt; $c = 10^{-4}$mol/ℓ: $V = -0.88$ volt; $c = 10^{-2}$mol/ℓ: $V = -0.82$ volt; $c = c^\ominus = 1$ mol/ℓ: $V = -0.76$ volt. **6.15** Reduction potential: $V^\ominus = -\Delta G^\ominus/(z\mathcal{F})$; ($Ca^{++}$): z=2; $V^\ominus = -2.86$ volt; (Ag^+): z=1; $V^\ominus = +0.80$ volt; (Na^+): z=1; $V^\ominus = -2.71$ volt; (Cr^{++}): z=2; $V^\ominus = -0.91$ volt; (Al^{+++}): z=3; $V^\ominus = -1.68$ volt; and (Ni^{++}): z=2; $V^\ominus = -0.24$ volt. Electrochemical series: $Ca^{++} < Na^+ < Al^{+++} < Cr^{++} < Ni^{++} < Ag^+$.

6.16 $Fe(s) + \frac{1}{2}O_2(g) + 2H^+(aq) \rightarrow Fe^{++}(aq) + H_2O(\ell)$; $\Delta G^\ominus_{298} = -316.1$ kJ/mol; (z=2); $\Delta V^\ominus = +1.64$ volt; $K_a = 2.24 \cdot 10^3$; $\Delta V = 1.54$ volt. The galvanic process is spontaneous, cf. eqn. (6.50), since $\Delta G < 0$ and $\Delta V > 0$. **6.17** $Sn(s) + 2H^+(aq) \rightarrow Sn^{++}(aq) + H_2(g)$; $c = 10.9 \cdot 10^{-2\mathrm{pH}}$ mol/ℓ. **6.18** (Cu^{++}): $G^\ominus_{298} = 66.0$ kJ/mol; (Cu^+): $G^\ominus_{298} = 50.3$ kJ/mol; $[Cu^+] = 2.9 \cdot 10^{-4}$ mol/ℓ. **6.19** $[Fe^{++}] = 1.44 \cdot 10^{-13}$ mol/ℓ. **6.20** A-B: $V = -0.93 + 0 \cdot \mathrm{pH}$ volt; B-C: $V = -0.42 - 0.059 \cdot \mathrm{pH}$ volt; C-D: $V = -0.09 - 0.088 \cdot \mathrm{pH}$ volt.

As the first in the world, H.C. Ørsted succeeded in making a specimen of the element aluminium in 1824. The picture shows Ørsted's chemical laboratory which was established in a former stable in the northern Copenhagen.

APPENDIX D

Subject index

To facilitate look-ups in this book, a subject index is included; this gives reference to the descriptive text in the theoretical sections of the book and to the examples, tables, exercises, and figures. The following markings and codes are used in this index:

- All self-contained key words are indicated by an *upper-case* first letter (example: *Reinforcement corrosion*).
- Some important key words are provided with a number of subordinate entries; these are indented and printed with a *lower-case* initial letter (example: Atmospheric air / *composition*).
- A reference, for example *1.23*, denotes (*chapter.page number*) in the book; references to particularly important *definitions* or *descriptions* are emphasized in bold-face letters (example: *the Avogadro constant,* **1.5**).
- The references are supplemented with information of how the subject has been treated in the book, shown as: **ex**: *example*; **f**: *figure*; **t**: *table*; **exc**: *exercise*.

Q

R

S

T

APPENDIX E

Acknowledgements for illustrations

Cover: Electron microscopy of some of the reaction products which are formed by hydration of Portland cement. The picture shows hexagonal plates of portlandite, rods of ettringite and clusters of calcium-silicate-hydrate. Magnification approx. 22.000 X. Paul Stutzman, Nat. Inst. of Sci. and Techn., USA

Many photos are taken by photographer Lizzi Allesen-Holm in cooperation with Per Freiesleben Hansen. These are referenced as LAH & PFH in the following.

1.4 Karel Jakubec; 1.6 Paul Stutzman, Nat. Inst. of Sci. and Techn., USA; 1.7 LAH & PFH; 1.10 LAH & PFH; 1.14 Murerfagets oplysningsråd; 1.15 John Kerstholt; 1.16 LAH & PFH; 1.19 AGA A/S; 1.23 LAH & PFH; 1.27 LAH & PFH; 1.28 Press Photo Siemens AG, Energy Sector; 1.30 Danish National Metrology Institute; 1.31 LAH & PFH; 1.32 Murerfagets oplysningsråd; 1.33 LAH & PFH; 1.34 H+H International A/S; 1.35 H+H International A/S; 1.37 Aalborg Portland A/S; 1.38 LAH & PFH; 1.41 Polyteknisk Forlag; 1.42 LAH & PFH; 1.43 Aalborg Portland A/S; 1.44 PFH; 2.3 Aalborg Portland A/S; 2.5 LAH & PFH; 2.7 Ole Mejlhede Jensen; 2.8 LAH & PFH; 2.11 LAH & PFH; 2.19 LAH & PFH; 2.20 Rosenkilde & Bagger; 2.23 Aalborg Portland A/S; 2.30 Beton- og konstruktionsinst.; 2.33 D.C. Hoffmann, Caltech; 2.36 LAH & PFH; 2.40 Swedish Research Council Formas; 2.41 Beton- og konstruktionsinst.; 2.42 LAH & PFH; 2.43 Andrew Dunn; 2.44 Beton- og konstruktionsinst.; 2.46 Aalborg Portland A/S; 2.47 LAH & PFH; 3.1 LAH & PFH; 3.2 LAH & PFH; 3.5 LAH & PFH; 3.9 Fg2; 3.11 LAH & PFH; 2.12 Folketinget; 3.14 DAfStb Heft 170 Berlin (1965); 2.23 LAH & PFH; 3.24 LAH & PFH; 3.27 Zement-Kalk-Gips 28 (7) 1975; 3.32 Aalborg Portland A/S; 3.33 Knauf Danogips; 3.34 LAH & PFH; 3.36 Aalborg Portland A/S; 3.37 Polyteknisk Forlag; 3.41 Parr Instrument Company 6755 Solution Calorimeter; 3.42 Beton- og konstruktionsinst.; 3.45 U.S. National Oceanic and Atmospheric Administration; 3.47 LAH & PFH; 3.48 Paul Stutzman, Nat. Inst. of Sci. and Techn., USA; 3.49 Aalborg Portland A/S; 4.1 Ole Mejlhede Jensen; 4.2 PFH; 4.15 Aalborg Portland A/S; 4.16 Wilson Bentley; 4.19 LAH & PFH; 4.26 Y. Kariya, Shibaura Inst. of Techn., Japan; 4.28 Aalborg Portland A/S; 4.29 Zement-Kalk-Gips 30 (6) 1977; 4.30 Aalborg Portland A/S; 4.31 Earth Sci. and Geogr., School of Physical and Geogr. Sci., Keele Univ.; 4.32 Aalborg Portland A/S; 4.36 NETZSCH-Gerätebau GmbH; 4.37 Aalborg Portland A/S; 4.39 LAH & PFH; 4.41 LAH & PFH; 4.44 LAH & PFH; 4.45 LAH & PFH; 4.47 H+H International A/S; 5.1 LAH & PFH; 5.2 LAH & PFH; 5.4 LAH & PFH; 5.7 LAH & PFH; 5.12 LAH & PFH; 5.13 LAH & PFH; 5.18 Kraft Energy Systems; 5.19 Aalborg Portland A/S; 5.20 Aalborg Portland A/S; 5.21 Aalborg Portland A/S; 5.26 W.A. Côté, S.U.N.Y. and P. Hoffmeyer, Technical University of Denmark; 5.28 Byggefejlregisteret; 5.30 Byggefejlregisteret; 5.36 Byggefejlregisteret; 5.38 United States Geological Survey & Mineral Information Institute; 6.1 Stork Veco BV; 6.2 LAH & PFH; 6.3 Ole Mejlhede Jensen; 6.8 Ole Mejlhede Jensen; 6.10 Aalborg Portland A/S; 6.14 Roger McLassus; 6.15 Skarpenord Corrosion a.s.; 6.16 LAH & PFH; 6.23 LAH & PFH; 6.33 Ole Mejlhede Jensen; 6.34 Byggefejlregisteret; 6.35 Malene Thyssen; 6.42 LAH & PFH; 6.43 Datamax; 6.46 Ole Mejlhede Jensen & Torben Jacobsen; 6.47 Fontana & Greene, Corrosion Engineering, McGraw-Hill 1978; 6.51 R. C. Frank, General Motors Res. Lab. (Currently Physics at Augustana College, (IL) USA); 6.53 Polyteknisk Forlag; 6.54 Rosenkilde & Bagger; 6.55 Mila Zinkova; 6.56 Polyteknisk Forlag.

Line figures: All by Per Freiesleben Hansen except A.1 by Ole Mejlhede Jensen.